PHYSICS RESEARCH AND TECHNOLOGY

THE WAY TO ETIS

APPLIED GRAVITATIONAL MASS SPECTROSCOPY

PHYSICS RESEARCH AND TECHNOLOGY

Additional books in this series can be found on Nova's website
under the Series tab.

Additional e-books in this series can be found on Nova's website
under the e-book tab.

PHYSICS RESEARCH AND TECHNOLOGY

THE WAY TO ETIs

APPLIED GRAVITATIONAL MASS SPECTROSCOPY

KRISTINA ZUBOW
ANATOLIJ ZUBOW
AND
VIKTOR ZUBOW

nova publishers
New York

Additional color graphics may be available in the e-book version of this book.

Library of Congress Cataloging-in-Publication Data

Zubow, Kristina, 1976- author.
The way to ETIs : applied gravitational mass spectroscopy / Kristina Zubow, Anatolij Zubow and Viktor Zubow (R&D Department, Aist H&C GmbH, Germany, and others). -- First edition.
pages cm
Includes bibliographical references and index.
ISBN 978-1-62808-091-9 (hardcover)
1. Mass spectrometry. 2. Gravitational waves. 3. Matter--Properties. I. Zubow, Anatolij, author. II. Zubow, Viktor, author. III. Title.
QD96.M3Z83 2013
543'.65--dc23

2013019731

Published by Nova Science Publishers, Inc. † New York

CONTENTS

Chapter 1

INTRODUCTION

The principle of material analysis using shock waves is well known and widely accepted in nature. For example, a fish generates acoustic signals in water and analyze the replies it receives; analogously, the woodpecker employs this method to find larvae in trees. The algorithm of this technique is based on the interaction of acoustic waves with the substance. The larger the geometric object, the stronger and longer must be the shock waves. On the other hand, increasing shock wave energy worsens the quality of the information. Therefore, the woodpecker varies its hit power on the wood to get exact locational information concerning where the larvae are or their sizes and quality. Thus, there is a natural model describing a device for scanning the heterogeneity in matter using the analysis of reflected shock waves. This model was observed to concentrate into itself physical and mathematical models as well as neural brain networks together so that they create a system for the generation and analysis of signals with the needed characteristics. Its realization without modern computer techniques and program design, however, is impossible. The next challenge was to understand and explain the huge information flow. Our brain is not able to process a sequence of five to seven signals and even less umpteen variable ones. Nevertheless, nature gives examples for living organisms (e.g. dolphins) that are able to process many more signals. The problem is more complex because the reflected signals are very weak and "dirty" with noises from their surroundings. Simply hitting an object weakly informs us about its hardness and sound. With a relatively high probability we learn, whether the object is a porous or monolithic and whether it is made of metal, plastic, wood, stone, sand, or liquid.

In the physical chemistry of polymers, the term "free volume" is often used when referring to the inside of the polymer structure. Mass concentrations at the level of clusters/domains should be their reverse. The sizes of the domains are in the nanometer area because their masses are umpteen billion Daltons and they are characterized to determine the most important properties of materials. The clusters in molecular matter are considered seed crystals at the point of transformation into the crystalline state. While the crystalline state is well studied by many methods, the analysis of the amorphous unstable one is problematic even though this state is the most important. It determines the direction of the biochemical processes, crystallization of substances, and destruction of materials. It is not right to believe that there is absolute chaos in the amorphous part. Our world has been built up by nature harmonically and each building stage exists as a compromise between array and chaos. Here, the empirical rule is applicably: 70±15% (array) to 30±15% (chaos), mainly with regard to liquids, polymers, and biochemical processes (mutagenesis).

To get an idea of the structure of the amorphous molecular matter, a new construction of devices is needed by which the recording of mass concentrations in stationary gravitational energy clusters is possible. The new method could be based on the influence of dosed energy impulses on the sample and the simultaneous analysis of the reflected ones. Here, the energy impulses are characterized as shock waves that are easily recorded. Technically the following tasks have to be solved: the source of the shock waves, receiver of reflected the shock waves, interface, and the administration system and service.

We attempt to inform a broad circle of readers on the essential details and the possibilities for the suggested method, therefore, we do not stop at difficult mathematical calculations, physical and chemical models, or electronics and computer design. The method we are introducing is called gravitation mass spectroscopy (GMS). Only after reading the monograph as a whole it is possible to understand the key aspects of this method and its huge potential. Examples of investigating some substances, processes, and events will be presented as they are helpful. The interaction of weak shock waves with pure substances will be discussed as well as their long-range order at the level of mass concentrations (clusters). Furthermore, the monograph covers the analysis of processes in which clusters are involved (physical chemical processes, remote cluster interaction). The clusters are characterized to be both wave and particle (wave-particle dualism)—with new revolutionary perspectives in natural science that could therefore be opened.

Chapter 2

WEAK MOLECULAR INTERACTIONS

Weak intermolecular interactions encompass the following:

- Fluctuation of charge densities on atoms and the induction of opposite charge densities on adjacent ones. These are the so-called dispersion interactions (Van der Waals force, widespread in the 20th century). They will be realized in inert gases and substances with a zero dipole moment (CH_4, CCl_4). The whole interaction energy of electrically neutral molecules can be calculated according to the Mie equation:

$$E = - A \cdot d^{-6} + B \cdot d^{-12}$$

where A and B are empirical constants and d is the inter atomic distance [1, p. 472].

- Dipole-dipole interaction between molecules with dipole moments or with one dipole and one induced dipole on another molecule. For instance, substances with polar molecules like water, ammonia, halogenated hydrocarbons [1].
- Hydrogen bridges between molecules. Here the hydrogen atom is linked with an atom of strong electro negativity (halogen, oxygen, nitrogen, etc.). The energy of hydrogen bridges are low, around 4…40 kJ/mol.
- Donor-acceptor interactions (coordination bonds). In this case, one atom is characterized to be the owner of an electron pair while the other one is the provider of an electron-free orbital (nomenclature of the 20th century) [2]. Hydrogen bridges are classified as a case of donor-acceptor interaction. Some problems, which arise at the investigation of weak donor-acceptor interactions, for instance, in mixtures of organic liquids, can only be understood after reading [3].
- Gravitational interaction. This type of remote interaction between mass concentrations has been developed by Newton and is described by:

$$F = \gamma \cdot m \cdot M / r^2$$

where γ is the gravitation constant $6.672 \cdot 10^{-11}$ $N \cdot m^2 \cdot kg^{-2}$, and m and M are masses (kg) that interact at the distance r (m). The reasons for this interaction have yet to be found.

- Our discovery of remote interaction between ensembles of mass concentrations (domains/clusters) [4, 5] that we believe to cause the Newton forces.

All of the above mentioned interaction forces inside of an individual substance with the same long-range order will be classified as cohesion forces whereas those between different substances will be regarded as adhesion forces. Adhesion forces, however, can be developed inside of one and the same substance when clusters of different structures appear, e.g. isomers, clusters of different isotopic content, clusters with different conformation and configuration (isotactic, syndiotactic polystyrene). The terms "cohesion" and "adhesion" will be of importance in understanding the processes of cluster formation in individual substances, solutions, and polymers.

2.1. The Energy of Weak Molecular Interactions

Thus, molecular clusters are described as the product of weak energy, cooperation, and structural interactions that are not sufficient for the clusters' stability. Another factor in stabilizing the cluster structure was observed to be stationary waves in white noises [6, 7]. The lifetime of clusters is low, being in the diapason of ps and ns for liquids [8] and about some seconds for polymers [9]. Clusters interact with each other at a distance and form super cluster structures — ensembles. A long-range order in liquids and polymers will be built by these ensembles of clusters [10, 11].

According to Trouton's rule, in liquids, only 1% of molecules are freely moving [12], with the huge remaining part being in a bound state. The thermodynamic model of area-wide network bonds between all molecules cannot withstand the criticism because of strong internal tensions and the loss of flowability; a possible way out from this "dead end street" could be the acceptance of cluster formation in liquids and polymers. For example, while investigating methanol, Zakharov and colleagues found clusters in this alcohol and even calculated their free surface energy [13]. The structure of water clusters has been described by many authors [14, 15, 16], (see Figure 2.1.1) because the appearance of clusters in natural and synthetic polymers has been proven experimentally [4]. The formation energy of clusters in individual liquids and solutions is observed to be lower than the heat background (kT), < 2...3 kJ/mol [6], and for solvated clusters of ion pairs (SCIP), the number is even less at about a J or mJ per salt mole [11]. As we already know, the stability of an organized structure depends not only on the energy component but also on entropy being the determining factor in the case of SCIP. The formation energy dependency on the seed crystal size is discussed in [17] and only in two cases does it approve of low values around zero.

The analysis of periodical structures in the long-range order of liquids using GMS can be understood by applying a simple example. Assume that the periodical structure consists of *n* elements, whose lifetime is remarkably shorter than the observation time. That means the elements disappear as soon as they have arisen. At recording the time of elements' formation/destruction the observer will see only one part of them, because the other one has been destroyed. At the next moment, the former visible elements have disappeared and new ones are at another place appeared and so on. For the observer it looks like chaos. Accelerating the reaction of both the emergence and disappearance of elements means that they are visible to the observer; an impression arises and the observer is able to see more elements and a higher order. This scenario said can be compared with the principle of a screen where the display is a result of fast movements of an X-ray point and the visible "play of time" in the observer's brain. The observer now understands that his perception involves not only *n* elements but also their wave properties. The wave properties of clusters were discovered by the authors of the present monograph and described in their work [10].

The long-range order in liquids is believed to be built of three different kinds of clusters: skeletal, randomly appearing, and intermediary. All of these three cluster types form an ensemble whose stability depends on the cooperative interaction of the clusters inside the ensemble—like a unit—and on its ability to achieve the lowest potential energy. The size and stability of an ensemble are directly connected with the mass of liquids and destroying factors (temperature and pressure fluctuations, external energy flows, vessel walls, colloids, gases, ions).

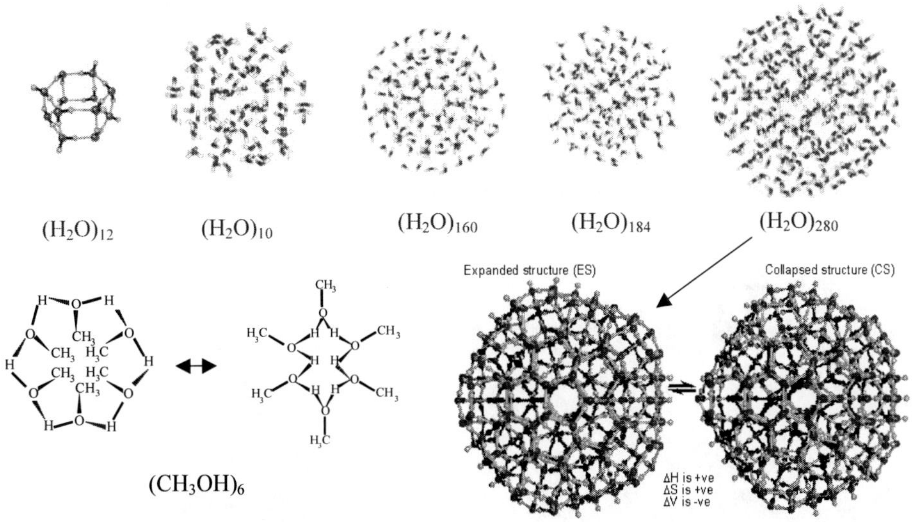

Figure 2.1.1. Calculated water cluster models (with the kind permission of Professors Chaplin, http://www.lsbu.ac.uk/water/index.html, and Lenz, $(H_2O)_{12}$ [14]) and clusters in methanol experimentally observed by the authors [10].

2.2. Identification Methods for Weak Molecular Interactions

Weak molecular interactions can be analyzed through classic physical chemical methods using modern computer techniques. Furthermore, modeling molecules and molecule processes with the help of computer design is based primarily on the Lennard-Jones potential and its variants [18]. It has to be mentioned that in the Lennard-Jones potential calculations, the Mie equation (§2) will be used. Lastly, the gravitation mass spectroscopy method suggested in this monograph is suitable in identifying weak molecular interactions, too, although it works at the level of molecular clusters/domains.

2.3. MOLECULAR CLUSTERS

Molecular clusters are described as the precursors of seed crystals. The authors of [19] have even managed to observe clusters of saccharose in water, as well as those of solvated ion pairs clusters ($KCl \cdot nH_2O$), and to take of a photograph of them directly. Clusters will be built in stationary gravitational waves and stabilized by gravitational noises where their formation energy will be provided by energy clusters of the physical vacuum [4]. As will be shown later in §21 and [4, 20], unbalanced processes of dissolving and condensation of baryonic matter in the physical vacuum, proceeding with super light velocity, are the main source of gravitational noises. On the other side, molecular clusters are characterized to emerge from weak gravitational radiation that is found to be individualized for each cluster and can be considered a "finger print." This weak gravitational radiation from a cluster and cluster ensembles changes the gravitation field of its surroundings. Anthropogenically changed white noises and their transfer in color noises lead to new formation conditions of clusters and cluster ensembles. This has consequences for the structure of seed crystals, the long-range order of liquids and polymers, their technological processes, and their product quality. This can schematically be represented as follows (Figure 2.3.1).

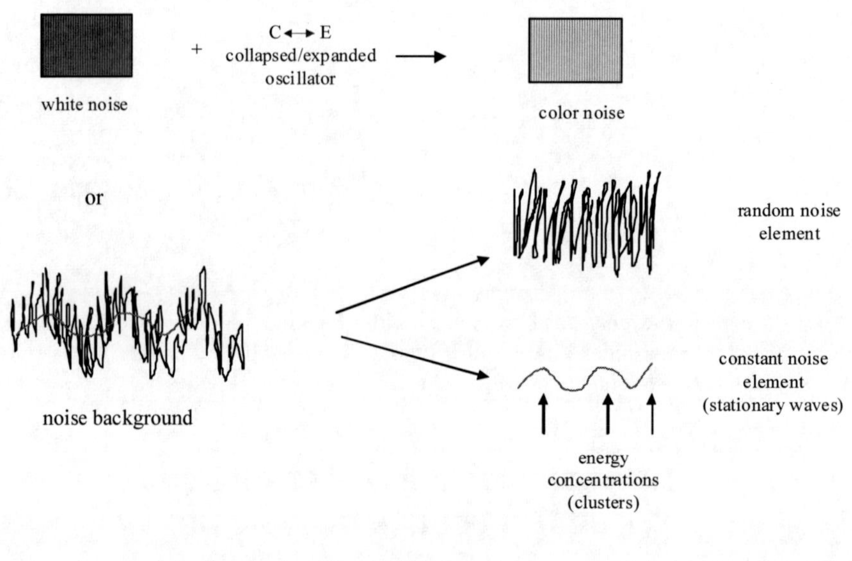

Figure 2.3.1. The interaction of background gravitational noises with weak gravitational noises of molecular clusters. C – collapsed (close-packed), E – expanded cluster form (loose clusters). The constant noise element is caused by stationary waves being the energy supplier for the formation of molecular clusters. The white noise can be seen on the TV screen when the antenna is switched off while color noise is the white noise that is disturbed by external energy flows, e.g. radio disruptions.

According to Figure 2.3.1, the stabilization of molecular clusters through the background noise leads to the formation of a new background noise, reflecting the wave character of the cluster and its structure. After forcefully shifting the C ↔ E balance (e.g. pressure on a polymer, stirring a liquid, or at beating), the wave character of a cluster will be changed. That means its gravitational "fingerprint" becomes different. This "fingerprint" is characterized as typical of the energy profile/histogram for the proton jumping between the cluster atoms and physical vacuum. In accordance with the resonance, the oscillation of a cluster ensemble remotely influences a smaller analog one, and large ensemble forces cause the other one to subordinate itself. Here a new ensemble consisting of the large ensemble and the small one will be built.

A number of processes in chemistry, biochemistry [6], and the hydrodynamics of liquids (change of viscosity under seismic waves) [21] as well as wall effects ([22], §10.9) and the influences of celestial bodies (§9.5.1, §10.10, §16) points to the fact that there is a resonance between cluster analogs being far away each other (from any number of meters to astronomic units). These cluster analogs are characterized by equal or at least similar energy forms of gravitational noises, which arose from protons solved/condensed in/from the physical vacuum. Understanding the interactions at such large distances is possible only when large masses are involved in this resonance process. Here, as will be shown later, the oscillating masses have to be seen as clusters of matter in which a "transmission of colors" between the dissolution and condensation part of protons takes place. The "transmission of colors" could be compared with a flock of bird or shoals of fishes changing their density, direction, and structure.

Dimers were found to be the simplest molecular clusters; their formation scheme is shown below:

$$\mathrm{A} + \mathrm{A} \leftrightarrow 2\mathrm{A}$$

this means that it is probable at a given time and in a defined space to find a dimer characterized by a concrete lifetime and a concrete energy formation even when it is remarkably lower than the destroying heat background (kT). Even if the life and observation times are very different, the dimer will be recorded by the device in any case. For very small lifetimes, however, the talk is on the highest statistical probability (Ψ) between two limit cases, only:

$$[\text{no signal}] \leftrightarrow [\text{yes signal}]$$

Stable clusters will be observed when this balance is shifted to [yes signal] and then the clusters are able to be extracted and characterized by conventional physical chemical methods, e.g. a unique cluster in polystyrene [23]. In the reality, $0 < \Psi < 100\%$ so it never amounts to zero or 100%. For dimers with similar cohesion forces, Ψ will be postulated to always be larger than zero where the postulate can be applied to mass concentrations of two or more molecules with nearly the same cohesion energies, as well as to mass concentrations (domains) within a large molecule, e.g. proteins or synthetic polymers.

The adhesion-cohesion interactions and structural factors, which stabilize the clusters/domains with the lowest potential energy, were found to be the driving forces of mass concentration formation. For example, in the formic acid dimer, the cohesion interaction between two equal molecules will be enforced by the structure factor (ring formation) and approach of the dimer rotating form to a spherical one with a minimum potential energy.

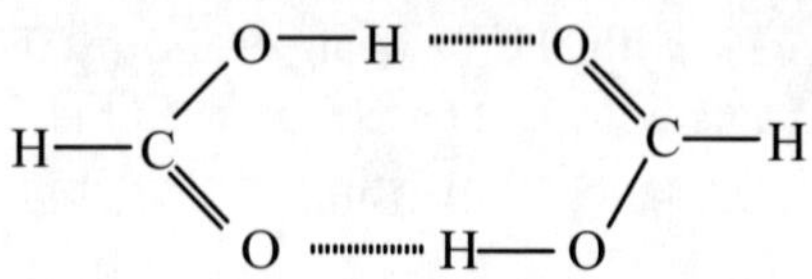

On the contrary, in the unstable expanded benzene dimer, the cohesion forces are assumed to be the essential component of linking forces whereas the structuring factor belongs to the destroying parameter. On the other side, in the stable collapsed benzene cluster, the weak cohesion interaction will be enforced by the structure component (the rings are perpendicular to each other). In the context of the atom molecule theory of the 20th century, the stability of the two benzene dimers could not be explained. Applying the new understanding of the molecule as an ensemble of nuclei and of electrons as stationary shock waves, however, means that the conflict will be solved (§22 and §23):

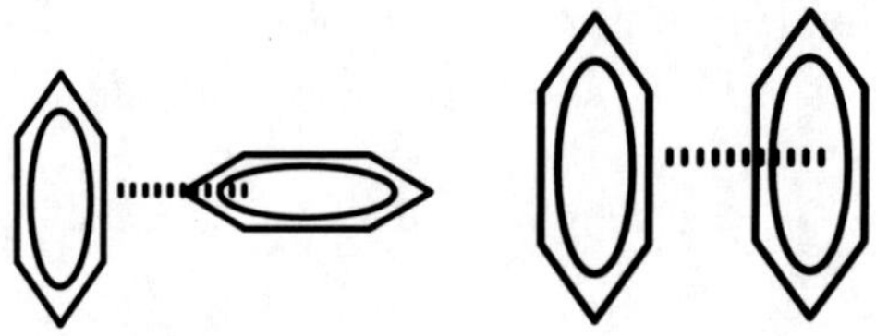

(collapsed cluster) (expanded cluster)

The existence probability of the expanded and collapsed clusters in liquid benzene was observed to be different, but under the influence of external structure factors, e.g. polymer chains (polystyrene, §9), the existence of the one or other dimer gets more probable and depends on the chain configuration.

The adhesion factors of cluster stabilization are characterized to play an important role in the long-range order formation in liquid mixtures (§10.8), solvated cluster ion pairs (§10), and in the formation of proteins (§11). To get some idea on clusters, computer models simulating the molecular dynamics of a defined number of molecules have been applied [24, 25, 26]. As will be shown later, stationary gravitational waves approve only determined mass concentrations to be built. Here molecular clusters as an organized matter obtain the energy for their existence from energy clusters in the stationary gravitational waves of a physical vacuum.

In Figure 2.1.1, some clusters in individual liquids that are statistically most probable are given. The theoretical cluster models for water and methanol were proven experimentally by the authors. It has to be mentioned that computer models only describe the quantifiable clusters, but do not reflect the qualitative properties or the fast changing surface structure and mechanism of cluster interaction with the surroundings they are part of. On the other side, the internal cluster structure is not less important, e.g. solvated clusters of ion pairs (cluster $HCl \cdot 98H_2O$ [27]) and clusters in styrene are characterized by a layer structure that gets even more complicated in the area of large masses and their ensembles (§5 and §8).

The reader should know that every mass concentration is characterized by an individual histogram of gravitational energy differences between the proton dissolution in the physical vacuum and its condensation from that one (§21, §22, §23). The two last processes are not discussed in detail because they are not the object of the monograph. The histogram problem has been investigated for more than 50 years in Russia, however, without yielding an explanation of the observed effects [28].

As shown below, the cluster structures (ensembles) in individual liquids and their mixtures are not constant and cannot be explained by the polymorphism of liquids. Cluster structures were found to be highly sensitive to weak energy fields both at the level of singular clusters and super clusters where heat flows, mechanical fields, seismic shock waves, resonance fields of cluster analogs, and gravitational radiation mainly belong to them. Clusters in polymers (domains) are slightly different from those in low molecular substances. The lifetime of clusters in liquids is around ps but those in polymers are usually nearly one second or more. In polymers, the talk is on a low-frequency movement of mass groups (macromolecule fragments), which do not have a defined surface. Here the small clusters (up to 200 Dalton) have to be understood as fast building and fast destroying (fluctuating) mass concentrations while larger, more stable domains exist as spirals, coils, sub-micelles, micelles, and as their associates. The modeling of domain structures in polymers is shown in Figure 2.3.2. The heterogeneities in solid polymers, polymer solutions, and gels can be ascribed to domains, too. Additional information about domains in proteins and synthetic polymers can be found in the chapters §11 and §9. Concerning the topic, only a few publications were researched by the authors. For example, the water entropy and the formation of hydrogen bridge bonds inside polymer coils were concluded to be the main factor determining the native biomolecule structure [29]. It is not sufficient to explain the domain formation only with the hydrogen bridges, though. All of the questions that have arisen will be discussed in this monograph and the answers will become a new base for the investigation of biomatrices and biopolymers.

After using the Zubow gravitation spectrometer for space observations (§24), it was shown that the idea of the cluster structure can be applied to the whole universe. Furthermore, based on the latest achievements in nuclear physics, the idea can successfully be assigned to the atomic nucleus. To understand the cluster ensembles as a form of existent matter means that it could become a unique platform within natural science.

Large clusters (> 1 kDa) ask for a new nomenclature because of a qualitatively new hierarchy in the molecular matter structure. These clusters themselves are ensembles of smaller clusters. Thus, in the following, different terminologies with a corresponding explanation shall be used.

One of the most important parameters for the stabilization of weakly organized structures has been ignored until the present time. This is the white noise, which penetrates the planet Earth [6, 7]. White noise was shown to be the external energy source assuring the entropy stabilization of molecular clusters. The effect is visible at the influence of mechanical fields (noises) on dispersed systems, e.g. sand with a broad particle size that is distributed on mechanical sieves. With vibration, not only does a particle size distribution take place, but also the formation of new regular structures on the surface of the sieve (http://www.youtube.com/watch?v=nO0bSSXmr1A&feature=related) occurs. Analog behavior was observed at the heating of a water layer with the formation of a six-member ring; here the role of the noise field was played by the heat flow [30].

Anthropogenically changed white noise causes changes in the resonance field (natural background field) of molecular clusters, unstable and metastable states in low-frequency movements of large masses in synthetic polymers, mutagenesis in proteins, etc. A conflict therefore arises between the molecular cluster structures under natural and anthropogenic conditions. Using computer simulation, an array of the most probable clusters with the lowest energy can be obtained, but with consideration to the stabilization by gravitational (white) noises, it should be acknowledged that new clusters could appear while simulated ones disappear. For example, the trimers of the benzene cation are not recorded under normal conditions [31]. In the following, some factors will be called, which either stabilize or destroy the cluster structure (Figure 2.3.3). It has to be mentioned that under real conditions, summary effects of stabilization/destruction are observed. The sum of all factors usually determines the Ψ value.

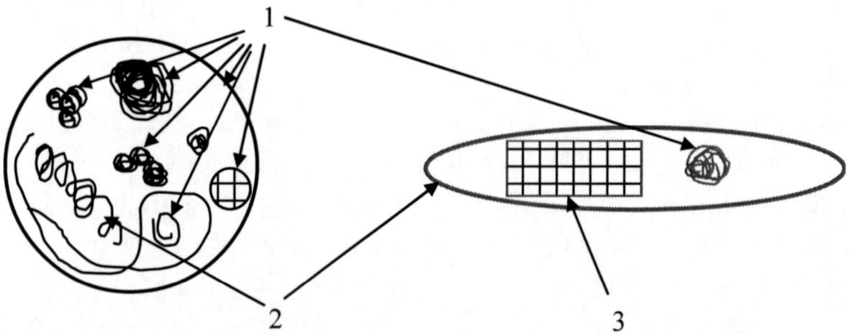

Figure 2.3.2. Models of mass concentrations (clusters/domains) in polymers. 1 – spherical coils, short spirals, spirals' associates, loops, spherical-like seed crystals, and compact members of chains; 2 – forms to be recorded as difficult (cluster with non-spherical forms); and 3 – forms that will not be recorded (crystals, crystallites).

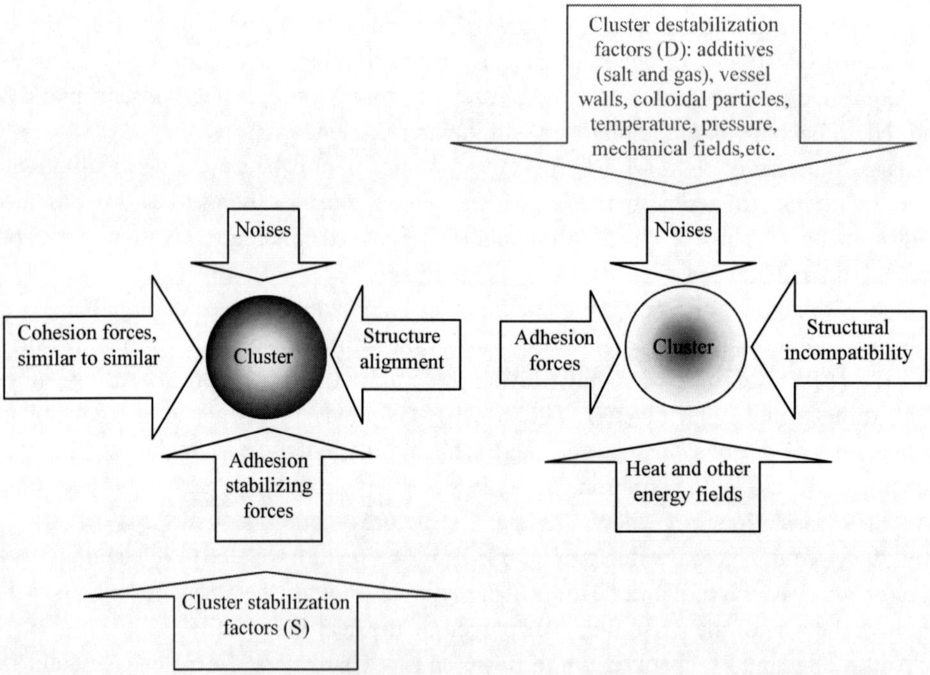

Figure 2.3.3. Factors of stabilization (S) and destabilization (D) of mass concentrations.

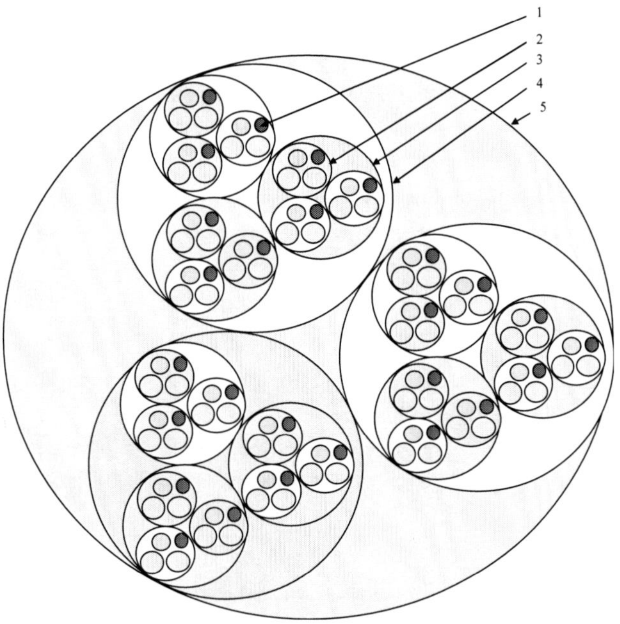

Figure 2.3.4. Architectural model of molecular cluster ensemble formation arising from the appearance/disappearance of mass concentrations. 1 – simple molecule dimers and molecule complexes, 2 – molecule associates and spirals in polymers, 3 – simple coils of folded macromolecules, 4 – sub-micelles of coils, and 5 – micelles and super coils.

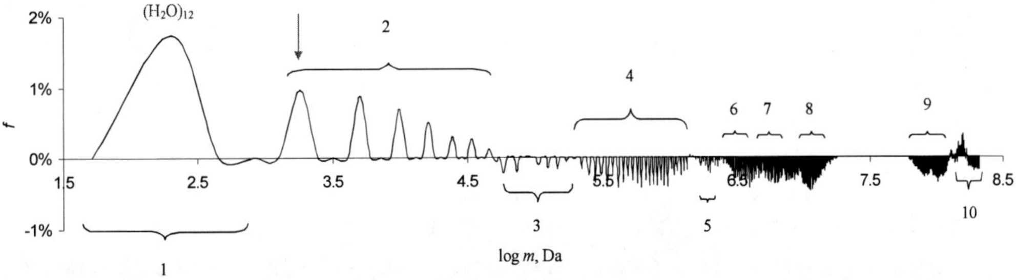

Figure 2.3.5. Refolded GMS-spectrum of potatoes amylopectin (genetically modified potatoes, extract of the pure amylopectin #1332), weak shock waves ($p <$ N/m^2), and the Zubow force constant (Z = $6.8 \cdot 10^{-15}$ N/m), 291 K. The base mass concentration types are marked with brackets.

Applying the Zubow gravitation mass spectroscopy, it is possible to look inside the "architecture" of the matter structure (Figure 2.3.4). The structure model can be developed further from atomic nuclei to molecular clusters, micelles, super micelles, colloidal particles and their associates, bio-populations, and even to concentrations of stars and galaxies. Regarding this model, a review GMS-spectrum of pure amylopectin extracted from genetically modified potatoes [32] will be shown in Figure 2.3.5 as an example of a cluster ensemble. The main mass concentrations in the interval between 200 and 100 million Daltons are labeled with brackets. Corresponding to the mass hierarchy given in Figures 2.1.1; 2.3.2; 2.3.4; and 2.3.5, at least 10 kinds of mass concentrations can be chosen in the first approximation. These are the signals of: (1) members of polymer chains and low molecular substance inclusions in polymers, e.g. water $(H_2O)_{12}$ [14, 33]; (2) spirals and spirals' associates in coils, e.g. spirals of the base amylopectin cluster consisting of 12 α-D-glycopyranose members [34] (signal is marked with an arrow); (3) area of spirals' associates and sub coils; (4, 5) area between sub coils and super coils/super micelles, and (6–8), (9), and (10) areas for micelles' associates and larger particles.

Conclusion

Thus, mass concentrations are concluded to be built in molecular matter, although the reasons for that still have to be researched.

On the other side, molecular mass concentrations (clusters, domains) are assumed to be formed in stationary gravitational waves of white noises, which will be proved in the present monograph.

Stationary waves in gravitational noises form the structure of the physical vacuum from which molecular clusters probably obtain their energy.

References to 2

[1] Campbell J.A. Chemical Systems. Energetics, Dynamics, Structure. W.H. Freeman and Company. San Francisco. 1970.

[2] Minkin V. I. Co-ordination link. Dictionary of Chemical Enciclopedy. Moscow. Soviet Enciclopedy. 1983, p. 275.

[3] Beatriz Giner, Santiago Martin, Hector Artigas, Maria C. Lopez, Carlos Lafuente. Study of Weak Molecular Interactions through Thermodynamic Mixing Properties. J. of Phys. Chem. B. 2006, vol. 110, pp.17683-17690.

[4] Zubow K.V., Zubow A.V., Zubow V.A. Ensemble of Clusters – New Form of Molecular Matter, Risks and Chances. Zubow Equations. In Advances in Chemistry Research, 2010, vol. 5, pp. 107- 145. Editor J.C. Taylor. Novapublisher. NY.

[5] Zubow K., Zubow A.V., Zubow V.A. Water Clusters in Plants. Fast Channel Plant Communications. Planet Influence. Journal of Biophysics Chemistry, 2010, vol.1, no.1, pp. 1-11.

[6] Horsthemke W., Lefever R. Noise-Induced Transitions. Theory and Applications in Physics, Chemistry and Biology. Edit. Hermann Haken, vol. 15. Springer-Verlag, Berlin, Heidelberg, NY, Tokio 1984.

[7] Halberg Franz; Cornelissen Germaine; Regal Philip et al. Chronoastrobiology: proposal, nine conferences, heliogeomagnetics, transyears, near-weeks, near-decades, phylogenetic and ontogenetic memories /Biomedicine & pharmacotherapy. 2004, vol. 58 Suppl 1, pp. 150-187.

[8] Guang-Jun Guo, Yi-Gang Zhang, Ya-Juan Zhao, Keith Refson, Gui-Hua Shan. Lifetimes of cagelike water clusters immersed in bulk liquid water: A molecular dynamics study on gas hydrate nucleation mechanisms. J. Chem. Phys. 2004, vol. 121, no. 3, pp.1542-1548.

[9] Bartenev G. M., Zelenev J. V. Physics and Mechanics of Polymers. Moscow, Vyshaja Shkola. 1983, p. 21 (in Russian).

[10] Zubow Kr. V., Zubow A.V., Zubow V.A. Cluster structure of liquid alcohols, water and n-Hexane. J. of Appl. Spectr. 2005, vol.72, no. 3, pp. 321-328.

[11] Zubowa, K. V.; Zubow, A. V.; Zubow, V. A. Oscillation spectra of seed NaCl crystals in aqueous solutions. J. of Applied Spectroscopy. 2005, vol. 72, pp. 840-847.

[12] Campbell J.A. Chemical Systems. Energetics, Dynamics, Structure, vol. 2. W.H. Freeman and Company. San Francisco. 1970.

[13] Zakharov V.V., Brodskaya E.N., Laaksonen A. Molecular dynamics simulation of methanol clusters J. Chem. Phys. 1998, vol.109, pp. 9487-9494.

[14] Annika Lenz, Lars Ojamäe. On the stability of dense versus cage-shaped water clusters: Quantum-chemical investigations of zero-point energies, free energies, basis-set effects and IR spectra of $(H_2O)_{12}$ and $(H_2O)_{20}$. Chemical Physics Letters. 2006, vol. 418, pp. 361-367.

[15] Chaplin M. (the portal of south bank university of London) in http://www.lsbu.ac.uk/water

[16] Anderson P., Steinbach C., Buck U. Vibrational spectroscopy of large water clusters of known size. European Phys. J. D. 2003, vol. 24, pp. 53-64.

[17] Tshernov A.A. Crystallization. Physical Enciclop. Dictonary. Moscow. Soviet Encyclopedia. 1984, p. 319 (in Russian).

[18] Chemical Enciclopedy. Moscow. Soviet Encyclopedia. 1990, vol. 2, p. 1164 (in Russian).

[19] Lian Li., Ogawa T. Clusters and their properties in aqueous solutions of KDP, KCl and sugar. J. of Crystal Growth. 2000, vol. 211, pp. 286-289.

[20] Zubow K., Zubow A.V., Zubow V.A. The Phenomenon of Proton Dissolving in Vacuum and of Proton Condensation from Vacuum. Two Forms of Protons, Structure of Nuclei, Electrons and Atoms, J. of modern physics, vol.1, no.1, 2010. 33-43.

[21] Omori Soichi, Komabayashi Tetsuya, Maruyama Shigenori. Dehydration event and earthquake in the subducting stab. Frontiers Science Series, 44 (Physicochemistry of water in geological and biological systems), 241-257. Publisher: Universal Academy Press, Inc. 2004.

[22] Zubow K., Zubow A.V., Zubow V.A. The Phenomenon of Water Clusters Distribution in Water and Solvated Clusters of Ion Pairs of NaCl in Solution Near the Wall. The Wall Effect. J. of Chem. Industr. Sankt-Petersburg. 2006, vol. 83, no. 6, pp. 300-306 (in Russian).

[23] Zysman, Viviane; Nguyen, Tuan Q.; Kausch, Henning H. Degradation on freezing dilute polystyrene solutions in p-xylene. J. of Polym. Sci., Part B: Polym. Phys. 1994, vol. 32, no.7, pp. 1257-1269.

[24] Koneshan, S.; Rasaiah, Jayendran C. Computer simulation studies of aqueous sodium chloride solutions at 298 K and 683 K. J. of Chem. Phys. 2000, vol. 113, no.18, pp. 8125-8137.

[25] Siu, Chi-Kit; Fox-Beyer, Brigitte S.; Beyer, Martin K.; Bondybey, Vladimir E. Chemistry-A Ab Initio Molecular Dynamics Studies of Ionic Dissolution and Precipitation of Sodium Chloride and Silver Chloride in Water Clusters, NaCl(H2O)*n* and AgCl(H2O)*n*, *n* = 6, 10, and 14. Europ. J. 2006, vol.12, no.24, pp. 6382-6392.

[26] Peterson K.A., Puzzarini C. Systematically convergent basis sets for transition metals. II. Pseudopotential-based correlation consistent basis sets for the group 11 (Cu, Ag, Au) and 12 (Zn, Cd, Hg) elements. Theor. Chem. Acc. 2005, vol. 114, pp. 283-287.

[27] K.V. Zubow, A.V. Zubow, V.A. Zubow, Principles of gravitation spectroscopy. New form of molecular matter. Processes. Fields. “Aist Handels- und consulting gmbH”, dep. R&D, Berlin, 2011, electronic book (www.zubow.de), p. 900. (in Russian).

[28] Shnoll S. E. "The Arrow of Time" in the Experiments in which Alpha-Activity was Measured Using Collimators Directed at East and West. Progress in Physics. 2006, vol. 2, no. 4, pp. 39-45.

[29] Yuichi Harano, Ronald Roth, Yuji Sugita, Mitsunori Ikeguchi, Masahiro Kinoshita.Physical basis for characterizing native structures of proteins /Chem. Phys. Letters. 2007, vol. 437, pp. 112-116.

[30] Milton van Dyke. An Album of Fluid Motion. The Parabolic Press Standford California, 1982.

[31] Schriver K.E., Paguia A.J., Hahn M.Y., Honea E.C., Camarena A.M., Whetten R.L. Are clusters of nonpolar molecules icosahedral? J. Phys. Chem. 1987, vol.91, pp. 3131-3133.

[32] Reichmann M., Schwarzfischer A. Pure amylopectin starch from potatoes. A renewable raw material with novel quality. Kartoffelbau. 2006, no. 4, pp. 170-172.

[33] Bogdanov E.V., Manturova G.M. Equiclaster model of water. Biomedicinskaja Radioelektronika. 2000, no. 7, pp. 19-28. (in Russian).

[34] Zubow K.V., Zubow A.V., Zubow V.A. Using of flicker noise spectroscopy for non destroying analysis of nano structures. Zavodskaja Laboratorija. Diagnostics of materials. 2008, vol. 74, pp. 40- 45. (in Russian).

Chapter 3

PRINCIPLES OF GRAVITATION MASS SPECTROSCOPY (GMS)

The main principle of gravitation mass spectroscopy is based on recording the resonance energy between the weak gravitational radiation (GR) of the surrounding molecular matter (base gravitational field) and that of the sample (sensor). Here the weak gravitational radiation (WGR) is formed through nuclei concentrations in a molecular cluster where the nuclei concentrations are the product of gravitational fields in the structured physical vacuum (Figure 3.2.1) with regard to Zubow equations (§3.2.1), [1]. The energy balance of the proton with the physical vacuum (balance between proton dissolving in a physical vacuum and super light condensation of the proton from the physical vacuum) was found to cause WGR [2] (§22):

$$P \leftrightarrow E_{\Gamma}$$

Where P is the proton and E_{Γ} is the gravitational energy. In the world surrounding us, this balance is shifted a little to the proton dissolution side, although under the constant influence of celestial bodies, shock waves, and other energy factors, it has to be characterized as flexible and sensitive. The processes according to this balance proceeds with super light velocities (§22, §23) [3], generating WGR through atomic nuclei.

The proton concentrations (atomic nuclei) in molecular clusters can be described as follows (Scheme I) and is characterized by two kinds of energy histograms.

In nuclei/protons' concentrations, the WGR effect will be enforced through resonance and cooperative interactions between the same atomic nuclei that oscillate as one dimension of uniform pendula (ensemble with lowest potential energy). In contrast, in the dissipative state, these effects are not distinctive. Therefore, the histograms reflect the energy state of the nuclei concentrations in a molecular cluster as well as their balance with similar ensembles of the surroundings. The nuclei concentrations can be influenced in two ways: (i) through the change of the gravitational field (white noises) and (ii) through externally generated shock waves. These items will be discussed in the present monograph but in the reverse order.

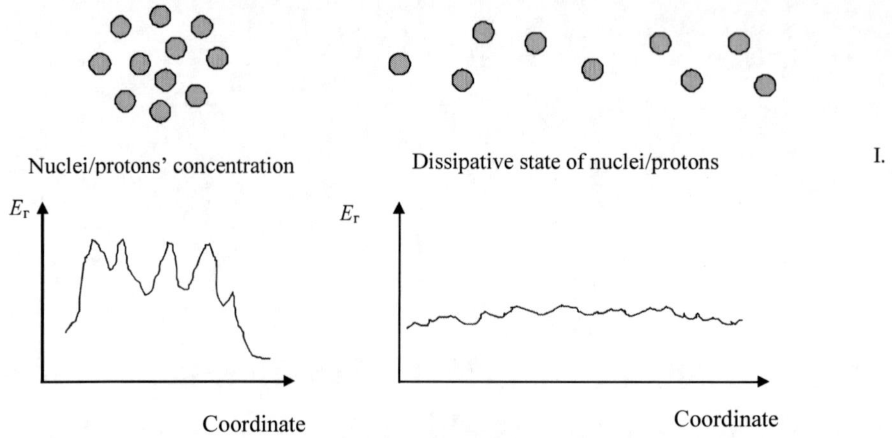

Our finding describes a new form of matter—molecular cluster ensembles, which were built under the influence of energy clusters in a physical vacuum (Figure 3.2.1). An information and energy balance was observed between the energy clusters and molecular ones. This model of the universal structure should be an indication for the heterogeneity of the physical vacuum (§18) and it will be discussed in detail in the present book.

The molecular cluster ensembles are characterized by their own energy capacity, which was found to be highly sensitive to weak energy flows. In the metastable state, the energy histograms of cluster ensembles or their GMS-spectra were even shown to be invertible, being one of the main properties of molecular clusters. Additionally, the inversion effect belongs to the most sensitive characteristics of the method and its understanding is the main principle of GMS.

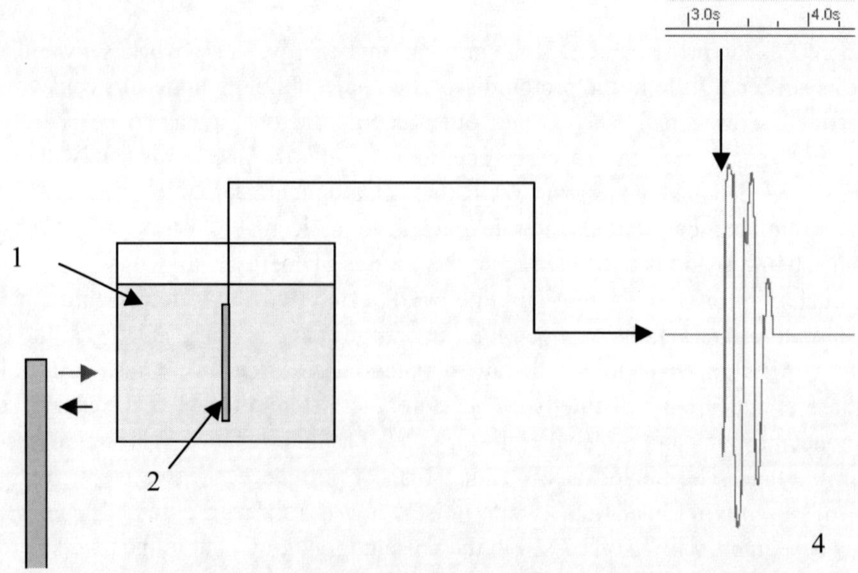

Figure 3.0.1. Scheme illustrating the principle of GMS-spectroscopy. 1 – vessel with water (100 ml), 2 – GMS-sensor, 3 – glass rod provided for beating on the vessel wall to generate weak shock waves in the water, and 4 – response signal from direct shock waves and reflected ones.

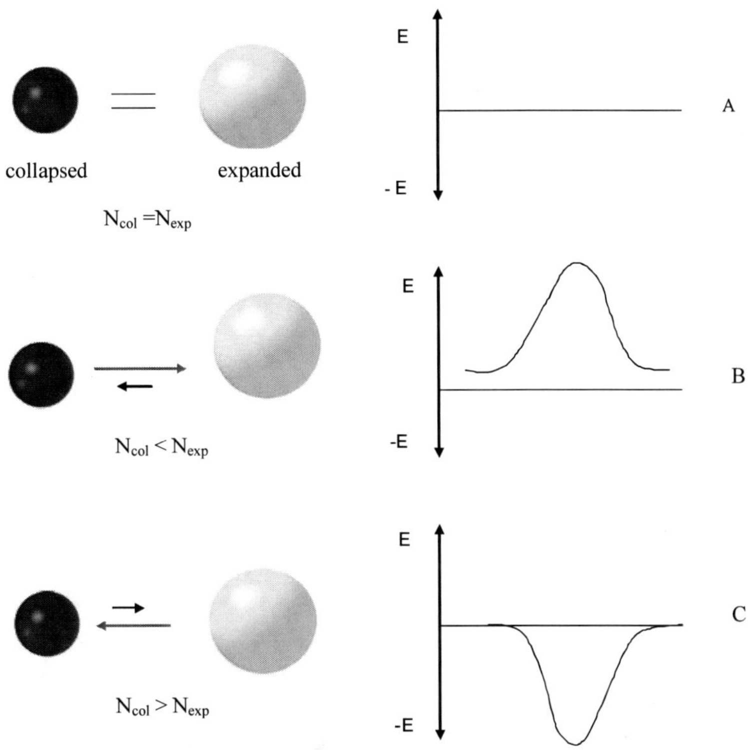

Figure 3.0.2. Three possible cases for molecular cluster interactions with shock waves. (A) is a concentration of both expanded and collapsed clusters ($N_{col} = N_{exp}$); the character of the interaction with the shock wave (absorption and release of energy) is the same. In this case, the GMS-signal is not registered; a mutual neutralization of signals is proceeding. If the balance (expanded – collapsed) is shifted to the expanded form, then a signal of an endothermic process appears in the GMS-spectrum, (B) indicating that energy shall be absorbed by the expanded cluster, which leads to its melting. In contrast, if in this balance the collapsed form of clusters is dominating, (C) then we have an exothermic process where the collapsed clusters are densified and crystallized (energy release). If there are no mass concentrations in the substance (theoretical case), then there are no GMS-signals either, e.g. a crystal monolith without any defects or amorphous regions. Here, however, signals from energy clusters shall be observed (§18).

In GMS, the interaction energy between weak shock waves and the sample will be analyzed where the shock waves can be man-made as well as generated by gravitational noises, e.g. from celestial bodies, earthquakes, etc. The practical principle of this spectroscopy will be shown using a simple experiment (Figure 3.0.1).

Upon beating the vessel wall, weak shock waves were found to be generated in the water; they approach the GMS-sensor and change its structure and electrical potential. Then in the sensor structure, the distribution of hydrogen bonds between the water molecules shall be changed as well as the conditions for proton jumping. The proton jumping is characterized to be of high importance and will be discussed in detail in a special chapter (§21). Mass concentrations in the substance are able to absorb shock wave energy as well as reflect it. All these events change the sensor potential, which can be recorded. Furthermore, every mass concentration is described by its own typical energy histogram of the proton jumping in the physical vacuum and back (§21). Weak gravitational radiation, generated by the array of proton jumping, influences the protons' state in the sensor and therefore reflects the structure of the molecular clusters. WGR has to be assigned to white noises because they are pulsating shock wave flows (Scheme I). Hence, the WGR of molecular matter is the base of practical GMS-spectroscopy.

In the next section, three possible cases for cluster interaction with shock waves will be shown.

3.1. Interaction Principles of Matter with Shock Waves

In paper [4] the principle of how weak shock waves with a spherical subject interact, when applied to GMS, shall be explained. Below, the interaction of a sample with weak shock waves shall be discussed.

3.1.1. The Influence of the Molecular Matter's Density of Molecular Matter and the Barrier's Form

In GMS, weak shock waves can be generated through a forced oscillation of a molecular film in the GMS sensor. Here a quasi-running wave will be produced, which moves through the liquid or solid sample. The wave movement is characterized by the wave resistance and does not depend on the wave form. On the other side, this process is described by the known equation for the impedance of acoustic media through which a plane wave is running:

$$p/\mathrm{w} = \mathrm{dc}$$

where p is the sound pressure in the running plane wave, N/m^2; w is the oscillation velocity of sample particles, m/s; d is the medium density, kg/m^3; and c is the wave velocity, m/s.

Due to the fact that the density of most liquids and polymers is nearly 10^3 kg/m^3, a simplified equation can be used to calculate the oscillation velocity of sample particles (mass concentrations):

$$\mathrm{w} = 10^{-3} \cdot p/\mathrm{c}$$

The sound pressure in a running plane wave (c ~ 1,500 m/s) at impact on a Chaplin water cluster (pendulum) of 280 molecules (4.4 kHz; diameter of 3 nm; 1,000 kg/m^3) amounts to ~1 N/m^2 at w = $6.6 \cdot 10^{-7}$ m/s (~10% pendulum deviation of the radius of 1.5 nm, $4400 \cdot 10\% \cdot 1.5 \cdot 10^{-9}$ m). In the next section, these waves shall be called "strong shock waves" ($p \geq 1$ N/m^2) while "weak shock waves" will be understood as waves with a lower pendulum deviation than 10% ($p < 1$ N/m^2). By modeling the oscillation process of this cluster, it becomes clear why cluster destruction takes place when the deviation is higher than 10%. Sure, the classification of shock waves into weak and strong ones requires a special chapter in the present monograph, however, for the first edition, this division is sufficient.

To generate shock waves, the GMS-device was used: $< 3 \cdot 10^{-2}$ V·A for strong waves and $< 5 \cdot 10^{-9}$ V·A for weak ones. When the medium is heterogeneously mixed, the wave resistance depends on the barrier form. For a barrier with the diameter, S, which is moving in a medium with the density, d, the wave resistance will be described by:

$$C_B = F/(dv^2/2)S$$

where F is the force and v is the velocity of direct movement [5].

Therefore, the wave propagation in a substance was observed to depend on the substance density as well as on existing barriers whose form and nature (absorption or release of energy upon interaction with shock waves) influence the resistance of waves reflected from the barriers. This is the information base of GMS.

We found the criterion of the globule cluster structure to be the harmonized physical model of an oscillating ball (moment of inertia of ball rotation $J = (2/5)mr^2$, where m is the mass, r is the radius, and J is the moment of inertia) with the computer data of the Chaplin models for clusters consisting of 280 and 1,890 water molecules (http://www.lsbu.ac.uk/water/index.html) while other geometrical rotation figures, like hollow or compact cylinder, cone, or randomly rotating bodies ($J = 0.5\pi \cdot \rho \cdot \int r^4 dx$, where ρ is the density of the rotating body and π = 3.14 and is contour integration from x_1 to x_2) were not suitable because of bad correlation. The criteria apply to the general thermodynamic laws concerning the minimization of the bodies' surfaces as well. Weak gravitational shock waves from white noises were assumed to interact with barriers according to the same physical and mathematical models. No literary data on this matter, however, has been researched up to 2011.

3.1.2. The Influence of the Substance's Durability

Object destruction by shock waves is known and is well described in military science. The jump of molecular densification (shock wave) is reflected by a barrier, where the shock wave and barrier are destroyed. Here, the destruction process depends on the force and pressure of the shock wave as well as the nature and structure of the barrier's heterogeneity. How a barrier absorbs shock wave energy is dependent on its nature, form, density, and structural stability (construction) [6]; it belongs to the platform of GMS, too.

3.1.3. Generation of Weak Shock Waves for GMS

Energy impulses compulsorily generated on the sensor surface raise changes in the sensor structure (Figure 3.1.3.0). In the absence of an electrical potential, the layers of a polar substance on the sensor surface, however, are orientated with increasing distance to the sensor surface and the alignment of the molecules is decreased. In contrast, an additional electrical potential increases the thickness of the orientated molecule layer. A compulsorily and periodically changed thickness of the film [7] could be applied to beat the sample surface that is in contact with the sensor. After the disappearance of the enforced potential, the system returns to its beginning state [4, p. 62] where the velocity of the return to the beginning depends on the sample nature, temperature, pressure, background noises, etc. Film beating raises a reaction in its molecular structure and therefore reflects the sample's assemblage. In the next part, the shock wave reflected from the sample returns to the sensor layer and changes its electrical potential. This can be illustrated using the scheme in Figure 3.1.3.0.

Theoretically, the structure of the reflected wave (dotted line) has to agree with that of the waves that strikes the sample, though practically, it is quite different. A real shock wave only penetrates into the sample some mm deep depending on its power. Here the sample actively interacts with the shock wave, and absorbs and dissipates its energy. In this depth of some mm, the waves to be reflected shall be formed. The reflected waves are directed contrarily to the incident ones and they differ structurally from each other. The reflected waves are found to hold the information on the sample's structural heterogeneity; additionally, they influence the generation of shock waves and decrease its performance. This decreased generator performance of broadband shock waves can be recorded to analyze the structure of the energy loss. Furthermore, shock waves can initiate an oscillation of mass concentrations (clusters), which for their part becomes a resource of different energy solitons (because of a changed distance between nuclei). Weak gravitational radiation like solitons spreads out in liquids up to some cm and in solids up to 2-3 mm. For gigantic cluster ensembles (tectonic masses, planet and star systems), however, the influence can spread over arbitrary distances. The creativity of the researcher consists in registering this radiation, the real signals' extraction from background noises, and in the signals' assignment for a suitable understanding. For measuring the energy loss, routine methods like potentiometry can be applied but the signals are subject to further processing: archiving, extraction, and cleaning for which a new computer design is needed.

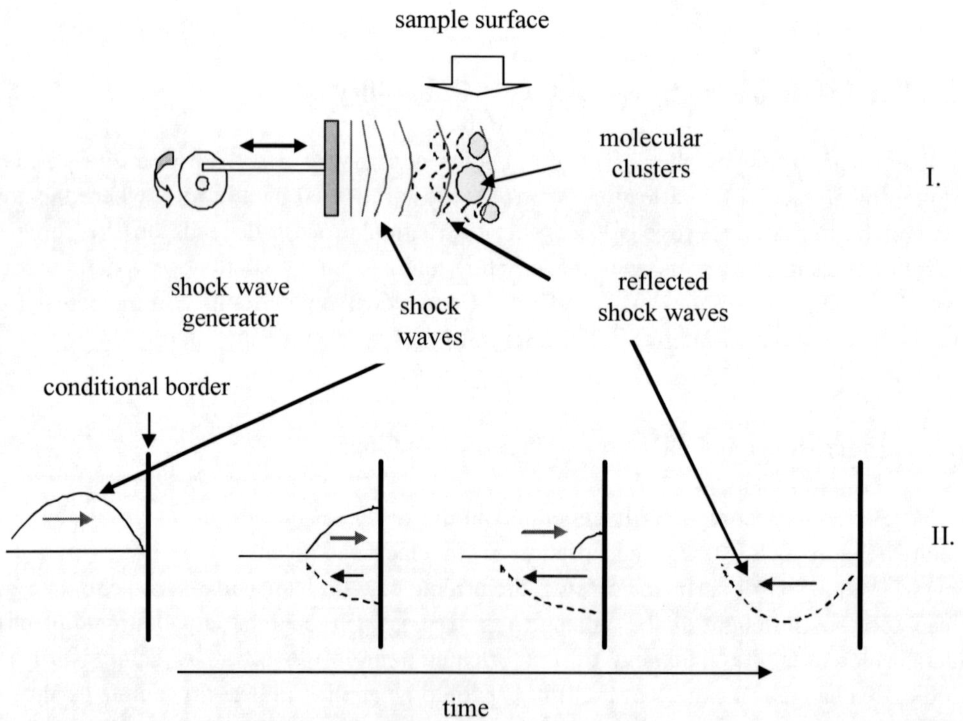

Figure 3.1.3.0. Scheme for the generation of shock waves and their influence on the sample (I) and the shock wave entrance into the sample and exit of the reflected wave from the sample (II). The reflected wave is marked with a dotted line.

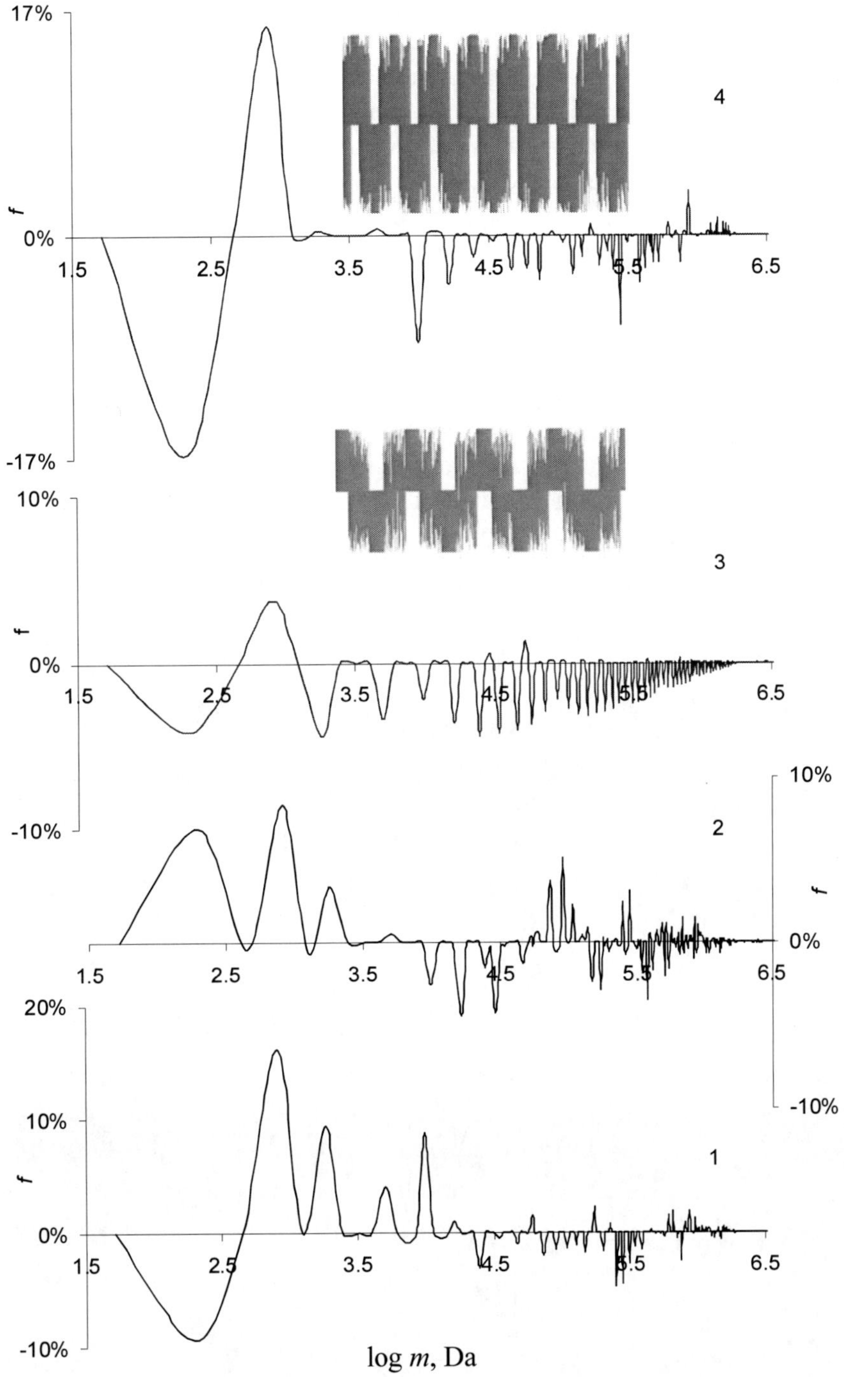

Figure 3.1.3.1. Review of GMS-spectra for an aqueous NaCl solution obtained after recording with two kinds of shock waves. 1 – 26 wt. % solution aged for 18 months, strong shock waves (comment in the text); 2 – 26 wt. % freshly prepared solution, strong shock waves; 3 – 3.5 wt. % freshly prepared solution, weak shock waves; and 4 – 3.5 wt. % solution aged for 18 months, strong shock waves. Waves before signal cleaning are given above spectra 3 and 4.

To investigate the cluster structure in molecular matter, shock waves are needed but their power should only marginally influence the sample structure. On the other hand, the anthropogeneous shock waves also have to be taken into account; their influence could be minimized by applying a special sample protection, weak shock waves, sensitive hardware, and intelligent computer programs. The use of standard methods for shock wave generation is one and the same—measuring cells and sensors are found to be the best alternative to increase reproducibility. The influence of earthquakes and celestial bodies, however, must not be forgotten. As shown below, the change of shock wave generation leads to that result under one and the same conditions in the base water cluster (11±1 molecules). It can dominate either the expanded form or the collapsed one. The reaction of polymer materials on shock wave changes is more stable than that of individual liquids and solutions. For salt solutions, the main signals of solvated clusters of ion pairs remain unchanged while the interaction of clusters with surroundings are highly changed and the number of weak signals arisen from non-skeletal clusters is decreased. The stabilization of weak shock waves is of high importance for the GMS-calibration as well as to understand the formation of the super cluster structures and the cluster's interaction with its surroundings.

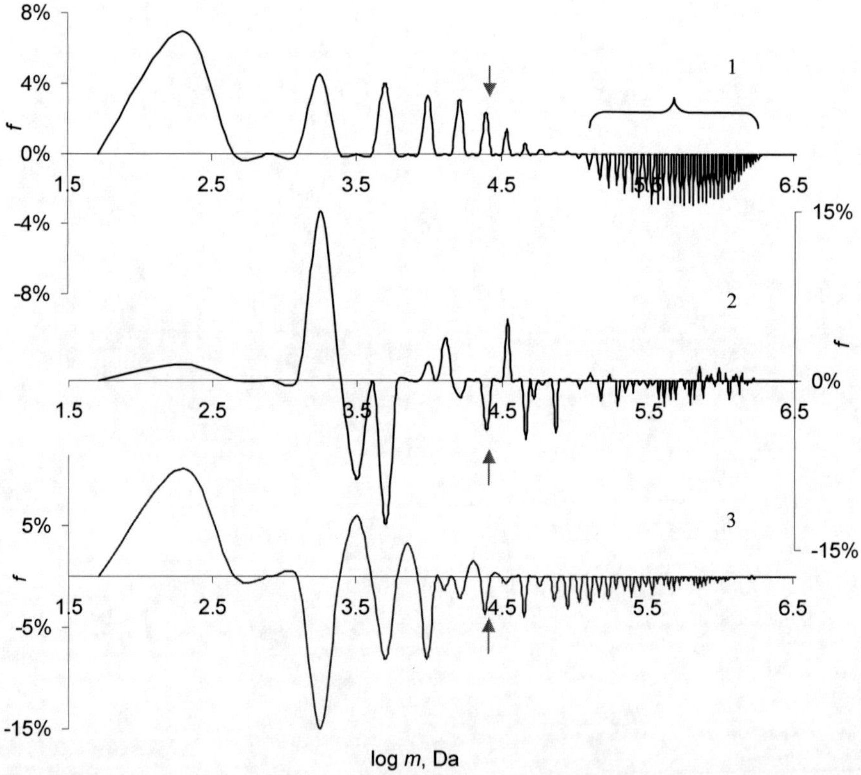

Figure 3.1.3.2. GMS-spectra of dried powders (< 9 wt. % H_2O, 1, 2) of casein and casein gel (60 wt. % H_2O, 3). Review spectra (2, 3) – strong shock waves spectrum (1) – weak shock waves ($p < 1$ N/m^2), Z = $6.4 \cdot 10^{-15}$N/m. The sub-micelle oscillation diapason is marked with brackets. Casein coils with a molecular mass of 24±0.5 kDa were taken as intern standard (described by an arrow), 293±1 K.

In Figure 3.1.3.1, there are two raw spectra given where the incident shock waves are overlapped by their reflected ones and the GMS-spectra after cleaning and signal extraction. There are differences found between the spectra of the salt solution. Stronger waves simplify the GMS-spectra with the loss of some information, though. In contrast, a wave's weakening maintains the super cluster structure of the sample. Thus, the lower the shock wave energy, the more the spectrum corresponds to the reality. On the other hand, the negative influence of a strong wave can be decreased by a shorter exposure time. The observer decides by himself what wave energy and measuring time have to be used to gain optimal information on small molecular clusters and the concentration of small clusters, super cluster structure, and super cluster agglomerates. For example, it is possible to obtain the most information on the globular casein structure (globule structure, sub-micelle, micelle/distribution, and micelle interaction with surroundings) by varying the character of the shock waves (energy, form). Whereas for the analysis of the simplest clusters (molecular dimers, seed crystals), a maximal lowered pressure at the shock wave front is needed. With the best GMS-spectrometer the shock wave energy can be reduced until only the solitons' registration and the natural noises caused by cluster oscillations remain.

In Figure 3.1.3.2, the GMS-spectra of the globular protein casein using two kinds of shock waves are shown. There are strong differences in the signal intensities, even leading to reversed signals for the very same oscillators. The quality of a shock wave, e.g. its front pressure and form, is the main factor that determines the signal reproducibility, the right signal interpretation, and energy calibration.

In Figure 3.1.3.3, how the impact time of the shock wave influences the average molecular mass of clusters (M_{GMS}) is shown.

Independent of its power, the shock wave influences the long-range order in any case. Two extreme cases will be shown in Figure 3.1.3.3. M_{GMS} change usually is not more than 10 to 20% for the most of analyzed molecular systems.

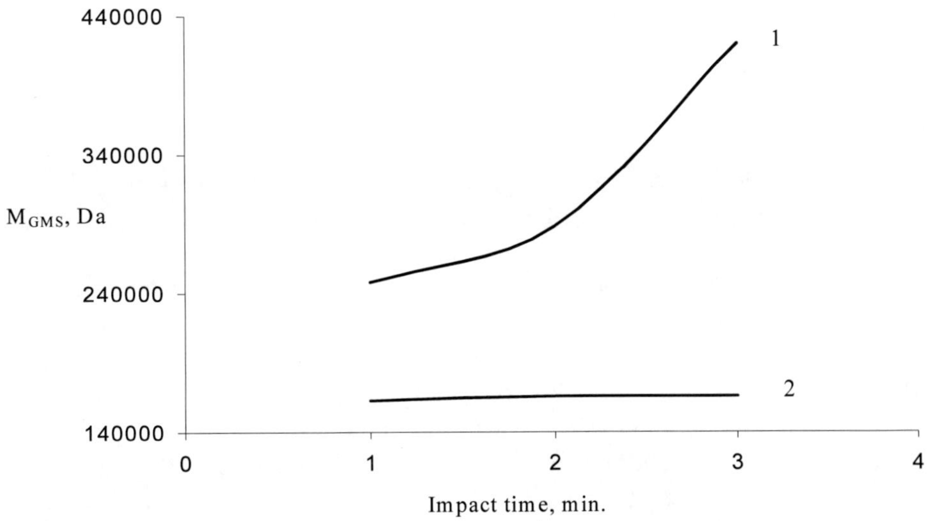

Figure 3.1.3.3. Average molecular cluster mass versus impact time of strong shock waves (M_{GMS}). 1 – vitamin C (keto form) melted and cooled to 293 K and 2 – bitumen melted and cooled to 293 K.

The choice of experimental conditions depends on which goal shall be accomplished. For instance, to analyze very weak effects of the cluster formation by small masses, the shock wave influence has to be decreased (impact time, pressure, and form). On the other side, to improve the signal resolution, modern signal processing and computer design will be applied. Moreover, different mathematical procedures like dividing, subtraction, multiplication, etc., of spectra are possible but only in the frame of one experimental series.

Weak gravitational radiation from the sample will be analyzed, recording the resonance energy between the sensor, WGR (energy clusters), and the source of gravitational radiation, e.g. celestial bodies, molecular clusters, etc. In the case of molecular clusters, weak shock waves will be generated as the result of the appearance/disappearance of molecular clusters in the sample. Here the electrical potential of the sensor being in direct contact with the sample will be recorded.

The excitation of the gravitational field is observed to be the result of the remote resonance interaction between WGR of different substances (Figures 3.1.3.4 and 3.1.3.5, §14), where WGR is caused by an unbalanced dissolution/condensation process of protons in a physical vacuum (§21).

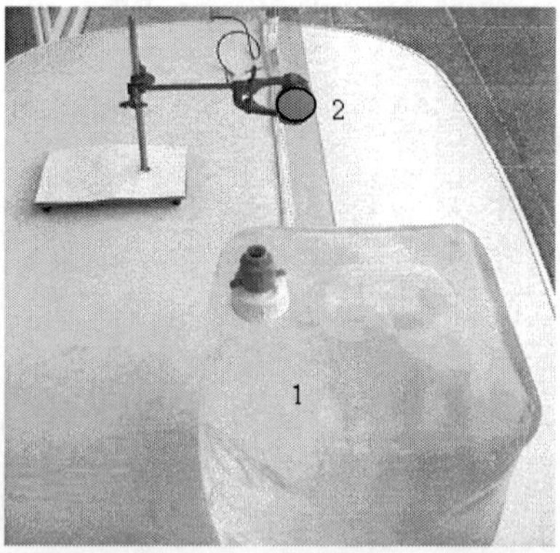

Figure 3.1.3.4. Equipment for the investigation of the remote interaction between sample WGR and sensor WGR. Example for the GMS-sensor approaching the plastic bag filled with water.

According to this experiment, as well as to the experiments in §10.10, §14 and §23, molecular matter (water, concrete, iron, etc.) is characterized by WGR. As is visible in Figure 3.1.3.5, approaching the sensor with a small mass (little level of WGR radiation) in relation to the bag containing the larger mass (water, concrete, iron, etc.), the gravitational field in the sensor space will be excited. The cluster structure e.g. of water in the GMS-sensor will be influenced most strongly by this excitation. All these events change the electrical potential of the GMS-sensor, which can be measured.

To not complicate the monograph material unnecessarily, the reasons for proton dissolution/condensation in/from a physical vacuum are renounced (§10, [2]); only the consequences are discussed. Clusters or mass concentrations reflect the energy state of stationary waves that were caused by gravitational radiation in white noises. On the other side, the presence of molecular clusters indicates that proton jumping is not a random process but rather an ordered one.

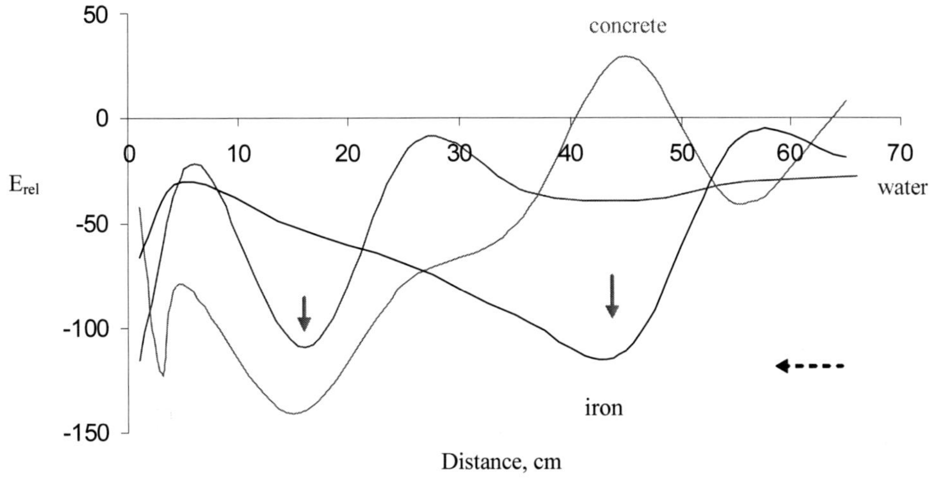

Figure 3.1.3.5. GMS-sensor potential (in relative units of energy arising from an energy cluster ensemble, pendulum I) that is lost to larger masses (molecular cluster ensemble with a larger potential energy, pendulum II) upon approaching (described with an arrow) pendulum I and pendulum II (Figure 3.1.3.4). The extremes of the curves are marked with vertical arrows.

Thus, there are two principal directions to analyze the long-range order in molecular matter:

1. Compelled generation of incident shock waves and analysis of shock waves reflected from heterogeneities in the sample
2. Analysis of white noises (gravitational radiation) and their structurally stable shock waves (H-soliton), which are generated by proton jumping in the sample

Whereas the first direction should be interesting for high energy technologies, the second one is more applicable for non-destructive analysis of the long-range order in molecular matter.

3.2. A New Form of Molecular Matter – Cluster Ensembles: Particles and Waves

A new form of molecular matter—cluster ensembles—has been discovered by the authors; understanding clusters as molecular mass concentrations is essential to this new discovery. A cluster ensemble is characterized as an energy carrier for a unit of clusters. Ensembles were found to be unique sensors that react on all energy flows known up to now, from seismic waves of earthquakes to the gravitational radiation of planets. For their existence, molecular clusters in the ensemble get their energy from energy clusters (EC) in the stationary gravitational waves of their surroundings. The possibility to form molecular clusters and their ensembles is strongly determined by EC (Figure 3.2.1) [1].

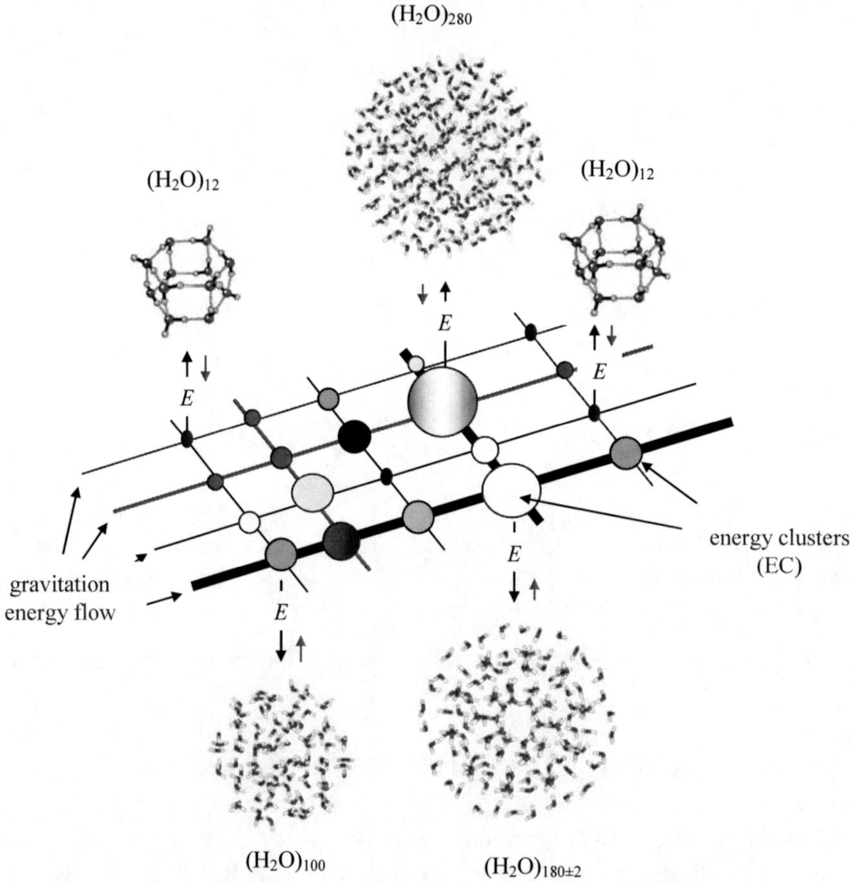

Figure 3.2.1. Model describing how molecular clusters are formed from energy clusters in stationary gravitational waves of white noise (understanding the energy field is shown in only one surface network). At nodes of the network, the energy clusters are shown as colored spheres with sizes correlating to the potential gravitational energy level caused by different cosmic sources. Above the network are some water clusters that get their energy for their self-organization from the energy network of EC. Cluster models were kindly provided by Chaplin [8] and Lenz [9]; the base cluster $(H_2O)_{12}$ first was discovered by Bogdanov and Manturova [10].

3.2.1. Zubow Equations

The cluster distribution in an ensemble shall be described by Zubow equations (I-III) even though the existence of clusters in molecular matter is already well known. Clusters, which should be understood as molecular mass concentrations, are experimentally proofed in liquids [11, 12]. The formation of a cluster ensemble where each cluster takes a certain energy level, however, is not completely clear even at present. To extract a single cluster from this ensemble without destroying the cluster or ensemble is not possible. For instance, for every polymorph state of liquid water there is its own ensemble. The authors of the present work found an equation connecting masses of clusters (m, Dalton) with the frequencies of their oscillations (ω, Hz) [1, 12]:

$$m = (Z \cdot N_A \cdot 1000)/(4\pi^2 \cdot \omega^2) \quad \text{(I)}$$

here Z is the Zubow constant $(7\pm2)\cdot10^{-15}$ N/m, [12]) and N_A is the Avogadro number $(6.02\cdot10^{23})$. The equation can be simplified until:

$$m = 10^{11}\cdot\omega^{-2} \quad \text{(II)}$$

The clusters were characterized to be thermodynamically stable in a pico- second interval [13]. The energy needed for their stabilization was taken from noises [1, 14] because the energy for their destruction was caused by heat fluctuation (E = kT). The dependence of the cluster number (K = *1, 2, 3, 4, 5...*) in the molecular matter on its mass was described as follows:

$$m = (250\pm20)\cdot K^{(2.2\pm0.2)} \quad \text{(III)}$$

Therefore, Zubow equations make it possible to get some idea of the nature of clusters and their ensemble as well as their purposeful search in chemical technologies.

Due to the fact that clusters are characterized by oscillation frequencies that correspond with their masses and interaction degrees with their surroundings, a resonance oscillation field can be supposed with whose help one particle dimension influences the oscillation of its analogous being in a certain distance. Here the oscillation of a singular molecule (nuclei ensemble) or a molecular cluster can be described with a pendulum (Pendulum I), and a variety can be described as pendula (Pendula II, ensemble of clusters), that oscillates in a uniform rhythm to minimize the potential energy, as shown in the scheme below.

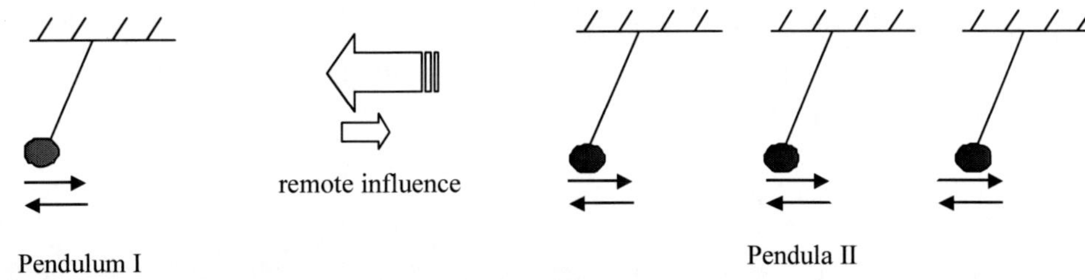

In this scheme, the pendula oscillation is dependent on the nature of both the pendulum and its surroundings.

The minimization of the potential energy is observed to be the moving strength for assembling the singular Pendulum I together with Pendula II to form a new pendulum that is a uniform and stable system. Based on resonance, the oscillation of other pendula shall be influenced by Pendula II, too, where stronger pendula compel weaker ones to oscillate in their rhythm. The mutual reaction between the hydrogen giant, Jupiter (Pendula II), and the dwarf, Earth (Pendulum I), serves as an example for this phenomenon (§18). Furthermore, the results given in Figure 3.1.3.4 confirm this proposition. The information transfer from Pendula II to Pendulum I and the attraction between both (Newton force, §18, §23) proceed through a physical vacuum.

Signals that are cleaned from the noises of the measuring cell (without the sample) will be saved (C_m); the measuring cell with the sample will be prepared analogously (C_{ms}). Moreover, the difference $C_{ms} - C_m = C_s$ is postulated to represent the real sample signal (C_s). The value C_s can be larger or smaller than zero and reflects the different interaction mechanisms of the shock waves with loose or dense barriers in the sample. The sum of all the observed signals ($\sum C_s$) corresponds with the whole energy effect that has arisen from the interaction of the shock wave with all clusters in the sample and shall be applied for the device calibration. For one experimental series, it is possible to calculate the part of equal cluster kinds (f_i) as the ratio of these signals to the absolute sum of cluster signals in the ensemble $f_i = C_s/(\sum|C_s|)$. The value f_i is usually indicated in percent. Negative values are ascribed to the part of the collapsed clusters, the positive to that of the expanded clusters. The average cluster mass (M_{GMS}) in the sample can be calculated according to the next equation:

$$M_{GMS} = \Sigma(m \cdot |f|)$$

where $\sum$ is the mathematical symbol for the sum of all cluster fractions and M_{GMS} is given in Daltons.

Additionally, it is interesting to calculate the part of the collapsed clusters ($Д_c$) by applying the following equation:

$$Д_c = \sum(|-f|)/\sum(|f|)$$

where $\sum(|-f|)$ is the absolute sum of all negative f values for all collapsed clusters and $\sum(|f|)$ is the absolute sum of all f values for all clusters.

Often, the value $F_{95\%}$ shall be used to get an idea of the numeric part of clusters—those signals are reproduced with more than 95% (at least three experiments). $F_{95\%}$ has to be understood as the ratio of signals recorded with a reproducibility of more than 95% ($N_{95\%}$) in relation to the whole signals' number (N – cluster kinds):

$$F_{95\%} = N_{95\%}/N$$

$F_{95\%}$ is characterized as describing the state of skeletal clusters in the long-range order in liquids and polymers, which shall be discussed in the present monograph. Furthermore, an important characteristic of the long-range order in the sample or in a single cluster is the summary signal instability (SI) inside of one experimental series. SI has to be seen as the maximal deviation of the signal intensity ($^{max}S_i$) from that of the average value (S_{av}), e.g. for a chosen signal, i, the signal instability (SI_i) in an experimental series is calculated using:

$$SI_i = (^{max}S_i - S_{av})/ S_{av}$$

This parameter is observed to be of high importance in characterizing remote cluster and cluster ensemble interaction.

3.3 The Physical Impact of the *F* Value

The mathematical sense of the f value has already been described above (§3.2). Due to the fact that the GMS-signals reflect the energy interaction of cluster ensembles with their surroundings, the f value describes the potential energy part of the cluster formation/destruction in a concrete ensemble; it has to be understood as a constant force for the cluster's interaction with its surroundings. For small clusters, the contact interaction (per surface unit) with one another is found to be higher than for large clusters, therefore, the f value for small clusters mainly reflects its interaction with its surroundings and secondly, its part inside the cluster ensemble. The character of clusters' interaction with their surroundings (f) is an energy characteristic whose significance is lost with increasing oscillator mass. For large clusters, it reflects the relative cluster part in the ensemble (Figure 3.3.1).

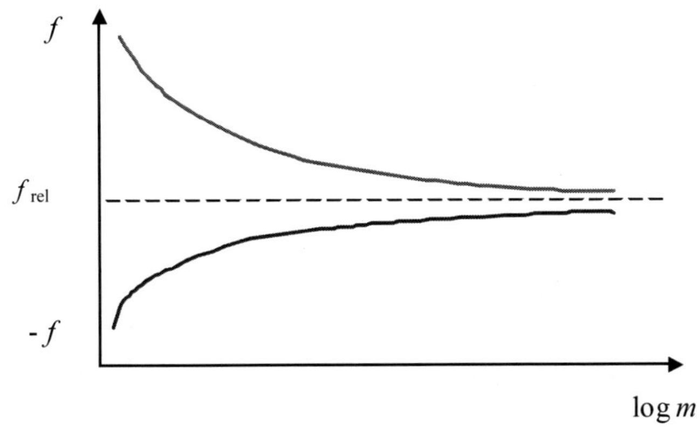

Figure 3.3.1. Change of the f value with increasing cluster molecular mass.

The two curves in Figure 3.3.1 show how f approaches f_{rel}, where f reflects the cluster part in the ensemble. For weak shock waves in the range of small clusters (up to log m = 4…5), f is a very sensitive characteristic regarding the cluster interaction with its surroundings. Additionally, it reacts on external influences, which are directed to this interaction, e.g. weak external energy fields; light and shock waves from earthquakes, mechanical fields, and resonance fields of neighboring analogous clusters; and the gravitational radiation from celestial bodies. The approach of f to f_{rel} is shown for expanded ($+f$) and collapsed ($-f$) clusters. It has to be mentioned that this approximation is a mathematical model; it is not monotonous in reality and depends on the cluster nature and its surroundings. Within an experimental series, f can be applied to get general spectra as well as to understand the cluster mass distribution and cluster interaction with its surroundings. The comparison of f values from different substances and processes, however, is doubtful but if the substances are very similar, it is possible to do so by applying proofs. The f values are found to be concrete for each molecular ensemble to get information on other parameters, e.g. energy of cluster formation, number of interaction contacts, and surface energy of clusters. Therefore, they can be used for device calibration. The approach of f to f_{rel} with increasing cluster molecular mass has been proven for the distribution of sub-micelles in casein and super coils in Aerosil and amylopectin [16]. As will be shown later, the intensity of the real signals (f) reflects the energy of proton dissolution in a physical vacuum (§21) [2, 3].

Chapter 4

EQUIPMENT DESCRIPTION FOR THE GRAVITATION MASS SPECTROSCOPY (GMS)

The principal Scheme for GMS is shown in Figure 4.0.

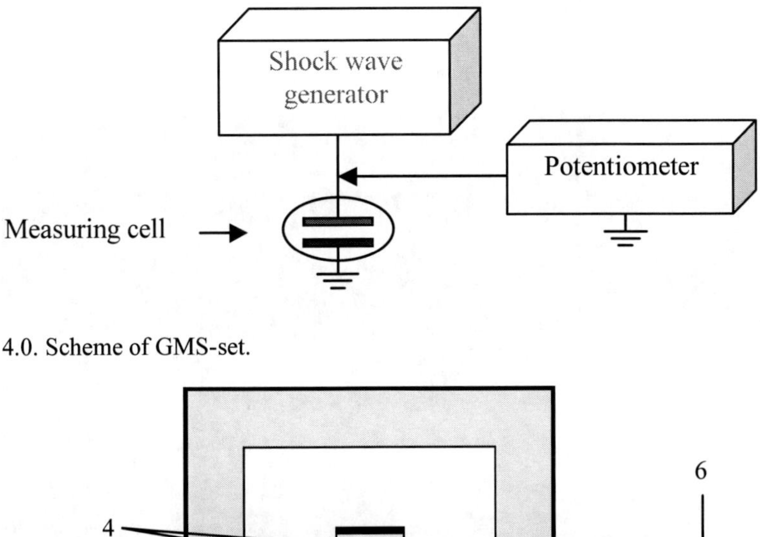

Figure 4.0. Scheme of GMS-set.

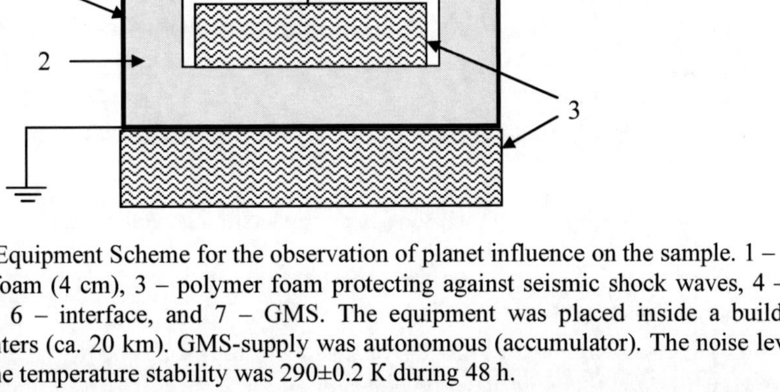

Figure 4.0.1 Equipment Scheme for the observation of planet influence on the sample. 1 – steel box, 2 – polystyrene foam (4 cm), 3 – polymer foam protecting against seismic shock waves, 4 – sample, 5 – GMS-sensor, 6 – interface, and 7 – GMS. The equipment was placed inside a building far from industrial centers (ca. 20 km). GMS-supply was autonomous (accumulator). The noise level in the box < 5 dB and the temperature stability was 290±0.2 K during 48 h.

Figure 4.0.2. Photo of the first GMS-model with internal design in tower.

Figure 4.0.3. Photo of the GMS-model "Agro" 2007 with external design for farming. 1 – potato tuber, 2 – GMS-sensor and interface, and 3 – external shock wave generator and notebook.

The GMS-spectrometer is based on personal computer standards so that it can be integrated into the hardware architecture using an internal or external design. For the internal design, the microelectronics for the shock wave generation shall be installed inside the computer block, while for the external one, the generator is provided through an external block that is linked with the computer by an interface. Samples were placed in a box that is maximally protected from outer energy flows (Figure 4.0.1)

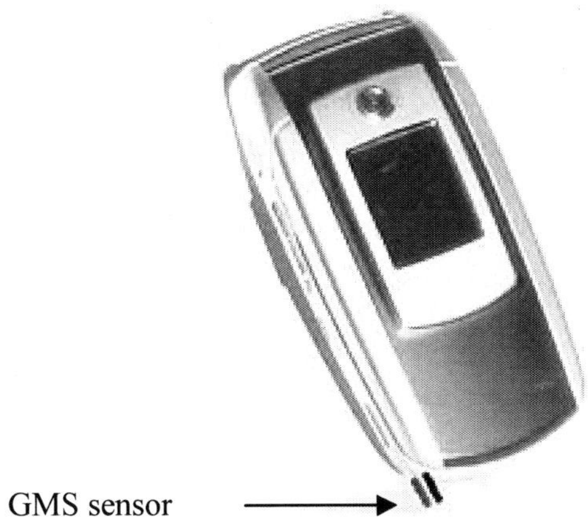

Figure 4.0.4. General view of GMS of the 3rd generation (in planning).

On the photo (Figure 4.0.2), one of the first GMS-models (2004) with internal design can be seen. It works under stationary conditions and is applicable in laboratories. A portable model with external design is shown in Figure 4.0.3. Here the mobility is the advantage; the accumulator is suitable for only 3 h of working time, though. Both models possess the same computer programs for internet access and for data processing/saving. For the 2007 model, the external shock wave generator is installed in the hardware of a notebook, which is useful for field tests and is linked with the global positioning system and the wireless protocol IEEE 802.

Further device minimization down to mobile phone size is possible (Figure 4.0.4).

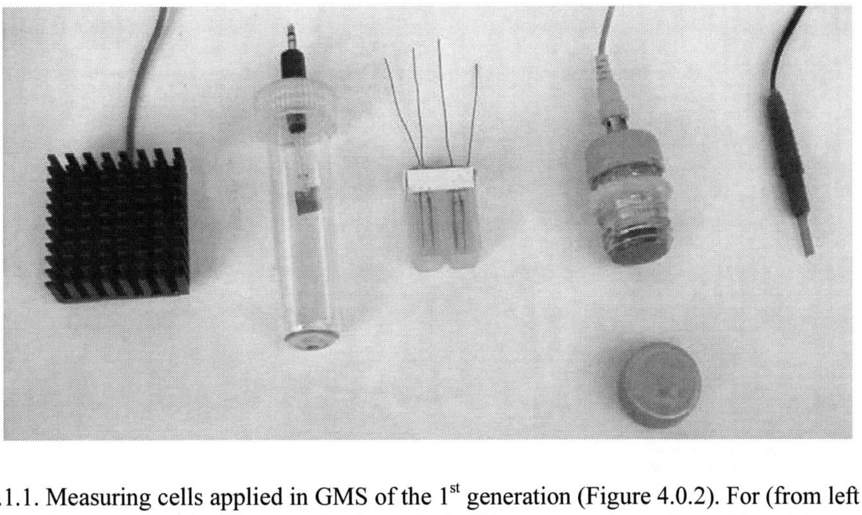

Figure 4.1.1. Measuring cells applied in GMS of the 1st generation (Figure 4.0.2). For (from left to right): gases (long-range order in adsorption layers), liquids, gels, powders, films, biomatrices, and minerals.

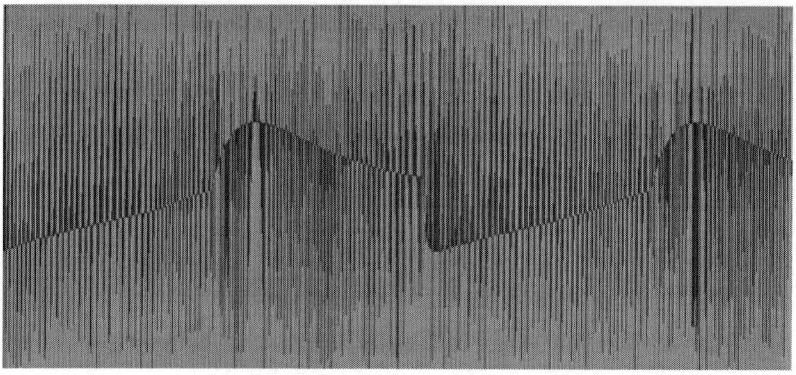

Figure 4.2.1. Oscillograph photo of strong shock waves together with reflected ones from water.

4.1. Measuring Cells for GMS-spectroscopy

In Figure 4.1.1, there are some types of measuring cells for solid and liquid substances that are used in GMS of the first generation (Figure 4.0.2). The most important consideration in the measuring process is to protect the cells from the external energy fields' maximal influences: mechanical fields, electromagnetical/heat flows, anthropogeneous noises, and changes of mass position around the measuring place. The measuring cells are characterized to be easily installed into the existing system of temperature stabilization (thermostats, Dewar vessel, etc.) and technological and controlling equipment.

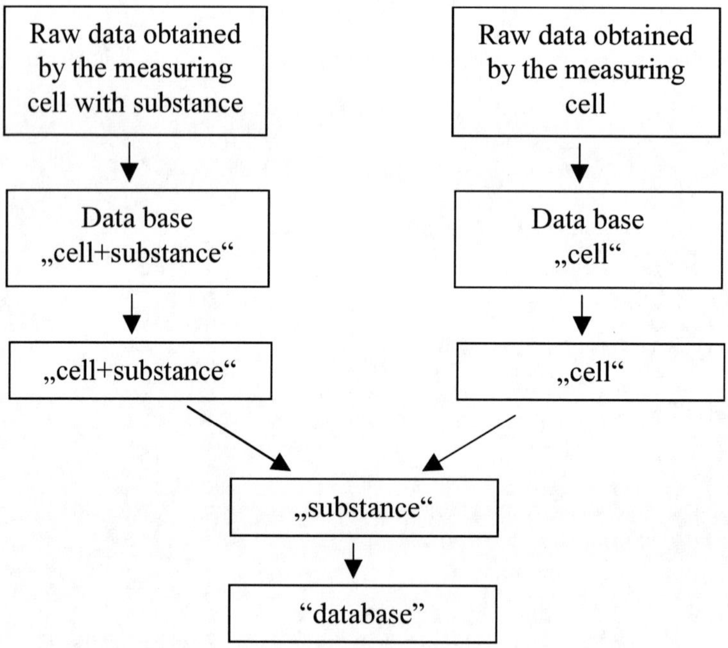

Figure 4.2.2. Secondary data extraction of GMS-signals.

4.2. Algorithm and Programs for GMS

In Figure 4.2.1., an oscillograph photo of shock waves regenerated by GMS (Figure 4.0.2) is shown. Here both the incident shock waves and reflected ones from an anisotropic medium (water) are shown. Additionally, the noises before the cleaning procedure are also featured. The algorithm is shown in Figure 4.2.2.

The GMS is based on known methods of statistical analysis of very weak signals in noises.

4.3. An Experiment (An Example)

Samples were placed in a box that was maximally protected from outer energy flows (Figure 4.0.1). The following materials were chosen as investigation objects: biomatrices of hen's eggs (egg white) and fresh potato tubers, potato starch (powder), distilled water, water solution of NaCl (3.5 wt. %), industrial polyethylene (PE), atactic polystyrenes, collagen (air bladder of a roach), melted quartz (glass), and methanol. The GMS-sensor was placed directly into the biomatrices, weakly pressed into the powders/films (1...2 kPa), or placed into liquids that were aged for six months before the experiment. Masses and frequencies of clusters were calculated by Zubow equations using for biomatrices and water, the constant $6.4 \cdot 10^{-15}$ N/m; for starch, $6.8 \cdot 10^{-15}$ N/m; and for PS, $5.2 \cdot 10^{-15}$ N/m. Atactic polystyrenes (aPS) prepared by a standard method of anionic polymerization in THF, in the presence of butyl lithium as a catalyst, [15] were chosen as investigative objects. The aPS samples had a molecular weight (M_w) of 15, 33, 190, and 450 kDa with a narrow molecular-weight distribution of 1.29, 1.81, 2.03, and 1.35. The molecular weight calculated from the oscillations of the macromolecule coils was equal to 16.7 kDa, 35 kDa, 199 kDa, and 458 kDa (the nearest signal in the GMS-spectrum to the M_w value). In the following section, the polymers will be briefly described as aPS-17, aPS-35, aPS-199, and aPS-458. Distilled styrene had $n_D^{20} = 1.5470$. The GMS-spectra were measured with the spectrometer of the Aist Handels und Consulting GmbH, Germany (www.zubow.de). Weight calibration of the device was carried out with the oscillation of the cluster $(H_2O)_{280}$ (5,040 Da) in bi-distilled water at 4,350±100 Hz for small clusters, with the oscillation of lysozyme coils at 2,451 Hz (15,754 Da) for the middle weight area, and with the oscillation of the cluster in polyethylene at 1,002 Hz (94,261 Da) for large oscillators. The energy calibration of the GMS-spectrometer for clusters in liquid styrene in the temperature interval 293...400 K was made by interpolating the formation energy of liquid styrene (15.3 kJ/mol, formation of all clusters) to the evaporation energy at 413 K (38.3 kJ/mol, destruction of all clusters). The sum of all heat effects observed in the measurement cell (0.125 ml) was the base for this calibration in this temperature interval. The measuring error was 10±5%.

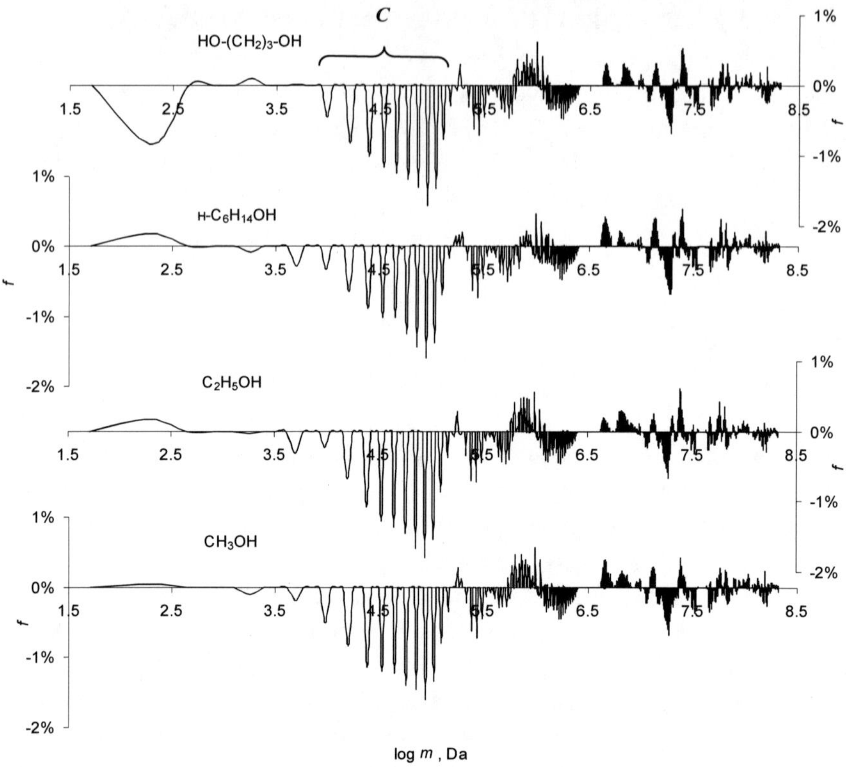

Figure 4.3.1. GMS of cluster ensembles in several alcohols: 291 K, $Z = 6.4 \cdot 10^{-15}$ N/m, and $p < 1$ N/m^2.

4.3.1. Typical GMS-Spectra of Liquids

In Figure 4.3.1, GMS-spectra of cluster ensembles (up to 200 million Dalton) in several alcohols are given, and in Figure 4.3.2, we have presented GMS of cluster ensembles in inorganic liquids.

As visible in Figure 4.3.1, the ensembles are similar; the cluster signal in an ensemble is negative when it is in a collapsed form that is enriched with energy in contrast to the expanded one, which is characterized by a positive *f*-value. The mass interval, ***C***, is typical of sub-micelle structures and appears in every substance, which is influenced by white noises [16]. On the other side, with increasing thermal energy or energy from anthropogenic noises (colored noises), the interval ***C*** can be destroyed.

In Figure 4.3.2 GMS-spectra of cluster ensembles in inorganic liquids are given for comparison with those in alcohols.

Visible clusters in each ensemble are formed by strongly determined masses according to the Zubow equations. The cluster ensembles given in Table 4.3.1 were observed experimentally by the authors with the help of the GMS-device (Figure 4.0.2).

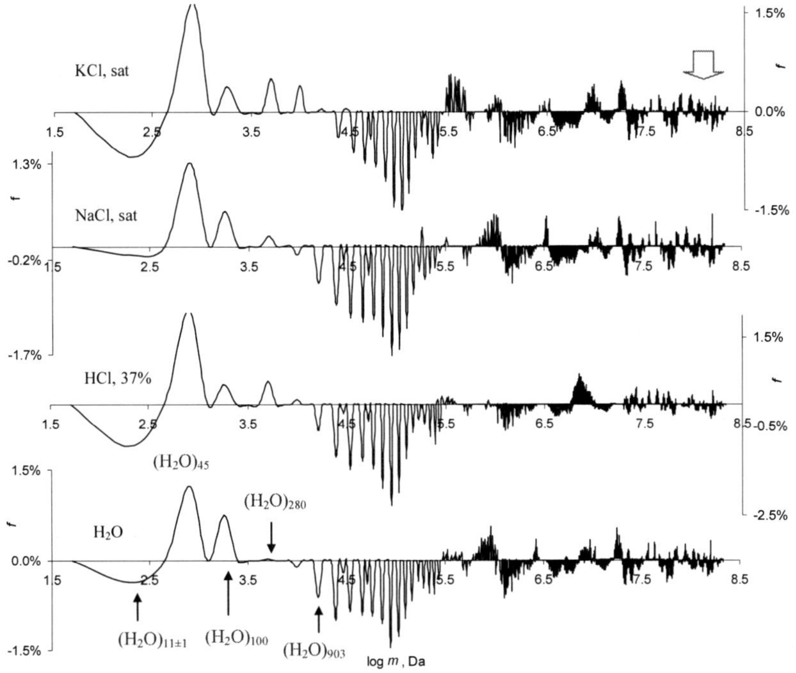

Figure 4.3.2. GMS-spectra of cluster ensembles in water and saturated salt solutions in white noises at 295 K ($Z = 6.4 \cdot 10^{-15}$ N/m, $p < 1$ N/m^2).

Table 4.3.1. Selected molecular cluster ensembles. The number of molecules (members) in a cluster for the first seven members of the Zubow array shall be given. Non-excited ensembles under normal conditions in white noises are shown.

Substances	*1*	*2*	*3*	*4*	*5*	*6*	*7*
H_2O	11±1	45	100	280	545	903	1,351
MeOH	6	55	152	297	492	736	1,030
Acetone	3	28	77	151	250	374	522
n-Hexane	2	20	36	57	81	111	183
p-Xylene	2	17	30	47	93	153	229
Styrene*	2	18	24	32	40	50	72
aPS	2	18	50	97	161	241	287
PEHD	8	73	204	397	658	984	1,376
Casein**	198	1,807	5,034	9,817	16,262	24,325	34,006
Starch***	1	10	27	52	86	129	180
Aerosil, 380 м^2	3	28	77	151	198	250	373
SiO_2 (glass)	3	27	76	148	246	367	514
H_2SO_4	2	8	18	51	100	166	248

* freshly distilled, ** bovine (mass in Dalton), and *** potato ("Agria"), members of α-D-glucopyranose.

The clusters in an ensemble were found to compete for energy sources in the ensemble, which was visible as a steady rearrangement of the cluster structures [17] where the rearrangement time could last up to some months or even years under normal conditions [18]. Ensembles were able to form mirror-isomers and ensemble-"oboroten" ("wryneck"); they are characterized by inversion of density (collapsed–expanded) and energy capacity. Furthermore, they compete with each other for the energy potential of white noises ("wall effect" [19]). Large ensembles grow at the expense of a small one. The reason for this is the potential energy minimization of the new ensemble. It was concluded too that ensembles attract each other at distances. To the authors, this phenomenon is the background of Newton's Law of Gravitation. In mixtures, cluster ensembles of different substances are not mixable but within one ensemble, a virtual mass transfusion goes between the clusters that reflect the gravitational energy transfusion (fluctuation) on the level of energy clusters in white noises (Figure 3.2.1).

Cluster ensembles were found to be characterized by the number of cluster kinds (*N*), ratio of expanded and collapsed forms (D_c - ratio), average cluster molecular mass (M_{GMS}), average ensemble energy, ratio of skeletal clusters (clusters with signal reproducibility higher than 95%, $F_{95\%}$), and the ratio of random clusters (with signal reproducibility lower than 20%).

Since clusters and cluster ensembles are sensitive to seismic waves of earthquakes, they are risks for chemical technology [20, 21]. Furthermore, cluster ensembles can influence and do interact with each other at distances. Therefore, they are sensitive chemical and physical processes. This means that when planning chemical productions (chemical engineering), such factors must be taken into account, e.g. distance between chemical reactors and the influence of the biomatrix of the staff and people in the surroundings. Concerning living organisms, this influence changes cluster ensembles in biomatrices, for example in water, and finally in all bio-chemical processes in organisms. Therefore, nano-technologies have led us to a new critical point at which risks are heavily predictable, especially on a nano-electronics and computer administration basis.

CONCLUSION

Weak shock waves are suggested to be applied for the analysis of mass concentrations including size, form, density, structural inhomogeneity, and their interaction with the surroundings.

Clusters are found to form an ensemble.

Between the clusters of an ensemble exists a hierarchy.

The larger an ensemble, the higher its potential energy.

Large ensembles are characterized to dominate and control small ones while the last ones will be integrated into the larger clusters, compulsorily forming a new ensemble with minimal potential energy.

This process is the basis of Newton's Law of Gravitation.

REFERENCES TO 3 AND 4

[1] Zubow K.V., Zubow A.V., Zubow V.A. Ensemble of Clusters – New Form of Molecular Matter, Risks and Chances. Zubow Equations. In Advances in Chemistry Research, vol. 5, 2010, pp. 107-145. Editor J.C. Taylor. Novapublisher. NY. USA.

[2] Zubow K., Zubow A.V., Zubow V.A. The Phenomenon of Proton Dissolving in Vacuum and of Proton Condensation from Vacuum. Two Forms of Protons, Structure of Nuclei, Electrons and Atoms, J. of modern physics, vol.1, no.1, 2010 pp.175-184.

[3] Zubow A., Zubow K., Zubow V.A. New Experimental Method for Determination of Distances to Celestial Bodies and their Masses. In Horizons of Physics. Editor Richard Schortemeyer III. Nova Science Publishers, Inc. NY. Vol. 279. 2012, in print.

[4] Kedrinsky V.K., Shokin J.V., Vshivkov V.A., Dudnikova G.I., Lasareva G.G. Generation of shock waves in a fluid spherical bubble clusters / Reports of the Academy of Sciences. 2001, vol. 381, no. 6, pp. 773-776, in Russian.

[5] Physical Encyclopedic Dictionary. Moscow. Soviet Encyclopedy. 1984, p. 84.

[6] Physical Encyclopedic Dictionary. Moscow. Soviet Encyclopedy. 1984, p. 778.

[7] Frolov J. G. Kurs kolloidnoi chimii. Poverchnostnie jawlenia i dispersnie systemy. Moscow.: Chimia. 1982, p. 54.

[8] Chaplin M., SBU London, http://www.lsbu.ac.uk/water/index.html.

[9] Annika Lenz; Lars Ojamäe. On the stability of dense versus cage-shaped water clusters: Quantum- chemical investigations of zero-point energies, free energies, basis-set effects and IR spectra of $(H_2O)_{12}$ and $(H_2O)_{20}$. Chemical Physics Letters. 2006, vol. 418, pp. 361-367.

[10] Bogdanov E. V.; Mantrova G.M. Equicluster water model. Biomedical Radioelectronics. 2000, no. 7, pp.19-28. (in Russian).

[11] Lian Li; Ogawa T. Clusters and their properties in aqueous solutions of KDP, KCl and sugar. Journal of Crystal Growth. 2000, vol. 211, pp. 286-289.

[12] Zubow Kr. V.; Zubow A.V.; Zubow V.A. Cluster structure of liquid alcohols, water and n- hexane. Journal of Appl. Spectr. 2005, vol. 72, no. 3, pp. 321-328.

[13] Guang-Jun Guo; Yi-Gang Zhang; Ya-Juan Zhao; Keith Refson; Gui-Hua Shan. Lifetimes of cagelike water clusters immersed in bulk liquid water: A molecular dynamics study on gas hydrate nucleation mechanisms. Journal of Chem. Phys. 2004, vol. 121, no. 3, pp. 1542-1547.

[14] Horsthemke W.; Lefever R. *Noise-Induced Transitions. Theory and Applications in Physics, Chemistry and Biology*. Edit. Hermann Haken, vol. 15; Springer-Verlag: Berlin, Heidelberg, NY, Tokio, 1984.

[15] Gausepohl H, Gellert R. (Eds.). Kunststoff Handbuch: Polystyrol. Wien: Hanser Verlag München, 1996.

[16] Zubow Kr. V.; Zubow A.V.; Zubow V.A. Use of flicker noise spectroscopy for undestroyed analysis of nano structures. Zavodskaja Laboratorija. Diagnostika Material. 2008, vol. 74, no. 9, pp. 40-46 (in Russian).

[17] Zubow Kr. V.; Zubow A.V.; Zubow V.A. Investigation of the distribution of water clusters in vegetables, fruits and natural waters by flicker noise spectroscopy. Biofizika. 2007, vol. 52, no. 4, pp. 585-592 (in Russian).

[18] Zubow A.V.; Zubow Kr. V.; Zubow V.A. The Investigation of the Salt Dissolving Mechanism and of the Ageing of the NaCl Solution in Water by Flicker Noise Spectroscopy. Journal of chemical Industry (Saint.-Petersburg, Russia), 2007, vol. 84, no. 1, pp. 40-48.
[19] Zubow K., Zubow A.V., Zubow V.A. Molecular Cluster Distribution in Sea Water Near the Wall. Antifouling aspect, Journal of Clusters, 2011, v.22, no. 4, 603-619. (DOI) 10.1007/s10876-011-0402-6.
[20] Zubow A.V.; Zubow Kr. V.; Zubow V.A. The Phenomenon of Water Clusters Distribution in Water and Solvated Clusters of Ion Pairs of NaCl in Solution Near the Wall. The Wall Effect Journal of chemical Industry (Saint.-Petersburg, Russia), 2006, vol. 83, no. 6, pp. 300-306.
[21] Zubow Kr. V.; Zubow A.V.; Zubow V.A. Phenomenon of Influences of natural disasters (hurricanes and earthquakes) on the long-range order in liquids. Risks and chances. Russian Journal of Applied Chemistry. 2007, vol. 80, no. 11, pp. 2001-2004.

Chapter 5

CLUSTER STRUCTURE OF LIQUID ALCOHOLS, WATER, AND *n*-HEXANE

Different views of the structure of liquids exist. Prominence is given to the structures of liquid alcohols and water [1-4] at the level of formation and rupture of intermolecular hydrogen bonds. For example, hydrogen bridges with lengths of 0.16 and 0.20 nm, representing intermolecular structures of two types, have been detected in [2]. In this work, bifurcation models of intermolecular interactions in liquid methanol are discussed and a hypothetical liquid fragment consisting of 26 molecules is considered. In [1], the average lifetime of the hydrogen bridge between two methanol molecules at 300 K has been estimated at ~5–7 ns. The super molecular structure of alcohols at the level of cluster formation remains an open question [2, 4]. In [4], methanol clusters containing 64, 128, 192, 256, and 512 molecules at 200 and 300 K are considered and their thermodynamic characteristics (surface tension, potentials, and energies of cluster surfaces) are described. The data from computer simulation performed in [5] suggests the existence of two types of water clusters with close-packed and loose structures. The problem of the detection of clusters and determining their energy characteristics, however, remains to be solved. In [6], it was suggested that water possesses weak thixotropic properties. The times of its decomposition and reduction range from 3–4 s (local times) to 1 h (total time), and water experiences restructuring when it approaches the freezing point (temperatures lower than 281 K).

Since different molecules of one and the same liquid have different masses (isotopic composition) and interact differently with the analogous molecules of the environment, the impurities, and the vessel walls, it may be suggested that molecules with close thermodynamic characteristics associate into clusters. Such cluster formations should possess specific physical properties. Clusters are not stable; they are formed and break down constantly. Therefore, the statistical pattern of their existence can be detected only by modern electronic devices with the use of high-speed computers and computer programs.

Due to this fact, we attempted to investigate the structure of liquids in water, methanol, *n*-hexanol, and *n*-hexane through the method of GMS, which is described in detail in [7]. The GMS method makes it possible to detect statistical ensembles of structures that are formed and break down rapidly in liquids.

As is seen from the GMS-spectra of the liquids studied (Figure 5.1), clusters of two types oscillate in them: clusters with negative energies that interact with the spectrometer shock wave with energy release and, as a result, are packed and crystallized (close-packed clusters) and clusters with positive energies that take part in the melting and decomposition processes (loose clusters).

The experiment has shown that the energies of formation of clusters in alcohols are much higher than the energies of formation of clusters in *n*-hexane and are lower than the energies of formation of clusters in water. This is explained through the action of the alkyl group (+*I* effect) on the oxygen of the hydroxyl group as well as the concentration of a hydrogen-bond potential on this group and the decrease in the number of surrounding molecules competing for the hydrogen bonds. In the case of water, additional possibilities for the formation of hydrogen bonds appear and the number of competing neighboring water molecules increases, with the result that the clusters become weaker and the energy of their formation decreases. In our opinion, clusters are formed in *n*-hexane due to the cooperative-structure effect arising in it as a result of the weak Van der Waals interactions.

Close-packed clusters of mass 43,760 Da are formed in *n*-hexane, *n*-hexanol, and methanol. In Figure 5.0.1, a cluster of this type is denoted with an asterisk. It may be suggested that the cohesive forces of the alkyl groups are responsible for the formation of these oscillators. With an increase in temperature to 300 K, the signal intensity of them decreases most strongly when they are measured in alcohols. Oscillators of mass 3,100 Da (denoted by a cross) are formed in methanol (clusters of 97 molecules) and *n*-hexanol (cluster of 30 molecules). These clusters manifest themselves clearly in the spectra of water at temperatures higher than 274 K. It is possible that the hydrogen bonds of the hydroxyl groups are responsible for their formation.

The GMS method makes it possible not only to detect oscillations of individual clusters but also to reveal regularities in the behavior of oscillators representing sets of organized structures (statistical ensembles) in liquids similar in structure. Analysis of tabular data shows that "islands" of clusters with close masses appear in such liquids. In these "islands," clusters are detected with a reproducibility of 99%; they form the skeleton of the cluster structure of liquids. For example, aggregates of clusters close in size with masses falling within the range of 960,000–1,500,000 Da were reliably detected in *n*-hexane. In passing to *n*-hexanol, methanol, and water, the set of clusters decreases. We failed to explain the reason for the formation of such cluster islands. Analogous effects were detected by us at other temperatures.

Figure 5.0.2 shows the GMS-spectra of methanol at different temperatures. It is seen that, in the region of small and medium-sized clusters ($\log m > 4.5$), the masses of the oscillating methanol clusters change insignificantly but their energy changes markedly. As the temperature increases, the set of large clusters and their energies decrease significantly. In this case, the intensities of the peaks in the region of medium and small clusters increase. It may be suggested that an increase in the temperature of an alcohol leads to the disruption of its cluster structure and the homogenization of the alcohol. A small loose methanol cluster ($\log m = 2.28$) interacting with the spectrometer shock wave with energy absorption is presented in Figure 5.0.2. This cluster consists of six alcohol molecules. Using the Cochrane pivotal models, we obtain two possible structures for it (Scheme I):

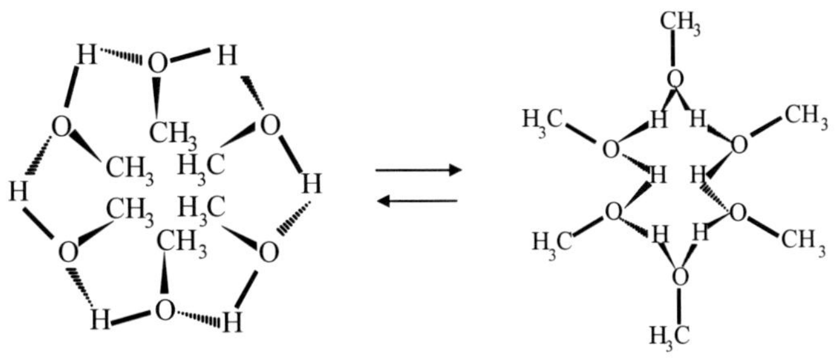

(I)

These structures manifest themselves in the GMS-spectra as single oscillators. They are typical of micelles. The question of which hexamer structure dominates in the GMS-spectrum still remains open.

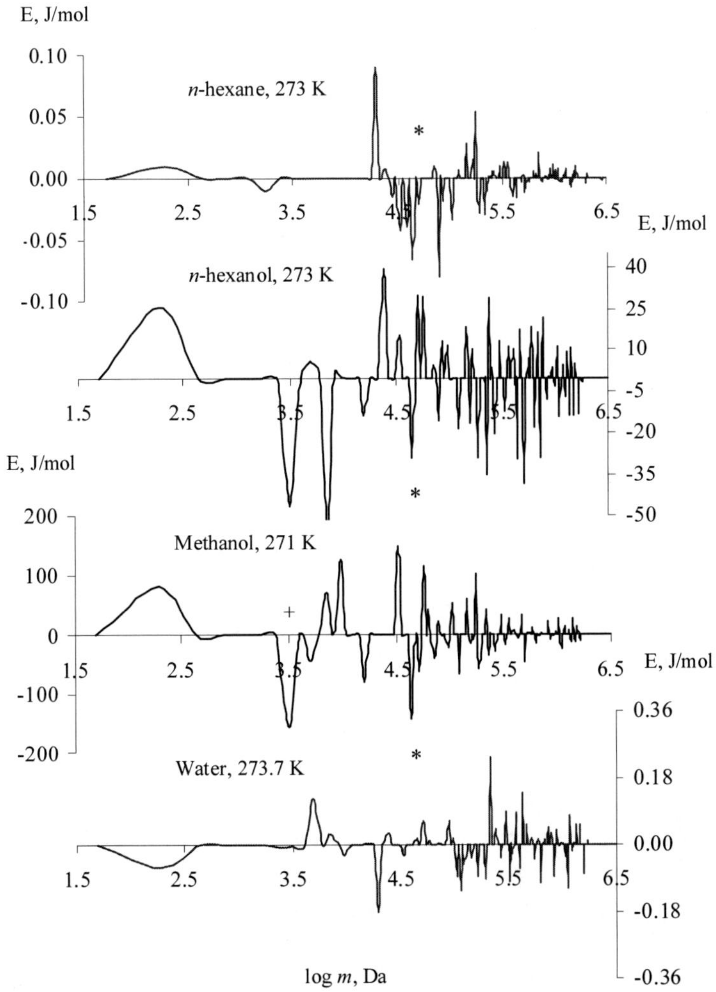

Figure 5.0.1. GMS-spectra of the liquids studied: $Z = 6.4 \cdot 10^{-15}$ N/m and $p > 1$ N/m^2. (For GMS-spectra for $p < 1$ N/m^2, see in Figure 4.3.1).

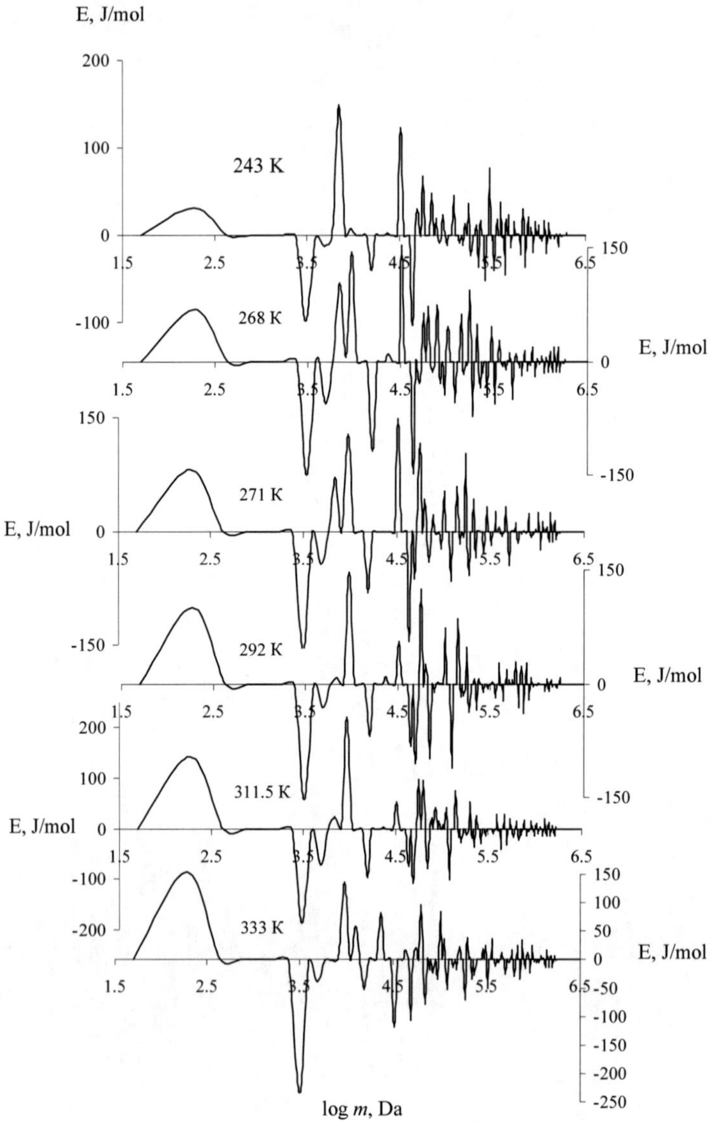

Figure 5.0.2. GMS-spectra of methanol, $p > 1\ N/m^2$.

Clusters with a mass equivalent to 97 methanol molecules (log m = 3.49) manifest themselves in the spectra of all alcohols at different temperatures. They are close-packed clusters that, when interacting with a shock wave, release energy and, as a result, are packed and crystallized. When the temperature of the indicated clusters decreases from 330 to 240 K, the energy of their formation decreases by 60% (from −250 to −100 J/mole). This, in our opinion, is explained by the fact that the energy of formation of a cluster as an element of a statistical ensemble of clusters depends on its interaction with the environment. The denser the environment and the larger the number of intermolecular interactions of the medium with the surface of the cluster, the lower the energy required for its destruction. Thc mass of the indicated oscillating clusters is close to the mass of the methanol cluster consisting of 128 molecules that were detected in [4]. It has been shown in this work that the energy of the surface tension of this cluster increases with the decrease in the temperature from 300 to 200 K, which is evidence of the enhancement of the surface effects of the cluster's interaction with the environment.

The fact that methanol clusters can be characterized not only by their mass and size but also by their surface tension allows for the conclusion that certain liquids (nano emulsions) have a colloidal structure. In this case, long-range order liquids can be investigated by the GMS method. Dimers of *n*-hexanol and *n*-hexane can be represented as oscillators in the following expanded form (we give their expanded structures for clarity):

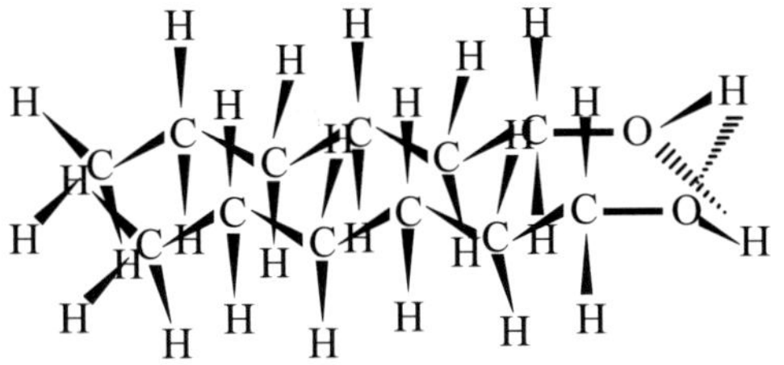

n-hexanol

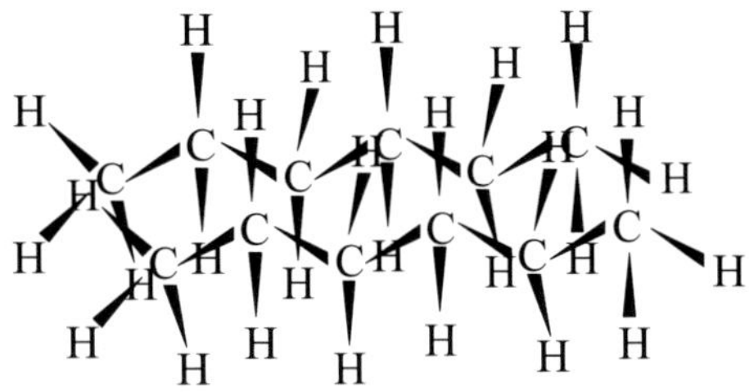

n-hexane

The Cochrane pivotal models of larger clusters allow us to suggest that they have a micelle structure. It should be noted that the micelle model of methanol clusters contradicts their bifurcation representation made in [2].

The energy characteristics of different clusters in *n*-hexanol at 341 K behave differently. Clusters consisting of 1,496; 15,803; and 17,219 molecules are relatively stable at all temperatures; some clusters (n = 323) are strongly dependent on the temperature, and the energy characteristics of the other clusters (n = 552) change monotonically. Analysis of the table data on the distribution of clusters in *n*-hexanol also points to the existence of stable cluster islands. It is conceivable that these islands are responsible for the macro oscillations detected in liquids.

In [5], two states of the water cluster $(H_2O)_{280}$, in which it has a close-packed (ρ = 1 g/cm^3) and a loose (ρ = 0.94 g/cm^3) structure, are described. The oscillations of the masses of these clusters manifest themselves in their GMS-spectra at log m = 3.7 (Figure 5.0.3). At temperatures lower than 280 K, the indicated cluster has a close-packed structure. It is packed and crystallized as a result of interaction with the spectrometer shock wave. At temperatures higher than 285 K, this cluster has a loose structure. It can be seen from Figure 5.0.3 that, in the temperature range 280-285 K, the energy of formation/decomposition of clusters in water is the lowest. This phenomenon agrees with the data of [6], where it has been established that the shear deformation of water changes sharply at 281 K. Consequently, it may be suggested that clusters in water substantially influence its rheological properties.

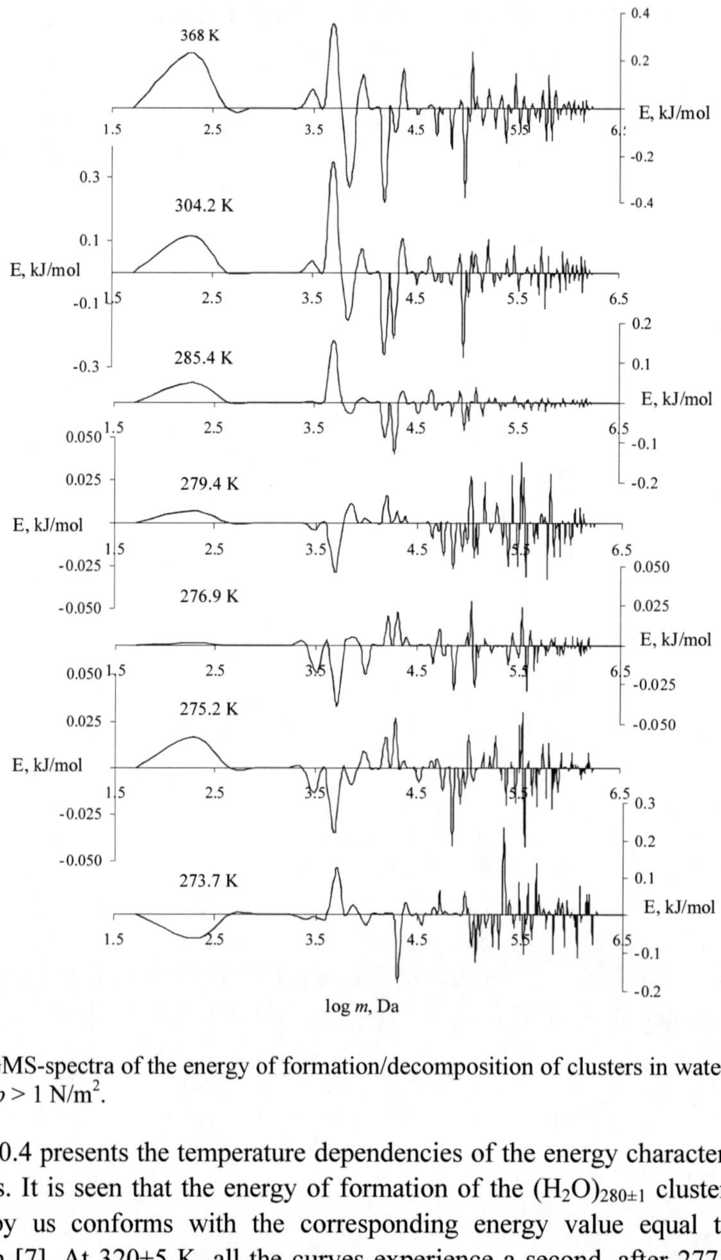

Figure 5.0.3. GMS-spectra of the energy of formation/decomposition of clusters in water in the process of its heating, $p > 1$ N/m^2.

Figure 5.0.4 presents the temperature dependencies of the energy characteristics of small water clusters. It is seen that the energy of formation of the $(H_2O)_{280\pm1}$ cluster (~1 kJ/mole) determined by us conforms with the corresponding energy value equal to ~1 kJ/mole determined in [7]. At 320±5 K, all the curves experience a second, after 277 K, deflection, which points to the fact that the whole water structure is transformed. It is possible that this restructuring leads to the known collapse of biological systems at a temperature higher than 42°C.

The symmetry of the temperature dependencies of the energy of formation and decomposition of clusters in water (see Figure 5.0.4) is favorable to the dissipation of the internal stresses in the liquid and their stabilization. Otherwise, the hydrogen bonds formed in the water network gave rise to giant internal stresses that make the existence of the liquid impossible. We detected analogous regularities for alcohols and *n*-hexane.

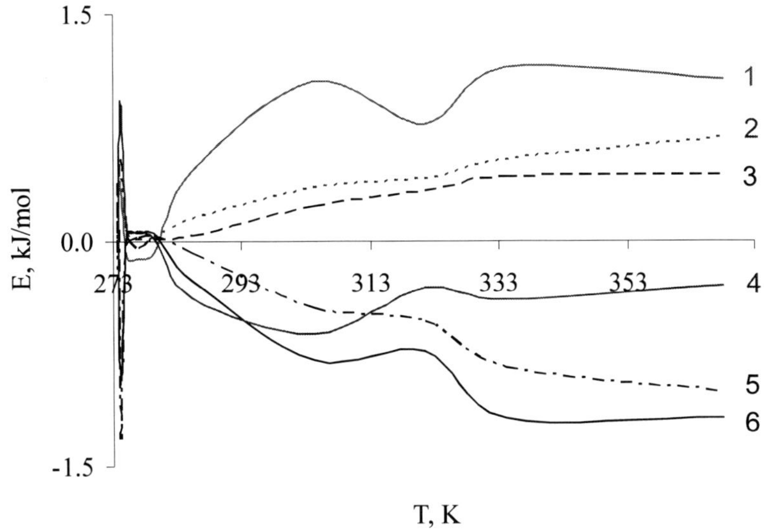

Figure 5.0.4. Energies of formation of clusters in water at different temperatures: 1 – $(H_2O)_{280}$, 2 – $(H_2O)_{12}$, 3 – $(H_2O)_{545}$, 4 – $(H_2O)_{1,119}$, 5 – $(H_2O)_{402}$, and 6 – $(H_2O)_{903}$.

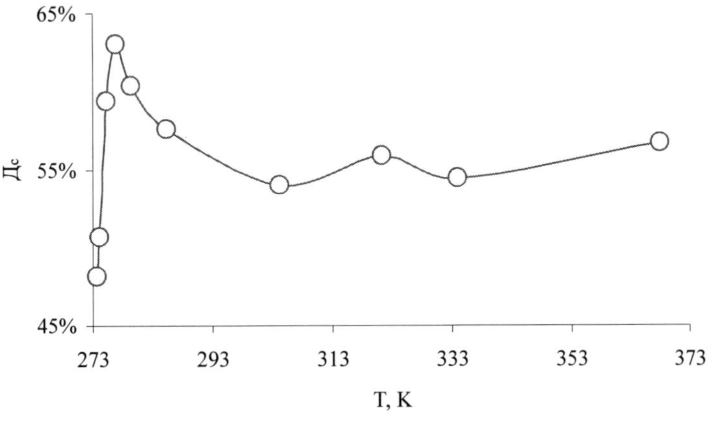

Figure 5.0.5. Dependence of the fraction of collapsed clusters ($Д_c$) in water on its temperature. The fraction of clusters was calculated, using the data presented in Figure 5.0.3, as the ratio between the total intensity of the signals of collapsed clusters and the total intensity of all the signals in the GMS.

Analysis of the GMS-spectra presented in Figure 5.0.3 shows that the oscillating clusters of the statistical ensembles in water can be divided (according to the terminology of Professor S. F. Timyshev [8]) into close-packed and loose clusters. The fraction of an individual cluster or the fraction of closely packed or loose clusters can be calculated as the ratio of the sum of their signals to the total sum of all the signals of clusters detected at a given measurement temperature. Figure 5.0.5 presents the results of such calculations for close-packed clusters in water at different temperatures.

It is seen that at a temperature of maximum water density (277 K), the fraction of the close-packed clusters is maximum. It is reasonable to suggest that the anomalous density of water at this temperature is due to its cluster structure. The temperature hysteresis of the $(H_2O)_{280}$ cluster was detected by the GMS method in the process of heating and cooling of water in the temperature range 273-280 K (Figure 5.0.6). As is seen, the $(H_2O)_{280}$ cluster is formed in two ways: in the process of heating water and in the process of its cooling. The cluster fraction formed in the process of cooling water at the highest point of 275.4 K, and this maximum shifts to 277 K in the case where the cluster is formed in the process of heating. The energy of formation of this cluster reaches a maximum at 275.4 K only in the process of cooling water (Figure 5.0.6b). These facts indicate that the cluster can be transformed from the close-packed form into the loose form.

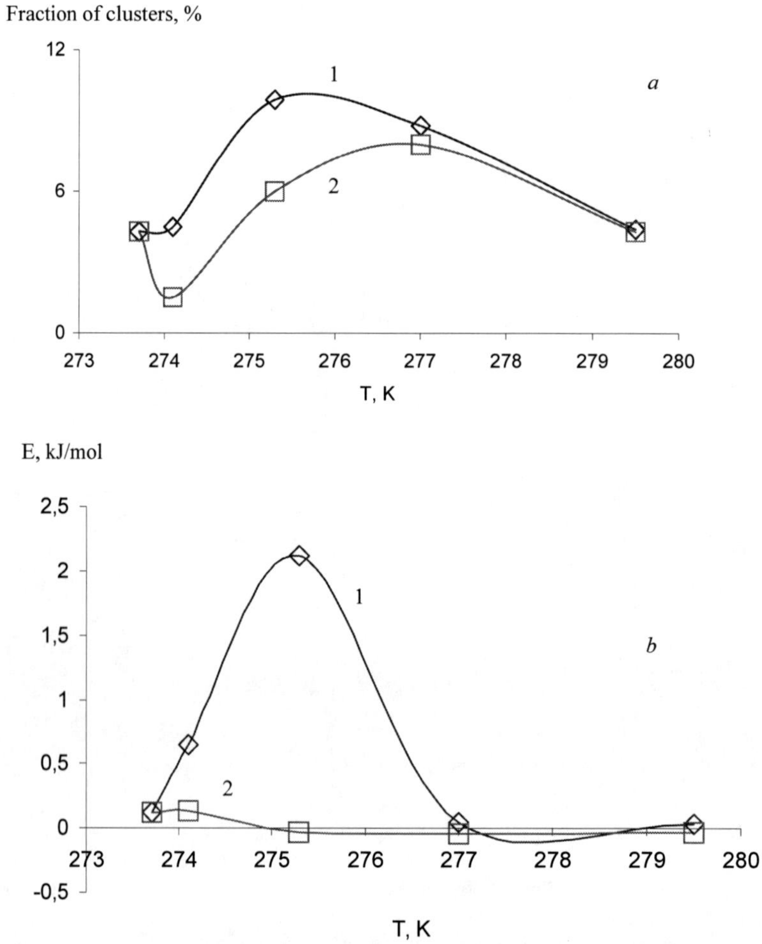

Figure 5.0.6. Hysteresis of formation of the $(H_2O)_{280\pm1}$ cluster in water: a – dependence of the fraction of the $(H_2O)_{280\pm1}$ cluster (fraction of this cluster signal of the total sum of all the signals in the GMS-spectrum) on the temperature, b – dependence of the energy of formation of the cluster on the temperature, 1 – curves obtained in the process of cooling, and 2 – curves obtained in the process of heating.

The distribution of clusters in water depends not only on its temperature but also on its nature: cooled boiled water or water from melted snow. It is possible that the phenomenon of different tastes of water and its different technological properties depending on the season (flotation) and medical properties ("living" and "dead" water) are determined by the cluster structure of water.

5.1. Water Clusters and Celestial Bodies

In [9] we described how cluster ensembles interact at a distance where the interaction is based on the resonance of their weak gravitational radiation. WGR shall be explained by a proton dissolving in a physical vacuum with its following condensation from the vacuum (influence on the gravitational field, §21). WGR of samples was supposed to be in resonance with the analogous gravitational radiation of planets at their culmination. Therefore, it should be of interest to find out how molecular cluster ensembles on Earth are influenced by the gravitational radiation of the planets. If this suggestion is right, one could expect planetary influence on chemical technologies through the changing of cluster ensembles (long-range order) in substances in reactors leading to changed technological conditions, product qualities, and new unknown risks.

Analyzing molecular clusters, there was observed a trivial correlation with the masses of the planets and of the sun additionally with their hypothetical gravitational radiation (Table 5.1). In Table 5.1 the base for such resonance (equal frequencies) between water cluster ensembles and planet clusters shall be shown.

Table 5.1. The main noise frequencies (ω) of the planets and of the sun that were calculated according to $m = N_A \cdot 10^{11} \cdot \omega^{-2}$, where compared with the frequencies of water clusters in a potato biomatrix, $m = 10^{11} \cdot \omega^{-2}$

Characteristics			Water clusters*	
Celestial bodies	*m*, g	ω, Hz	ω, Hz	Formula
Mercury	3.40E+26	13,309±2,000	11,111±2,000	$(H_2O)_{45}$
Venus	4.87E+27	3,517±200	3,431±200	$(H_2O)_{472}$
Earth	5.97E+27	3,176±200	3,193±200	$(H_2O)_{545}$
Mars	6.40E+26	9,700±1,500	7,454±1,500	$(H_2O)_{100}$
Jupiter	1.90E+30	178±20	178±20	$(H_2O)_{174,420}$
Saturn	5.69E+29	325±30	333±30	$(H_2O)_{50,124}$
Uranus	8.70E+28	832±50	827±50	$(H_2O)_{8,131}$
Neptune	1.03E+29	765±40	769±40	$(H_2O)_{9,385}$
Moon	7.35E+25	28,624±4,000	21,517±4,000	$(H_2O)_{11\pm1}$
Sun	1.99E+33	5.5±1	5.5±1	$(H_2O)_{36,453,036}$

* The mass of a cluster corresponds to the number of water molecules.

The Chaplin [5] clusters $(H_2O)_{178}$, $(H_2O)_{280}$, $(H_2O)_{903}$, $(H_2O)_{1351}$, $(H_2O)_{1601}$, and $(H_2O)_{1889}$ that were formed in the noises from the galaxy and the sun or that where caused through the interference of noises from single planets do not fit in the table.

It has to be mentioned that the mass of water clusters is proportionate to the number of hydrogen bridges in the cluster.

CONCLUSION

Liquids, having a long-range order, can be described by a statistical ensemble of clusters. Clusters with close-packed and loose structures are formed in liquids.

The mean statistical energy of formation/decomposition of clusters in a liquid depends on the nature of this liquid. This energy falls within the range −0.4 to +0.4 kJ/mole when the indicated processes occur in heated water and can be half the order of magnitude in cooled water.

The distribution of clusters in a liquid depends on the temperature and the method of its attainment. Liquids can be considered nanodisperse emulsions of clusters, i.e., as thermodynamically stable lyophilic colloidal systems.

Liquids exist on the condition that they have a cluster structure in which the mutually opposite vibrations of clusters dissipate the internal stresses and sustain the liquid state.

Clusters can be transformed from the expanded state into the collapsed state and vice versa.

The GMS method allows one to rapidly and reliably determine the presence of clusters in liquids.

REFERENCES TO 5

[1] Matsumoto M. and Gubbins K. E. Hydrogen bonding in liquid methanol. Journal of Chem. Phys., 1990, vol. 93, no. 3, pp.1981–1994.

[2] Pagliai M., Cardini G., Righini R. and Schettino V. Hydrogen bond dynamics in liquid methanol. Journal of Chem. Phys., 2003, vol. 119, no. 13, pp. 6655–6662.

[3] Riccardo Chelli, Sonia Ciabatti, Gianni Cardini, Roberto Righini, and Piero Procacci. Calculation of optical spectra in liquid methanol using molecular dynamics and chemical potential equalization method. Journal of Chem. Phys., 1999, vol. 111, no. 19, pp. 4218–4229.

[4] Zakharov V. V., Brodskaya E. N. and Laaksonen A. Molecular dynamics simulation of methanol clusters. Journal of Chem. Phys., 1998, vol. 109, no. 21, pp. 9487–9493.

[5] Chaplin M. SBU London: http://www.lsbu.ac.uk/water/clusters.html

[6] Stebnovskii S. V., About Shift Durability of Water. Zhyrnal of Tekh. Fiz., 2004, vol. 74, no. 1, pp. 21–24 (in Russian).

[7] Lee H.M, Suh S.B, Kim K.S. Erratum: Structures, energies and vibrational spectra of water undecamer and dodecamer: An ab initio study. [J. Chem. Phys. 114, 10749 (2001)]. Journal of Chem. Phys., 2001, vol. 115, pp. 7331-7334.

[8] Deshcherevsky A. V., Lukk A. A., Sidorin A. Y., Vstovsky G. V. and Timashev S. F. Flicker- noise spectroscopy in earthquake prediction research. Natural Hazards and Earth System Sci., 2003, vol. 3, pp.159–164.

[9] Zubow K.V., Zubow A.V., Zubow V.A. Ensemble of Clusters – New Form of Molecular Matter, Risks and Chances. Zubow Equations. In Advances in Chemistry Research, vol. 5, 2010, pp. 107-145. Editor J.C. Taylor. Novapublisher. NY.

Chapter 6

GMS-SPECTROSCOPY OF NONANOIC ACID ($n\text{-}C_8H_{17}COOH$)

The ability of organic acids to form the simplest clusters—dimers—has already been known for a long time. Dimerization proceeds with the help of hydrogen bonds between carboxylic groups:

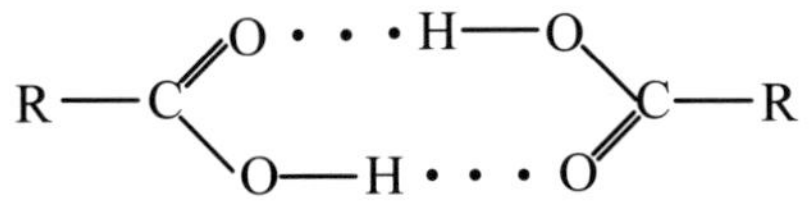

It should be of interest to understand how the long-range order in liquid organic acids will be developed. For example, for nonanoic acid, the GMS-spectrum of its long-range order shall be discussed. The acid is of the following formula:

$$H_3C{-}CH_2{-}CH_2{-}CH_2{-}CH_2{-}CH_2{-}CH_2{-}CH_2{-}COOH$$

Due to the cooperative cohesion effects between similar CH_2 –groups, the aliphatic part can be expected to be wound into clusters like in hexane (§5). Furthermore, the interaction of carboxylic groups (-COOH) probably leads to the formation of large micelle and sub-micelle structures. In Figure 6.1, two GMS-spectra of this acid obtained are shown after the impact of weak and strong shock waves.

Differences in the GMS-spectra of nonanoic acid after weak and strong shock waves' influences are shown in the Table 6.1.

Table 6.1. Influence of shock waves on the main long-range order characteristics of nonanoic acid at 293 K.

p, N/m^2	M_{GMS}, Da	$Д_c$, %	$F_{95\%}$	N
>1	484,020	59	65	52
<1	357,136	15	20	45

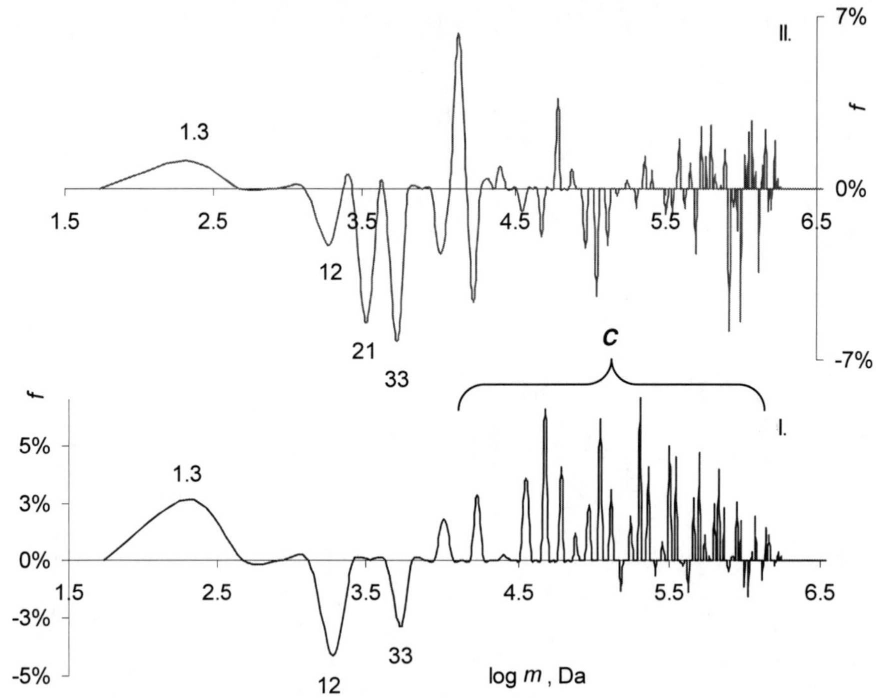

Figure 6.1. GMS-spectra of nonanoic acid (Merck, 0.904…0.907 g/cm^3) at 293 K. I – $p < 1 N/m^2$ and II - $p > 1 N/m^2$. Ensembles of masses up to 3 million Dalton, Zubow constant amounts to $6.7 \cdot 10^{-15}$ N/m. Numbers above signals – number of acid molecules.

Visible weak and strong shock waves are characterized to differently influence the main long-range order characteristics of nonanoic acid in the liquid state. Under strong shock wave influence, a destruction of clusters connected with their densification takes place where the average cluster mass (M_{GMS}) increases while the part of the collapsed ($Д_c$) and skeletal clusters ($F_{95\%}$) sharply rises and the number of cluster kinds (*N*) gets greater, too.

At the level of small clusters the liquid is described by a distinctive long-range order (Figure 6.1), which will be discussed in detail in the next. For the strong shock wave, there are clusters with masses equally to 1.3; 12; 21 and 33 acid molecules while for the weak one 1.3; 12 and 33 molecules are observed, only. The cluster consisting of 21 molecules (collapsed form) seems therefore rather to be a product of cluster destruction. Under these conditions the other small clusters are stable and do not change their density.

The signal of the smallest cluster corresponds to the mass of 1.3 molecules, which can be illustrated as follows (Scheme I):

Cluster-1.3

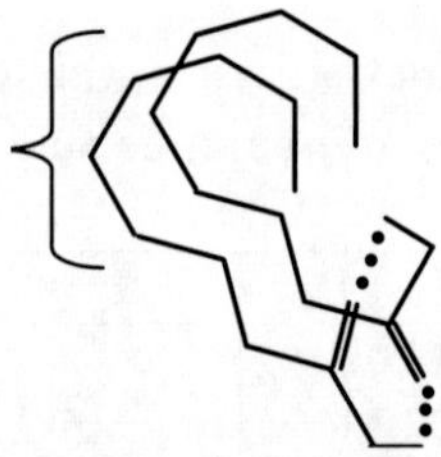

(I)

Here are the skeletons of two molecules (dimer) that are both linked by hydrogen bonds between carboxylic groups (marked with dotted lines) and by aliphatic fragments through weak cohesion forces (van der Waals linkage). The prototype of a spiral meander is found to be the structuring factor of aliphatic chains' interaction. Thus, the expanded cluster-1.3 is represented by these aliphatic fragments. Using the Cochrane models, the structure of this dimer has been proved.

The cluster structure of 12 molecules is given below (Scheme II):

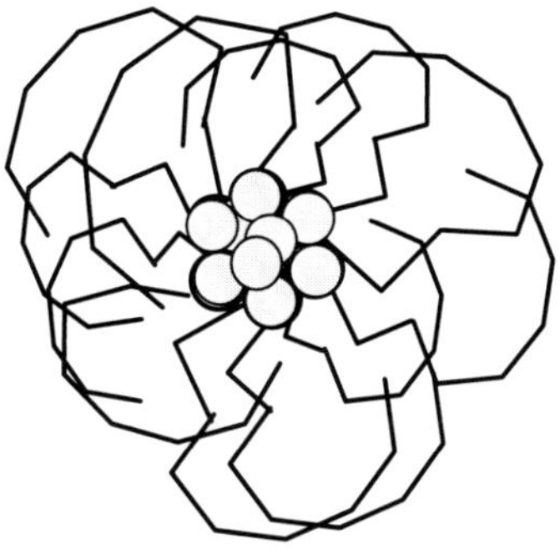

(II)

As is visible, the polar carboxylic groups (described as circles) are concentrated in the cluster center while the aliphatic chains are located as half-folded ones on the outer side (bracket lines). The cluster remembers the micelle structure of surface-active substances [1, 2] (the area of the micelle mass in Figure 6.1 is described with a horizontal bracket).

The cluster of 33 molecules seems to consist of two clusters, namely, the cluster-12 and the cluster-21, because they appear only after it was subjected to strong shock waves (Figure 6.1).

In the area of large clusters (log m > 4), a broad Gauss distribution of expanded clusters (marked with brackets) is found to be under weak shock waves only. It remembers the signal concentration in the mass interval C that was observed in liquids (§4.3, §7), polymers (§9) [3], and biomatrices (§15), too. This signal concentration has to be ascribed to micelles. Under strong shock waves modeling technological noises, this harmony in the long-range order of liquid nonanoic acid shall be destroyed.

Analogous regularities are found for other monocarboxylic acids.

CONCLUSION

The long-range order in liquid monocarboxylic acids is characterized by cluster and super cluster structures.

The structure of small clusters in liquids is observed to be a preliminary one to micelles.

Clusters in organic acids are characterized to be formed in stationary waves of white noises that penetrate the Earth.

REFERENCES TO 6

[1] Frolov J. G. Colloid chemistry. Surface effects and disperse systems. Moscow. Chimija. 1982, p. 298 (in Russian).
[2] http://en.wikipedia.org/wiki/Micelle
[3] Zubow Kr. V.; Zubow A.V.; Zubow V.A. Use of flicker noise spectroscopy for non-destructive analysis of nano structures. Zavodskaja Laboratorija. Diagnostika Materialow. 2008, vol. 74, no. 9, pp. 40-46 (in Russian).

Chapter 7

GMS-SPECTROSCOPY OF SULFURIC ACID

According to some authors [1, 2, 3, 4] liquids were characterized by a long-range order being represented by statistical ensembles of clusters and super cluster structures. Furthermore, clusters could be described by thermodynamic properties such as surface tension and potential and surface energy [1]. These clusters were formed in stationary waves of white noises permeating our planet [5, 6, 7]. The energy of the cluster formation for individual liquids was in the range of kJ/mol and even below [3, 8] that for solvated clusters of ion pairs. It amounted, however, to J/mol or mJ/mol. A thermal field (kT, k – Boltzmann constant, T – temperature, K) permanently destroys clusters of molecules but it is in equilibrium with the enforced formation of clusters. Energy clusters in a physical vacuum [9] are the driving force behind the molecular cluster formation from which molecular clusters get the energy for their organization. The lifetime of molecular clusters is in the picoseconds' time [10]. Individual liquids and mixtures thereof as well as polymers can be as seed crystal nanosuspensions [8]. The currently known standard physical and chemical methods for the isolation and analysis of molecular clusters can only be applied for polymers [11]. For liquids, this is not yet currently possible, although by using statistical analysis methods of weak signals, the long-range order (LRO) was investigated at the cluster ensemble level [2, 8].

To understand LRO in the "bread" of the chemical industry, sulfuric acid will open new applications, improvements of known technologies, the rational use of resources, and know-how in development. Publications concerning LRO in liquid sulfuric acid have not been researched by the authors but there are some reports on the simplest acid clusters with water ($(H_2SO_4)_y{\bullet}(H_2O)_x$, up to hexamers) under the formation of nano aerosols (Ø ~ 3 nm) in the upper atmosphere [12, 13, 14, 15]. These works allowed us to understand the formation of the smallest acid clusters with water with minimal exposure to other molecules. Cluster formation should proceed in liquid acid but its solutions will be somewhat different than in aerosols.

Applying computer simulation of water clusters, a number of hypothetical structures was calculated (Figure 2.1.1). A lot of models for clusters consisting of up to 20 molecules were suggested [16] whereas for large clusters, the number of cluster kinds was really low [17]. Computer modeled large clusters selected as the most probable structures were shown in Figure 2.1.1; the two structure types, collapsed and expanded, which are often in equilibrium with each other, were given. In relation to the surroundings, the equilibrium can be shifted to one or the other structure.

The cluster structure of water is not constant [2]. It depends on temperature [18], energy fields, impurities, salt ions, and pressure. In water, molecules are not identical [19]; they differ in their isotopic composition, interaction with surroundings, impurities (colloidal particles, salt ions), and external fields. Furthermore, they will be influenced by vessel walls [21]. Clusters of water molecules with similar physical characteristics, however, can be formed while being the background for water with constant properties. According to this logic, a similar LRO formation process in sulfuric acid and its water solutions can be expected. LRO in liquids can be experimentally determined using statistical methods such as gravitational mass spectroscopy (GMS) [2, 8, 20].

The aim of the present study was to investigate molecule clusters in liquid sulfuric acid and its aqueous solutions by GMS.

Sulfuric acid (Merck) and distilled water ($\chi = (20\pm10)\cdot10^{-6}\ \Omega^{-1}\cdot cm^{-1}$) served as the investigative objects. Both H_2SO_4 and water were aged for 0.5 years at 288±2 K in a dark box, free of mechanical influences. A part of the acid was preheated to 520 K and cooled to 290 K in a natural way (LRO destruction). The solutions were prepared through the successive dilution of the acid and storing for four weeks at 286 K (excluding external energy fields to reach the thermodynamic balance [9]). The temperature deviation during the measurement (290 K) did not exceed ±0.1 K and the atmospheric pressure remained constant during the experiment (6 h). For the measurement, the GMS spectrometer belonging to the firm, Aist Handels und Consulting GmbH (Germany), was applied where the sensor was placed in the center of a glass cell (20 ml) [2, 8, 20]. The wall impact was taken as a systematic error and ignored [21]. The measurement time was equal to 30 s. The GMS-spectra were documented in the coordinates of cluster masses (m) and their relative shares in the cluster ensemble (f). The calibration of the energy spectra was performed using the following formula: $E_i = \Delta S_i \bullet E_{290K}/|\Sigma\ (\Delta S)|$, where ΔS_i is the signal difference between the sample in the cell and the empty cell for the cluster i, $|\Sigma(\Delta S)|$ is the absolute sum of the differences of cleaned signals for all clusters from $i = 1$ to N, E_{290K} is the energy of cluster formation in sulfuric acid at 290 K (interpolated value between the energy of the destruction of all clusters at acid boiling at 609 K (50.1 kJ/mol), and the energy of cluster formation at crystallization at 283.6 K (10.7 kJ/mol). Two kinds of shock waves were used ($1\ N/m^2 > p > 1\ N/m^2$).

The method relies on digital multidimensional signal processing [9, 22, 23], which is caused by the weak interaction of shock waves with molecular clusters in liquids where the masses and oscillation frequencies of the clusters were calculated according to the Zubow equations [9] using the Zubow constant $6.4\cdot10^{-15}$ N/m for water and for water solutions. To divide between the different cluster forms, the collapsed clusters were described with $-f$ and the expanded ones with f, according to the modeled cluster $(H_2O)_{280}$ (http://www.lsbu.ac.uk/water/index.html, Figure 1). The bigger the absolute value of f the more individual is the cluster, and the smaller is its interaction with the surroundings. Its oscillations become freer, and on the contrary, the lower the f, the greater the oscillator interaction with the surroundings and the less the cluster is pronounced as a real particle. The average energy for the cluster formation in water was given in [2]. The orgraph for the cluster signals' cleaning was given in [24].

The GMS-spectra of sulfuric acid and water were shown in Figure 7.0. As visible, there is a sequence of cluster signals whose hypothetical formulas are presented, too.

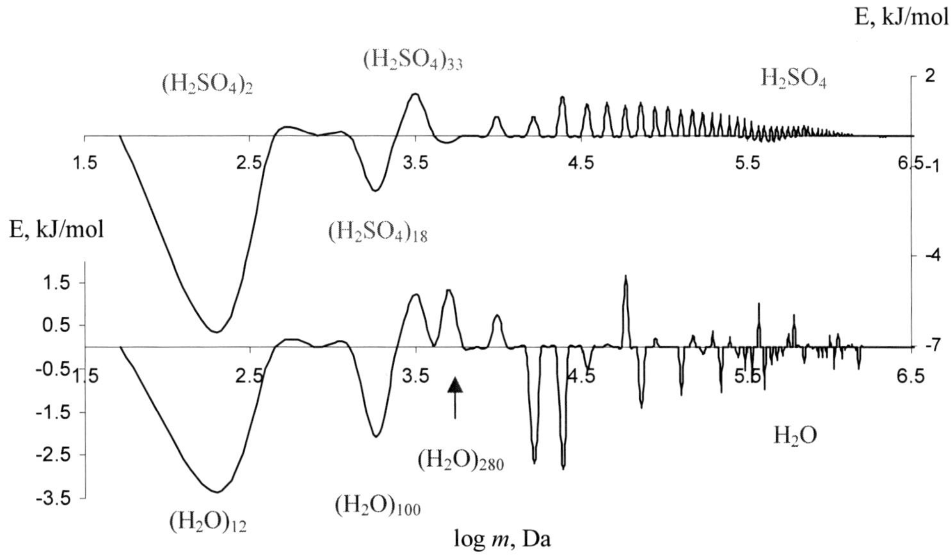

Figure 7.0. GMS-spectra of fresh H_2SO_4 and aged water. For H_2SO_4, average cluster mass = 191,487 Dalton (M_{GMS} = $(m\bullet\Sigma|f|)$, where f is the ratio of cleaned signals to the absolute sum of all cluster signals); part of collapsed clusters = 68% $(\Sigma(-f)/\Sigma|\pm f|)$; part of skeletal clusters ($F_{95\%}$) with reproducibility of 95% = 3%; and the number of all cluster signals (N) = 138. For water, M_{GMS} = 226,189 Da, part of collapsed clusters =70%, $F_{95\%}$ = 77%, and N = 48 at 290 K.

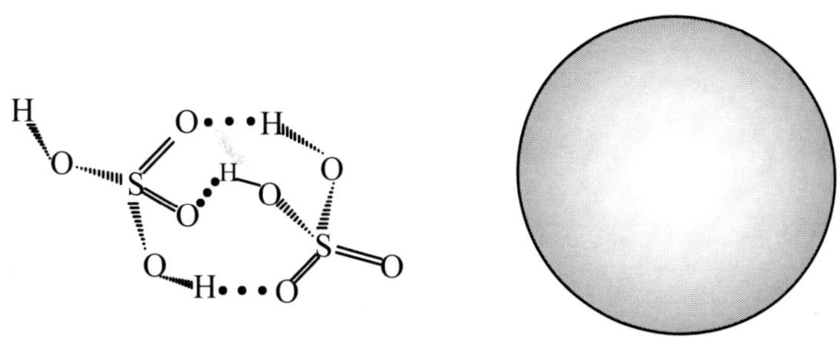

(I)

The isomers of sulfuric acid dimers were calculated by Chang as well as their formation energies [12]. The experimentally determined value was equal to 6.6 kJ/mol (collapsed cluster, Figure 7.1, scheme I), being in good agreement with that of 10.5 kJ/mol calculated by Chang. The difference of 3.9 kJ/mol is due to the impact of surrounding molecules in the liquid destroying the dimer. The cluster of 18±2 molecules (cluster-18) seems to be built from 9 dimers, $\{(H_2SO_4)_2\}_9$ (Figure 7.1), in such a way that the hydroxyl groups are concentrated in the center (cluster core) while the oxygen atoms are on the cluster surface. The hydroxyl groups in the cluster core are united by hydrogen bonds, which thus stabilize the cluster (a circle with grid). Due to the pre-micelle structure, the cluster-18 is able to oscillate as a spherical statistical unite.

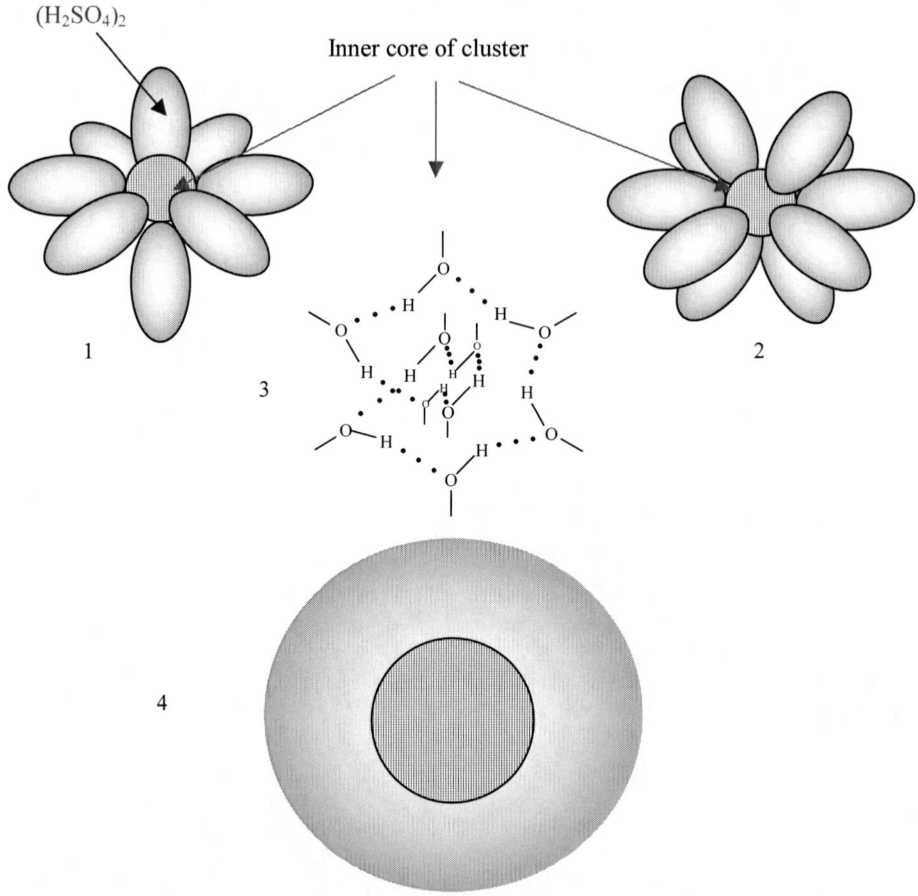

Figure 7.1. Most probable statistical structures 1 – $\{(H_2SO_4)_2\}_8$, 2 – $\{(H_2SO_4)_2\}_{10}$, 3 – model for the inner cluster core formed by hydroxyl groups, and 4 – common model of the average cluster-18 $\{(H_2SO_4)_2\}_9$.

The cluster distribution in an ensemble is not uniform (Figure 7.2). There are at least five cluster concentrations in the acid and only one in water.

For sulfuric acid, the cluster fraction distribution was satisfactorily described by the equation:

$$W_K = (-2E\text{-}06)x^6 + 9E\text{-}05x^5 - 0.0021x^4 + 0.0225x^3 - 0.1296x^2 + 0.4091x + 0.2703$$

where $x = m \bullet 10^{-5}$ Daltons. The dependence of the cluster number (n) on its oscillation frequency (ω, Hz) obeyed the first Zubow equation $n = 10^9/\omega^2$ [9]. For water, the cluster fraction distribution followed the equation:

$$W_w = 0.123\ln(x) + 0.6341$$

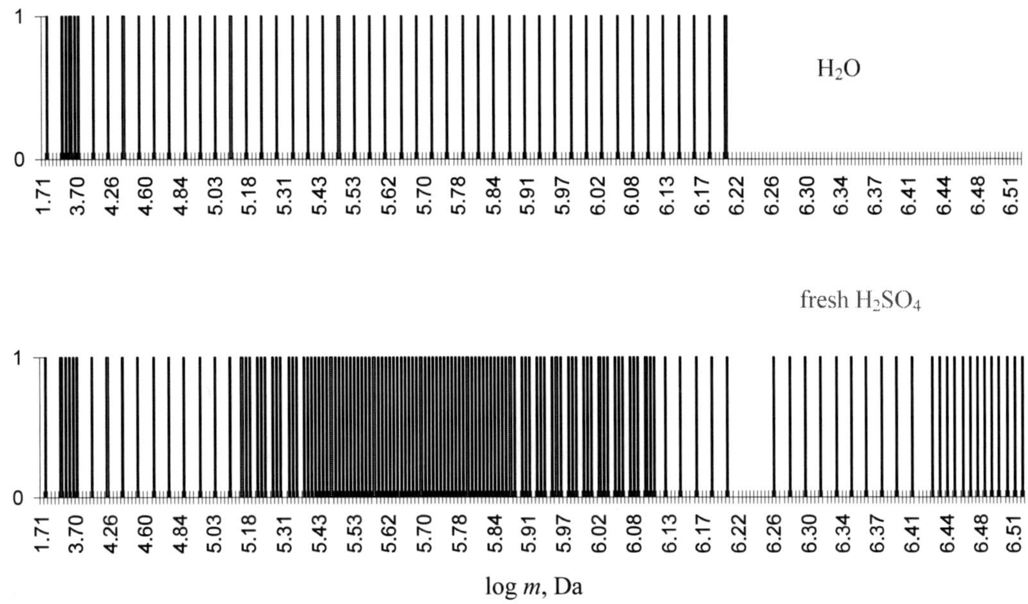

Figure 7.2. Bar GMS-spectra of fresh sulfuric acid and aged water At 290 K.

Remarkably, the cluster masses of $(H_2SO_4)_{248}$ and $(H_2O)_{1351}$ were equal to that of a polystyrene cluster, which was extracted by Zisman [11]. As the cluster mass does not depend on the nature of the liquid or polymer, the cluster formation was concluded to proceed in stationary waves of white noises serving as a forced stabilizer for the organized structures [9]. Here the destroying role was played by the thermal background.

At the liquids' storing, the LRO changed where the energy state shifted to a new thermodynamic balance as it did in different GMS-spectra for fresh and aged water (Figure 7.1, [2]). For the Chaplin cluster $(H_2O)_{280}$ (Figure 2.1.1), interaction with the surroundings was found to weaken, leading to cluster individualization, e.g. the formation of energy in fresh and aged (six months) distilled water was around 0.4 kJ/mol [2] and 1.32 kJ/mol, respectively. The cluster formation energy calculated by Chaplin was around 1 kJ/mol [3]. For the reproducibility it is therefore necessary to consider such properties of liquids.

The GMS-spectra of H_2SO_4 and its aqueous solutions in the concentration range of 80 to 0.001 wt. % were shown in Figures 7.3 and 7.4. Additionally, before measurement, the solutions were kept for 24 hours at a place isolated from seismic shock waves.

From Figures 7.3 and 7.4, it is visible that at the acid concentrations 10 wt. % $< C <$ 20 wt. %, the cluster formation energy changed as a result of LRO restructuring and changes in the density of clusters to the opposite. A puzzle for many is the behavior of the acid with the concentration of 18 wt. % [25]. We explained this anomaly with a special solvation of a H_2SO_4 molecule by 8 H_2O. LRO in both acid and water was assumed to be destroyed at the dilution of H_2SO_4 with water. In concentrated acid, the H_2SO_4 molecules formed clusters in which water can be included leading to their LRO destruction. As the water concentration increases, the roles changed and at ~18 wt. %, water highly influenced the solution LRO. At up to C ~ 0.1 wt. %, however, its influence is not yet dominant (Figure 5). At concentrations below 0.1 wt. %, LRO of the solution is completely determined by water and for C < 0.001 wt. %, the effect of H_2SO_4 is even neglected.

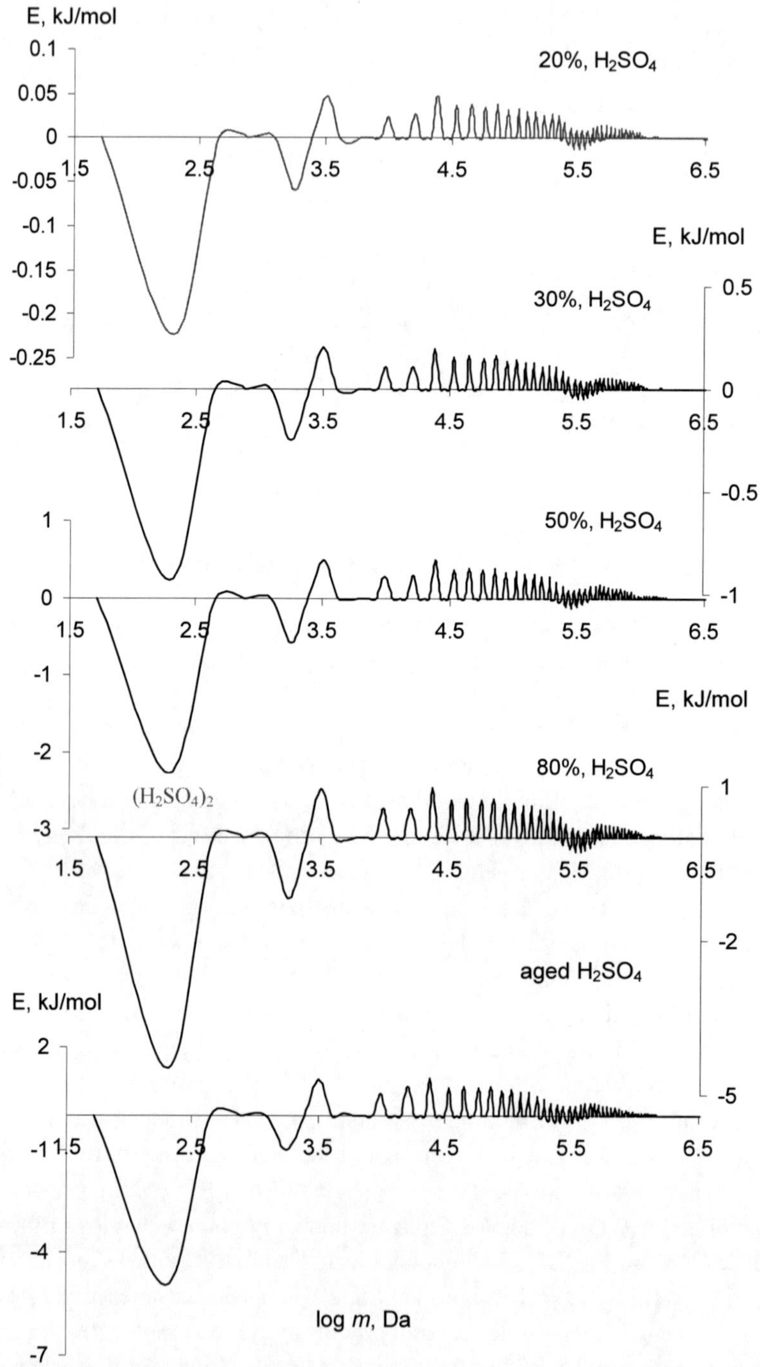

Figure 7.3. GMS-spectra of H_2SO_4 and its aqueous solutions (20 wt. % and higher).

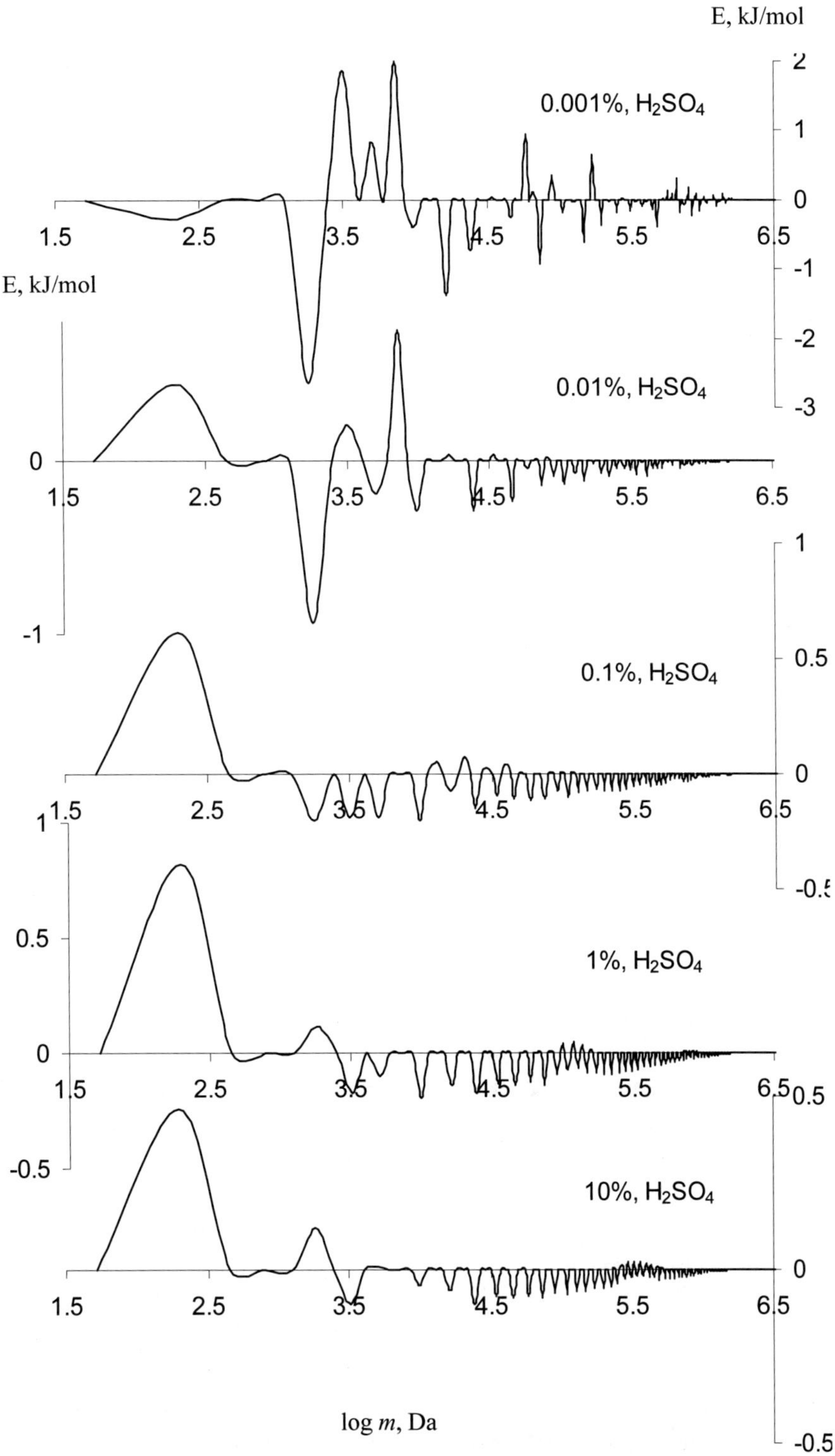

Figure 7.4. GMS-spectra of aqueous H_2SO_4 solutions from 10 down to 0.001 wt. %.

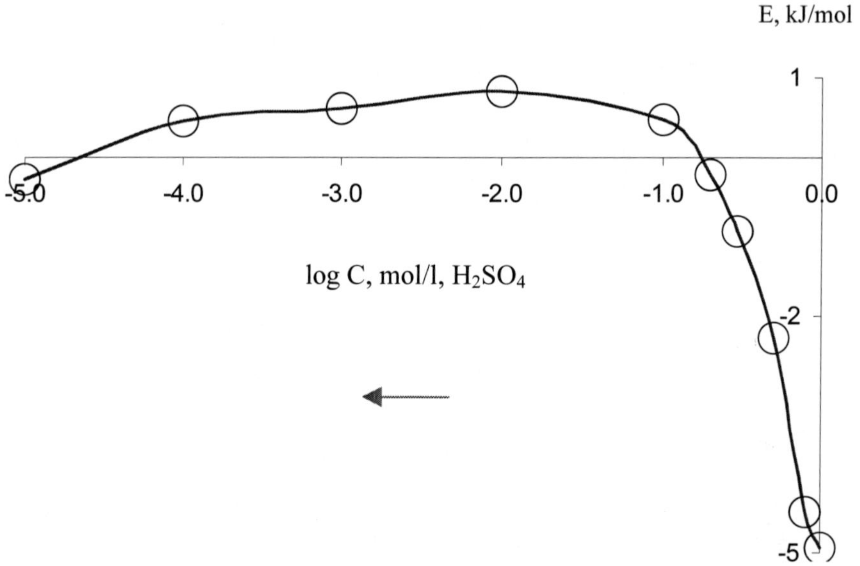

Figure 7.5. Formation energy of the cluster 198±5 Da at acid dilution (direction of the dilution process is indicated by the arrow).

For the cluster with the mass of 198±5 Da, with increasing acid dilution, the cluster formation energy is first strongly decreased and then stagnated. Finally at very high dilution, it is again negative (Figure 7.5).

In H_2SO_4, the mass of the above cluster corresponded to the dimer $(H_2SO_4)_2$ and in water, it corresponded to the base water cluster $(H_2O)_{12}$ [4,5], which was really $(H_2O)_{11\pm1}$ [8]. The little changing character of the curve (0.4…0.8 kJ/mol) in the concentration ranged from 0.01 to 10 wt. % and could evidence the existence of stable molecular clusters consisting of acid and water. These clusters could be $(H_2SO_4)\bullet(H_2O)_4$ [5, 26] and $(H_2SO_4)_2\bullet(H_2O)_1$ [12]; they were densified and crystallized under the influence of the spectrometer shock waves with energy release. In concentrated or pure H_2SO_4, the interaction of the dimer $(H_2SO_4)_2$ with shock waves was different, however, because it melted and absorbed energy. That means it was expanded. Two possible structures with three hydrogen bridges (HB) and different formation energies were developed by Chang [12]. As visible in Figure 7.5, the curve is divided into two events with a break point at ~ 18 wt. % due to the domination of two different cluster types, expanded clusters with two HB, and collapsed ones with three HB. Hence there is a specific mechanism of successive cluster destruction by water. From 100 to ~18 wt. %, water highly attacks the sulfuric acid LRO destroying first the inter cluster and later the intra cluster HB. Here the stable dimer $(H_2SO_4)_2$ with three hydrogen bonds between molecules (C > 80 wt. %) is transforming to the less stable one with two HB (~1 kJ/mol, C ~ 30 wt. %). The experimentally determined formation energy of the sulfuric acid dimers was found to be almost twice lower than the one calculated by Chang (10.5 kJ/mol [12]). This is understood as the result of the cluster interaction with its surroundings causing weakened intra cluster HB in the acid.

CONCLUSION

A long-range order in sulfuric acid and its aqueous solutions was found to be formed by molecular clusters having formation energy from 2 to -7 kJ/mol.

The cluster distribution in all investigated liquids was influenced by white noises of the gravitational field.

REFERENCES TO 7

[1] Zakharov V.V., Brodskaya E.N, Laaksonen A. Molecular dynamics simulation of methanol clusters. J. Chem. Phys. 1998. V. 109. №21. pp. 9487-9493

[2] Zubow Kr. V., Zubow A.V., Zubow V.A. Cluster structure of liquid alcohols, water and n- Hexane. Journal of Appl. Spectr. 2005, v.72, №3, pp. 321-328.

[3] M. Chaplin in http://www.lsbu.ac.uk/water/index.html

[4] Annika Lenz, Lars Ojamäe. On the stability of dence versus cage-shaped water clusters: Quantum-chemical investigation of zero-point energies, free energies, basis-set effects and IR spectra of $(H_2O)_{12}$ and $(H_2O)_{20}$. Chemical Physics Letters. 2006, v. 418, pp. 361-367.

[5] Bogdanov E.V., Mantrova G.M. Equicluster water model. Biomedical Radioelectronics. 2000, no. 7, pp. 19-28 (in Russian).

[6] Horsthemke W., Lefever R. Noise-Induced Transitions. Theory and Applications in Physics, Chemistry and Biology. Edit. Hermann Haken, vol. 15. Springer-Verlag, Berlin, Heidelberg, NY, Tokio 1984.

[7] Zubova, K. V.; Zubov, A. V.; Zubov, V. A. Low frequency movement of cluster-12 in potato amylopectin during growth. Influence of white noises. Khimiya Rastitel'nogo Syr'ya. 2009, no., 2, pp. 77-87 (in Russian).

[8] Zubow Kr. V., Zubow A.V., Zubow V.A. Oscillation spectra of seed NaCl crystals in aqueous Solution. Journal of Appl. Spectr.2005, v.72, №6, pp. 840-847.

[9] Zubow K.V., Zubow A.V., Zubow V.A. Ensemble of Clusters – New Form of Molecular Matter, Risks and Chances. Zubow Equations. In *Advances in Chemistry Research*, vol. 5, 2010, pp. 107-145. Editor J.C. Taylor. Nova Science Publishers, Inc. NY.

[10] Guang-Jun Guo; Yi-Gang Zhang; Ya-Juan Zhao; Keith Refson; Gui-Hua Shan. Lifetimes of cagelike water clusters immersed in bulk liquid water: A molecular dynamics study on gas hydrate nucleation mechanisms. J. Chem. Phys. 2004, v., 121, no., 3, pp. 1542-1547.

[11] Zysman, Viviane; Nguyen, Tuan Q.; Kausch, Henning H. Degradation on freezing dilute polystyrene solutions in p-xylene. Journal of Polymer Science, Part B: Polymer Physics. 1994, v. 32, №7, pp. 1257-1269.

[12] Chang-Geng, Kari Laasonen, Ari Laaksonen. Two sulfuric acids in small water clusters. J. Phys. Chem. A. 2003, v. 107, № 41, pp. 8648-8658.

[13] Hanson D.R., Lovejoy E.R. Measurement of the Thermodynamics of the Hydrated Dimer and Trimer of Sulfuric Acid. J. of Phys. Chem. A. 2006, v. 110, №31, pp. 9525-9528.

[14] Floyd K.B., Lovejoy E.R. Experimental Thermodynamics of Cluster Ions Composed of H_2SO_4 and H_2O. 2. Measurement and ab Initio Structures of Negative Ions. J. Phys. Chem. A. 2003, v. 107, № 46, pp. 9812-9824.

[15] Noppel H., Vehkamäki H., Kulmala M. An improved model for hydrate formation in sulfuric acid-water nucleation. J.of Chem. Phys. 2002, v. 116, № 1, pp. 218-228.

[16] Shruti Maheshwary, Nitin Patel, Narayanasami Sathyamurthy, Anant D. Kulkarni, Shridhar R. Gadre. Structure and stability of water clusters $(H_2O)_n$, n=8-20: An Ab Initio Investigation. J. Phys. Chem. A. 2001, v. 105, № 463, pp.10525-10537.

[17] Andersson P., Steinbach C., Buck U. Vibrational spactroscopy of large water clusters of known size. Eur. Phys. J. D. 2003, № 24, pp. 53-56.

[18] Yves A. Manz, Bin Chen, Glen J. Martyna. Temperature-dependent water structure: Ab initio and empirical model predictions. Chem. Phys. Letters. 2005, v. 405, pp. 294-299.

[19] Campbell J.A., Chemical Systems. Energetics, Dynamics, Structure. W.H. Freeman and company, San Francisco, 1970.

[20] Zubow A.V., Zubow K.V., Zubow V. A. The Phenomenon of Water Clusters Distribution in Water and Solvated Clusters of Ion Pairs of NaCl in Solution Near the Wall. The Wall Effect. J. of Chemical Industry (Saint Petersburg). 2006, v. 83, №6, pp. 300-306 (in Russian).

[21] Zubow K., Zubow A.V., Zubow V.A. Molecular Cluster Distribution in Sea Water Near the Wall. Antifouling aspect. Journal of Clusters, 2011, v.2 2, no. 4, pp. 603-619.

[22] Dan E. Dudgeon, Russell M. Mersereau. Multidimensional Digital Signal Processing. Prentice-Hall,Inc., Englewood Cliffs 1984.

[23] Müller E.. Digitale Signalanalyse in der Hydroakustik (Teil 2).Angewandte Akustik. Band 3. S. VEB Verlag Technik, Berlin, 1989, 1 Aufl.

[24] Zubow K., Zubow A.V., Zubow V.A. Water Clusters in Plants. Fast Channel Plant Communications. Planet Influence. Journal of Biophysics Chemistry. 2010, vol.1, no.1, pp.1-11

[25] Valeev A. Ch., Smirnov P.R. X-ray study anionic hydration in aqueous sulfuric acid from the difference of the experimental structure functions J. Phys. Chem. 1995, v. 95, №2, pp. 228- 231 (in Russian).

[26] Lotnik S.V., Chamiddulina M. A., Kasakov V.P. Effect of structure frozen solutions of H_2SO_4 at Radioluminescence. Izw. RAN. Ser. Chem. 1996, №6, pp. 1566-1568 (in Russian)

Chapter 8

SUPER CLUSTER STRUCTURE IN AROMATIC LIQUIDS

The understanding of the long-range order in aromatic hydrocarbons is of direct practical importance for solution processes, organic synthesis, drying, and polymerization. In the present chapter, the base characteristics of the long-range order at the level of cluster ensembles will be discussed for *n*-xylene and styrene. Concerning the formation of dimers in liquid aromatic hydrocarbons, little information has been researched by the authors, only [1, 2, 3, 4, 5], however, their structures have to be represented as follows (Scheme I)

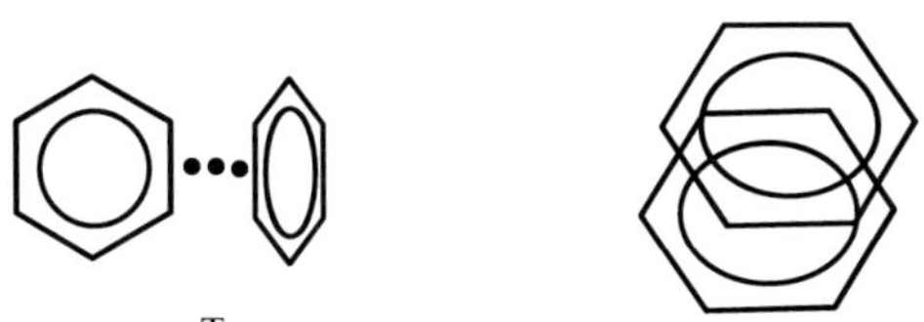

T

(I)

(perpendicular (T) and parallel position of benzene rings in dimers)

These two types of dimers can be identified with GMS using the right shock waves. Up until now it is not clear which of the forms is stable, but according to Janowski [5] and Lee [6], it must be the T type dimers. Their suggestion, however, is based on pure calculations where the influence of the surroundings is ignored.

As discussed by Mine [7], the formation of benzene trimers is also possible, too, whereas Janowski [5] even informs about sandwich structures.

In the next, the GMS-spectra of the two individual aromatics, *p*-xylene and styrene, which are of high practical importance, will be discussed. The GMS-spectrum of *p*-xylene under weak shock wave influence is given in Figure 8.0.1. As visible, there is a sequence of clusters in the collapsed and expanded forms. There are signals in the area of the small clusters being equal to 2, 17, 47, 93, and 153 *p*-xylene molecules where the collapsed forms dominate in the first two clusters but the expanded forms dominate in the last ones. In the area of cluster masses from $\log m = 3.8$ to $\log m = 5.6$, all clusters will be transferred into the expanded form; similar behavior is found for other organic liquids and water (§§5, 6, and 7).

More large mass concentrations ($\log m > 5.6$) are characterized as cluster associates, which are formed under the influence of white noises [8]. Their formation mechanism needs further investigation and will not be discussed in this chapter.

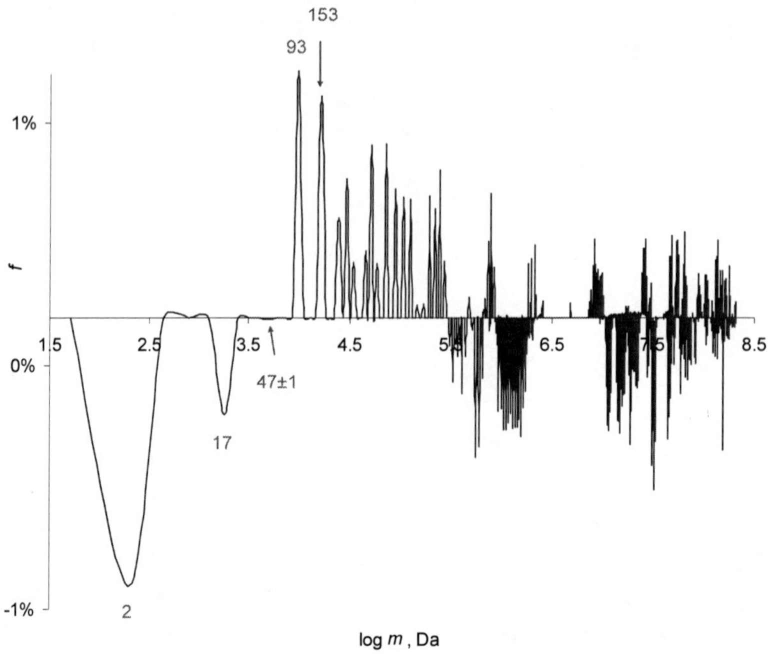

Figure 8.0.1. GMS-spectrum of *p*-xylene, 297 K, $p < 1$ N/m^2. Ensemble up to 210 million Daltons, $6.4 \cdot 10^{-15}$ N/m. Numbers above signals – number of *p*-xylene molecules. M_{GMS} = 44,664,869 Da, $Д_c$ = 61%, and $N = 987$.

For small clusters, the following sequences of clusters are possible to separate (in the range of the measurement error):

$\sim 153 \approx 93+47$; $93 \approx 47+47$; $47 \approx 17+17+17$ and $17 \approx 2 \times 8$

The correlation between the cluster masses is similar to that of other investigated liquids, e.g. acetone (§7, Scheme IV) and water (§5). The structure of *p*-xylene-153 will therefore be represented as follows, where the clusters, e.g. 47+47+47+2+2+2+2+2+2, appear and disappear permanently (irisation effect) according to the principle of the "Russian matroshka" (Scheme II).

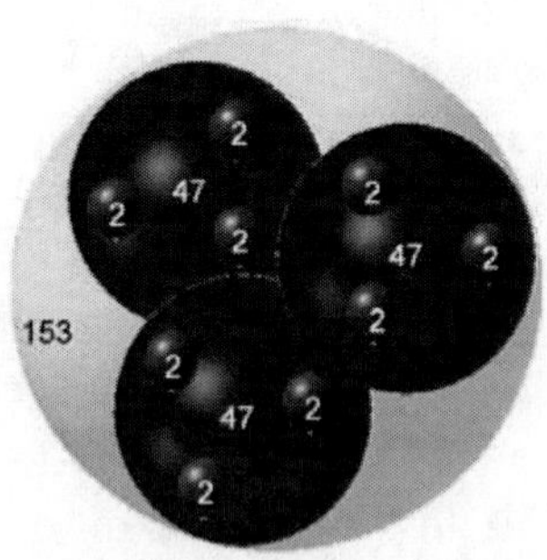

(II)

The base cluster of two molecules is in the center of every cluster-17 (not shown), which builds the cluster-47 where two clusters-47 assemble the cluster-93 and so on. Here the clusters can oscillate synchronously as well as asynchronously, leading to one (153) or several (2, 17(18), 47, 93) GMS signals.

The irisation effect can be illustrated using Scheme III:

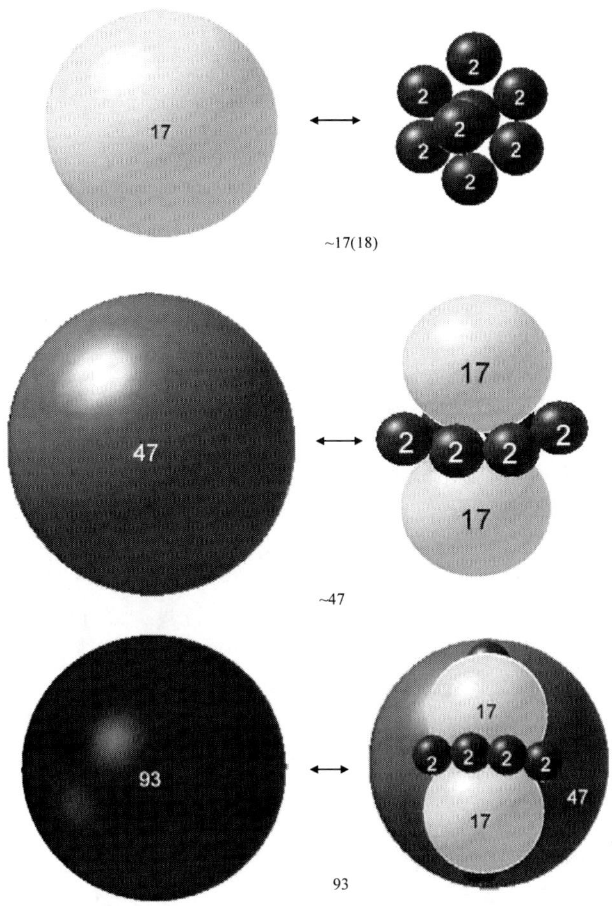

III.

(III)

The base *p*-xylene cluster is found to be a dimer, whose rotating figure has to be seen as a sphere (Scheme IV).

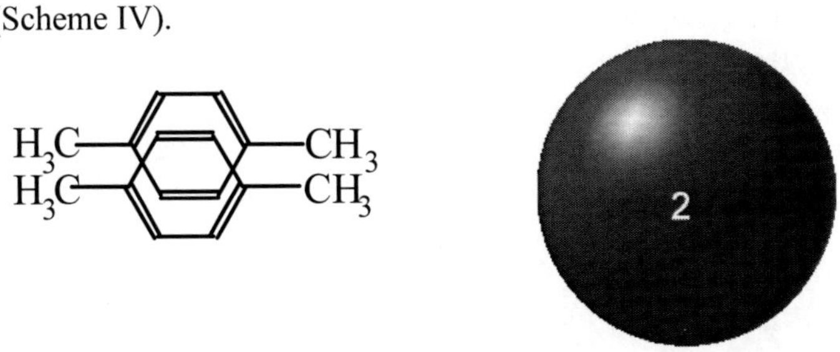

Hypothetical figure of a rotating dimer

(IV)

The integral cluster fraction distribution in liquid *p*-xylene in the mass interval shown in Figure 8.0.1 is given in Figure 8.0.2. As is visible, the curve does not rise monotonously; it is characterized by three steps.

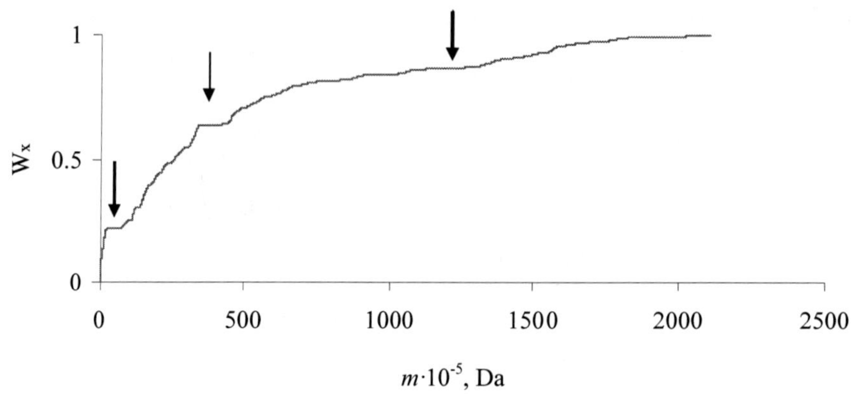

Figure 8.0.2. Integral cluster fraction distribution in liquid *p*-xylene. For a cluster ensemble up to 210 million Da, see Figure 8.0.1.

In the next section, we will analyze how strong shock waves model real conditions, for example, in the chemical technology (heating, stirring, mechanical fields' influence, etc.) change of the long-range order in *p*-xylene (Figure 8.0.3). The influence of strong shock waves is distinctive. While the small clusters (2 and 17) still exist, the middle (mass interval from log m = 3.5 to log m = 4.2) disappears completely and the signals that arise from the super large clusters (log $m > 4.5$) are not even reproducible.

At heating, the dimer changes its signal prefix, which could be an indication for a balance between the collapsed and expanded structures (Scheme I). At 286.6 K it is like the cluster-17 in its collapsed form. Mainly however, at 299.4 K, both clusters are found to be expanded. While cluster-2, upon further heating up to 368 K, practically disappears, the signal intensity of the expanded cluster-17 sharply increases. The heating of *p*-xylene is connected with the increased Brownian motion of the molecules, and the clusters-2 are assumed to disappear (individuality loss) and to be involved in the structure of the cluster-17. Therefore, the cluster-17 seems to be a rival for the base cluster-2 and to take its functions inside a cluster ensemble in color noises.

The GMS-spectra calibration (Figures 8.0.1 and 8.0.3), according to the energy, is difficult due to the etalon absence. Additionally, different cluster ensembles with different energy capacities and different surroundings' dependencies are possible. The formation energies of the benzene dimers are known [1, 2, 3, 4] and can be applied as an internal etalon but only for the smallest clusters.

It could be of interest to understand how cluster formation proceeds in monomers, for example, styrene being the base raw material for the production of polystyrenes (Scheme V).

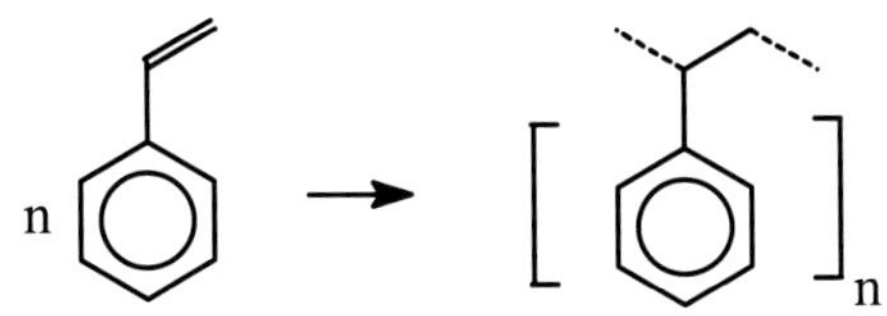

(V)

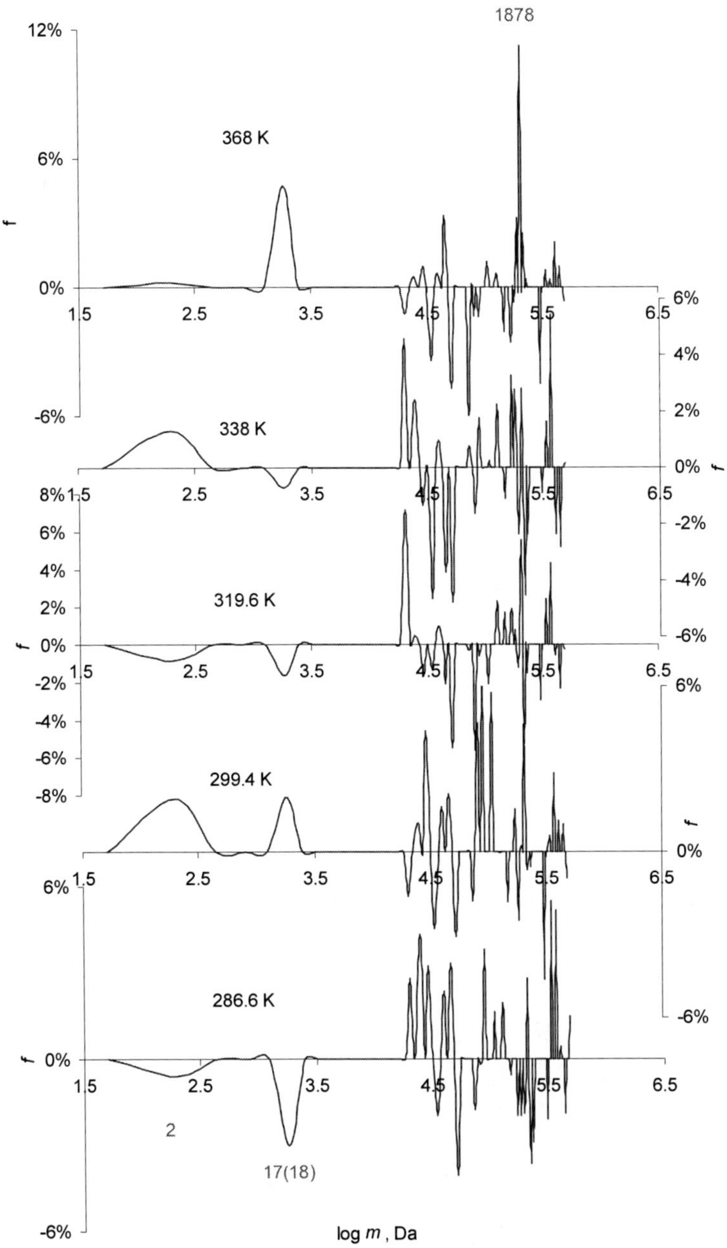

Figure 8.0.3. GMS-spectra of *p*-xylene at heating (6 K/min). Ensemble up to 3 million Daltons. $Z = 6.4\cdot10^{-15}$ N/m. Numbers above signals are the number of *p*-xylene molecules, $p > 1$ N/m^2.

The idea of cluster formation in liquids is based on the fact that there are no two equal molecules in liquids. From the authors [1, 9, 11], it was found that clusters are stable in the whole temperature interval in which the liquid exists. Cluster formation in liquid styrene influences the structure of the synthesized polystyrene, especially when the polymerization process proceeds in the presence of other compounds, e.g. solvents, catalysts, etc., which increase the cluster formation in the monomer. It cannot be discounted that at the polymerization structure, elements of the statistical cluster ensemble in monomers were transferred to the structure of polymer chains, which then reflects the cluster structure of the monomers.

Clusters of 50 monomer molecules and more introduce a new challenge for molecule ordering, therefore, their modeling is difficult (Figure 8.0.4). Its structure can be represented as one cluster consisting of two base clusters with 18 and 32 styrene molecules, which oscillate as one unit (merging of structures, Scheme VI). This unit is built as a layer structure probably with internal oscillation of collapsed and expanded elements [10]. The larger clusters, consisting of 72, 97, etc., molecules also seem to be built by these base clusters with 18 and 32 styrene molecules. Their stability should be explained by the layer structure, too.

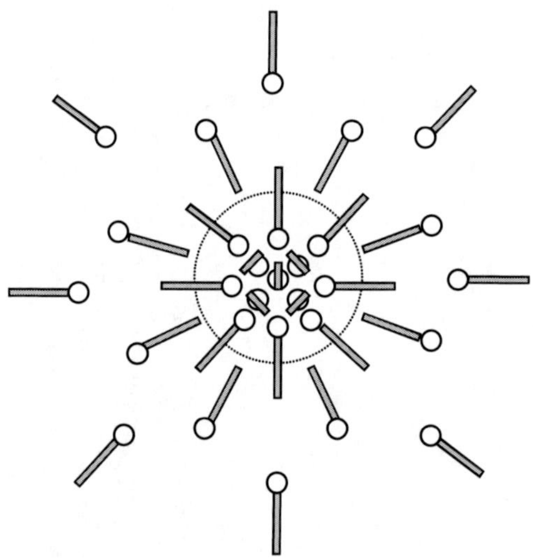

Figure 8.0.4. Projection of styrene molecules of the cluster-50 (slice plane) upon a plane. The nucleus (marked with a dotted circle) is the cluster-18 containing 18 styrene molecules. The vinyl groups (circles) are concentrated in the center while the phenyl groups (little rods) are on the cluster surface, which interacts with the molecules of the next layer through the phenyl groups. The molecules assembling in the next layer proceeds via the vinyl groups.

In Figure 8.0.5, the GMS-spectra of styrene at different temperatures are given. As shown, the structure of the cluster distribution already changed remarkably at 343 K so the clusters consisting of dimers collapsed and the ratio of the collapsed clusters to the expanded one, consisting of 18 (7,353 Hz), 32 (5,525 Hz), or 50 (4,405 Hz) molecules approached just one.

As was shown in [11, 14] the three clusters consisting of 18, 32, and 50 molecules are the base elements for the long-range order in liquid. The smallest oscillating cluster was observed at 22.2 kHz (200±20 Da). The number 2 (dimer) in Figure 8.0.5 marked the signal of this oscillator. The styrene dimer structure can be imagined as a vibration of expanded [2] and collapsed form [11, 14]. The presence of dimers in liquid aromatic hydrocarbons was already discussed by Zacharopoulos [3]. The energy parameters of benzene dimers were used for the analysis of the cohesion interaction in polymers containing phenyl groups. Additionally, styrene dimers were found with mass spectrometry (200 μJ) at 301 K by Zensen [1].

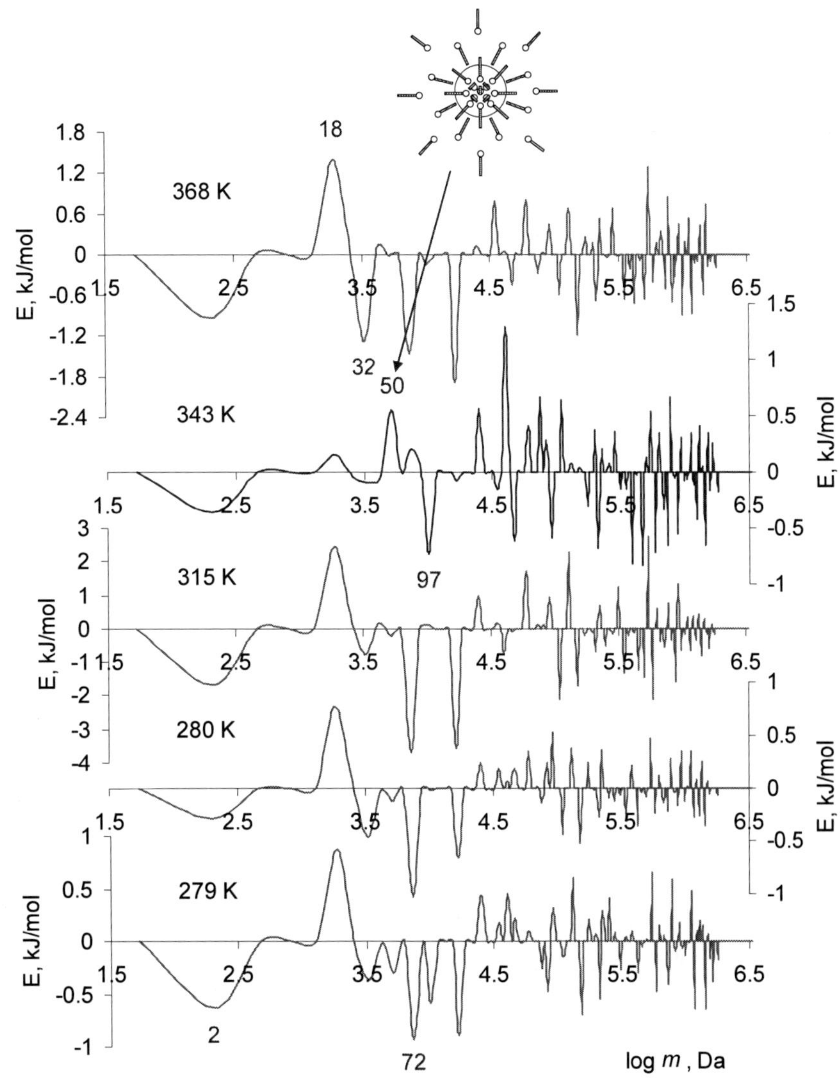

Figure 8.0.5. GMS-spectra of styrene at different temperatures. For the temperature stabilization, 30 minutes were given. Strong shock waves ($p > 1$ N/m^2, $Z = 6.6 \cdot 10^{-15}$ N/m). Color noises – industrial conditions. The numbers over the signals correspond to the number of styrene in a cluster.

The cluster of 18 molecules shall be described with the Cochrane model where 14 molecules fit well in the model but 4 molecules only satisfactorily. In the sphere model for this, the vinyl groups are directed to the center. The clusters consisting of 32 molecules were modeled so that the vinyl groups of styrene were partially on the cluster surface and the aromatic rings in the second layer between the vinyl groups. In this model, an approaching of the vinyl groups is prevented, therefore, 26 molecules fitted well into the model but only six fit satisfactorily. This structure model prevents the formation of isotactic polystyrene (iPS) blocks and promotes syndiotactic configurations (§9).

As shown in Figure 8.0.5, the styrene clusters are sufficiently stable structures and they are present in the whole investigated temperature interval. Through the analysis of these spectra, the amount of collapsed clusters in the liquid monomer can be calculated as the absolute sum of all negative signals (*-f*). This value is not constant and has two minima at 293 and 343 K. The liquid monomer has two structure transitions at heating. The authors [12] found an analogous structure for water that transitions at 277 and 315 K (§5).

As shown in Figure 8.0.6, the density of styrene clusters is different from those of clusters with the same mass in polymers.

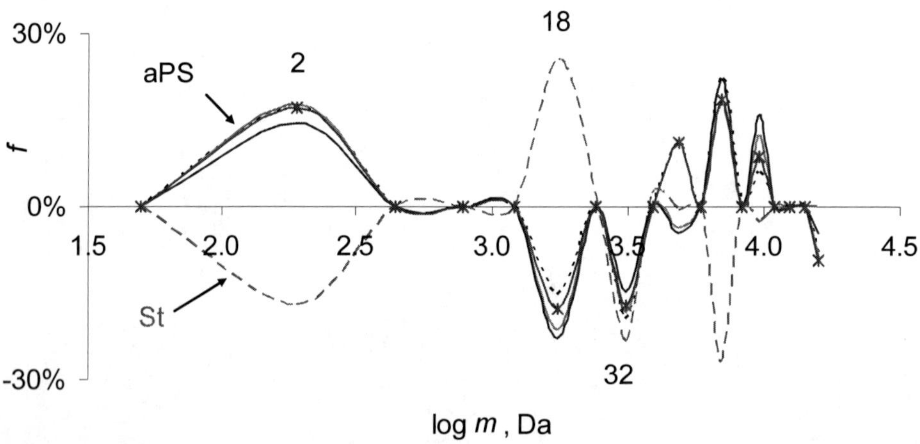

Figure 8.0.6. GMS of aPSs and styrene (dotted line) at 366 K for a cluster ensemble up to 15 kDa, aPSs with 17 kDa, 35 kDa, 199 kDa, and 458 kDa were used (according to GMS).

The dimer in styrene is represented in the collapsed form but in aPSs in the expanded one (diads). The chain is curved (spiral state) leading to no parallel aromatic rings in these diads (Figure 9.1.1). The cluster consisting of (17±1) ~18 members in aPS oscillated in a collapsed form but in the liquid monomer in an expanded one. The cluster with 32 styrene units in the polymer and the cluster consisting of 32 styrene molecules in the liquid monomer possess the same density.

The formation energies of clusters in organic liquids are very small. Clusters could be stabilized by the entropy factor (cooperation effect of the intermolecular interaction) and by a special super cluster structure in an ensemble that is built by three clusters (18, 32, and 50 styrene molecules/members) where two of them (18 and 32) oscillate in an opposite direction (collapsed and expanded) but the third one (50) operates in a layer. The first two clusters are connected by the opposite oscillation that proceeds in the third cluster. Here the third cluster acts as a binder (see Scheme VI for aPS below, [13]). Generally, according to Zubow equations, only certain molecular mass concentrations are permitted by the gravitational field (§3.2).

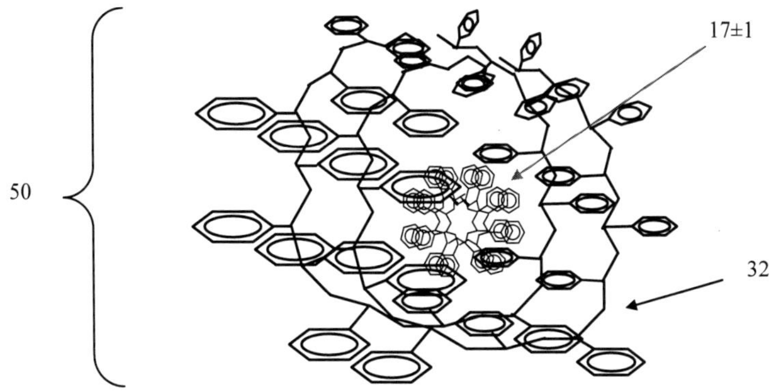

(VI)

Oscillations of layer structures on the cluster surface in liquid water were observed by Yoshiteru Yonetani [10], who ascribed to the oscillations of liquids different densities and adhesion energies. A similar effect probably takes place in liquid styrene, too. It is assumed that these effects stabilize the super cluster structure of the liquid at the level of primary building units consisting of these three clusters. Analogous conclusions about the stabilization of the cluster structure in liquids were made for water, methanol, *n*-hexanol, *n*–hexane [13], and styrene [1].

CONCLUSION

The long-range order in liquid aromatics is characterized by clusters, cluster associates, and super cluster structures.

Under white noises, the base cluster consists of two molecules.

The energy clusters of the physical vacuum are found to be the energy supplier for molecular cluster formation.

REFERENCES TO 8

[1] Zensen T.A, Jarski P, Eaton J.G. A REMPI study of styrene(Ar)$_n$, n=4–12 clusters. Chemical Physics Letters. 2000, vol. 326, no.5, 6, pp. 389-394.

[2] Robert A., Di Stasio Jr., Gert von Helden, Ryan P. Steele, Martin Head-Gordon. On the T- shaped structures of the benzene dimer. Chemical Physics Letters. 2007, vol. 437, pp. 277-283.

[3] Zacharopoulos N, Vergadou N, Theodorou D.N. Coarse graining using pretabulated potentials: Liquid benzene. Journal of Chem. Phys. 2005, vol.122, p. 244111; doi:10.1063/1.1948370.

[4] Erin R. Johnson, Gino A. DiLabio. Structure and binding energies in van der Waals dimers: Comparison between density functional theory and correlated ab initio methods. Chemical Physics Letters. 2006, vol. 419, pp. 333-339.

[5] Tomasz Janowski, Peter Pulay. High accuracy benchmark calculations on the benzene dimer potential energy surface. Chemical Physics Letters. 2007, vol. 447, pp. 27-32.

[6] Lee E.C., Kim D., Jurečka P., Tarakeshwar P., Hobza P., Kim K.S. Understanding of Assembly Phenomena by Aromatic−Aromatic Interactions: Benzene Dimer and the Substituted Systems. Journal of Phys. Chem. A. 2007, vol. 111, no. 18, pp. 3446-3457.

[7] Mine M., Mori H., Ogata M., Tanaka S., Tsutsui T., Miyoshi E. Computational research of the electronic structure of benzene trimer cation by ab initio method. Chemical Physics Letters. 2007, vol. 438, pp. 157-161.

[8] Horsthemke W., Lefever R. Noise-Induced Transitions. Theory and Applications in Physics, Chemistry and Biology. Edit. Hermann Haken, vol. 15. Springer-Verlag, Berlin, Heidelberg, NY, Tokio 1984.

[9] Kawanishi, Shunichi; Sasuga, Tsuneo; Takehisa, Masaaki;Takasaki Radiat. Existence of a novel liquid phase of styrene. Journal of Physical Chemistry. 1981, vol. 85, no. 9, pp. 1271- 1274.

[10] Yoshiteru Yonetani. A severe artifact in simulation of liquid water using a long cut-off length: Appearance of a strange layer structure. Chem. Phys. Letters. 2005, vol. 406, pp. 49-53.

[11] Zubow K.V., Zubow A.V., Zubow V.A. Ensemble of Clusters – New Form of Molecular Matter, Risks and Chances. Zubow Equations. In Advances in Chemistry Research, vol. 5, 2010, pp. 107-145. Editor J.C. Taylor. Novapublisher. NY. USA.

[12] Zubow KV, Zubow AV, Zubow VA. Cluster structure of liquid alcohols, water and n-Hexane. Journal of Appl. Spectr. 2005, vol.72, no. 3, p. 321-328.

[13] Zubow, A.V. Zubow, V.A. Zubow, "Principles of gravitation spectroscopy. New form of molecular matter. Processes. Fields. "Aist Handels- und consulting gmbH", dep. R&D, Berlin, 2011, electronic book (www.zubow.de), p. 900 (in Russian).

Chapter 9

POLYSTYRENES

9.1. STATISTICAL ENSEMBLE (LONG-RANGE ORDER) IN APS SOLUTIONS

In Figure 8.0.6, the GMS-spectra of polymers solved in *p*-xylene are given. These spectra were achieved by subtracting the signals of oscillating clusters in a "good" solvent from the signals of the clusters in the aPS solution. In earlier experiments it was found that at polymer concentrations higher than 0.02 mol/l, the spectra become more complicated; this is caused by the concentration's effects. For a better understanding of the oscillations in the coils, a concentration lower than 0.018 mol/l was chosen at which the highest sensitiveness of the device and the lowest concentration effects could be observed.

After analyzing the spectra interval up to 15 kDa and comparing the oscillator signals in the polymer coils (diluted solution) with each other and with the small clusters in styrene, an agreement of the mass oscillators was found (Figure 8.0.6) for aPS even at the density level. We assume that in the small cluster interval all aPS possess the same structure. This is only possible if the polymer chains in small clusters are of the same mass as well of as the same form that depends on the chain configuration. Since the signal intensities of the clusters are comparable, the same interaction mechanism with the surroundings can be accepted. This is only possible, however, when the clusters have the same surface, which is built by cohesion forces of chains with the same configuration.

It was observed by Natta [1] that the elementary unit in crystalline isotactic polystyrene consists of 18 styrene members that build a spiral. Thus, in the polymerization process of liquid styrene from expanded clusters consisting of 18 styrene molecules in the polymer the same clusters of 18 styrene members were found though, in the collapsed form. For the formation process of small clusters the Cochrane model was applied (Figure 9.1.1). With this modeling the building of expanded diad clusters at the curving of the isotactic chains in PS was confirmed. The formation of the collapsed polymer cluster of 18 members leads to fragments with isotactic chains and the diad clusters oscillate as one unit, which was supported by the results of GMS.

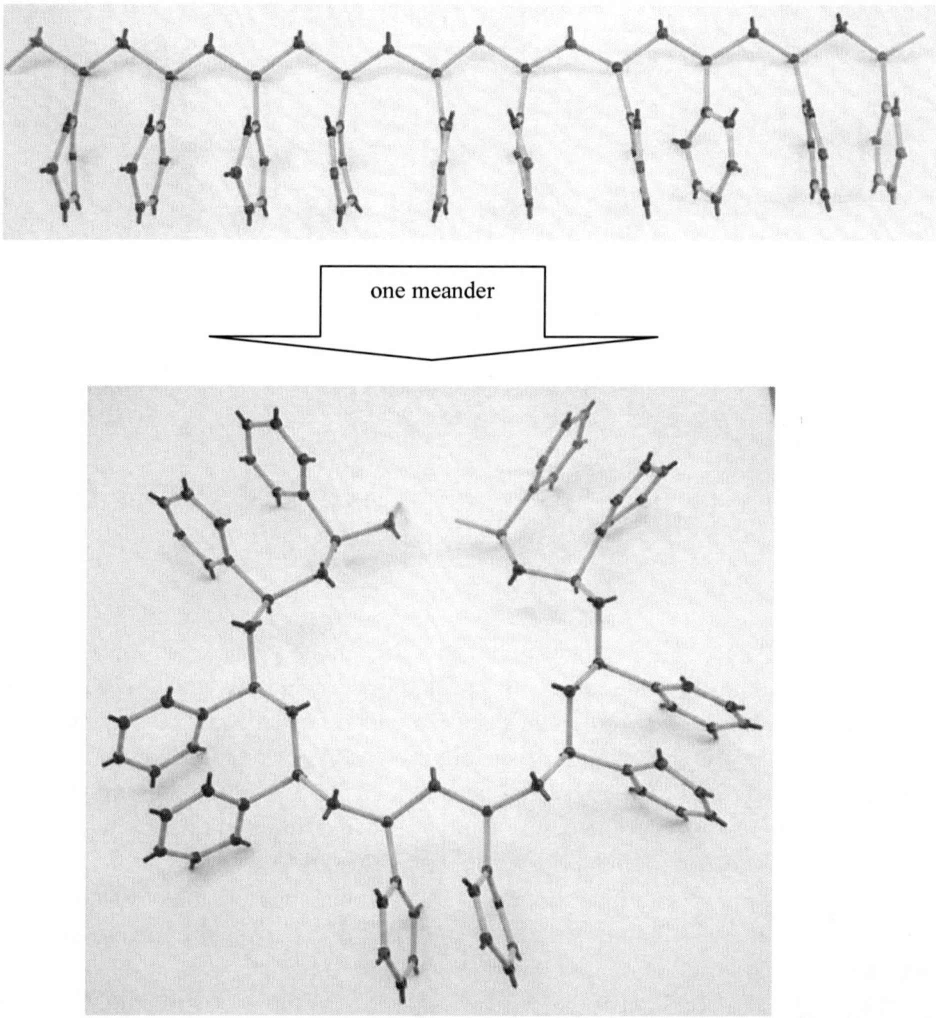

Figure 9.1.1. Cochrane model of an open chain and of a chain coiled in an isotactic configuration.

Using GMS of diluted aPS solutions with different molecular weights, it was shown that the cluster structure of polymers reflects the long-range order in the liquid monomers. For small clusters this regularity is especially visible, §8, Figure 8.0.6 [2].

It should be of great interest to investigate the super cluster structure of aPS with different molecular weights. For this purpose the GMS of the "good" solvent (*p*-xylene) was subtracted from the GMS of the aPS solution (0.018 mol/l, Figure 8.0.6). As result of this mathematical subtraction a GMS of a "naked" polymer in a diluted solution was obtained. Such spectra were recorded for all investigated polymers. Dividing the GMS of the high molecular aPS (S_x) by the GMS of aPS-17 (S_{17}, 15 kDa) permits us to see differences in the super cluster structure with increasing molecular weight.

In Figure 9.1.2, the results of these calculations are given; as is visible, nearly all signals are at the noise level besides the signal ratio of clusters-50 with a mass of 50 styrene units (4,405 Hz, Figure 8.0.4) and of the new cluster-60 (50 styrene units + 10 *p*-xylene molecules, 4,016 Hz). They strongly differ from the general level. This new cluster-60 appears in all aPS with a molecular weight that is higher than that one of aPS-17. For solid polymers, this effect was not found, therefore, the new oscillator with a mass of 60 styrene units in aPS solution is an "old" cluster of 50 styrene units, which adsorbed 10 *p*-xylene molecules and is built from polymer chains with atactic configuration. According to results of Figure 9.1.1 and 9.1.2, it has been concluded that the aPS chains are built from blocks consisting of chain fragments with different configurations. Only the blocks of clusters with 50 styrene units adsorbed the solvent that leads to the polymer dissolution in *p*-xylene.

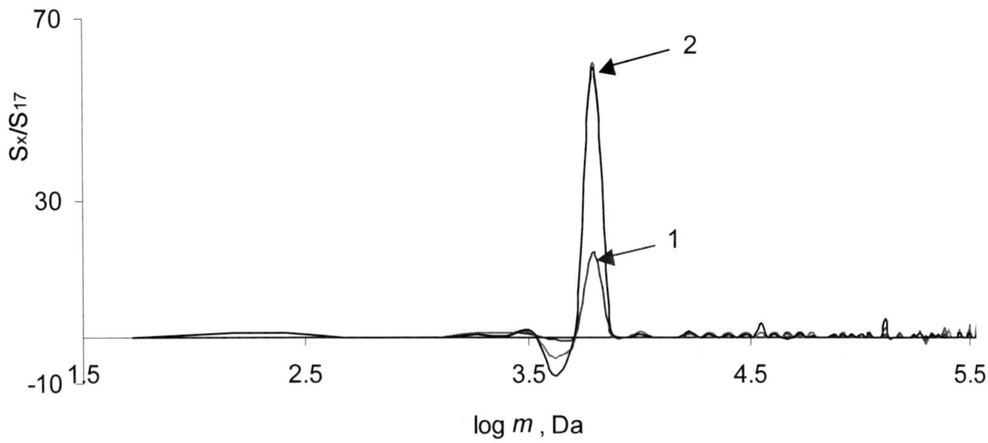

Figure 9.1.2. Signal dividing of GMS of high molecular aPS (S_x) by GMS of aPS-17 (S_{17}). At 297 K, 0.018 mol/l. 1 – S_{35}/S_{17}, 2 – S_{199}/S_{17}, and 3 – S_{458}/S_{17} (curves 2 and 3 are identical), here S_{35} – signals in aPS with 35 kDa, S_{199} – 199 kDa, and S_{458} – 458 kDa.

9.2. The Cluster Structure of aPS Coils in Solution

In a solution, a coiled chain is most likely an open one that can be explained with the thermodynamic minimization of the inner energy (Figure 9.1.1). In the model, the coiling of an open isotactic polystyrene (iPS) chain leads to a spiral conformation [3, 4] with mainly isolated isotactic diads. The phenyl groups in the diads are ordered parallel and opposite [5]. Accordingly, a model of the cluster of 18 members is built as a spiral of two meanders where each meander consists from eight to nine styrene units. This model agrees well with Yozo [6]. According to him, the iPS unit cell is formed by isotactic chains consisting of 18 styrene units. Oscillators of isotactic and syndiotactic triads, tetrads, pentads, etc. (up to 18 units), which vibrate as one cluster were not found by GMS or by Cochrane models; it was shown that they cannot exist (Figure 8.0.6).

The modeling of the polymerization of the 18-styrene molecule cluster has shown that the cluster shall be completely transferred from a monomer to a polymer with an isotactic chain configuration. In this cluster, the polymer chain is coiled in a spiral of collapsed form. Each of the chain ends has a block consisting of 18 units in isotactic configuration, followed by a second block of 32 units in syndiotactic configuration (modeling). These clusters are antagonists and they are forced to exist besides each other. Only at high pressures can they be separated [2], Figure 9.4.1.

Even in the crystalline state, the chain coiling of polystyrene (Figure 9.1.1) has been [7, 8], which supports our results obtained in the modeling of the aPS chains. Based on the GMS results, it was concluded that the polymer chain consists of polystyrene blocks with different configurations. These blocks contain clusters with different cohesion energies. The clusters 18 and 32 are linked by 2-4 styrene units. The third cluster consisting of 50 styrene units is of atactic configuration (§8). The next clusters in the polymer chains are in atactic configuration, too. In the solution, the polymer chain consisting of clusters (blocks) coils by itself due to cohesion forces. This coil has to be discussed as a cluster ensemble. To calculate the content of clusters in an aPS-17 coil, the maximum GMS-signals (minimal interaction with the surroundings) has been applied (Table 9.2.1).

Table 9.2.1. Mass content of clusters (M) in an aPS-17 coil. 0.018 mol/l polymer solution in *p*-xylene. Average values at 297 and 373 K. M = abs (S_i)/ Σ (abs (S_{18})+abs (S_{32})+abs (S_{50})+abs (S_{72})+abs (S_{97})), here S_i is the average value of the signal intensity of the cluster, sum (Σ (S_n)) is the absolute sum of all cluster signals in the polymer coil. is the number of styrene units in a cluster.

Clusters in aPS-17	Frequency, Hz	N	M, wt. %
iso-diads (i)	22,222	2	
spiral (iso)	7,353	18	21±0.5
spiral (synd)	5,525	32	26±1
collapsed cluster, atactic	4,405	50	16±1
expanded cluster, atactic	3,676	72	27±3
expanded cluster, atactic	3,155	97	9±1
coil, collapsed cluster	2,451	161	

As is visible, the mass concentration of the isotactic block (21%) is comparable with that of the syndiotactic one (26%), which is in satisfactory agreement with the IR results obtained by Pokrovsky [9].

The signal intensities of the small clusters in aPS-17 change only a little with a temperature increase from 297 to 373 K—with the exception of the cluster with 97 styrene units. Heating the polymer solution leads to an increase of its signal intensity by 20-30 %, which is caused by a partial reduction of its interaction with the surroundings and its individualization. The stability of these signals also shows that the interaction between a "good" solvent and a polymer is not strong.

Many authors investigated the coil conformation of polystyrene molecules of different molecular weights in solution with different methods [10, 11] without any clear idea of the coil formation and the transition of the coil to a spiral. Nevertheless, an answer for this question can be found with the help of a clear idea on the configuration of polystyrene chains, on their interaction with the solvent, on temperature influence, and on cluster formation in coils.

The model about the heterogeneous structure of aPS coils suggested by us is different from Picarra's homogeneous model [12]. Although Picarra confirmed the model experimentally, it cannot serve as a basis for the formation of the aPS coil consisting of chains with different configurations.

9.3. COIL ASSOCIATES OF APS IN SOLUTION

As GMS-spectra of aPS-17 in solution have shown, the polymer coils build associates (Table 9.3.1).

Table 9.3.1. Analysis results of oscillating associates (up to nonamers) in the GMS-spectrum of aPS-17 in 0.018 mol /l solution in *p*-xylene. The mass, *m*, is given in Da. At 297 K. Deviation = $(m_{obs}-m_{exp})/m_{obs}$

F, Hz	associates	m, observed	form	m, expected	deviation
2,451	coil (monomer)	16,770	collapsed	15,000	11%
1,695	dimer	35,069	expanded	30,000	14%
1,377	trimer	53,100	expanded	45,000	15%
1,160	tetramer	74,858	collapsed	60,000	20%
1,103	pentamer	82,877	collapsed	75,000	10%
1,002	hexamer	100,342	expanded	90,000	10%
882	oktamer	129,553	expanded	120,000	7%
817	nonamer	150,933	expanded	135,000	11%

The associates consisting of five and six coils are the most stable and they oscillate as collapsed and expanded units, respectively. Other oscillators (clusters) of coil associates are represented as expanded structures containing up to 13 wt. % included solvent. The signals of associates are considerably weaker than the oscillator signals of clusters in coils. In the case of aPS-17, they are a multiple of the molecular weight of 16.7 kDa. For a tetramer, the collapsed structure of an oscillator can be given as a model of four spheres (each sphere is a coil) that was placed in the corners of a tetrahedron. Details on larger associates (pentamer, hexamer, etc.) shall be given in [13]. Oscillations of new clusters were observed that arose at the contact points of the coils. These inter molecular clusters oscillate both as collapsed and expanded structures. They are probably formed from macromolecule chains with atactic configuration. Their analysis could give an idea about the real form of the coils.

It was observed by Zysman [14] that when freezing diluted solutions of polystyrene in *p*-xylene, the crystals of the solvent cut the macromolecule coils with a weight of 29 kDa. His results and our analysis data (Table 9.3.1) could be evidence of the block formation of super clusters in aPS. The coil associates are mass concentrations, too, and they belong to another level of mass concentrations hierarchy in our universe.

9.4. THE CLUSTER STRUCTURE OF APS COILS IN THE SOLID STATE

To understand the spatial arrangement of clusters in the super molecular and super cluster structure in solid aPS we investigated the interaction of clusters with each other in the moving process under pressure. The background for this is that the oscillation intensity depends both on the concentration of the clusters and on their interaction with their surroundings. The signal intensity approaches zero if the clusters are smeary, for example, after their active interaction with surroundings and their disappearance. It reaches the highest value for free cluster oscillation if there is no interaction with the surroundings [15]. For the solid aPS-17, oscillations of clusters with a mass of 18, 32, 50, 72, and 97 styrene units were observed.

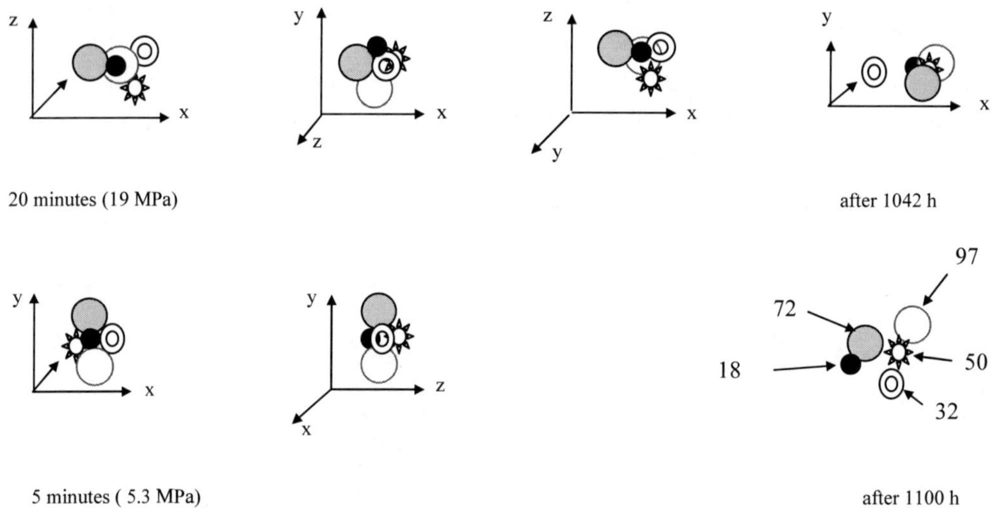

Figure 9.4.1. Dynamics of spatial arrangement of clusters in aPS-17 coils at a pressure of the solid polymer, 291±2 K.

Based on this idea and on some experiments under pressure between 13 kPa and 747 MPa, the spatial arrangement of clusters in the aPS-17 coil was characterized (Figure 9.4.2). It was found that the spatial arrangement of clusters in the coil is not always constant but depends on time, pressure, and temperature. The spatial ordering is usually constant for high pressure but at low pressures the clusters permanently change their spatial coordinates. Low pressures stimulate a spatial rearrangement, but high pressures limit it. Only at middle pressures (some MPa) can the primary super cluster structure of the solid polymer be investigated. It was observed that the clusters consisting of 18 styrene units with isotactic configuration and of 32 styrene units with syndiotactic configuration antagonize each other. They are the first that shall be divided from the other. In contrast, the clusters of 18, 50, and 72 styrene units form the most stable super cluster structures, which shall be explained by the similar cohesion energy and their high adhesion to each other. We think that the clusters of 50 and 72 styrene units have many similarities on the chain configuration level with the cluster consisting of 18 styrene units with isotactic configuration.

In diluted solutions, the cluster number (N_a) in coils weakly depends on the molecular weight of the aPS and only after $M_w > 29$ kDa; it can be described as follows $N_a = 0.8168 \cdot \ln (M_w) + 77.914$. In the solid state, the number is reduced by increasing M_w from 65 for aPS-14 to 52 for aPS-199. The complete number (N_g) of clusters both in coils and between them is up to M_w as described with the equation $N_g = 8.31 \cdot \ln (M_w) - 14.0$; here M_w is the average molecular weight polymer determined with viscometry. The polymer block with the mass of 29 kDa ($(PS)_{280}$) is therefore an element that is present in all aPS structures as secondary building unit (SBU). The analysis of the intensity change of the oscillating SBU and of small clusters under pressure influence showed that the small clusters are components of the SBU. SBU consists of five clusters in this model (Figure 9.4.2). The chains in these clusters enriched with isotactic configurations are near the central cluster of 18 styrene units. The central cluster is entirely built of an isotactic chain and could act as a seed crystal. In contrast, the chains that for the most part have a syndiotactic configuration are on the periphery. In the brackets, the values for non-spherical oscillators are given. These conclusions resulted from the investigations on cluster movements in aPS under pressure influence [13]; thus, the structure of SBU in aPS reflects the super cluster structure of the monomer in the liquid state.

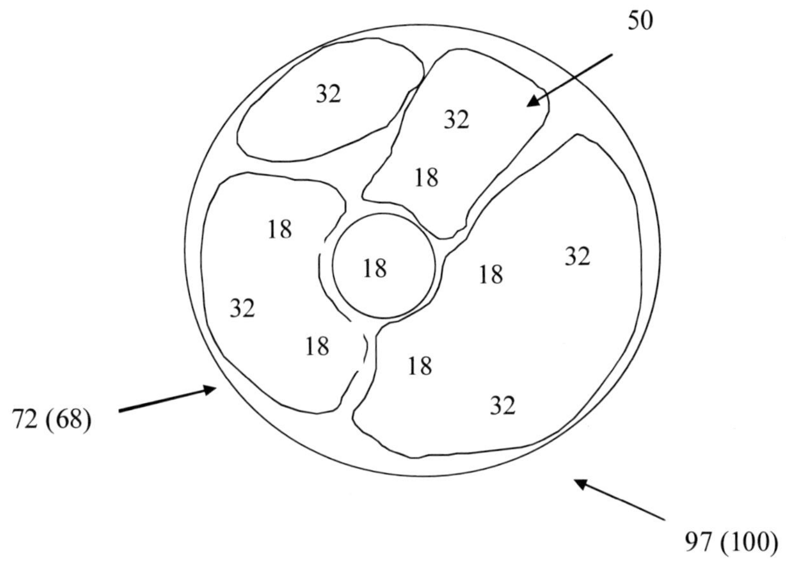

Figure 9.4.2. Model for SBU in aPS, see comment in the text.

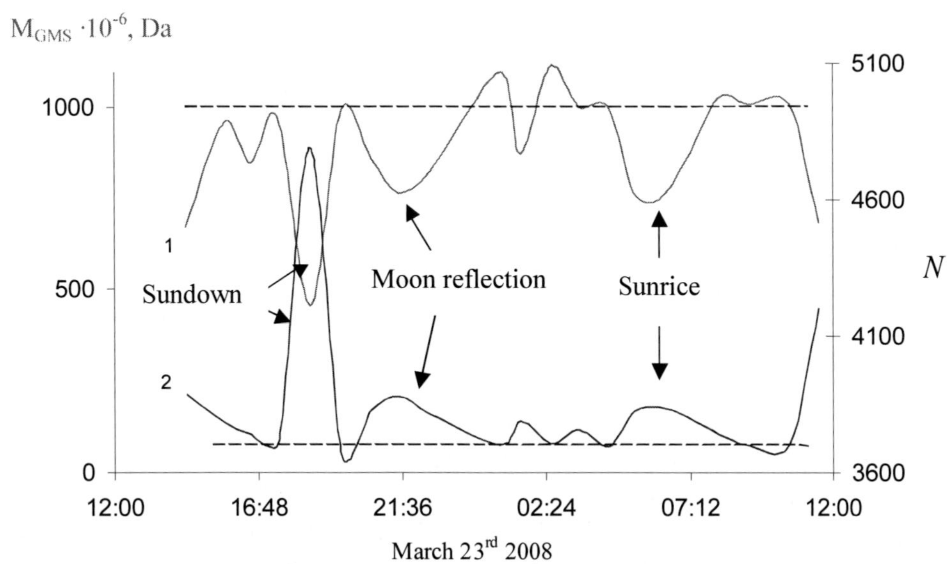

Figure 9.5.1. Change of the average molecular cluster mass (1, mass interval up to 4 billion Da) and of the number of PS cluster kinds (2) in 25 μm PS film at full moon period, sunrise 06:03, moonset 06:06, sundown 18:24, and moonrise 20:52 (CET, North Germany). Dotted lines – base lines for both signal levels. The sample is placed in a metallic box, Figure 4.0.1.

9.5. PROPERTIES OF POLYSTYRENE CLUSTERS

9.5.1. The Planet's Influences on the Cluster Structure of Polystyrene

It was observed that molecular cluster signals clearly correlate with the position of celestial bodies (Figure 9.5.1, [2, 16, 17]). In Figure 9.5.1, the change of the average molecular cluster mass and of the number of PS cluster kinds is shown. The sun shows its greatest effect at midday (12:02) and during sundown. It has to be mentioned that during the full moon, the moon's gravitational field also influences the long-range order in PS [13]. In Figure 9.5.2 it is visible how diads and clusters of 18 members interact with the surroundings at full moon on August 16, 2008.

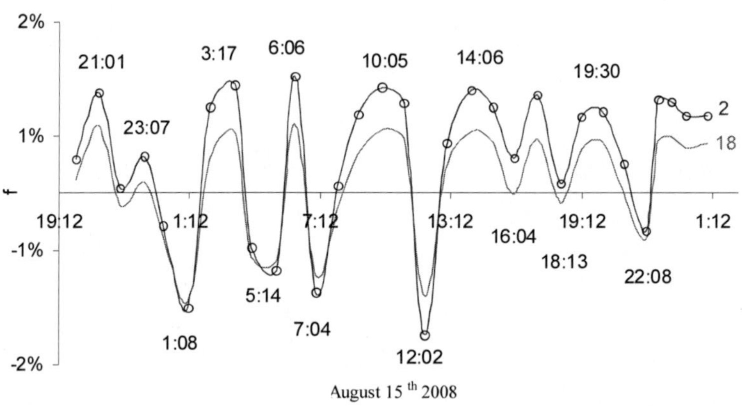

Figure 9.5.2. Interaction of diads (2) and clusters consisting of isotactic chains (18) with their surroundings in industrial PS at full moon period. Cluster ensemble up to 4.3 billion Da.

The interaction of both clusters with their surroundings proceeds synchronously. For both clusters, this character of PS chain conformations is confirmed by Cochrane models (Figure 9.1.1). The energy removal of weak gravitational radiation from the sample to the planets causes an expanding of the clusters (*+f*), however, in the absence of planets, they return into the collapsed form (*-f*). Thus, the planet's gravitational radiation was observed to reversibly destabilize the cluster interaction with its surroundings in polymers. The planet's gravitational radiation seems to influence the condensation-dissolution process of protons in a physical vacuum that leads to a reversible destruction of energy clusters (Figure 9.5.3). It is reminded that the energy clusters supply the organized molecular clusters in PS with energy.

It is interesting to investigate the phenomenon of energy cluster inversion under the influence of celestial bodies (Figure 9.5.3). This inversion should be connected with a changed direction of gravitational radiation flows that briefly destabilized the energy balance between energy clusters in an energy ensemble. After that the energy supply in the molecular cluster ensemble shall be interrupted leading to its changed stability. As is visible in Figure 9.5.3, the planet's GR inverts the cluster ensemble energy in polystyrene at 10:05 and 12:02 and at 19:59 the cluster harmony in the ensemble is fully changed (see Figure 9.5.2, too). The signals of cluster-18 remain unchanged, although at 19:59 a new balance between the expanded and collapsed clusters seems to adjust where the expanded spirals dominate about the collapsed one.

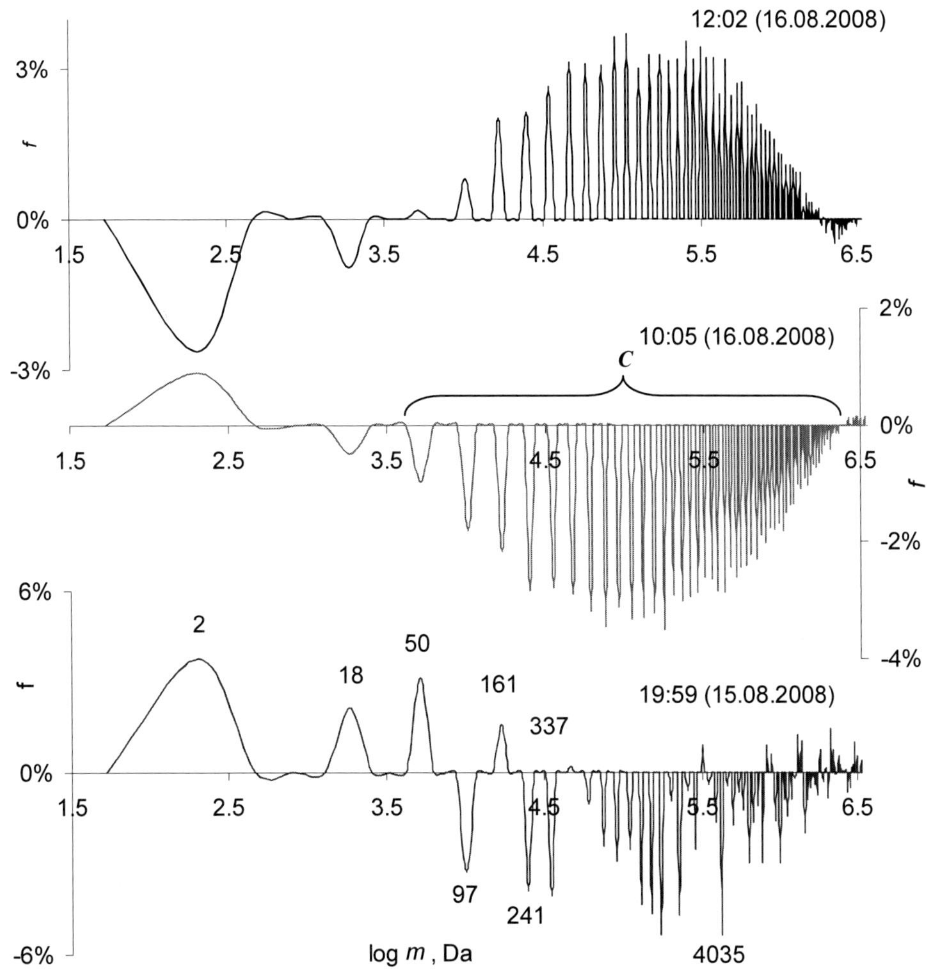

Figure 9.5.3. GMS of industrial PS films. Cluster ensemble up to 2 million Da. For some clusters, the number of styrene members is written over the signal.

CONCLUSION

The long-range order in polymers is formed by clusters, cluster associates, and large domains.

The cluster formation in polystyrenes is determined by styrene clusters.

Weak gravitational radiation arising from polystyrene clusters is observed to be in resonance with the gravitational radiation from celestial bodies.

REFERENCES TO 9

[1] Natta G., Corradini P., Bassi I.W. Crystal structure of isotactic polystyrene. Nuovo Cim. Suppl. 1960, vol. 15, pp.68-82.

[2] Zubow K.V., Zubow A.V., Zubow V.A. Ensemble of Clusters – New Form of Molecular Matter, Risks and Chances. Zubow Equations. In Advances in Chemistry Research, vol. 5, 2010, pp. 107-145. Editor J.C. Taylor. Novapublisher. NY. USA.

[3] Natta G, Pasquon I. The Kinetics of the Stereospecific Polymerization of α- Olefins, in Advances in Catalysis and Related Subjects, Eley D.D., Selwood P.W., Weisz P.B. (eds.), vol. XI, pp. 2-67, Academic Press, Inc., NY, 1959

[4] Itagaki, Hideyuki; Takahashi, Ikunori. Helix-Coil Transition and Physical Gelation of Isotactic Polystyrene in cis- and trans-Decalin Monitored by Fluorescence Measurements Macromolecules. 1995, vol.28, no.16, pp. 5477-5486.

[5] Ayyagari C, Bedrov D, Smith GD. Structure of Atactic Polystyrene: A Molecular Dynamics Simulation Study. Macromolecules. 2000, vol. 33, no.16, pp. 6194-6199.

[6] Chatani, Yozo; Nakamura, Norio. Coiled-coil molecular model for isotactic polystyrene gels. Polymer. 1993, vol. 34, no. 8, pp.1644-1648.

[7] Frank A. Bovey. High Resolution NMR of Macromolecules. NY. Academic Press. 1972.

[8] Greis O, Xu Y, Asano T, Petermann. Morphology and structure of syndiotactic polystyrene Journal of Polymer. 1989, vol. 30, no. 4, pp. 590-594.

[9] Pokrovsky E.I, Fedorowa E.F. Determination of stereoregularity of polystyrene by infrared spectroscopy. Vysokomolekulyarnye Soedineniya, Seriya A. 1964, vol. VI, no. 4, pp. 647-651.

[10] Lee E.C, Muller S.J. Flow Light Scattering Studies of Polymer Coil Conformation in Solutions in Extensional Flow. Macromolecules. 1999, vol.32, no. 10, pp. 3295-3305.

[11] Menasveta M.J, Hoagland D.A. Light scattering from dilute poly(styrene) solutions in uniaxial elongational flow. Macromolecules. 1991, vol. 24, no. 11, pp. 3427-3433.

[12] Picarra S, Duhamel J, Fedorov A, Martinho J.M.G. Coil−Globule Transition of Pyrene-Labeled Polystyrene in Cyclohexane: Determination of Polymer Chain Radii by Fluorescence. Journal of Phys. Chem. B. 2004, vol. 108, no. 32, pp.12009-12015.

[13] Zubow, A.V. Zubow, V.A. Zubow, "Principles of gravitation spectroscopy. New form of molecular matter. Processes. Fields. "Aist Handels- und consulting gmbH", dep. R&D, Berlin, 2011, electronic book (www.zubow.de), p. 900 (in Russian).

[14] Zysman, Viviane; Nguyen, Tuan Q.; Kausch, Henning H. Degradation on freezing dilute polystyrene solutions in p-xylene. Journal of Polym. Sci., Part B: Polym. Phys. 1994, vol. 32, no.7, pp.1257-1269.

[15] Drago RS. Physical Methods in Chemistry. Philadelphia, London, Toronto. W.B. Saunders Company. 1977.

[16] Zubow K., Zubow A.V., Zubow V.A. Water Clusters in Plants. Fast Channel Plant Communications. Planet Influence. Journal of Biophysics Chemistry. 2010, vol.1, no.1, pp.1-11.

[17] Zubow K., Zubow A.V., Zubow V.A. Scanning of the Sun and other celestial bodies with help of gravitation spectroscopy. Optic and Photonics Journal. 2011, no. 1, pp.15-23.

Chapter 10

INTERACTION OF WEAK SHOCK WAVES WITH SOLUTIONS

In the twentieth century it was believed that ideal salt solutions are homogeneous systems [1, 2], although solvated ions in water were also acknowledged [3, 4]. Only recently has new data on the structure of solutions been appearing, which characterizes the liquids as long-range order (LRO) cluster constructions [5]. The so-called ideal solutions are then considered through the prism of non-ideal solutions. Using different analysis methods, solvated clusters of ion pairs of salts (SCIP), which reach sizes > 500 nm, are observed [6]. Then the ideal solutions are understood as nanodisperse systems of SCIP. The authors of [7] photographed SCIP and even calculated its distribution and concentration. On the other hand, it has become clear that the LRO formation is a long process that can take up to several years [8]; it is a process in which water "plays the first violin." Furthermore, it is sensitive to a weak energy exposure. In solutions of inorganic salts, water and SCIP seem to compete for the free energy of white noises. Water forms its clusters apart from SCIP even in concentrated salt solutions. While LRO in diluted solutions is determined by water in concentrated ones it depends on SCIP.

Here, with the example of SCIP in KCl and NaCl aqueous solutions, we show which interesting events take place during the dissolution and storing of salts recorded by GMS. To understand LRO, two types of shock waves are applied: strong shock waves for solutions used in technology (color noise) and weak ones for the fine structure of clusters and their interaction with surroundings (white noise). To get the energy characteristics of SCIP is also not an easy task because of a missing reliable internal standard and an incorrect calibration of the whole system without taking into account the individual characteristics of each cluster.

10.1. GMS-SPECTROSCOPY OF INORGANIC SALT WATER SOLUTIONS (NACL, KCL)

GMS-spectroscopy of some chloride solutions is of interest to understand the structure of: (i) individual saturated solutions (technology), (ii) 3.5 wt. % NaCl solution—the analog of sea water, and (iii) 0.1 wt. % NaCl solution, simulating physiological solutions in medicine.

The aim of the study is to get an idea about the cluster formation, the equilibrium for seed crystals' formation [9], the formation and decomposition of associates [4], and the cooperative interaction in organized biological systems [10, 11]. The energy effects of these processes are often lower than the sensitivity of the most measuring instruments.

The literature information about the structure of solutions is different [3, 4]. Traditionally it is assumed that the salt solution is represented by solvated chlorine and sodium ions in a network of water molecules being linked by hydrogen bonds [4, 12]. To clarify what is correct or incorrect, GMS-spectra of water and saturated solutions (HCl, KCl, and NaCl) are analyzed, Figure 10.1.1.

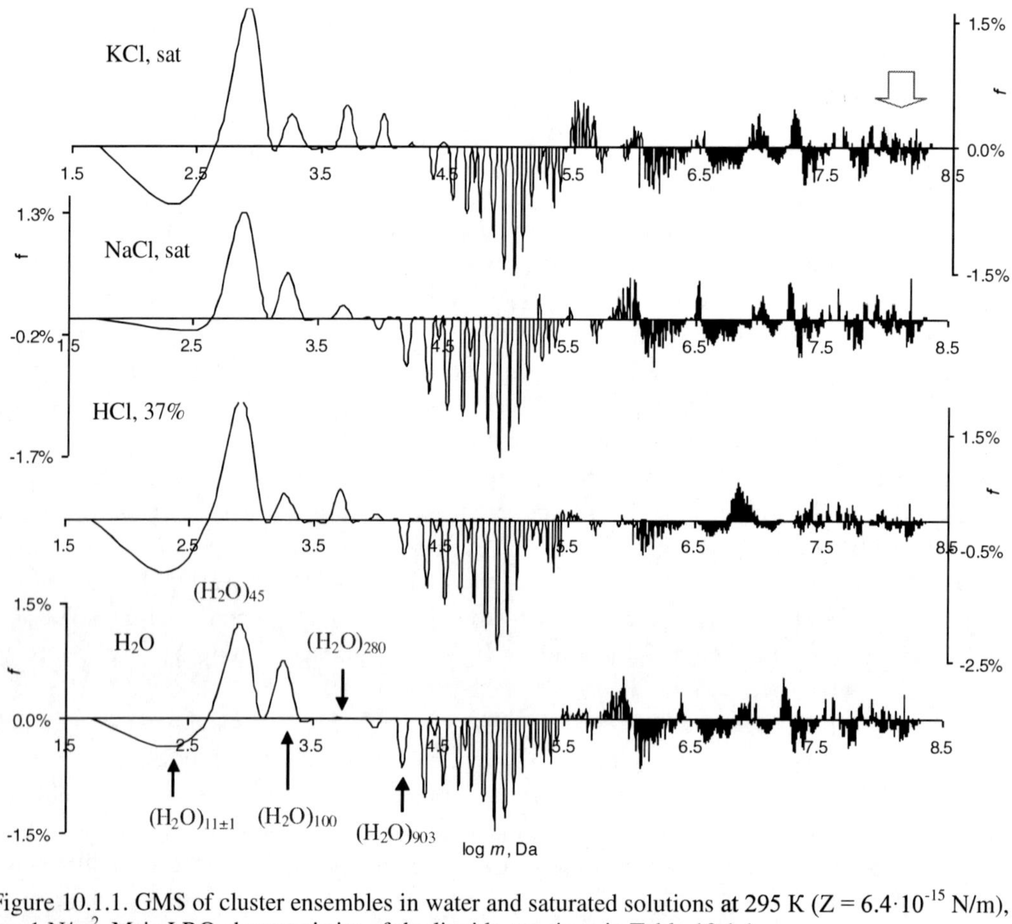

Figure 10.1.1. GMS of cluster ensembles in water and saturated solutions at 295 K ($Z = 6.4 \cdot 10^{-15}$ N/m), $p < 1$ N/m^2. Main LRO characteristics of the liquids are given in Table 10.1.1.

As is visible, each liquid is characterized by a cluster ensemble, and in accordance with the Zubow equations, a cluster is formed only at a strictly determined mass. The cluster ensembles given in Table 10.1.1 are recorded experimentally by the authors with a GMS-spectrometer. The weak shock wave, however, is not able to destroy the small base water cluster $(H_2O)_{11\pm1}$ and transform it into the expanded form it consistently takes in all solutions. Up to a mass of log m = 5.5 the spectra differ only slightly. The weak shock wave seems to be unable to differentiate between water and solutions, which is explained by the fact that the formation energy of SCIP is lower than that of clusters in water [5], §5. First, a spectrum of water (see also §5.3.1) will be considered given that LRO characterizes the polymorph state of water under test conditions. Unexpectedly, the rare water cluster of 45 molecules (Mercury cluster, Table 5.1) appears. During the entire measurement period from 2003 to 2006, the cluster could never be registered. In 2007, however, it was recorded in the same NaCl solution and water. A possible explanation could be the influence of the planet Mercury [13], Table 5.1.

Table 10.1.1. Some basic characteristics of LRO in water and aqueous solutions. For experimental conditions see Figure 10.1.1. The demineralized water, aged for 6 months in a plastic sealed container in a dark and quiet place, is characterized by a conductivity of (20±5)·10^{-6} cm^{-1} ohm^{-1}.

Remarks	Liquids	M_{GMS}, Da	$Д_c$	*N*
demineralized	H_2O	44,988,884	71%	1,362
"old solution," 12 months	NaCl-sat.	47,492,975	73%	1,287
freshly prepared (4 h)	KCl-sat.	44,180,169	66%	1,338
"old solution," 36 months	HCl, 37%	40,933,951	62%	1,208

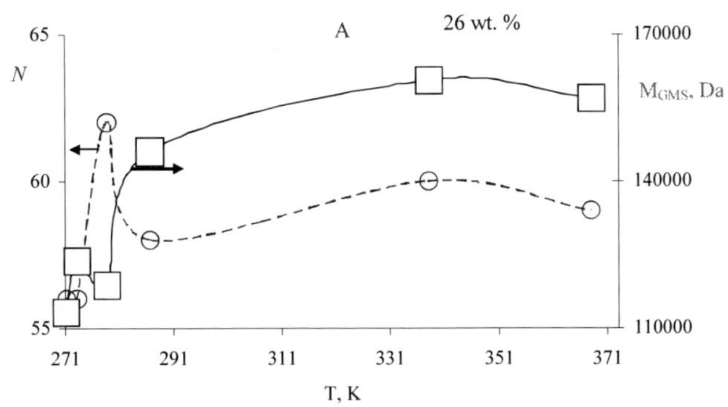

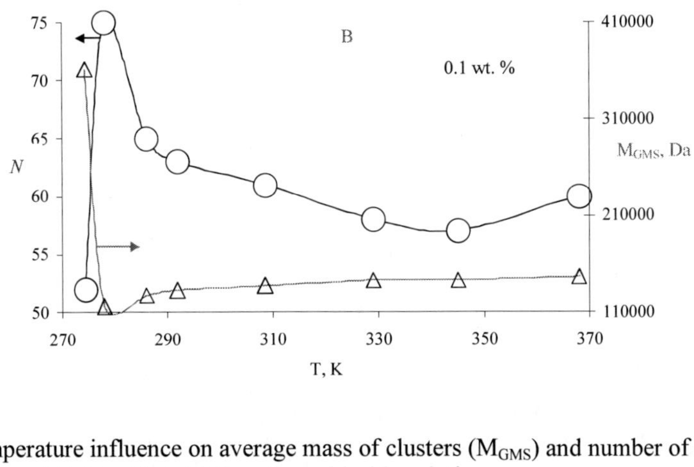

Figure 10.1.2. Temperature influence on average mass of clusters (M_{GMS}) and number of cluster kinds (*N*) in saturated (A) and highly diluted (B) sodium chloride solutions.

Let us consider the data in Table 10.1.1. Concerning all LRO parameters, HCl, 37% is described with the minimum values while for the saturated NaCl solution (NaCl-sat.), the highest part of the collapsed clusters and the largest average molecular weight are observed. In all the solutions, however, the number of cluster kinds is less than in water. That means salts and HCl partially destroy LRO of water clusters. Under the impact of weak shock waves it is difficult to observe significant differences between water and solutions, although one might speculate about minor deviations. The following is, therefore, an attempt to study LRO under strong shock waves.

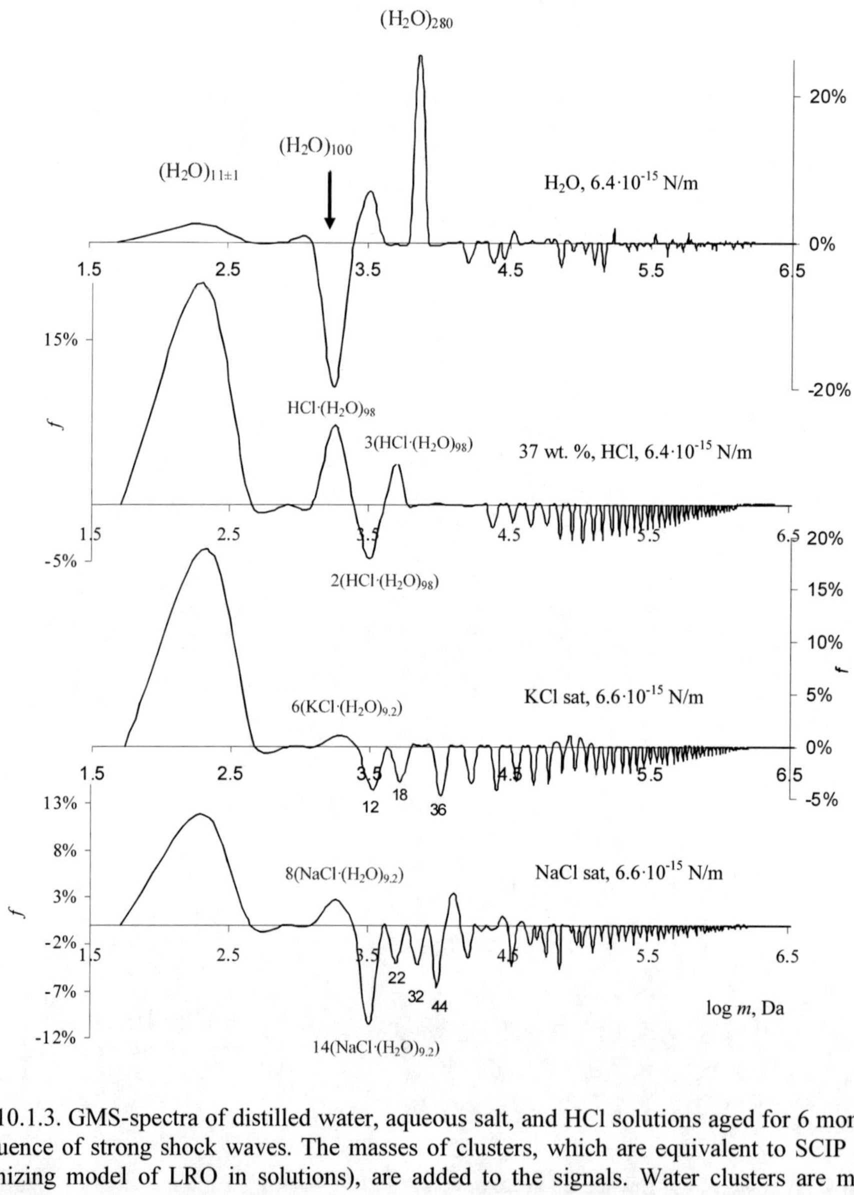

Figure 10.1.3. GMS-spectra of distilled water, aqueous salt, and HCl solutions aged for 6 months under the influence of strong shock waves. The masses of clusters, which are equivalent to SCIP structures (harmonizing model of LRO in solutions), are added to the signals. Water clusters are modeled in Figure 2.1.1.

The GMS-spectra of the salt solution, simulating sea water, shows, under strong shock wave influence, significant differences to the spectra that are exposed to weak ones. According to Figure 10.1.2, the main LRO parameters of saturated and highly diluted sodium chloride solutions change at heating. As is visible, the curves are characterized by some extremes; they are mostly in the low temperature range and shall be ascribed to reorganization processes in the solvent at cluster level.

Both for saturated and diluted solutions, the number of cluster kinds reaches a maximum with 277 K being an indication for restructuring of water's LRO outside the SCIP structures. Concerning the average cluster mass, it strongly increased at T > 273 K in the saturated solution while it decreases in the highly diluted one. For the concentrated solution, this can be understood as a larger individuality of SCIP; it has been released due to a melting process. In the highly diluted solution, the reduced M_{GMS} of the water clusters is explained with lower cluster individuality because of the hydrogen bonds between the clusters.

The liquids were aged for 6 months in a dark, isothermal place, which was protected from mechanical fields and changes to their LRO.

Figure 10.1.3 presents the GMS-spectra of water and solutions when they are exposed to strong shock waves. As is visible, the base water cluster, consisting of 11±1 water molecules and being in the expanded form, is present in all liquids, however, under weak shock wave impact, this cluster is mostly found in the collapsed form (Figure 10.1.1). It is noticeable that the Zubow constant for each liquid varies. Due to the fact that it is a force constant, it can also be used for understanding the inter- and intra-cluster interactions; it reflects the nature of a substance. While the water cluster of 45 molecules is observed under weak shock waves, it cannot be detected under strong ones due to their destruction. These results allow us to define more accurately and selectively the formation/destruction energies of individual clusters by varying the shock wave strength. The latter is solved by using special electronic systems.

As shown in Figure 10.1.4, with increasing salt concentration, the number of expanded cluster kinds decreases sharply. GMS-spectra of saturated salt solutions are observed to be less informative than those of diluted ones due to the fact that SCIP in complex polymer structures are not spherical [6, 14, 15]. In diluted solutions, however, there are more freely oscillating SCIP and less disturbances from the polymer-polymer interaction.

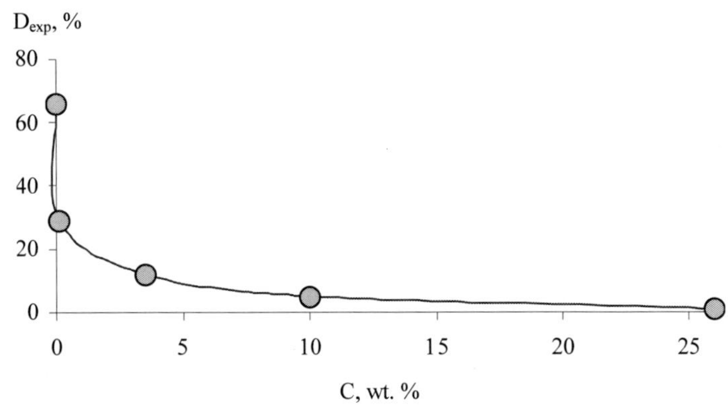

Figure 10.1.4. Part of expanded clusters, D_{exp}, versus the NaCl concentration in water.

First the GMS-spectra of a 3.5 wt. % aqueous NaCl solution simulating water in oceans shall be considered (Figure 10.1.5). At temperatures above 300 K, the spectra change significantly. This refers to signal intensities, signal shifting to larger clusters, and to the appearance of new clusters. As is visible, all spectra are characterized by oscillators with a negative energy formation, indicating an energy losing system that means the clusters become more densified; they crystallize and go into a more stable state. Increased signal intensity at a higher temperature is explained by cluster individualization and higher formation/destruction energy whereas the signal shifting to larger clusters (indicated by arrows) could be caused by the fact that the formation of clusters no longer proceeds in the white noises but in the color ones [16] due to an increased collision number of molecules in the solution and due to the transfer of the liquid into a new thermodynamic state. This phenomenon can be compared with the effect of vibrating sieves, where an increase in the frequency of vibration leads to the special agglomeration of particles and to a new particle size distribution in agglomerates.

Understanding of the structure of solutions is a topical problem of chemistry and physics. This especially concerns nanostructures based on weak cooperative interactions with the formation/destruction energies at the level of a few kJ/mole. Among them are the processes of cluster formation, equilibrium of seed-crystal formation, decomposition and formation of associates, and the cooperative interactions in organized biological systems. The energy effects of these processes are often below the sensitivity thresholds of most instruments used. Signals from weak interactions are masked by electronic instrumentation noise and energy fluxes from the outside, which complicate their identification and understanding. As we see it, this problem can be solved by using modern computers and familiar mathematical methods for "clearing" signals and representing large bodies of information in a form convenient for comprehension.

Interacting objects of investigation can be weak interactions in a liquid, e.g., in water. It is known that the structure of water is an aggregate of tiny volumes with the structure of ice, polymer formations $(H_2O)_n$, and individual molecules [12, p. 78], which are easily formed and destroyed under the action of very weak energy fluxes. For example, a water cluster (Figure 2.1.1) composed of 280 molecules is a typical example of the problem of revealing and establishing the nanostructure of a liquid.

Analogs of larger clusters are known in physics and chemistry [4, 6], and the problem of their role and identification is topical. With the use of the methods of X-ray spectroscopy of substances in the liquid state, periodic formations whose spectra are partially identical to the spectra of these substances in the crystalline form were revealed. It may be suggested that such correlations are determined by the cluster structure of liquids.

Conceptions of the cluster structure of liquids have become dominant for many researchers, who are beginning to form data banks on the structure and energy of clusters, e.g., small clusters of water $(H_2O)_{2...22}$ (see also: www.TheCambridgeClusterDatabase.htm).

The structure of the water solution of sodium chloride is an important object for understanding biological systems and the hydrochemistry of the sea. In the literature, there are different notions about the solution structure. It is traditionally considered that a salt solution consists of solvated chlorine and sodium ions in the water structure, whose molecules are bound to one another by hydrogen bonds. Due to the difference between these approaches to the comprehension of the structure of liquids, we attempted to reveal weak signals of water clusters and solvated clusters of salt ions in solutions through the GMS-method developed by us.

The principle of operation of the GMS-spectrometer is the same as that of the standard spectrometer, namely, the recording of weak oscillations of regular structures linearly interacting with the weak shock waves passing through the sample. Wave distortions at the output from the cell are recorded by a computer program, which clears real signals of random noise through familiar methods of mathematical treatment of digital signal packages.

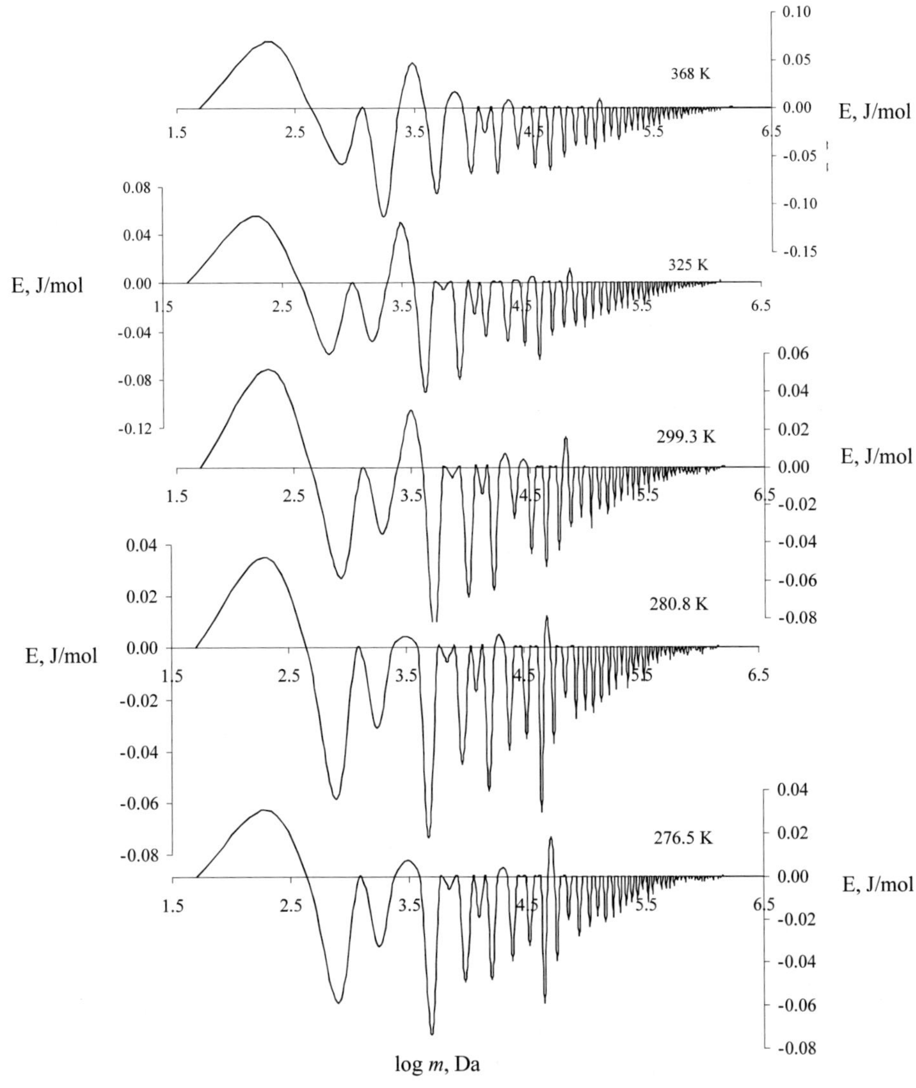

Figure 10.1.5. GMS-spectra of the SCIP in the aqueous solution of salt (NaCl 3.5 wt %).

Figure 10.1.5 shows the spectra of oscillations of statistical ensembles of clusters in the salt solution. They widely differ from the spectra of oscillations of water clusters in water [12]. It is seen that in the oscillation spectra of the clusters in the salt solutions, as the temperature is increased to above 300 K, marked changes in the intensity of the bands, their shift towards the formation of large clusters, and the appearance of bands of new oscillators are observed. All spectra are represented mainly by oscillators with negative values of the energy of their formation, which points to a system energy loss, i.e., to a densification of the structures of clusters, their crystallization, and the transition of oscillators to a more probable state. The increase in the band intensity with increasing temperature of the solution may be due to the increase in both the number of oscillators and the energy of their formation/destruction. The shift of the bands toward oscillators with large masses can be explained by the increase in the number of collisions of molecules in the solution and the transition of the liquid to a new thermodynamic state with a different distribution of the new oscillators. This phenomenon can be compared to the riddle effect, in which an increase in the riddle vibration frequency leads to an agglomeration of particles and their new mass distribution.

Table 10.1.2. Comparison of the hypothetical masses of the models of SCIP $(NaCl \cdot xH_2O)n$ to the found f values of the oscillator masses of clusters in the GMS-spectra (for the part of the spectrum in the region of small clusters) and the development of SCIP models for various concentrations of salt in water at 365 K

n*, 0.1 wt. %	n, 0.1 wt. % (found)	n*, 3.5 wt. % (calculated)	n, 3.5 wt. % (found)	n*, 26 wt. % (calculated)	n, 26 wt. % (found)
1	-	1	1.0 (-4.6 %)	-	-
2	2.2 (3.0 %)	2	2.2 (-9.0 %)	-	-
4	4.0 (-8.7 %)	4	4.0 (3.8 %)	-	-
6	6.3 (-3.3 %)	6	6.3 (-7.0 %)	7	6.9 (4.6%)
9	9.0 (-3.4 %)	9	9.0 (1.3 %)	-	-
12	12.2 (-6.0 %)	12	12.2 (-5.4 %)	12	12.2 (-8.8%)
16	-	16	16.0 (-2.0 %)	-	-
20	20.2 (-5.0 %)	20	20.2 (-5.4 %)	20	19.3 (-3.3%)
25	25.1 (-1.6 %)	25	25.1 (0.6 %)	-	-
30	30.3 (-0.3 %)	30	30.3 (-3.4 %)	28	27.7 (-3.4%)
36	36.0 (1.6 %)	36	-	38	37.6 (-5.9 %)
42	42.3 (-3.4 %)	42	42.3 (-4.9 %)	-	-
49	49.0 (1.1 %)	49	-	-	-
56	56.2 (-2.1 %)	56	56.2 (-5.1 %)	-	-
64	64.0 (-0.3 %)	64	-	62	62.3 (-5.4%)
72	72.3 (-3.0 %)	72	72.3 (-4.0 %)		
80	80.9 (0.5 %)	80	-	77	77.1 (-1.3%)
90	90.3 (-3.7 %)	90	90.3 (-3.1 %)	93	93.2 (-1.3%)
110	110.2 (-1.4 %)	110	110.2 (-2.9 %)	110	110.7 (1.1%)
121	121.1 (-1.6 %)	121	-	-	-
132	132.2 (-3.1%)	132	132.2 (-3.3 %)	130	130.3 (-3.7 %)

*Note: In brackets, the percentage of a given SCIP of salt in the sum of all revealed clusters is given. The sign before the percentage denotes energy absorption (–) or release (+) as a result of the interaction between the cluster and the spectrometer shock wave.

Table 10.1.3. Base models of the SCIP of ($(NaCl \cdot xH_2O)n$)

n	MODELS OF THE SCIP	Scheme of an oscillators
1	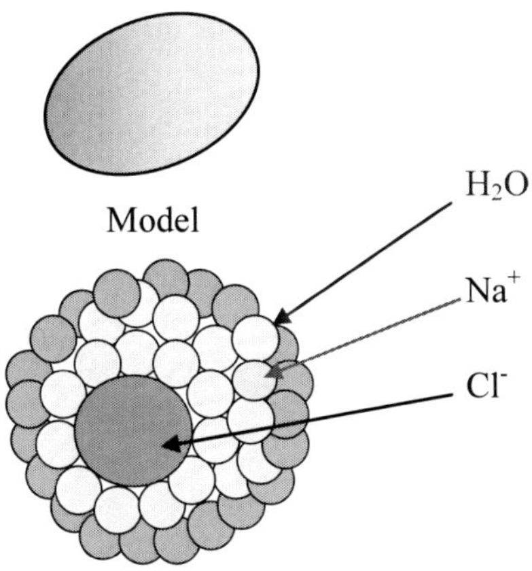 $NaCl \cdot 40H_2O$	
2	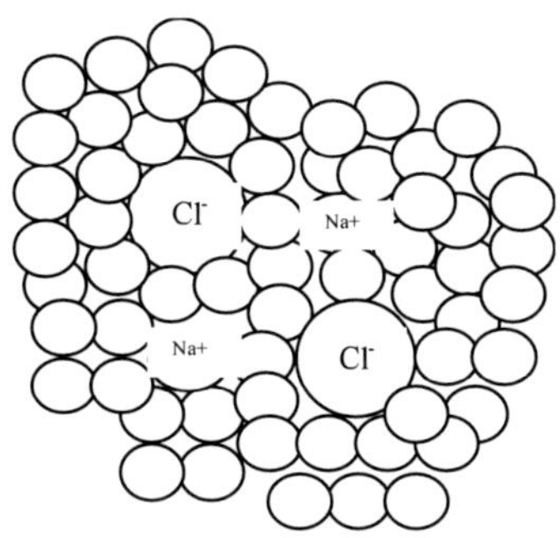	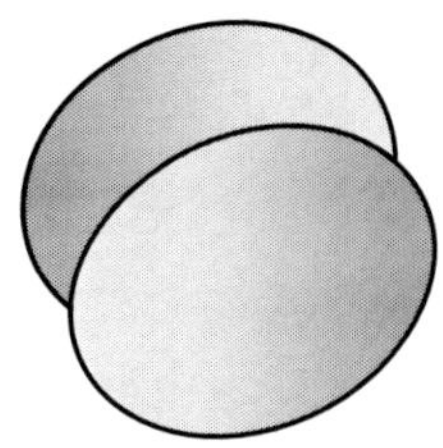
4	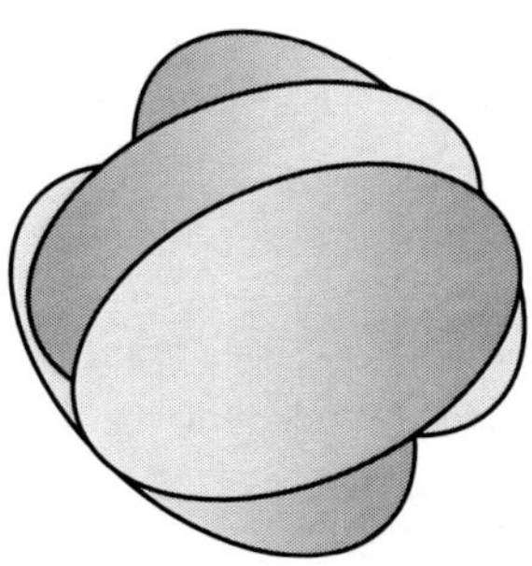	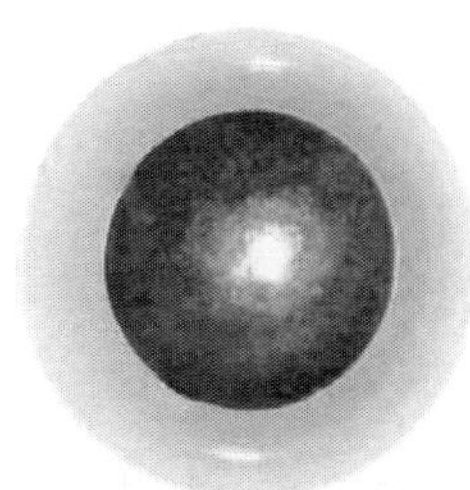

Unlike the oscillation spectra of water clusters (§5), in the GMS-spectra of the salt solution, harmony in the distribution of large oscillators ($n > 4.5$) is observed. The explicit law observed in this region is indicative of the polymeric character of the clusters formed.

Selecting the ratio of sodium chloride to solvate water $(NaCl \cdot xH_2O)_n$ and analyzing the literature data [17], we have found a good correlation between the SCIP models and the established values of the oscillation masses of clusters in the salt solution. For 26 wt. % NaCl, we chose the structure of the SCIP monomer (Table 10.1.2) in the form of $NaCl \cdot 9.2H_2O$, (theor. $NaCl \cdot 9.2H_2O$), and for 3.5 and 0.1 wt. % of solutions—an SCIP monomer of the form $NaCl \cdot 40H_2O$, which is consistent with the logic of the results is obtained. Such wide differences can be attributed to the "squeezing" of some water out of the SCIP structure in these solutions into the small clusters of water (oscillations at log $m = 2.28$). In so doing, the portion of small water clusters reaches 15% in the solution. Table 10.1.2 gives the correspondence of the main structures of oscillators of seed NaCl crystals in the form of SCIP to their models (see Table 10.1.3).

From Table 10.1.2 it is seen that with the increasing concentration of salt, the set of SCIP structures sharply decreases. The monomer SCIP is reliably detected only for salt concentrations from 1 to 5 wt. %. In diluted solutions of salt, this monomer is probably "squeezed out" by water to form its polymers. At high concentrations of salt, the monomer-polymer equilibrium is also shifted towards the formation of SCIP polymers. Thus, at 3.5 wt. % in the solution, advantageous thermodynamic conditions for the existence of the monomer are realized.

The SCIP tetramer in diluted solutions is represented by a loose structure, and in the sea water model (3.5 wt %), on the contrary, it is represented by a structure that under the action of a shock wave is densified (crystallized) with energy release. In the concentrated solution this cluster has not yet been revealed (Table 10.1.2).

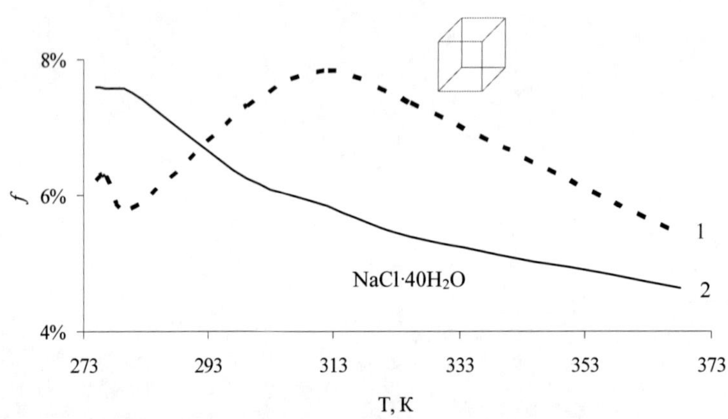

Figure 10.1.6. Ratios of the signals of the SCIP represented by a cube ($n = 12$) (1) and the monomer SCIP (2) to the sum of all signals of oscillators revealed in the spectrum versus the solution temperature.

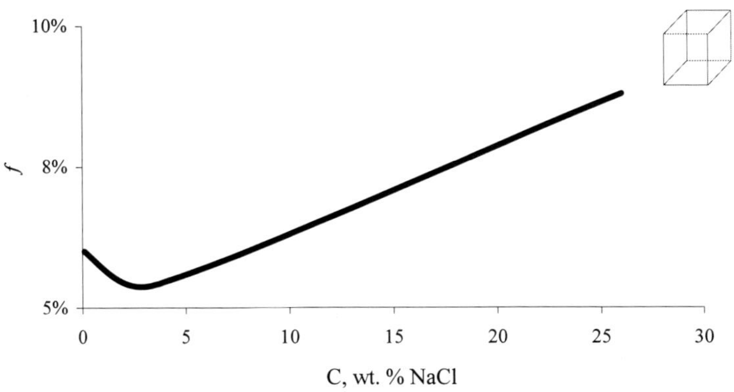

Figure 10.1.7. Fraction of the SCIP of salt dodecane ($n = 12$) versus the solution concentration at 368 K.

Figure 10.1.6 shows the temperature dependence of the fractions of the monomer SCIP and the salt dodecane SCIP to be similar in shape to the crystal cell of sodium chloride in the solution. It is seen that the fraction of the monomer SCIP signal monotonically decreases with increasing temperature and the fraction of the dodecane SCIP (n =12, cube, Table 10.1.3) has maxima at 277 and 313-315 K. The extremum at 313-315 K was observed by us in water (§5), [18] and we explained its appearance by the transformation of the cluster structure of the solvent. This transformation also possibly influences the salt SCIP in the solution and leads to the known collapse of biological systems at temperatures above 42°C.

Figure 10.1.7 shows the concentration dependence of the fraction of the SCIP signal represented by a cube. The presence of the minimum on the curve at 3.5 wt. % is so far difficult to explain.

Conclusion

Using the GMS-method, we have revealed oscillations of seed crystals of salt.

The basic element for the nucleation is an oscillating cluster with an internal skeleton—a prototype cube formed from 12 monomer SCIP of salt $NaCl \cdot 40H_2O$.

It has been established that the salt solution is not a perfect solution, as it was considered before; it can be treated as a nanodisperse system of salt SCIP in statistical ensembles of water clusters.

References to 10.1

[1] Dikarev S.N. Laboratory study of resonant regimes of motions of homogeneous fluid with free surface in a tilted rotating container. Izw. Ak. Sci. USSR Fisika Atmosphery i Okeana. 1990, vol. 26, no. 9, pp. 982-992.

[2] Manin D. Yu., Chernous'ko Yu., L. An experimental study of the stability of a quasi two- dimensional jet flow produced in rotating fluid by sinks and sources. Izw. Ak. Sci. USSR Fisika Atmosphery i Okeana. 1990, vol. 26, no. 5, pp. 483-492.

[3] Campbell J. A. Chemical Systems. Energetics, Dynamics, Structure. W.H. Freeman and Company. San Francisco. 1970. Russian edition, Moscow, Mir. 1975, vol. 3, p. 165.
[4] Izmai'lov N.A. Electrochemistry of solutions. 3th edition, Moscow, Chimia, 1976, p. 488, in Russian.
[5] Zubow K., Zubow A., Zubow V. A. Spectra of seed crystals oscillation of NaCl in water solution. Journal of Applied. Spectroscopy. 2005, vol. 72, no. 6, pp. 766-772.
[6] Yannis Georgalis, Kierzek A.M., Saenger W. Cluster Formation in Aqueous Solutions Observed by Dynamic Light Scattering. Journal of Phys. Chem. B, 2000, vol. 104, pp. 3405- 3406.
[7] Lian Li, Ogawa T. Clusters and their properties in aqueous solutions of KDP, KCl and sugar. Journal of Crystal Growth. 2000, vol. 211, pp. 286-289.
[8] Zubow A.V., Zubow K., Zubow V. A. The Investigation of the Salt Dissolving Mechanism and of the Ageing of the NaCl Solution in Water by Flicker Noise Spectroscopy. Chem. Industry (Sankt Petersburg). 2007, vol. 84, no 1, pp. 40-48 (in Russian).
[9] Perzov A.V. Born of the new phase. In. Chem Encyclopedia. M. Soviet Encyclop. 1990, vol. 2, p. 318 (in Russian).
[10] Schuster P., Wolschmann P. Universität Wien: TBI. Hydrogen Bonding: From Small clusters to Biopolymers, No. Pks. 99-007. Institut für Theoretische Chemie und Molekulare Strukturbiologie der Universität Wien.
[11] Bartenev G. M., Zelenev Yu. V. Physics and Mechanics of Polymers. Moscow, Vishaja Shkola, 1983, p. 21 (in Russian).
[12] Zai'tev O. S. General Chemistry.M. Chimia, 1990, p. 242, in Russian.
[13] Zubow K.V., Zubow A.V., Zubow V.A. Ensemble of Clusters – New Form of Molecular Matter, Risks and Chances. Zubow Equations. In Advances in Chemistry Research, vol. 5, 2010, pp. 107-145. Editor J.C. Taylor. Novapublisher. NY. USA.
[14] Léo Degréve, Fernando Luis B. da Silva. Large ionic clusters in concentrated aqueous NaCl solution. Journal of Chemical Physics. 1999, vol. 111, no. 11. pp. 5150-5157.
[15] Ishii, Katsuhiro; Iwai, Toshiaki. Proceedings of SPIE-The International Society for Optical Engineering (1999), 3740 (Optical Engineering for Sensing and Nanotechnology (ICOSN '99)), pp. 549-552.
[16] Horsthemke W., Lefever R. Noise-Induced Transitions. Theory and Applications in Physics, Chemistry and Biology. Edit. Hermann Haken, vol. 15. Springer-Verlag, Berlin, Heidelberg, NY, Tokio 1984.
[17] Peslherbe G.H., Ladanyi B.M., Hynes J.T. Structure of Na^+ ion pairs in water clusters. Chem. Phys. 2000, vol. 258, pp. 201-224.
[18] Zubow Kr. V., Zubow A.V., Zubow V.A. Cluster structure of liquid alcohols, water and n-Hexane. Journal of Applied. Spectroscopy. 2005, vol.72, no.3, pp. 321-328.

10.2. MECHANISM OF SODIUM CHLORIDE DISSOLVING IN WATER AND THE PROCESS OF SOLUTION AGING

The methods of preparing solutions, the dissolving processes, and the solution storage are of great importance in the chemical technology processes. First of all this concerns the stabilization of quality, rational consumption of resources, knowledge of the role of the salt state in a solution, and an understanding of the dissolving and solution aging processes.

The mechanism of dissolution of sodium chloride and the state of the salt ions in solution are considered in many publications [1, 2, 3, 4, 5, 6]. There are no publications, however, denoted to understanding this mechanism in view of long-range order formation and evolution on the level of static assembles (clusters) in water and in the solution of the salt in the initial steps of the formation of a thermodynamically stable solution as well as on the dynamics of the process of solution aging. Understanding this mechanism would allow the attainment of a new quality in performing technological processes and in understanding the processes in geochemistry, biochemistry, and even in cookery.

This work is aimed at the study of the mechanism of sodium chloride dissolving in distilled water and solution aging using the GMS-method [7, 8].

This method allows us to measure weight fractions of cluster type formations by a short period action of weak energy fields, thus introducing a minimum level of possible distortion in the formation of static assembles in the liquid.

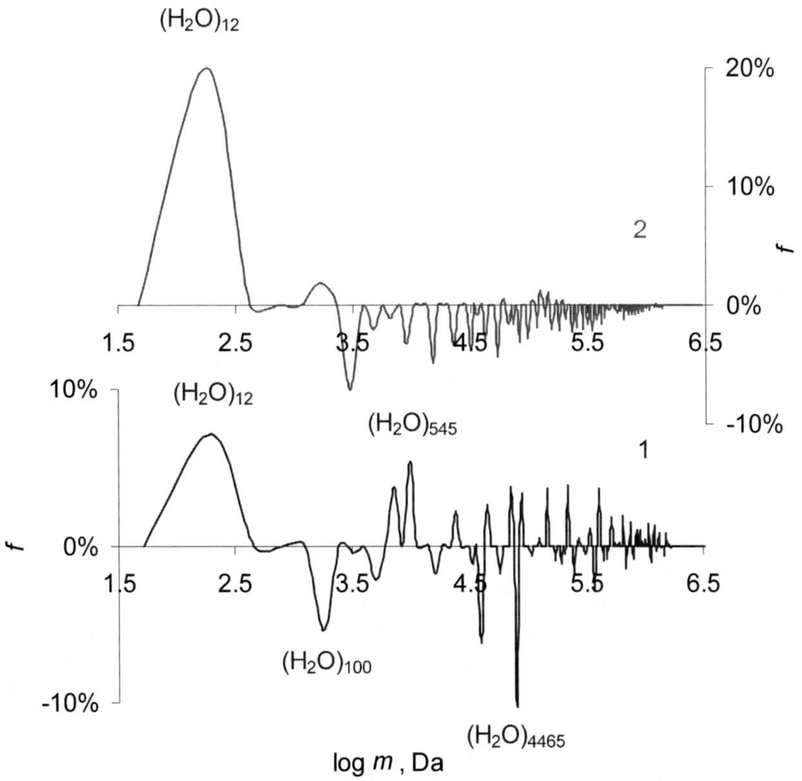

Figure 10.2.1. Survey GMS-spectra of distilled water (1) after fast cooling of vapor to 294 K and of 3.5 wt. % NaCl solution (2) after the addition of salt and three min of stirring, 294 K.

Figure 10.2.1 shows GMS data for water and a salt solution. As seen, both water and the salt solution are nanodisperse systems. The cluster distribution oscillating at the action of the weak shock wave is not homogenous. Adding the salt to water leads to the distortion of the cluster structure of the latter and to an almost complete disappearance of expanded clusters (+*f*), except for water clusters consisting of the molecular associates up to $(H_2O)_{12}$ [9, 10]. The salt ions, which are represented by a wide set of SCIP [7], allow water to form small clusters only, while medium and large statistical forms in the bulk solution are distorted. The spectrum of water shown in the figure differs slightly from that of distilled water obtained by water melting and heating to 300 K [8] due to the bigger number of unstable clusters (skeletal clusters) and due to difference in the history of the liquid. Such water is mostly used in chemical technology at the preparation of salt solutions.

Figure 10.2.2 shows the dependence of the average oscillator mass in water and in the salt solution in the process of its formation. It is seen that mechanical action in the process of stirring leads to a sharp decrease in the average cluster mass and the addition of the salt powder accelerates this process. In this case, the value of M_{GMS} falls down by a factor of more than three.

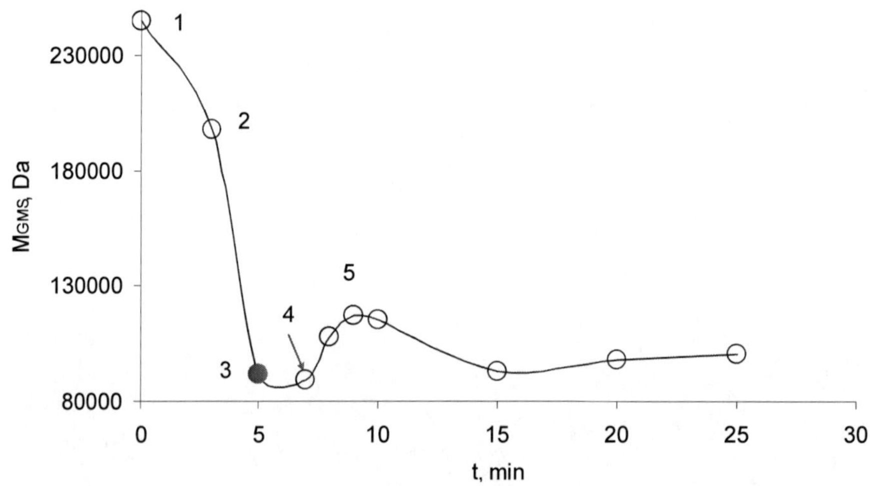

Figure 10.2.2. Plots of average mass variation of clusters (M_{GMS}) in water and in solution over time. (1) Parent water without stirring with magnet stirrer, (2) water at stirring with magnet stirrer, (3) solution after adding powdered salt and one min of stirring with magnet stirrer, (4) solution 3 after three min of stirring, and (5) solution without stirring (t = 0, start of aging).

Dissolving the salt stabilizes the process of strong change in M_{GMS} even while stirring the solution. When the stirring is stopped, after one min, there is an increase in the average oscillator mass, which then again decreases and stabilizes at the level of 96±5 kDa with a certain trend to a growing M_{GMS}. This running curve shows that in a process of redistribution in the cluster structure of solution, its evolution and further aging do occur. These processes can be visualized through the representation of GMS data as dash codes (Figure 10.2.3). The dash codes represent the presence or absence of a cluster with given mass regardless of its intensity of a corresponding signal in the GMS-spectrum.

From Figure 10.2.3 it is seen that in the process of salt dissolving, there also occur the processes of change in the cluster structure of the liquid. Adding the salt leads to the appearance in water of giant oscillators with a size comparable to colloid particles (> log m = 6.31). It can be suggested that they are oscillating colloid particles, the fragments of difficultly dissolving fragments of solid salt with minimum defects in their crystal structure. In the region of medium size clusters (log m from 4.05…5.25), we see the appearance of new oscillators that are not typical for water. They can be polymeric solvate clusters of the salt ion pairs (pSCIP) [7, 11].

Stirring the solution for another three min leads to the disappearance of the giant oscillators and to the establishment of a new thermodynamic equilibrium of SCIP in the regions of medium and big clusters. This equilibrium, however, is intermediate and not stable.

When stirring is stopped, the solution cluster structure evolution still continues. The dash spectrum (Figure 10.2.3) expands to the side of the appearance of very large oscillators with simultaneous compression of dashes in the region of medium and large clusters. Such a phenomenon continues for 2-3 min after the cessation of stirring and can be explained by the formation of instable SCIP (iSCIP) and their polymerization with the formation of polymer globular and oscillating structures.

After 15 minutes the iSCIP become quickly aged and destructive (Figure 10.2.3). The process of aging is accompanied with restructuring due to depolymerization of the most unstable globules and the formation of oscillating low molecular weight stable SCIP (sSCIP) that have a new structure which is in better harmony with the clusters of mass between log m = 4.62 and log m = 5.6. In Figure 10.2.3, this process is seen as disappearance of dashes in the region of giant clusters and strong increase in density of the dashes in the medium region (oligomeric clusters).

From seven min to 12.3 h, another trend occurs—a distribution process of dashes over the entire spectrum. This phenomenon we connect with the formation of a new thermodynamic equilibrium in the region of oligomeric clusters (numbers 77 to 100), leading to polymerization of iSCIP and the formation of new, stabler, high molecular oscillators with numbers 130 to 185.

After 0.8 h in the spectra, there again appeared giant clusters whose number increases with the solution aging. The reasons for their appearance can be the polymerization of new, low molecular, thermodynamically stable SCIP.

After 3.5 h aging of the solution, there occurs a certain decrease in the number of SCIP types in the medium area of spectrum and there is an increase in the number of oscillators in the region of big clusters. At 12.5 h, these changes become obvious and dominating. In the region of oligomeric (medium) clusters, the dashes' density decreases to the level that occurred at the moment of mixing salt and water, but in the region of giant clusters the dashes' density increased sharply.

Note that the cluster dashes No. 70 to 150 remain the most stable during the entirety of the experiment. They can be assigned to the carcass clusters of the solution, forming its long-range order structure.

At 500 h, the solution structure becomes mainly formed, but the process does not stop. In the solution, the processes of attaining thermodynamic equilibrium continue; they are accompanied by the dividing of the cluster structure into two big groups: with the masses < log m = 6.2 and > log m = 6.36 (Figure 10.2.3, progress vectors). The formation of two big groups of sub micro particles has been reported in [11]. This data confirms that there are no ideal solutions and the old theory of ideal solutions should be declined. These liquids cannot be considered homogenous systems. They are heterogenic on the nano level, the level defining sensible chemical and biochemical processes, which will be dominant in 21st century technologies.

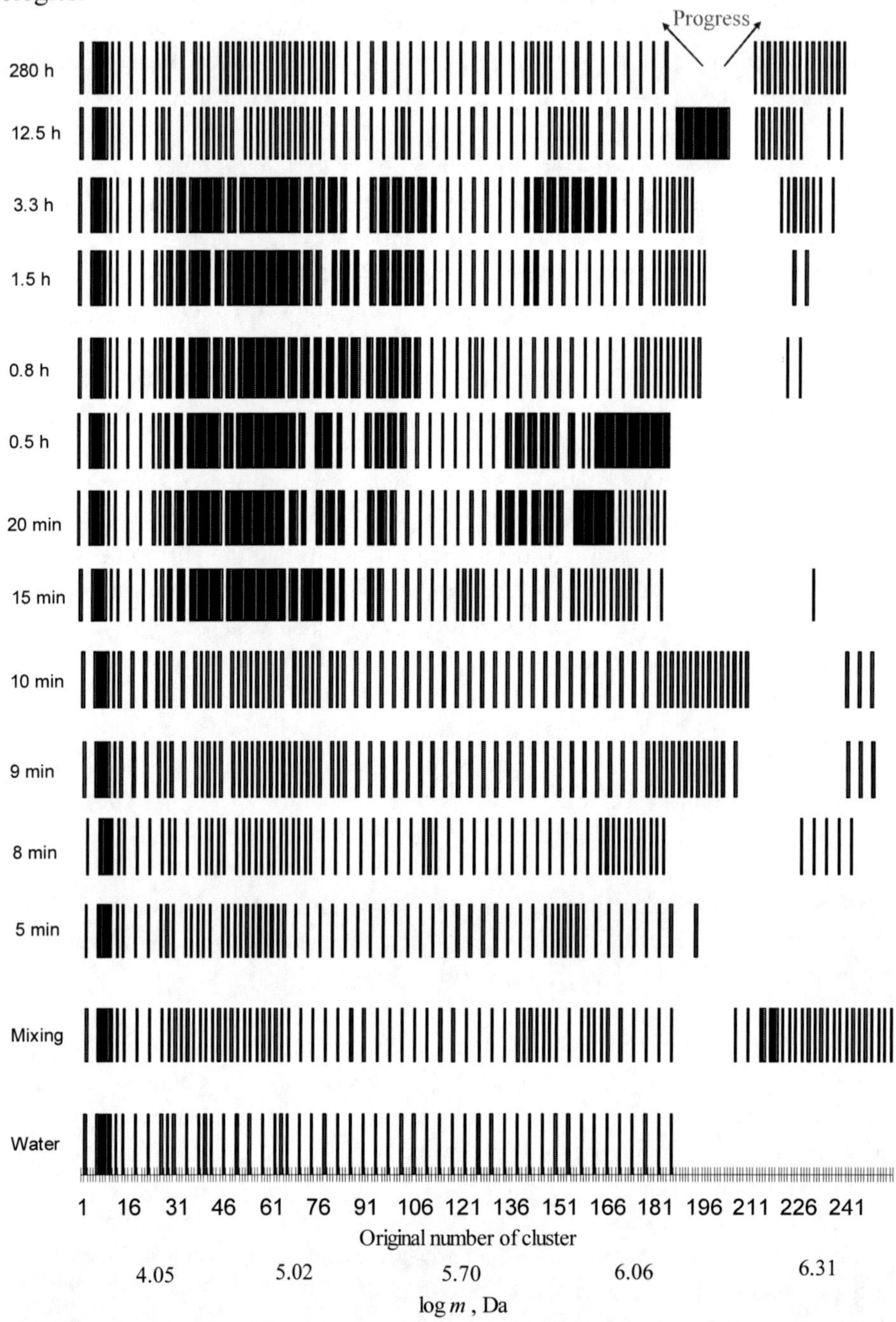

Figure 10.2.3. Dashed representation of the presence/absence of clusters in a liquid in the process of forming and aging its statistical structure (long-range order). The clusters' ordinal numbers tentatively correspond to their order in GMS-spectra (step of observation raster).

The processes of SCIP formation and their polymerization (polycondensation) pass a few steps in their evolution that can be represented as a dissolving of salt in water (Figure 10.2.4).

The evolution of dashes represented in Figure 10.2.3 also attests to the presence of fluctuations in the formation of a thermodynamically stable salt solution.

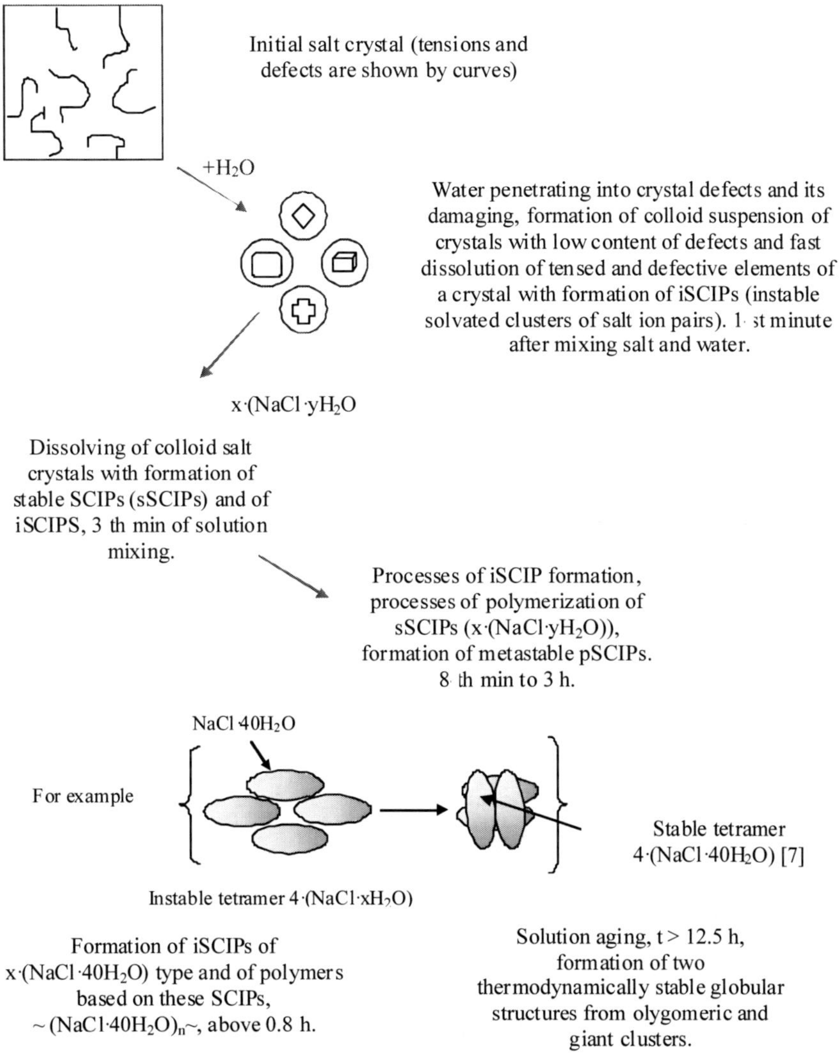

Figure 10.2.4. Model of dissolution of salt in water.

The signals of oscillating clusters in GMS prove the globular structure of the latter. Therefore, the structure of a completely formed solution of salt can be considered a thermodynamically stable nanodisperse suspension of polymer solvate cluster ion pair in water. Therewith, the salt solution long-range order is defined on the level of structural constructing of pSCIP and associates based on it.

The pSCIP should differ by anisotropy at least in its surface layers coming in contact with the bulk solution. In [5], it was found that a solvate shell of an alkali metal in water has layers in two levels, internal and external; the external layer is rather porous (expanded) and differs by structure from both the internal layer and surrounding solution. In [12], it was reported that in NaCl solutions, the H-bond bridges between water molecules are much longer than in pure water. These facts attest in favor of the presence in SCIP of an expanded surface layer consisting of water molecules.

It seems logical that upon the polymerization of SCIP, they approach each other up to the level of internal (condensed) layers of solvating water while the expanded layer will be formed on the surface of the polymer globule at the solution frontier. Similar phenomenon, which is characteristic of the cluster structure of water was described by Yoshiteru [13]. He found that 22% of water molecules have one H-bond bridge only and 77% have two bridges. We also modeled water clusters, which allowed us to understand that surface layer consists of the molecules with one H-bridge (22%) and internal layers of the molecules with two H-bridges (78%). In other words, the density of the SCIP surface layers should differ by density from the overall liquid density and from the density of the clusters inside the construction.

It seems possible to detect this layer by, for example, changing the pressure over the solution. If the layer exists, then according to the osmosis laws (van't Hoff equation, $\pi = icRT$, where π is osmotic pressure, R is gas constant, T is absolute temperature, and c is the solution molar concentration), it should respond to the change in pressure. In this equation, the i factor is connected with the electrolyte degree of dissociation (α) in a diluted solution ($i = 1 + \alpha(\nu - 1)$, where ν is number of ions formed at the dissociation of the electrolyte) [14] and can reflect changes in the SCIP structure. The influence of pressure on the clusters' distribution and their average mass can be reliably measured using the GMS-method.

We found that weak changes in atmospheric pressure correlate with the character of SCIP distribution in the solutions of salt, values of M_{GMS} parameter, and in the number of oscillators in the solution (Figure 10.2.5). Thus, the change in pressure of atmospheric air over the solution (+10%) for one sec results in significant changes of these parameters after ten minutes and brings the system to the new thermodynamic equilibrium on hours. These parameters vary adequately with the change in pressure. The correlation of solutions' physical properties (osmotic pressure, vapor pressure over the solution) with the ion hydration numbers was also reported earlier [15].

Thus, SCIP has a surface solvate shell that can play the role of membrane commutating with environment, e.g., for the tetramer $4(NaCl \cdot 40H_2O)$ it can be represented by the following model (this layer is dashed, Scheme I).

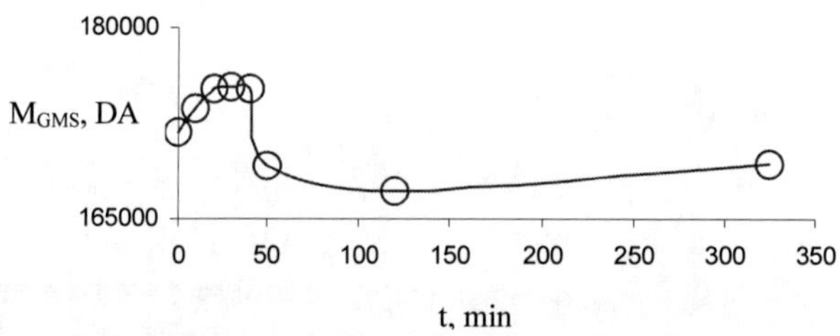

Figure 10.2.5. M_{GMS} response of 0.1 wt. % salt solution (analog of physiological salt solution) on sharp change in air pressure over the solution, $\Delta P = 11\%$. Number of oscillator types after 30 min increases by 35%.

We detected dependence of the cluster ordering hierarchy of the salt solution on the method of preparation of the solution.

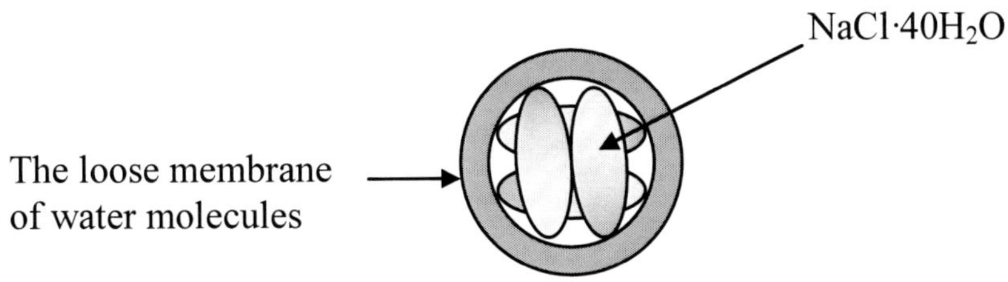

Scheme I.

In [7] the salt solutions were prepared by us by means of consecutive dilution of the concentrated (26%) solution and by keeping them for two weeks, excluding the careful action of light shaking and changes of temperature, while in the current work, the 3.5% salt solution (Figure 10.2.1) was prepared directly by dissolving salt in water. Comparison of GMS-spectra of the prepared 3.5% salt solution in the current work after 328 h of keeping it within the spectrum in [7] shows that they are not completely identical in the region of small clusters (log m from 2.6 to 3.6), but are instead identical with the spectrum of 26% salt concentration below log m = 3.7. This was also prepared by direct dissolution of the salt and keeping it for 320 h.

In the 3.5% solution obtained by direct dissolving of salt, dimeric SCIP $2(NaCl{\cdot}40H_2O)$ were found only in expanded form. The number of oscillator kinds in the solution obtained by salt dissolving is 56% higher than that in the solution obtained through dilution.

Thus, the GMS-method allows for the obtaining of some data on the solution origin, namely, was it obtained by either diluting or direct dissolving.

Technologists long ago noted that many processes of chemical technology with solutions depend not only on the order of mixing, conditions of stirring, duration of "ripening," etc., but also on weather situation (pressure jumping). We suggest that solutions of salts have long-range order analogous to that found in [7] and represented in the Scheme I structure of SCIP. Such SCIP are sensitive to change in pressure in apparatus and in vessels with open surface, and their changes can be used in fine chemical synthesis for enhancing the quality of products and certain reactions selectivity. The long time it takes to establish thermodynamic equilibrium in a salt solution can be used by technologists for the protection of inventions and formation of "know how."

Advance biological systems are known to be very sensitive to pressure changes. Atmospheric pressure probably affects solvate shells of not only SCIP, but also of dissolved compounds involved in biochemical processes as a sequence to distort metabolism in people (headache, joint-ache). Modern medicine still cannot explain the source of such diseases.

Evolutionary aging of salt solution according to the model shown in Figure 10.2.4 and discovered by us, is a phenomenon of the dependence of the cluster arrangement on the solution origin and can assist in the understanding of existence in oceanic stable layers and streams of water that have equal salt content but differ in temperature so they are not mixing together.

Conclusion

The process of salt dissolution in water is a multistep one and includes the following steps: formation of unstable colloid systems, metastable solvate clusters of the salt ion pairs, formation of unstable globular polymeric structures in the solution, and the formation of thermodynamically stable structures SCIP and globular polymers based on them. This process for 3.5 wt. % salt solution at 294 K requires basically hundreds of hours. The SCIP surface consists of an expanded layer of solvate molecules behaving as a membrane in the osmotic processes of informational interaction of clusters with the environment. Dash codes of long-range cluster organization in liquids can be used in the estimation of their quality and in the archiving and modeling of physical, chemical, hydro chemical, and biochemical processes. GMS-method allows for the obtainment of certain data on the solution origin history (it is obtained by diluting or by direct dissolving). The salt solution is not an ideal solution, as has been suggested earlier.

References to 10.2

[1] Zahn, D. Atomistic Mechanism of NaCl Nucleation from an Aqueous Solution. Physical Review Letters, 2004, vol. 92, no. 4, p. 040801/1.

[2] Water and aqueous solution, Horne, R.A., Ed., Willey-Interscience, 1972, p. 838.

[3] Karazhanov, N.A. Osnovy kinetiki rastvoreniya solei, Alma-Ata: Nauka, 1989, p. 189.

[4] Peslherbe, G.H., Ladanyi, B.M., Hynes, J.T. Structure of NaI ion pairs in water clusters. Chem. Phys., 2000, vol. 258, pp. 201-224.

[5] Karpov, G.V. On the structure of hydration shells of alkali metal ions in aqueous solutions. Chemical Physics Letters, 2005, vol. 402, pp. 300-305.

[6] Marañón, Di Leo, Marañón, J. Confined Na^+ and K^+ ions in an aqueous solution. Chemical Physics Letters, 2005, vol. 404, pp. 257-262.

[7] Zubow Kr. V., Zubow A.V., Zubow V.A. Oscillation spectra of seed NaCl crystals in aqueous solution. Journal of Appl. Spectr. 2005, vol.72, no. 6, pp. 840-847.

[8] Zubow Kr. V.; Zubow A.V.; Zubow V.A. Cluster structure of liquid alcohols, water and n- hexane. Journal of Appl. Spectr. 2005, vol. 72, no. 3, pp. 321-328.

[9] Bogdanov E. V.; Mantrova G.M. Equicluster water model. Biomedical Radioelectronics. 2000, no. 7, pp. 19-28. (in Russian).

[10] Annika Lenz; Lars Ojamäe. On the stability of dence versus cage-shaped water clusters: Quantum-chemical investigations of zero-point energies, free energies, basis-set effects and IR spectra of $(H_2O)_{12}$ and $(H_2O)_{20}$. Chemical Physics Letters. 2006, vol. 418, pp. 361-367.

[11] Yannis Georgalis, Kierzek A.M., Saenger W. Cluster Formation in Aqueous Solutions Observed by Dynamic Light Scattering. Journal of Phys. Chem. B, 2000, vol. 104, pp. 3405- 3406

[12] Masuda, K., Haramaki, T., Nakashima, S., Habert, B., Martinez, I., Kashiwabara, S., Structural change of water with solutes and temperature up to 100 degrees C in aqueous solutions as revealed by attenuated total reflectance infrared spectroscopy. Applied Spectroscopy. 2003, vol. 57, no. 3, pp. 274-281.

[13] Yoshiteru, Y. A severe artifact in simulation of liquid water using a long cut-off length: Appearance of a strange layer structure. Chemical Physics Letters, 2005, vol. 406, pp. 49-53.
[14] Ageev E.P. Osmos. Khimicheskii enciklopedicheskii slovar' (Chemical Encyclopaedia), Moscow: Sovetskaya Entsiklopediya, 1983, p. 419, in Russian.
[15] Spangberg, D., Wojcik, M., Hermansson, K. Pressure dependence and activation volume for the water exchange mechanism in NaCl(aq) from MD simulations. Chemical Physics Letters, 1997, vol. 276, pp. 114-121.

10.2.1. The Phenomenon of Water Polymorphism's Influence on the Dissolution Enthalpy of the NaCl Mechanism

In the chemical and biochemical technologies, where water is used as reactant or solvent, the water structure at the cluster level is ignored because of a general ignorance concerning how to control it. Models of water as a polymorphous substance were extensively studied in the 80s of the last century [1], however, only recently has their polymorphism really been proved [2, 3, 4]. Ensembles of water clusters are different in various ETIs [5]; they reflect the dynamics of the instability of gravitational fields [9] and can be a platform to understand chemical and biochemical processes on other planets. The practical use of this phenomenon on our planet opens interesting possibilities for process optimization and improving selectivity and reliability. For example, we know the sound effect of different frequencies and intensities on the formation of stable organized structures of powders (http://www.youtube.com/watch?v= nO0bSSXmr1A&feature=related), liquids, and solutions (http://www.youtube.com /watch? v=4shodbQMcmM&feature=related). The mechanism of this phenomenon is not yet understood. On the other hand, reference books give different values for the salt dissolution enthalpies in water, which can lead to complications in the calculation of the heat balance in chemical technologies [6, 7, 10].

The aim of this study is to investigate the influence of polymorphous water structures on the energy of the salt dissolution and to understand the interaction mechanism of a salt with a particular polymorphous water form.

Distilled water and powdered sodium chloride (99.9%) are used as investigation objects. Water is previously aged for six months under conditions protected from light and mechanical fields. In Figure 10.2.1.1, there is a scheme of a modified calorimeter for measuring heats of salt dissolution and GMS recording as well as for exposure with sound frequencies.

The standard method for the determination of the salt dissolution enthalpy is described in [10]. The calorimeter is calibrated according to the known heat of salt dissolution in distilled water equal to 4.16 kJ/mol [11]. The registration method of water in the LRO is described in [12, 13]. The distance between the sound frequency generator and the water surface is 3 cm and the sound power above the water surface $\leq 80\pm10$ dB. The GMS-spectrum (earlier called a "flicker noise spectroscopy," FNS) of water without energy impact is presented in Figure 10.2.1.3.

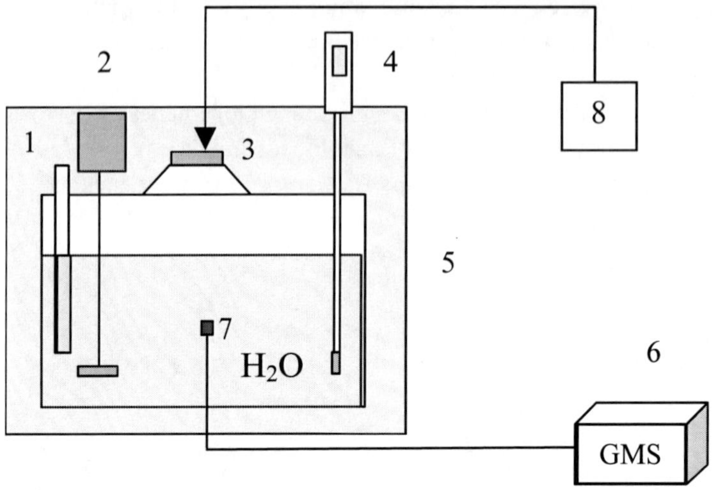

Figure 10.2.1.1. Scheme of calorimeter studying the influence of water polymorphism on the heat of salt dissolution. 1 – vessel filled with salt, 2 – stirrer, 3 – sound frequency generator, 4 – thermometer, 5 – box filled with polystyrene foam, 6 – GMS-spectrometer, 7 – GMS sensor, and 8 – frequency generator. To exclude earthquake shock waves, box 5 is placed on a foam layer (2 cm). To generate a specific polymorphous water structure with the help of a compulsory formation of a specific long-range order in the liquid at the cluster level, the following sound frequencies are used (Table 10.2.1.1).

Figure 10.2.1.2 shows the GMS-spectra of water exposed to sounds of different frequencies. It is seen that the cluster structure of water is different. These structures are found to be stable under sound exposure only when disconnecting the sound generator from the morphology immediately changes to an equal LRO. At low sound frequencies, the signal of the base water cluster $(H_2O)_{12}$ [14] gets a positive *f* value, which indicates its expanded structure [4, 9]. At the frequency of the Chaplin cluster of 100 molecules (7,353 Hz), its signals even disappear while the base water cluster becomes collapsed, -*f* [4, 9]. Since the dissolution of salts is accompanied by mechanical agitation, which is known to influence LRO [12], it is of interest to understand how the polymorphous water structures will be changed under mechanical stirring. Preliminary experiments showed that the origin of the water and its stirring affect the water LRO before sound exposure (Figure 10.2.1.3).

Four water forms (Figure 10.2.1.2) are applied to analyze their influence on the dissolution enthalpy of NaCl.

To get an idea on the cluster structure of solvated clusters of ion pairs in NaCl, the solution of the spectrum of the solvent has been subtracted from that of the solution [16], Figure 10.2.1.4. As is visible, SCIP dimers and tetramers disappear; they are the basic elements of the LRO formation in a salt solution at the level of seed crystals. On the other side, there are strong hexamer and nonamer signals indicating the destructive impact of the sound frequency of 7,353 Hz, probably. Furthermore, the frequency influence on the basic SCIP differs.

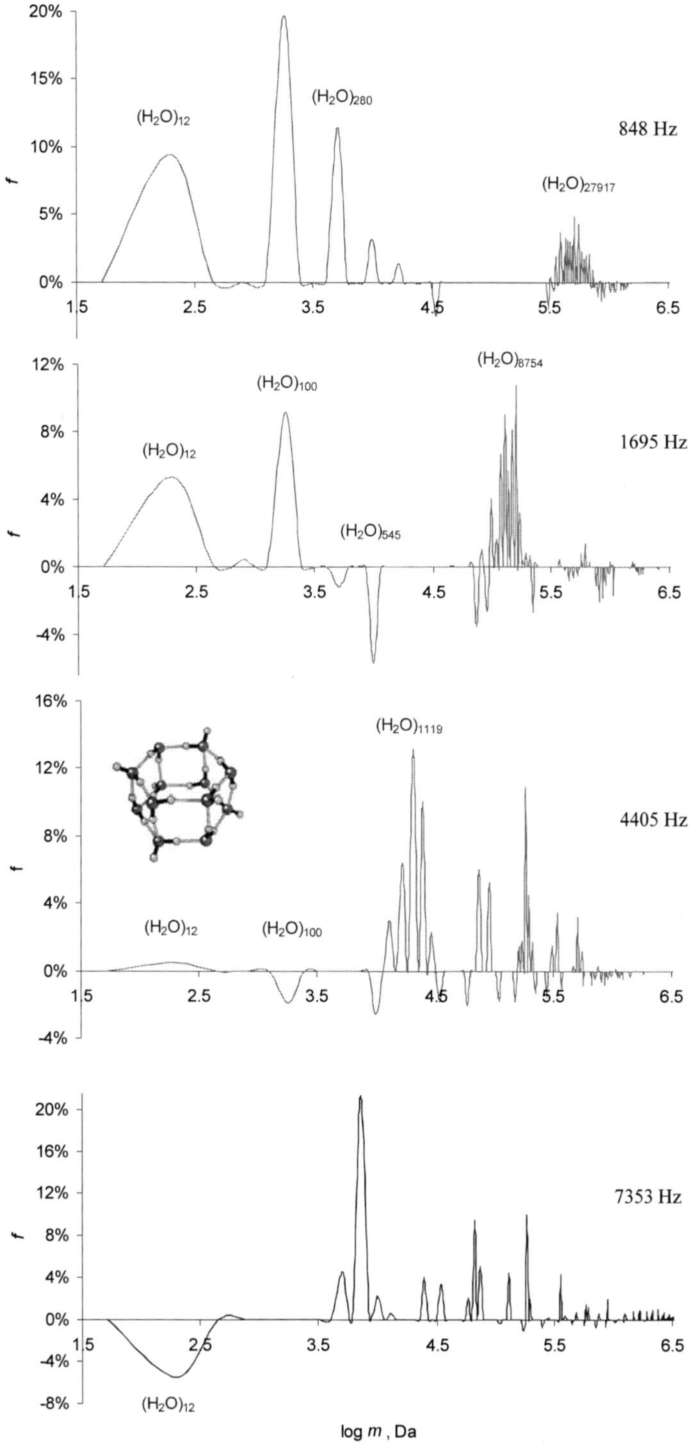

Figure 10.2.1.2. GMS-spectra of polymorphous water structures used for the study of the thermal effect from the salt dissolution. During the exposure to the appropriate frequency, the liquid is stirred at 293 K. The model of the base water cluster is kindly provided by Professor A. Lenz [15].

Table 10.2.1.1. Sound frequencies for the exposure of water and salt solutions and the corresponding Chaplin water clusters [8]. *The oscillation frequencies of water clusters are calculated according to the Zubow equations [9].

Frequency, Hz	Water cluster*
7,353	$(H_2O)_{100}$
4,405	$(H_2O)_{280}$
1,695	$(H_2O)_{1889}$
848	$(H_2O)_{7544}$

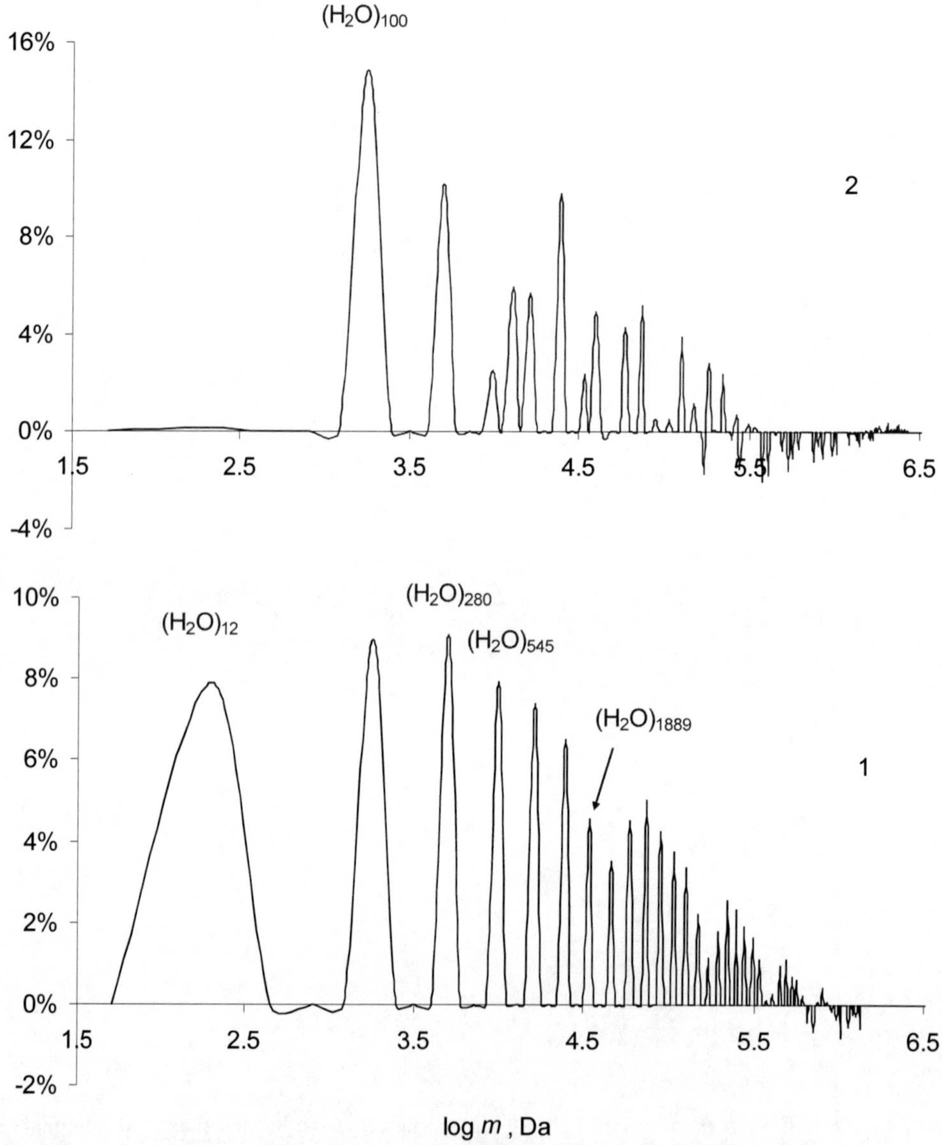

Figure 10.2.1.3. GMS-spectra of long-range order. 1 — distilled water and 2 – water (1) at stirring.

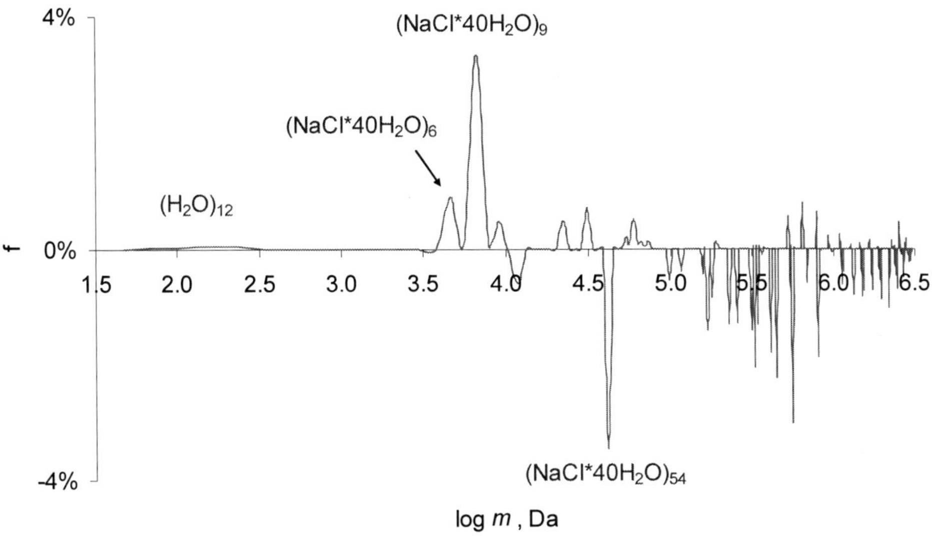

Figure 10.2.1.4. GMS-spectrum of SCIP in solution obtained by subtracting the water spectrum from the solution spectrum (2 wt. % NaCl). The dissolution process proceeds under stirring and sound exposure (7,353 Hz) at 293 K in a mass ensemble up to 3 million Daltons is investigated.

Table 10.2.1.2 presents the impact of polymorphous water structures on the dissolution enthalpy of salt. As is visible, the enthalpy highly depends on the specific polymorphous water structure. Reduced dissolving enthalpy by 14 times with decreasing sound frequency should not be attributed to that fact, which is connected with an easy energy pumping into the solution with the help of sound waves.

Table 10.2.1.2. Dissolution enthalpy of NaCl in distilled water with different LRO (different polymorphous structures) at 293 K

Frequency, Hz	Enthalpy of NaCl dissolution, kJ/mol
-	4.2
7,353	2.5
4,405	0.3
1,695	0.3
848	0.3

These events, however, should be considered from the standpoint of the specific cluster structure being characterized as a dynamic equilibrium between collapsed and expanded forms. For example, Professor Chaplin reports on this balance for the cluster consisting of 280 water molecules [8], Figure 2.1.1.

A strong enthalpy decrease is just observed for the oscillation frequency (4,405 Hz) belonging to this water cluster. Apparently for the smaller clusters, this effect is little or even unpronounced. The sound energy is considered to influence the protons in the hydrogen bonds (HB) leading to a changed distance between the oxygen atoms because of an enforced condensation of these protons from the physical vacuum [9, 17, 18, 19]. This has to be ascribed to the so-called effect of the balance shifting between baryonic protons and protons in the dissolved state in a physical vacuum. The clusters possess an additional energy stored in HB and upon water cluster destruction by the salt, they release it. Thus, the recorded dissolution enthalpies are effective values. Here two processes are running; the destruction of the salt crystal lattice cools the solution, and while the cluster interconverts, it heats.

CONCLUSION

The phenomenon can be successfully applied to the chemical industry for selectivity improvement, product yield enhancement, and production stability.

The phenomenon provides a more detailed understanding of the thermodynamics of the complex chemical processes and their control. Furthermore, technologists can take account of technological noises.

The water cluster ensemble impact on the thermodynamics of the dissolution and the interaction of an individual liquid with other substances can be a platform for a new understanding of the design of biological systems on other planets.

The phenomenon can be used for the fine structure analysis of the physical vacuum.

REFERENCES TO 10.2.1

[1] Ovchinnikov, L.N. Calculating the density of water at 0-340° based on the model of polymorphism. J. of Phys. Chem., 1982, v. 56, № 7, pp. 1613-1617 (in Russian).

[2] Mallamace, Francesco. The liquid water polymorphism. Proceedings of the National Academy of Sciences of the United States of America. 2009, v. 106, № 36, pp.15097-15098.

[3] Ricci, M. A.; Soper, A. K. Neutron diffraction study of water polymorphism. NATO Science Series, II: Mathematics, Physics and Chemistry. 2002, v. 81(New Kinds of Phase Transitions: Transformations in Disordered Substances), pp. 355-366.

[4] Zubow Kr. V., Zubow A.V., Zubow V.A. Cluster structure of liquid alcohols, water and n- Hexane. J. of Appl. Spectr. 2005, v.72, №3, pp. 321-328.

[5] Zubow K., Zubow A.V., Zubow V.A. Signals from Extraterrestrial Intelligences. Horizons in World Physics, editor, Albert Reimer. Nova Science Publishers, Inc. NY. 2013, vol. 280, pp. Chapter ID: _8738_Horizons in World Physics. Volume 280, 2013 in print. https://www.novapublishers.com/catalog/product_info.php? products_id =38323

[6] Drakin S.I. Heat of solution. Chemical Encyclopedia. Moscow, Soviet Encyclopedia, 1983, p. 564 (in Russian).

[7] Short handbook chemist. Ed. Perelman V.I., Moscow, Chimija, 1964, p. 98 (in Russian).

[8] Chaplin M. Structural forms in the icosahedral water cluster, aviable via South Bank University of London, 2009, in http://www.lsbu.ac.uk/water/index.html

[9] Zubow K.V., Zubow A.V., Zubow V.A. Ensemble of Clusters – New Form of Molecular Matter, Risks and Chances. Zubow Equations. In Advances in Chemistry Research, vol. 5, 2010, pp. 107-145. Editor J.C. Taylor. Novapublisher. NY. USA.

[10] Practical work on Physical Chemistry. Ed. N. K. Vorobev. Moscow. Chimija,1975, p. 127 (in Russian).
[11] Short handbook of physico-chemical values. Eds. K.P. Mishchenko, A.A. Ravdel. Leningrad. Chimija, 1967, p. 50 (in Russian).
[12] Zubow A.V., Zubow K.V., Zubow V.A. Mechanism of sodium chloride diossolving in water and the process of the solution aging. Russian Journal of Applied Chemistry. 2007, v. 80, №7, pp. 1249-1255.
[13] Zubow Kr. V., Zubow A.V., Zubow V.A. Oscillation spectra of seed NaCl crystals in aqueous solution. J. of Appl. Spectr. 2005, v.72, №6, pp. 840-847.
[14] Bogdanov, EV, GM Manturova Equicuster water model. Biomedical Electronics. 2000, № 7, p.19-28 (in Russian).
[15] Annika Lenz, Lars Ojamäe. On the stability of dence versus cage-shaped water clusters: Quantum-chemical investigations of zero-point energies, free energies, basis-set effects and IR spectra of $(H_2O)_{12}$ and $(H_2O)_{20}$. Chem. Phys. Letters. 2006, v. 418, pp. 361-367.
[16] Zubow K., Zubow A.V., Zubow V.A. Molecular Cluster Distribution in Sea Water Near the Wall. Antifouling aspect. J. of Clusters. 2011, v.22, no. 4, pp. 603-619.
[17] Zubow K., Zubow A.V., Zubow V.A. The Phenomenon of Proton Dissolving in Vacuum and of Proton Condensation from Vacuum. Two Forms of Protons, Structure of Nuclei, Electrons and Atoms. J. of modern physics. 2010, v.1, №1, pp.175-184.
[18] Zubow K., Zubow A.V., Zubow V.A. Experimental Platform for the Investigation of the Structural Heterogeneity of the Physical Vacuum. Vacuum Energy Risks and Chances. In *Horizons in World Physics*. Editor A. Reimer. NY. 2012, vol. 277, pp. 87-110. https://www.novapublishers.com/catalog/product_info.php?products_id=29310
[19] http://www.zubow.de/pdf/News.pdf

10.3. RIVERS

To investigate the cluster formation in natural waters e.g. rivers, lakes, rain, springs, glaciers, underground waters, water in biomatrices, as well as in seas and oceans is of great interest. There are some water cluster structures with a minimal potential energy that are modeled by computer programs; they are in the expanded or collapsed forms. Earlier it was concluded by the authors that in water solutions of salts (§10), alcohols and acids form water clusters first and then SCIP. After that they assemble with organics (§10.8). Furthermore, the item's "ideal solution" is argued to be incorrect and the structure of such liquids has to be understood rather as a nano-emulsion of solvent clusters or as nano-suspension of SCIP. This statistical cluster ensemble built in liquids is characterized by a long-range order—a compelled distribution of mass concentrations (clusters) under the influence of white gravitational noises. Here the stable long-range order is in balance with different destructive influences: heat/mechanical flows, resonance fields of surrounding cluster analogs, gravitational radiation from celestial bodies, and tectonic deformations. Hence, a changed long-range order in natural waters should influence hydro chemical, hydrological, technological, climatic processes, as well as communication processes of living organisms.

Dependencies are found to exist between the long-range order in natural water and seasons, time of day, temperature, sun radiation, water flows' depth and velocity, and celestial bodies' constellation [1], whereas the Sun and water flows' velocity are of the highest relevance.

The fluid flow structure of long-range order varies depending on the structure of the vortex, its duration and intensity of turbulence. In strong turbulence in a river, there are new signals of clusters, which typically are in a harmonious balance of collapsed-expanded in a laminar flow or in their absence (stagnant water). Depth has an effect on long-range order in river water, but it is still primarily determined by the structure of the flow and the structure of the Sun's rays penetrating into the deeper layers of water. In all cases, the Sun greatly affects the cluster structure of water in the surface layer (20…30 m). It should be noted that the planets exert a significant influence on the distribution of clusters in surface waters [1].

We chose the materials for this section from our archive for a small river, a creek, located in the interaction with a lake originating from rain. The river flows out of lake. Both river and lake most clearly highlight the essence of the problem with a simple example: a lake-river. Due to the fact that we have to do with multi-factor influences (engines, etc.), similar studies for larger rivers are not carried out. Figure 10.3.1 presented an overview of GMS-spectrum of clusters in the water of the Peene River, filmed in midsummer 2007. The spectrum contains the main signals of clusters shown in Figure 2.1.1. There was an appearance of weak signals from a cluster containing 45±1 water molecules found in the turbulent parts of the river, which could be a consequence of forced destruction of the cluster structure of water flow.

This figure was marked with the simplest water clusters and formed by the basic clusters consisting of 10, 11, and 12 molecules ($(H_2O)_{11\pm1}$). Note that clusters of 45, 100, 178, 280, etc. (§24, Table 5.1) interact resonantly with the gravitational fields of the celestial bodies and can be used as a new generation of gravitational sensors. Thus, the signals' instability of the Mercury cluster $(H_2O)_{45}$ in the river and the lake changes adequately at the approximation of Mercury to the Earth (Figure 10.4.7), and the signals $(H_2O)_{11\pm1}$ satisfactorily correlated with the full Moon. A correlation between the signals detected from the Venus, Mars, Jupiter, and Saturn clusters and the position of the planets (see Table 5.1) is observed. In §23, we tried to find an explanation for this phenomenon. It should be noted that the energy balance of the clusters in the river water varies seasonally and even during the day. It also has a dependency on weather conditions, which we attached to a number of factors.

For water in the river, the cluster composition is not identical. It depends on the flow velocity and turbulence. In strong turbulence a number of new clusters appear, which are not present in the calm water of the lake; these are the clusters of 60, 137, and 242 water molecules. The structure of large clusters changes dramatically. This charming section of the physical chemistry of natural waters is just beginning to take shape. There appears to be a lot of interesting work for hydrologists but because of the limited amount of space within this monograph, we are forced to leave this market segment for later on.

The bar code GMS-spectra enables us to identify the super cluster structure of water in the river. Some clusters are observed that occur in the lake water, too, (§10.4.3) though they have some differences. The clusters in the river are more like those in the surface layers of the lake than those in the layers near the ground (Figure 10.3.5). Apparently, the forced mixing of the river water flow and the wind exposure of the lake surface layers causes this.

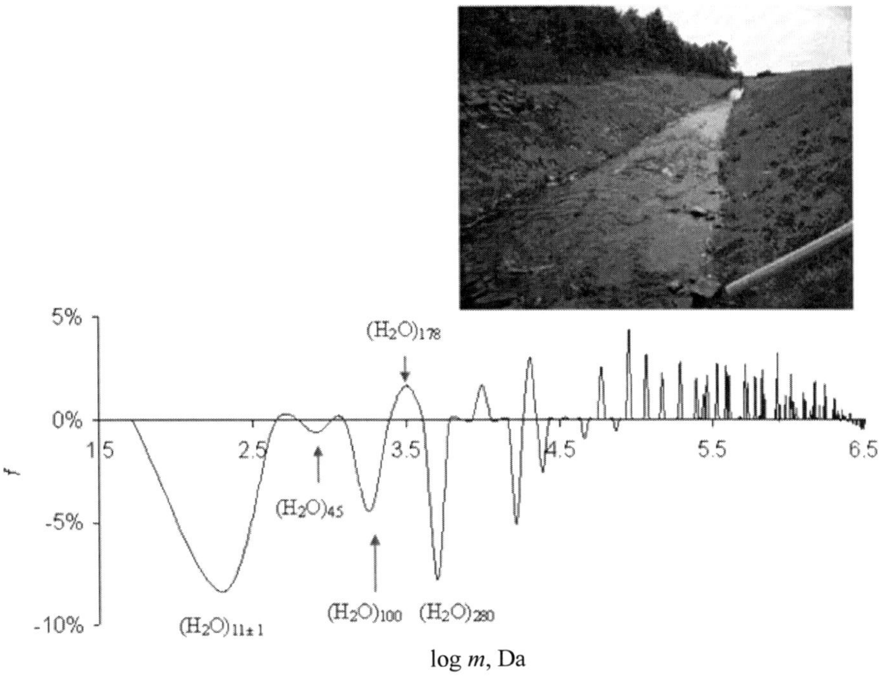

Figure 10.3.1. GMS-spectrum of water in the Peene River (Peene, North Germany), depth 0.15 m, 290 K. Overcast, rain. July 1, 2007, 11:00. 30 min. M_{GMS} = 519,885 Da, $Д_c$ = 34%, $F_{95\%}$ = 31%, N = 162, p < 1 N/m^2. The average results of three measurements of 30 seconds. The river flows out of the lake "Torgelower See" and originates from rain. Location of sampling is 300 meters from the lake (53.08°N and 12.75°E), picture in the right corner. The GMS-sensor was placed directly in the river water.

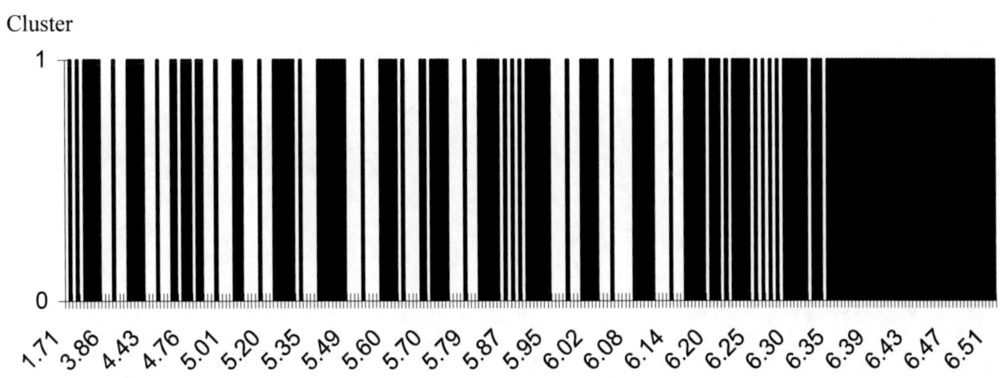

Figure 10.3.2. GMS of water in the Peene River displayed in a bar code (see Figure 10.3.1).

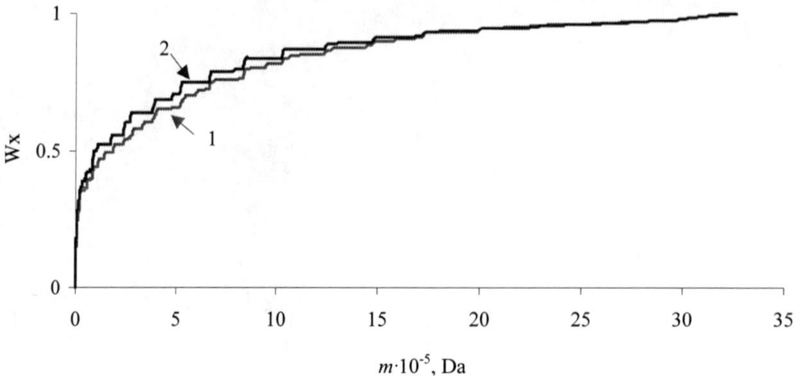

Figure 10.3.3. Integral distribution of cluster fractions in water from the river (1) (see Figure 10.3.1) and lake (of 1 m depth, 2). July 2007.

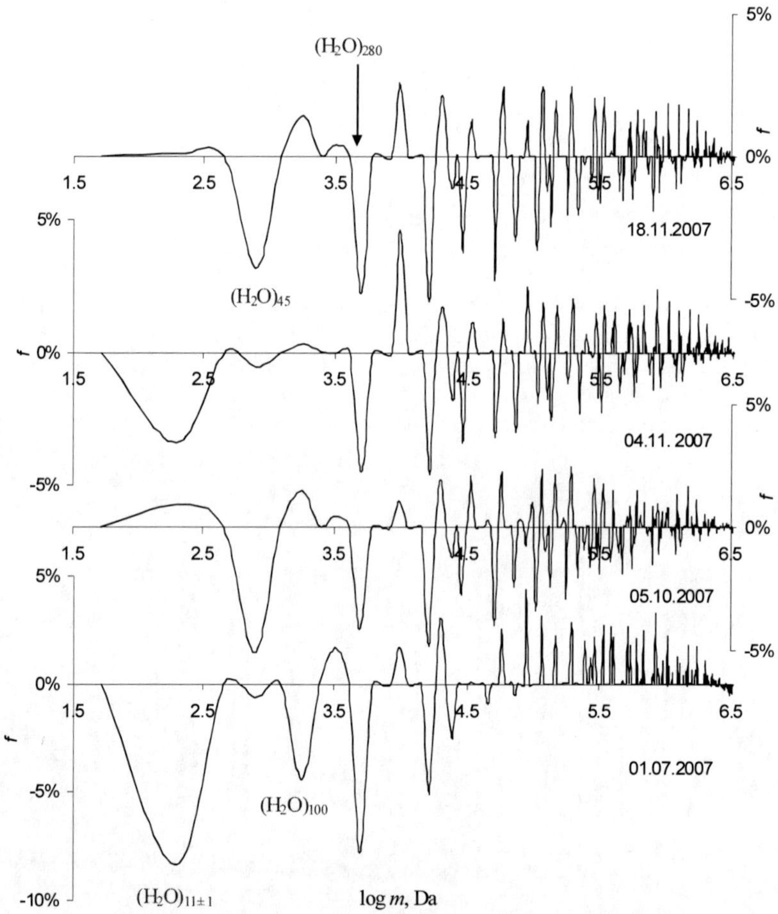

Figure 10.3.4. GMS-spectra of water in the Peene River (Peene, North Germany), depth 0.15 m, all three measurements made on a cloudy day: 01.07.2007: 11:30 CET (Central European Time), 290 K, M_{GMS} = 519,885 Da, $Д_c$ = 34%, $F_{95\%}$ = 31%, N = 162, $p < 1$ N/m^2. 05.10.2007: 08:12 CET, 286 K, M_{GMS} = 439,683 Da, $Д_c$ = 58%, N = 135, $p < 1$ N/m^2. 04.11.2007: 09:18 CET, 282 K, M_{GMS} = 510,303 Da, $Д_c$ = 53%, N = 135, $p < 1$ N/m^2. 18.11.2007: 08:06 CET, 279 K, M_{GMS} = 474,033 Da, $Д_c$ = 55%, N = 129, $p < 1$ N/m^2, Zubow constant taken equal to $6.4 \cdot 10^{-15}$ N/m.

The integral cluster distribution in water from the river and the lake is different (Figure 10.3.3). Differences can be seen in the small- and medium-mass interval, which is apparently due to the destruction of large water clusters by turbulence.

GMS-spectra of river water are not constant; they vary seasonally. Figure 10.3.4 shows a sequence of GMS-spectra of river water in the place shown in the photo in Figure 10.3.1. As is visible, the cluster structure of the river varies considerably only in the summer-autumn period. The most stable signal is that of the cluster consisting of 280 water molecules (Figure 2.1.1); the collapsed-expanded clusters' balance in the other small clusters changes in different ways. For large clusters (log $m > 4.5$), changes have occurred as early as October 5th. It should be noted that the background noise of the river murmur remained constant, but the structure of the noises changed adequately to the decreasing of the water temperature. With decreasing temperature in the river, the part of the collapsed clusters (seed crystals) increased while the number of cluster types decreased. The average mass of clusters remained almost constant.

It is interesting to trace the distribution of water clusters in the lake water in terms of its dependency on depth and to compare it with that found in the river flowing from the lake. In Figure 10.3.5, the GMS-spectra in the bar code for these waters on February 2008 are represented; the lake surface was covered with a thin layer of ice but the water in the river did not freeze due to its strong flow. One can see the difference in the distribution of the clusters. The long-range order in river water is characterized by a higher variety of signals compared with that found in the surface layer of the lake. The lowest number of cluster types is found in the bottom water of the lake.

Figure 10.3.5. GMS-spectra of water in the form of a bar code of the lake (different depths) and river. 17.02.2008: 08:00 CET. Water temperature: 275 K (lake), 276 K (river). Zubow constant = $6.4 \cdot 10^{-15}$ N/m, $p < 1$ N/m^2. Cloudy.

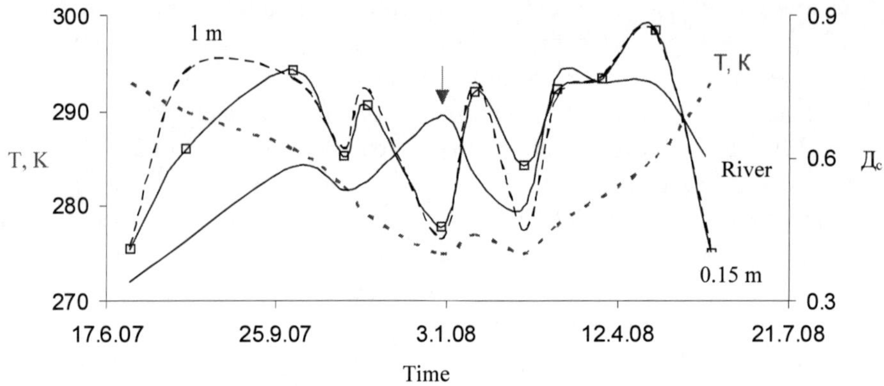

Figure 10.3.6. Part of collapsed clusters ($Д_c$) in river and lake water at different depths. All measurements were performed on a cloudy day from 08:00 to 10:00 in the morning. A mass range up to 2 million Daltons is taken. GMS-sensor was placed directly into the water.

Figure 10.3.6 shows how $Д_c$ changed during one year of observation. A clear correlation with the change in the water temperature is seen. The part of collapsed clusters in the surface waters varies with adequate temperature. Only on 31.12.2007 (indicated by an arrow) was a disharmony in the water long-range order for both the river and lake found. Since the value of $Д_c$ is an integral characteristic of the whole cluster ensemble, it only gives a general idea of the distribution of the collapsed clusters. At the level of individual clusters, the situation is more complicated. During the year, the individual clusters can repeatedly change their density, but it may remain relatively constant for several hours or even days.

Conclusion

Water from rivers consists of molecular clusters, which form a long-range order.

The natural flow of the river destroys the cluster structure of water, which is connected with the formation of several clusters; they are absent in calm water.

Temperature has a strong influence on cluster formation in natural water as well as on the energy hierarchy of individual clusters in the ensemble. Individual water clusters react differently to a weak energy field. The long-range order in water from rivers is concluded to be caused by white gravitational noises.

References to 10.3

[1] K.V. Zubow, A.V. Zubow, V.A. Zubow. Principles of gravitation spectroscopy. New form of molecular matter. Processes. Fields. Aist Handels- und consulting GmbH, dep. R&D, Berlin, 2010, electronic book (www.zubow.de), p. 854, (in Russian).

10.4. Lakes

The cluster structure of water in lakes is highly dependent on the origin of the lake (water coming into it), water dynamics, time of day and year, temperature, depth, water reservoirs, and the influence of the Sun. Below is material to study the formation of a water cluster in lake originating from rain (second rank according to the German classification). The Torgelower See is located in Northern Germany and is surrounded by agricultural lands. The maximum depth is 6…8 m and 80…90% of it is about 1…1.5 m. Figure 10.4.1 presents an overview GMS-spectra of the lake water at different depths recorded in a selected point. The procedure for obtaining the data is shown in the following figure (Figure 10.4.2).

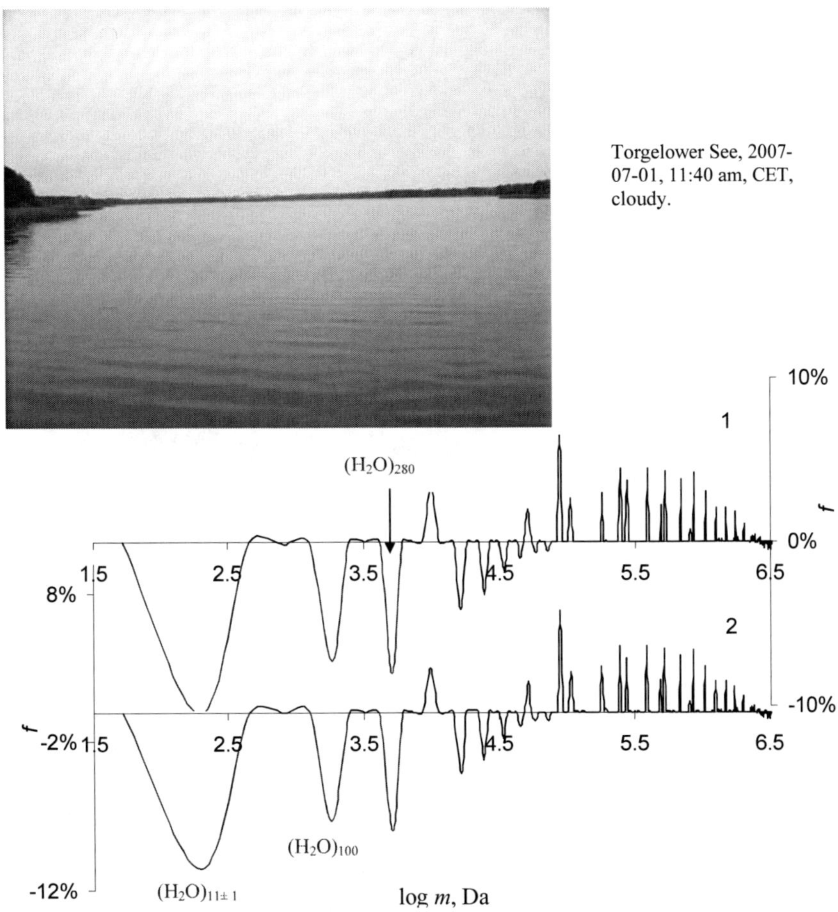

Figure 10.4.1. Overview GMS-spectra of water in the lake Torgelower See, Northern Germany, depth 0.15 m (1) and 1 m (2), 290 K. 01.07.2007, 11:40 CET. 1 – M_{GMS} = 472,529 Da; $Д_c$ = 41%; $F_{95\%}$ = 37%; and N = 159. 2 - M_{GMS} = 468,886 Da; $Д_c$ = 41%; $F_{95\%}$ = 23%; and N = 163. Average results of three measurements (30 s, $p < 1$ N/m^2).

These spectra were recorded in midsummer 2007, in cloudy weather, and in the absence of the Sun, which is the strongest factor affecting the building of a super cluster structure in natural waters. During this time, the water in the lake is clear and not burdened with the flowering process, which usually dominates in the second half of summer. Figure 10.4.1

shows the presence of the basic, small water clusters (Figure 2.1.1) in the freshwater lake. Attention is drawn to the absence of signals from the cluster consisting of 178 water molecules, which is present in the expanded form in the river flowing from the lake (Figure 10.3.1). The easiest clusters (up to 280 water molecules) are present in the collapsed form, but their signals are much stronger than in the river. This is an indication of a stronger individuality in the calm water of the lake. Since the salt content in the lake water is very small, in the GMS-spectrum, the signals of SCIP are overlapped by those of water clusters and their influence can be neglected.

Figure 10.4.2. Mobile GMS-spectrometer (1) and GMS-sensor (2) used for the analysis of the long-range order in natural waters.

A photo of GMS-spectrometer and a sensor applied to analyze the long-range order in natural waters is shown in Figure 10.4.2. The analysis procedure of the long-range order in the field is simple and involves direct sampling of the gravitational noise at the place of measurement. When the same measurement cell was first measured, the noises in the air and then in water (usually three times for 30 seconds) were measured, followed by the signals' cleaning and then subtracting the air signals from those in the water.

The analysis of the super cluster structure in the surface waters and deep layers of the lake did not reveal significant differences (Figure 10.4.3) in the warm season, however, in winter, there are differences (Figure 10.4.5). The differences are clearly visible when the spectra are set in relation to each other, for example, when the first spectrum is divided by the second, Figure 10.4.4.

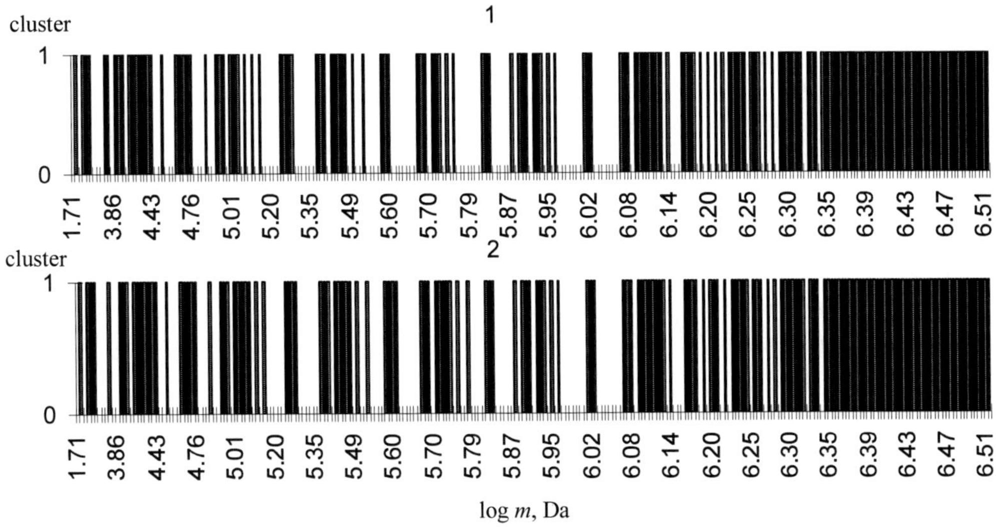

Figure 10.4.3. GMS-spectra of water in the lake in the form of a bar code (see Figure 10.4.1) for depths of: 1 – 0.15 m and 2 – 1 m.

The cluster with the mass of 337,611 Da (Figure 10.4.4) has been detected at a 1 m depth in the surface layers but its signal intensity decreases strongly (more than 160 times). On the other hand, we find it again in the river flow, where its signal intensity is about 8,000 times higher than in the surface layers of the lake. Hence its presence in some way is connected with the river flow.

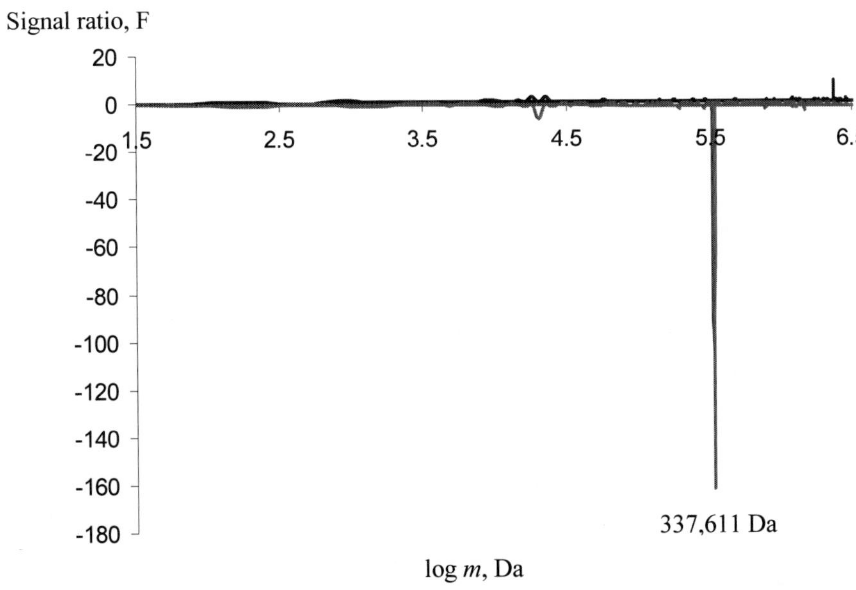

Figure 10.4.4. GMS-spectrum of signal ratios of clusters in the surface water area (0.15 m) to those in the deep zone (1 m) of the lake (F). The positive quadrant means the formation of new clusters. The inverse F (negative quadrant) means the disappearance of clusters. For conditions, see Figure 10.4.1.

It is interesting to study the long-range order changes in lake water during the day. Figure 10.4.5 shows the results of such research in the surface layers of the Torgelower See. It is visible that in the lake, during the day, there are strong changes in the long-range order. M_{GMS} achieves its maximum value at 11:00 and in the afternoon, it monotonically decreases by almost two times. The part of the collapsed clusters reduced from 06:00, but rises again at 15:00 when the sensor is shaded by tree crowns. Consequently, the Sun's radiation has a destructive effect on the water cluster structure in the lake. M_{GMS} growth in the first half of the day is due to the loosening of the cluster structure and the individualization of the clusters, leading to the weakening of the cluster interaction with its surroundings (an increase of the absolute values of *f*) under the influence of Sun's energy. In the afternoon, this process continued but at the level of destruction of the clusters.

It is visible from Figure 10.4.6 that at 19:00 in the lake there is no noticeable appearance of new clusters. Instead, there is the disappearance of a group of large clusters. It should be remembered that a balance between collapsed and expanded forms can only be discussed when the cluster mass is less than 5,000 Daltons, although, for large clusters, one should rather speak of cluster formation or decomposition. Therefore, the group of signals with log *m* > 5 shows the disappearance of large oscillators in the surface layers of the lake.

Table 10.4.1 summarizes the main characteristics of the long-range order in the lake water during June 22, 2005.

It is visible from Table 10.4.1 how the temperature influences the average cluster mass (M_{GMS}). The M_{GMS} varies inversely with temperature. That means its increase leads to a decrease of M_{GMS}. On the other hand, solar radiation affects the M_{GMS} and has an even greater impact on the maximum of the detected mass cluster (M_{max}) in this ensemble. The shading of the GMS-sensor through the trees by 15:00 leads to a strong increase of M_{max} as well as of the part of collapsed clusters by 23%. Thus, clusters are concluded to store solar energy, which accumulates in the structures of collapsed and very large clusters.

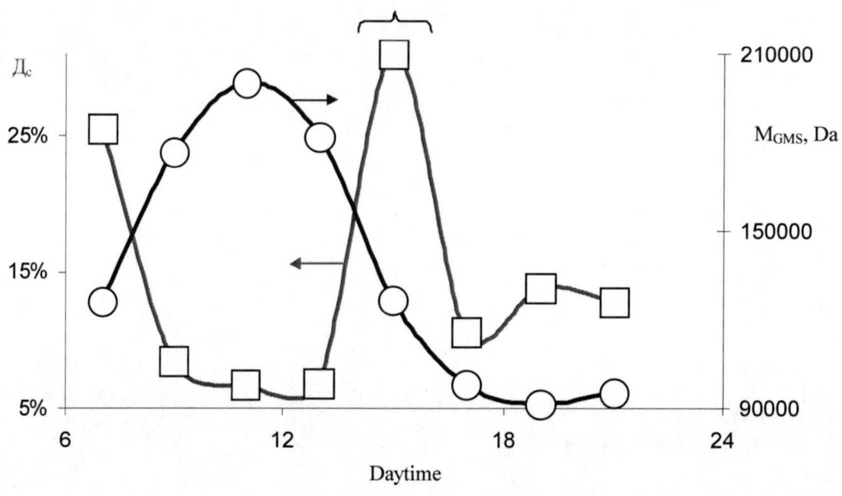

Figure 10.4.5. Change in average cluster mass (M_{GMS}) and the part of collapsed clusters (Дc) in the surface layers (0.15 m) of the lake Torgelower See during July 7, 2003. With a horizontal bracket, the time interval is marked in which the GMS-sensor was in the shade of the trees. 295±2 K. Sunny, cloudless weather.

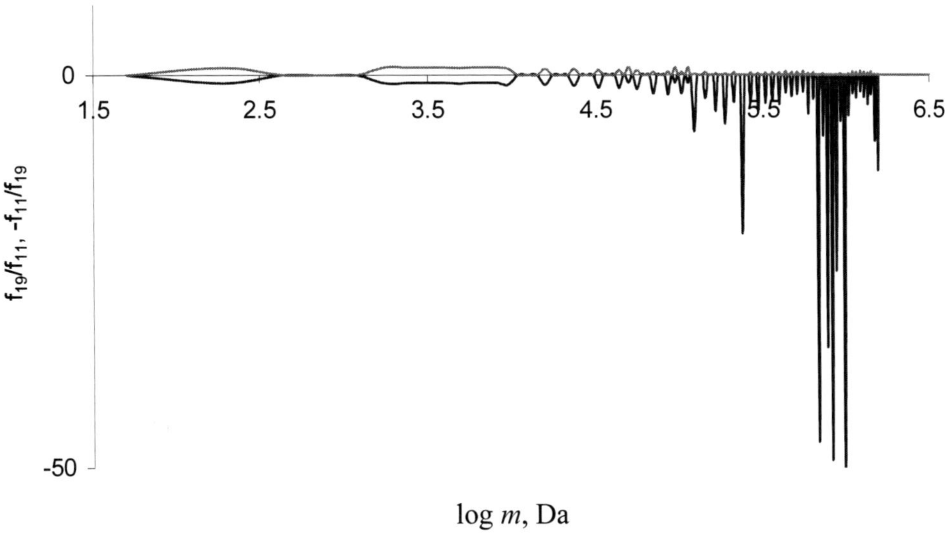

Figure 10.4.6. GMS-spectra of signal ratios (ratio of signals at 19:00 to those at 11:00 (f_{19}/f_{11})). The positive quadrant shows new clusters appearing (agenda line), and the negative one is for vanishing clusters—an inverse ratio ($-f_{11}/f_{19}$).

It is remarkable that the part of expanded clusters is very high ($Д_{exp}$, %, 74-93%) in the lake water, which means it is several times greater than that of the collapsed ones ($Д_c$, %). There are very little energy carriers at the cluster level, and thus the lake water is poor in energy. Comparison of the data given in Tables 10.4.1 and 10.5.1 shows a striking difference in the long-range order of rain and lake water. On average, the number of cluster kinds in the lake water is 8 to 10% higher than in rainwater (melt water).

In the period from June 2007 to July 2009, periodic changes in the interaction between the mercury cluster ($(H_2O)_{45}$, Table 5.1) and its surroundings at a depth of 1 m are recorded (Figure 10.4.7).

Table 10.4.1. Analysis results of GMS-spectra of lake water. Depth 0.15 m, slightly cloudy weather, $p > 1$ N/m^2, cluster ensemble up to 3 million Dalton.

Daytime	M_{GMS}, Da	M_{max}, Da	$Д_c$, %	$Д_{exp}$, %	Temperature, K	N	Note
7:00	130,051	2,131,459	26	74	292.7	61	shadow
9:00	182,747	1,813,045	8	92	292.8	57	sun
11:00	206,909	1,738,082	7	93	293.5	56	sun
13:00	187,997	1,813,045	7	93	296	57	sun
15:00	130,835	2,131,459	31	69	297	54	shadow
17:00	100,990	1,591,327	11	89	297.5	54	sun
19:00	94,487	1,591,327	14	86	297.1	54	sun
21:00	98,293	1,591,327	13	87	296	54	sun

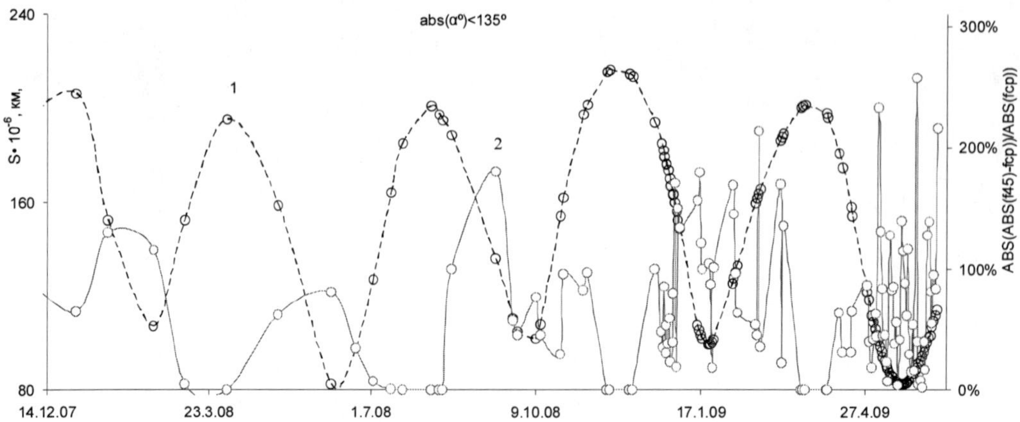

Figure 10.4.7. The distance (S) between Earth and Mercury (1) versus the signal instability of the mercury water cluster in the lake at a depth of 1 m (curve 2, the deviation from the average value). The analysis is done for a small cluster ensemble of up to 2.3 million Daltons and for the angle Mercury-Sun-Earth | α | < 135°. The distance data to Mercury was taken from the program "ZET 8" (www.astrozet.net), $p < 1$ N/m^{2}.

From Figure 10.4.7, it can be seen that the signal instability of the mercury water cluster changes periodically (88 days). When Mercury is approaching the Earth, the signal instability of the mercury water cluster increases by several orders of magnitude. When it is going away from the Earth, however, the signal instability is dramatically reduced. The cyclical nature of the instability agrees with the periodicity of Mercury orbiting around the Sun being equal to 88 days. The mercury cluster belongs to the whole ensemble of water clusters in the lake; it will be influenced by both the four moon clusters $(H_2O)_{11\pm1}$ and larger ones of the ensemble. All of this makes it difficult to interpret the signals correctly and requires complex mathematical processing.

Increasing the number of analyses at the moment of Mercury's approach to the Earth, a periodicity of the instability of the signals can also be found. The authors believe this can be explained by the fact that Mercury crosses the gravitational power lines between the Sun and the Earth.

This crossing has a shielding effect on the energy flow to the lake and to its biocenosis. Figure 10.4.9 shows the spectra of the super cluster structure (as a bar code) in a river and lake during summer and winter. As can be seen, the distribution of clusters in natural waters varies seasonally. Here, in summer, a harmonization of water cluster ensembles in the river and lake takes place whereas in the winter it is broken. The cluster distribution in the surface layer of the lake is excited by waves and in the river, it is similar but very different from that of the lake at a depth of 1 m. Consequently, temperature and mechanical agitation of the water structure affect its long-range order.

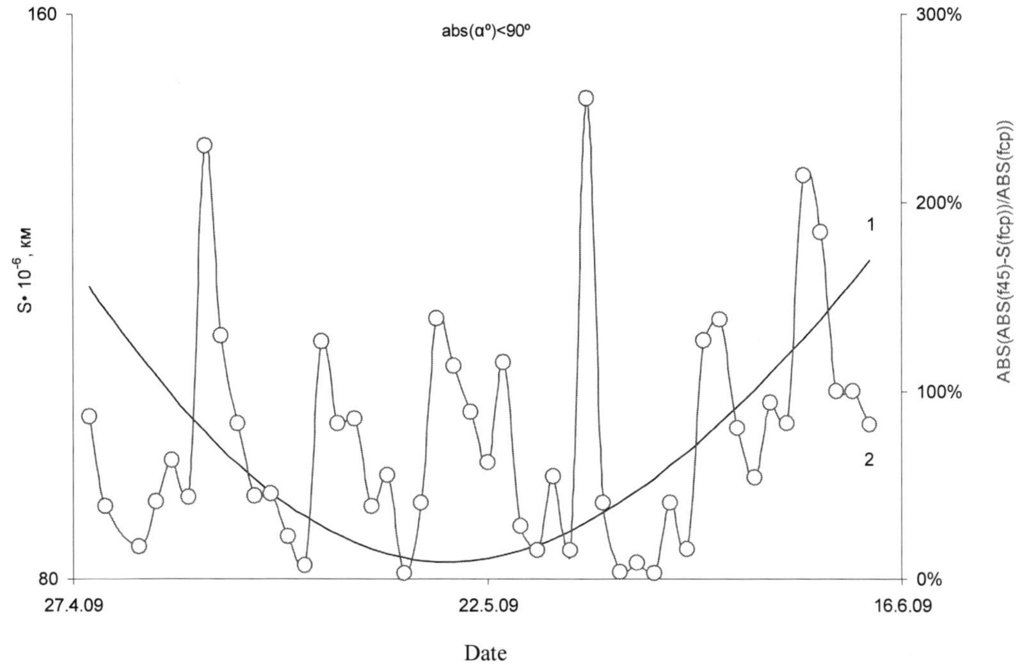

Figure 10.4.8. Fragment of Figure 10.4.7 for more frequent sampling of gravitational noise in the lake Torgelower See at the approaching of planets. For a legend, see Figure 10.4.7

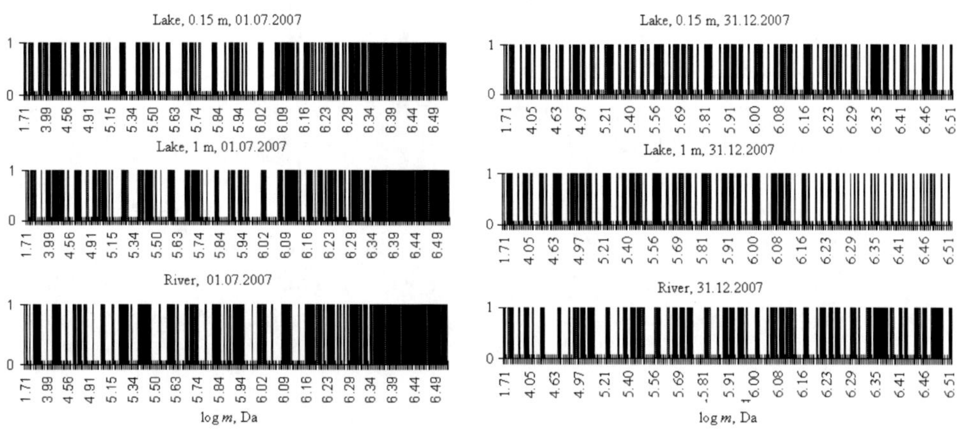

Figure 10.4.9. GMS-spectra of natural waters during summer and winter as bar codes. Summer at 291 K, winter at 275 K, $p < 1$ N/m^2, Zubow constant amounts to $6.4 \cdot 10^{-15}$ N/m.

CONCLUSION

White gravitational noises permeating the Earth have an organizing influence on the molecular structure of natural waters, destabilizing factors are energy flows from the Sun and forced movements of water.

The long-range order in natural waters is constant in small clusters, but varies depending on the temperature at the point of interaction amongst clusters and their surroundings.

Celestial objects influence the stability of cluster ensembles in the lake; they can therefore affect its biocenosis by changing the biochemical processes in which the water clusters are involved.

Natural water reservoirs are found to be applied as a new sensor for the analysis of events occurring in space.

10.5. Rainwater

It has long been known in agriculture that the harvest and its quality depend not only on the amount of used irrigation water, but also on the time of day, weather conditions, and water quality. The importance of these factors has also been observed by the animal breeders. The use of "distasteful" water dramatically reduces the immune response of animals to the action of pathogens. The general condition of the animals and their physiological stability in the herd deteriorate although the chemical composition of water is unchanged. There is no explanation for it. Similar problems are well known in chemical technology, for example, with seasonal changes in the flotation of minerals and the effect of the wall [1]. These influences are suspected to be caused by the differences in the long-range order in natural water. In the literature there is no information on this topic, so it has become the object of the present section.

Figure 10.5.1 shows a sequence of rainwater GMS-spectra, measured in a small sample (micro cells). This is done to understand the changes in the rainwater cluster structure inside the macropores of plowed land. As is visible in the spectra, these are not the signals typical of small water clusters (up to $\log m < 3$) where their absence was recorded even 23 hours after the rain. The changes in the intensities of the signals from medium and large water clusters are noticeable. The strongest signal change ended only 4 h after the rain. On the other hand, in samples of rainwater in large containers (> 100 ml), the signals of small water clusters are reliably detected (Figure 10.5.2). This phenomenon is due to the "wall effect" according to which at small distances from the wall (< 1 cm), the weak gravitational noises from protons jumping into the physical vacuum (§21) in both phases are overlapping. The new superposition of noises in the interphase space forms a new long-range order in water [1] (§10.9).

The practical interests of potato growers are taken into account, and the following experiments will be performed in small measuring cells that simulate the state of water in macropores of plowed land. In Figure 10.5.1, the behavior of the oscillation signal intensities of the clusters containing 100, 178, and 280 water molecules is unusual. These three clusters exist in rainwater (drop) in the collapsed form. While the signal intensity of the cluster-100 changes very little up to 17.5 h after the rain's beginning, the signal intensity of the cluster-178 increases 40-fold, from 0.2% to 8.3% and that of the cluster-280 reaches minimum values first after 1 h and after 17.5 h. It should be remembered that the increase in the signal intensity indicates an individualization of the cluster connected with the decrease of its interaction with the surroundings. In contrast, the signal stability of a cluster means that this is a skeletal cluster. Computer models of these clusters are shown in Figure 2.1.1. Analyzing the long-range order of water in fruits and vegetables and that of irrigation water yielded some differences [2]. It seems that only if the long-range order in the irrigation water and the biomatrix are similar can the quality of plants be increased.

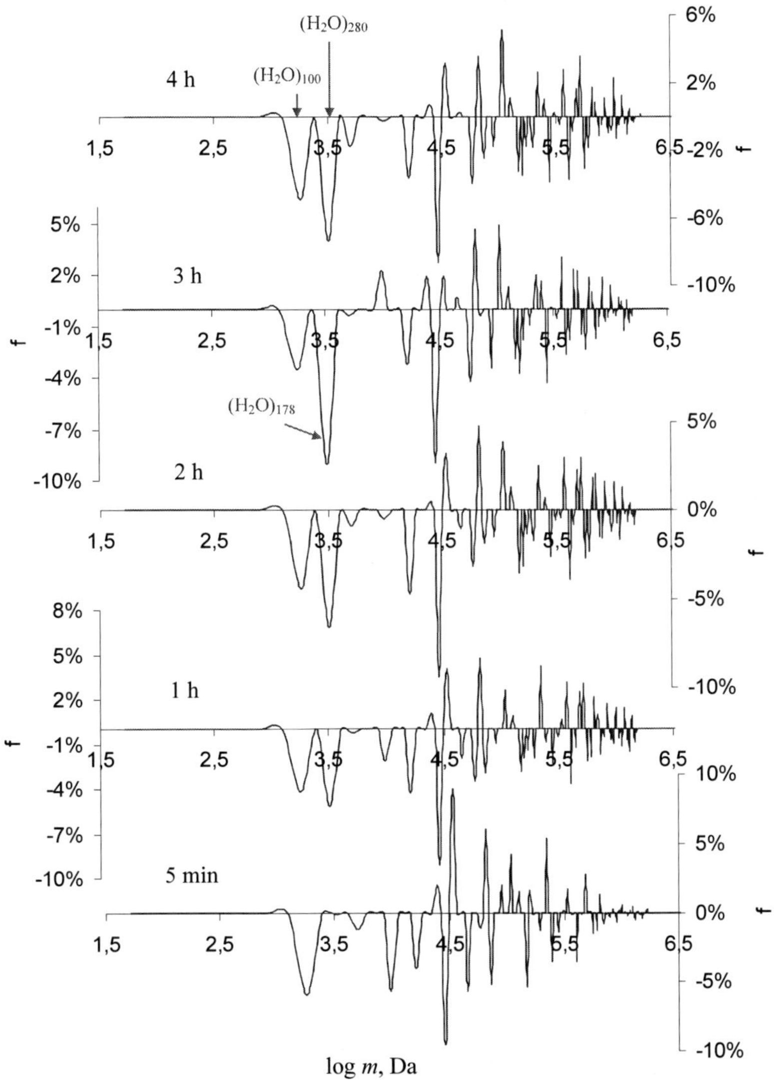

Figure 10.5.1. GMS-spectra of rainwater during aging, $p > 1$ N/m2. Sample volume 0.07 ml. August 30 2003, North Germany.

Table 10.5.1 shows the main parameters of the long-range order of rainwater (Figure 10.5.1) during aging.

Table 10.5.1 shows that the number of cluster kinds in this ensemble of rainwater practically does not change. The sum of the parts of collapsed clusters ($Д_c$), however, is unusually high during the whole period of measurements, and it is twice as high as that of the expanded clusters (100% - $Д_c$). This means that rainwater (melted snow) is enriched with energy that is concentrated in the structures of the collapsed clusters. The average mass of all clusters (M_{GMS}) found in the mass interval from 0 to log m = 6.5 increases in the first hour of aging, which is dramatically favored by increasing the cluster individualization (rise of $|f|$). After 7 h of aging, it reaches a constant value of ~213 kDa. Regarding the cluster with the largest mass (M_{max}), however, no systematic data is found. Taking into account the number of cluster kinds and parts of the collapsed clusters, the data in this table also indicates that the long-range order in rainwater is kept constant for a long time. With regard to the average mass of all the clusters though, we can talk about an aging of the cluster ensemble in the rainwater.

Table 10.5.1. Results of the analysis of rainwater GMS-spectra

t, h	M_{GMS}, Da	M_{max}, Da	$Д_c$, %	N
0.1	180,111	1,666,496	61	50
1	313,467	1,523,524	64	49
2	308,880	1,523,524	66	49
3	278,840	1,523,524	63	49
4	292,441	1,740,798	67	50
7	217,872	1,523,524	65	49
17.5	212,934	1,666,496	62	50
23	212,890	1,523,524	64	49

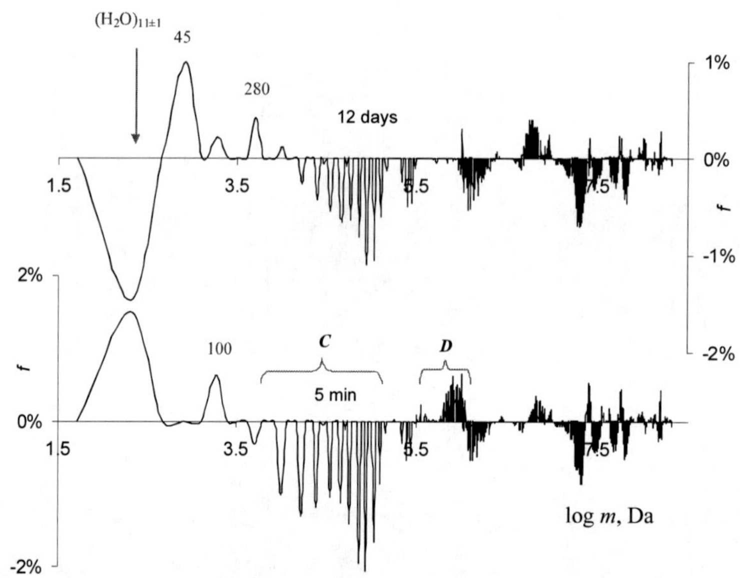

Figure 10.5.2. GMS-spectra of rainwater (July 22, 2007, North Germany) at 295±1 K, $p < 1$ N/m^2, Z = 6.4 · 10^{-15} N/m. Sample volume of 50 ml. It was observed that all clusters formed rows with a common sequence (see the first nine members in Table 10.5.2).

A second sample of rainwater (Figure 10.5.2) reveals not only a different picture of the long-range order, but also the fact that it is characterized by other types of aging of the cluster ensemble. For example, after 12 days of aging, the expanded cluster $(H_2O)_{11\pm1}$ receives permission to occupy a place in the new ensemble, although only in the collapsed form. The cluster of 280 molecules may be in the new ensemble only in the expanded form. Moreover, and quite unexpected, a new cluster ($(H_2O)_{45}$) appears. Significant changes in the oscillation intervals ***C*** and ***D*** are visible. The aging process is assumed to be connected with the cluster ensemble's slow approach to a new thermodynamic equilibrium.

The redistribution of the energy states of the clusters in their ensemble requires considerable time for small clusters, but for large clusters ($\log m > 5.5$), it needs even more time. This process seems to depend on the redistribution in small clusters. Thus, these results confirm the well-known phenomenon of water memory.

Table 10.5.2. Mass concentrations and their oscillation frequencies in liquids, polymers, and biomolecules

N	**m, kDa**	**kHz**
1	0.2±0.02	22.2
2	1.8± 0.1	7.4
3	3.1± 0.1	5.5
4	4.9±0.5	4.4
5	7.0±0.5	3.7
6	9.5±0.5	3.2
7	12.5±0.5	2.8
8	15.7±0.5	2.5
9	26±3	2.0

This row is well described by the following relation:

$$m = 268.19 \cdot N^{2.0576}$$

where *N* is the order number of a cluster in the spectrum, beginning with the basic cluster of 0.2 kDa, e.g. $(H_2O)_{11\pm 1}$ for water.

CONCLUSION

Gravitation mass spectroscopy is concluded to be applied for the determination of the long-range order in rainwater.

Fresh rainwater is enriched with collapsed, energy-rich clusters. Water clusters are formed in rainwater forcibly through stationary gravitational waves of white noises penetrating our planet.

REFERENCES TO 10.5

[1] Zubow A. V., Zubow K.V. and Zubow V.A. The Phenomenon of Water Clusters Distribution in Water and Solvated Clusters of Ion Pairs of NaCl in Solution Near the

Wall. The Wall Effect. Journal of chemical industry (Sankt Peterburg). 2006, vol. 83, no. 6, pp. 300-306.
[2] Zubow A. V., Zubow K.V., Zubow V.A. Investigation of the distribution of water clusters in vegetables, fruits and natural waters by flicker noise spectroscopy. Biophysics. 2007, vol. 52, no. 4, pp. 585-592.

10.6. Properties of Molecular Clusters in Water

10.6.1. Celestial Body Influences

Figure 10.6.1.1 shows how planets remove energy from a water sample at their culmination time. Thus, the Moon was found to strongly influence the long-range order in water, especially upon achieving a full moon (Figure 10.6.1.1 and 10.6.1.2). In addition, one can see how the Moon and the GR signal of Saturn overlap. On the other side, when the Moon disappeared, the GR signals of a "black hole" from the galaxy's center increased [1, 2].

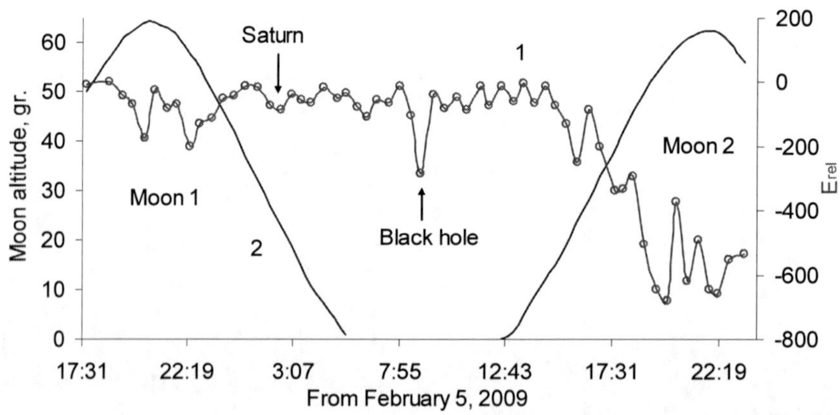

Figure 10.6.1.1. Drain of WGR energy from water (1) caused by the planets at their culmination between two Moon culminations, four (Moon 1) to three (Moon 2) days to the full moon on February 9, 2009 in Northeast Germany (2, Moon altitude), 290 K.

On curve 1, some signals regarding the drainage of gravitational energy from water by celestial bodies that lie on a line with the Earth disappeared. Their influence on water seems to be compensated in this constellation.

As shown in Figure 10.6.1.2, the ratio of collapsed clusters in water is very sensitive to energy drain from the planets. It already reaches a stable maximum when the Moon approaches its culmination.

In Figure 10.6.1.3, the behavior of a water cluster ensemble at full moon on February 9, 2009 is visible both in the number of cluster kinds and the average molecular mass of the cluster ensemble. They are strongly influenced by the GR of the Moon. Furthermore, the GR effects on the part of Saturn and the "black hole" at the galaxy's center were observed. Additionally, the GR of the Moon was found to overlap the GR from other planets. The two strong signals near to that of the "black hole" (sloping arrows, Figure 10.6.1.2) could be assigned to the GR of cosmic mass concentrations forming a ring around the "black hole." These three signals remind us of the Moon's influence, which occurred three days before the full moon. One may remark that in our galaxy center up to 20,000 small black holes with masses similar to that of our Sun are concentrated so that the black holes move further in the direction of the galaxy's center [1]. Consequently, an unstable GR signal from the Galaxy center should be expected; that is connected with unexpected changes in the long-range order of molecular matter, leading to new unknown risks [3].

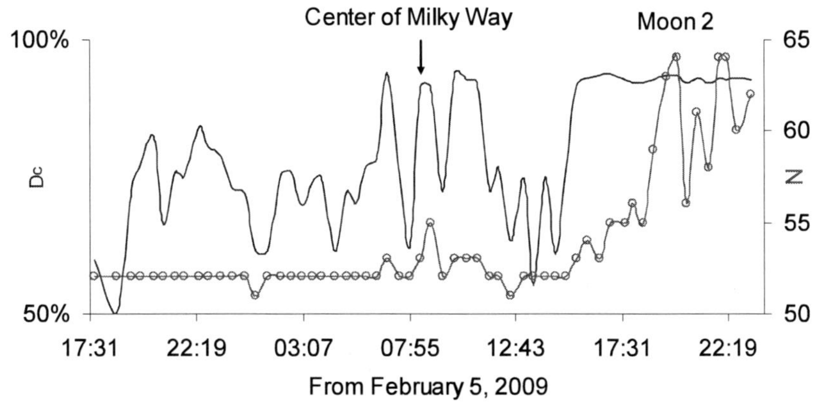

Figure 10.6.1.2. Change of part of the collapsed clusters (D_c) and number of cluster kinds (N, line with points) in water during observation time. Cluster ensemble up to 2 million Dalton (Figure 10.6.1.1).

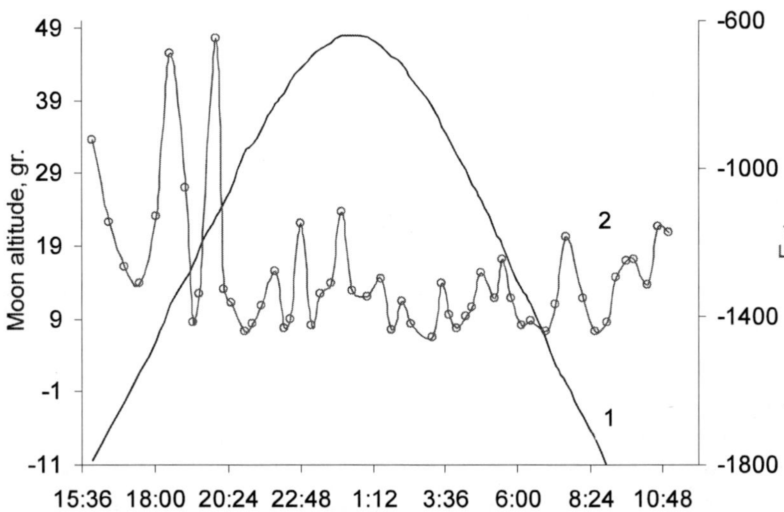

Figure 10.6.1.3. Resonance of water's WGR on the Moon's GR at full moon. 1 – Moon altitude (degrees) and 2 – resonance signals.

The GR of a celestial body influences the cluster ensemble energy in polystyrene, polyethylene, casein, collagen, starch, and water in vegetable and animal biomatrices as well as in salt solutions, liquid monomers (styrene), and even in glasses (melted quartz) [4], Figure 10.6.1.4.

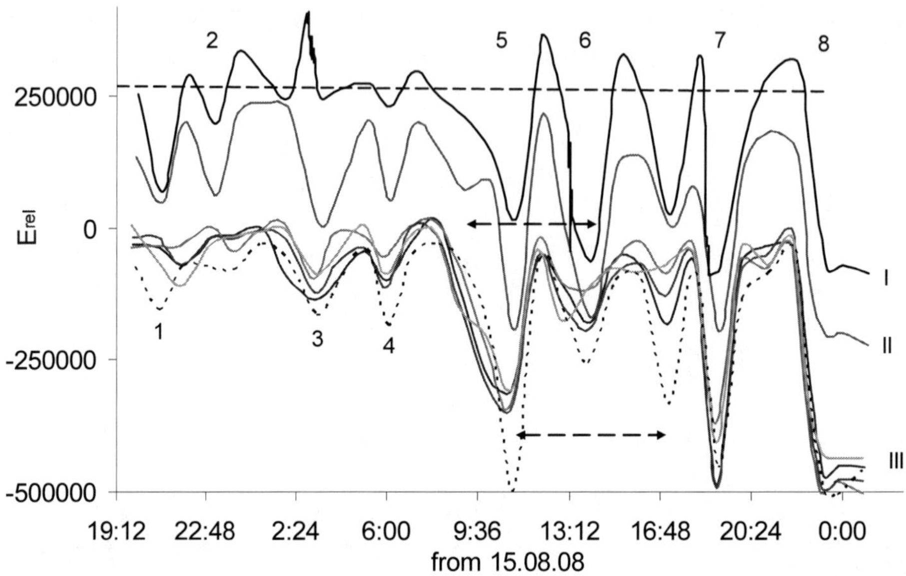

Figure 10.6.1.4. Energy flows in different substances: I – biomatrix of fresh potato tuber; II – melted quartz (glass); III – water, methanol, biomatrix of a hen's egg (egg white), polymers (dotted line), water solution of 3.5 wt. % NaCl; 1 – sundown, 2 – full moon before the Moon was covered by the Earth 24 h later (8); 3 – Uranus's culmination; 4 – sunrise; 5 – Mars's culmination; 6 – nearly simultaneous culminations of Mercury, Venus, and Saturn; 7 – Jupiter's culmination; and 8 – moon eclipse. Horizontal arrows denote the effects of interference of gravitational noises from the Sun (dotted line: from 09:00 until 15:00) and galaxy center (dotted line: from 11:00 until 17:00). The curves are built from points, which did not have one statistical value worse than 95%. Base line – horizontal line, 265,784 relative units (stated on 19:12, August 15, 2008).

On February 9, 2009, the full moon was less distinctive than on August 16, 2008. Therefore, the energy drain was also lower (300-500 times). Comparing Figures 10.6.1.3 and 10.6.1.4, it was found that if the long-range order was strongly destroyed by the Moon's GR, then the influence of the other planets shall also be great. Here we see a new spectrum of risks not only for chemical processes but for physical ones, too, e.g. in administration systems (computer networks).

The LRO in the samples was carried out at full moon in summer 2008 and winter 2009 in the northeast of Germany. Samples were put into a metallic box, which was maximally protected from light, anthropogenic noises, and heat and mechanical fields (Figure 4.0.1). In Figure 10.6.1.4, some energy curves reflecting energy flows (sum of the cleaned signals) in the samples shall be given. At full moon on August 16, 2008 the planets and constellations were in a unique formation and the moon was 80% in the Earth's shade. In order to acquire the base line, one has to do the following. It is necessary to register white noises at the sensor in a vacuum (10^{-4} N/m^2) that is maximally isolated from outer energy flows and extract the real signals. 265,784 relative units were calculated as a basic line. For this case, it represents the sum of all real EC signals for a hypothetical mass ensemble up to 4.3 billion Da. Every measurement lasted for 30 s.

In the next figure (Figure 10.6.1.5), it shall be shown how planets interacted with the LRO in water at full moon on February 9, 2009. Compared with Figure 10.6.1.2, the behavior of the curve shows a higher sensitivity to the events in the sky. The planets were observed to influence cluster groups in water differently. This shall open new risks as well as new opportunities for the chemical industry. The energy drain from the water cluster ensembles by the Moon's GR proceeds very specifically (Figures 10.6.1.2 and 10.6.1.3); first, the long-range order is quickly destroyed and then it is restored just as quickly. This means the energy platform quickly adapts to the new gravitational field (Figure 3.2.1) and the LRO of water clusters parallels the changes.

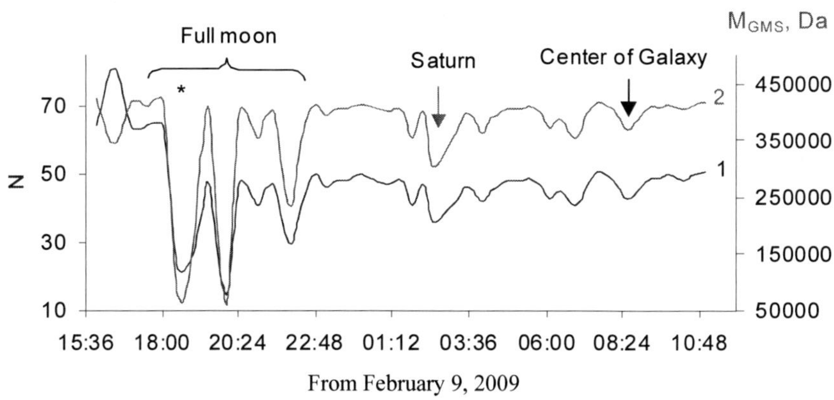

Figure 10.6.1.5. The behavior of the number of cluster kinds (1) and of the average cluster mass (2) in water at the culminations of the Moon, Saturn, and a "black hole" in Sagittarius.

The GR of celestial bodies was concluded to remove energy from hydrogen bonds and "switch off" the proton jumping process at the planet's culmination [5], §21. Switching off leads to a weakening of the energy content of the cluster ensemble as a whole and even to its disintegration. This phenomenon is of importance for the fine nano-technology of synthetic polymers, especially for the bio-technology of natural polymers, proteins, and polysaccharides. As we concluded above, the "black hole" is removing energy from the molecular matter samples in the direction of the galaxy's center. This center does not agree with that found in the literature [1]. For the observation of the "black hole," the literature describes the application of X-rays that do not react with mass concentrations in space. In contrast, the GR was found to be very sensitive to mass concentrations that probably cause the difference in the "black hole" position (§23).

Risks arise because of the destabilization of technology as well as of the bio-chemical processes in biomatrices connected with changes in the biological behavior (concentration loss, psychological instability, illness). Celestial bodies disintegrate water cluster ensembles (Figure 10.6.1.5) and remove energy from EC (Figure 3.2.1) into the water sample in the region of large clusters (shown as broad arrow). This cluster region is typical for sub-micelle and micelle structures in proteins, polymers, and solutions [4, 6]. Temporary disintegration of cluster ensembles causes important changes in the long-range order of water and finally in all chemical processes with water participation.

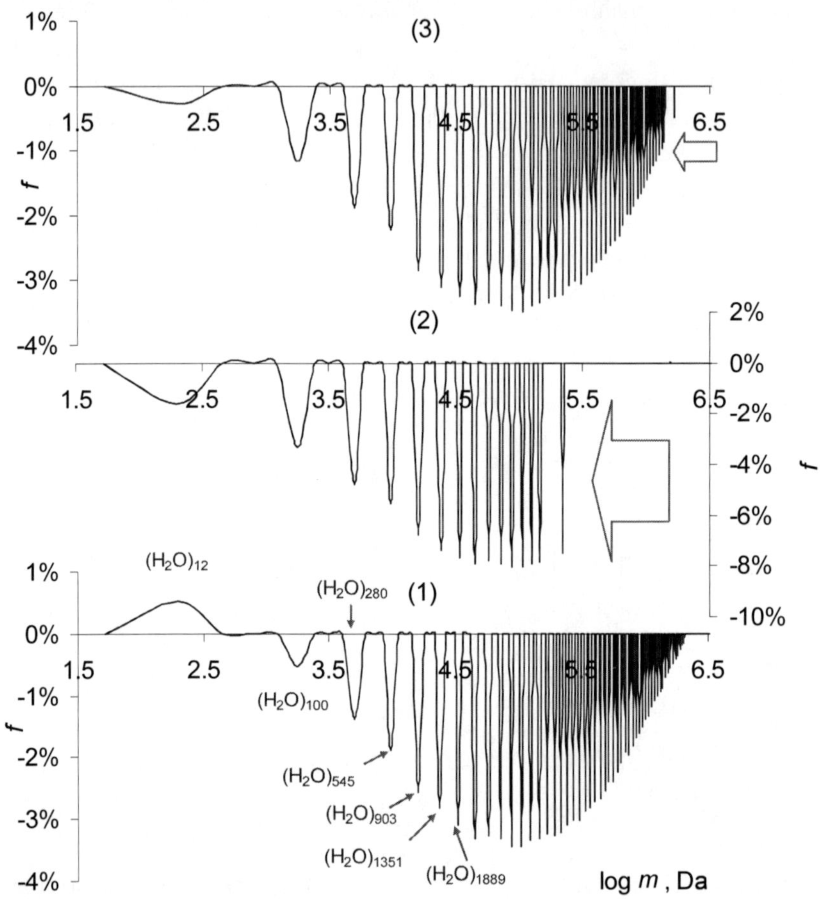

Figure 10.6.1.6. GMS of cluster ensembles (LRO) in distilled water (full moon period). Weak shock waves, $p < 1\mathrm{N/m}^2$, $Z = 6.4 \cdot 10^{-15}$ N/m. 1 – state before the Moon's culmination and after the culmination of the "black hole" (Figure 10.6.5, at 15:45), 2 – at the Moon's culminating moment (Figure 10.6.5, at 18:30, drawn with star), and 3 – at the moment of the "black hole's" culmination (Figure 10.6.1.5). For cluster structure models of water, see Figure 2.1.1. *f* – energy part of one cluster in the whole ensemble.

As is visible in Figure 10.6.1.6, these GMS of water are very different from the GMS measured earlier for other water polymorphs or in other times [7], §5. In our earlier works [4, 7], the LRO of water was concluded to be highly dependent on weak energy flows (Figure 10.1.1) that destroy the ensemble structure of EC (Figure 3.2.1). Therefore, water is a unique sensor for the energy state of an EC ensemble; it can be used to measure gravitational field excitations. In Figure 10.6.1.6, we see some GMS of distilled water at the culmination of selected celestial bodies. The state of the baryonic protons in hydrogen bridges of water, determining the inter cluster interaction, is assumed to be influenced by the neutrino flow to/from/to celestial bodies (§24), e.g. certain protons can be solved in the neutrino flow (like solvent) though, but this process is reversible and in the absence of it, they condense again to baryonic form.

10.6.2. Microgravity Influence

Microgravity strongly influences water cluster ensembles. Figure 10.6.2.1 and Table 10.6.2.1 show how the white noise in the water sample changes during its free fall and how the LRO of water varies.

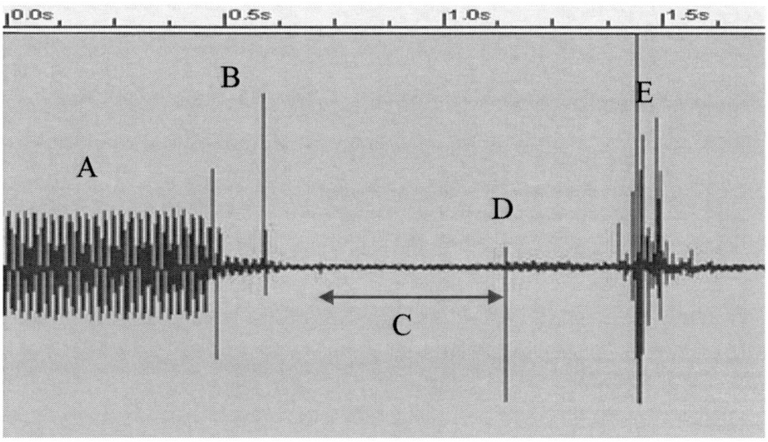

Figure 10.6.2.1. Spectrum of white noises in water (500 ml) aged for six months, at its free fall from 4.25 m height. Polyethylene vessel in grounded aluminum foil. A – white noises before the fall, B – acceleration, C – free fall (0.43 s), D – vessel hits the cement ground, and E – sensor destruction At 293 K, $p < 1$ N/m^2.

Table 10.6.2.1. Some LRO characteristics in water before and during its free fall (Figure 10.6.2.1). E_{rel} – relative energy of a water ensemble in the vessel.

Water state	M_{GMS}, Da	N	E_{rel}	f, 280H_2O	f, 1889H_2O
A	559,016	71	109	2%	4%
C	491,354	65	269	2%	2%

As is visible, the microgravity is strongly influencing the main cluster ensemble parameters while the Chaplin water clusters with 280 and 1,889 molecules are nearly unchanged. It has to be mentioned that the main events proceed in small clusters, where new clusters emerge and old ones disappear. Furthermore, to investigate the microgravity influence on molecular cluster ensembles, it is reasonable to apply the position of the GR-proton balance with their activated complex, §24.

CONCLUSION

The long-range order in water at cluster level is sensitive to gravitational radiation from celestial bodies.

Water cluster ensembles react to microgravity.

References to 10.6.

[1] Faustmann C. In Schwarze Löcher. Rätselhafte Phänomene im Weltall. 1 Auflage. Seifert Verlag, Wien, 2008. P. 187.
[2] Perkins D.H. Particle Astrophysics. Second Edition. Oxford University Press 2009. pp. 294- 297.
[3] Poisson, Eric; Sasaki, Misao. Physical Review D: Particles and Fields. 1995, vol. 51, no. 10, pp. 5753-5767.
[4] K.V. Zubow, A.V. Zubow, V.A. Zubow. Principles of gravitation spectroscopy. New form of molecular matter. Processes. Fields. Aist Handels- und consulting GmbH, dep. R&D, Berlin, 2010, electronic book (www.zubow.de), p. 854. (in Russian).
[5] Zubow K., Zubow A.V., Zubow V.A. The Phenomenon of Proton Dissolving in Vacuum and of Proton Condensation from Vacuum. Two Forms of Protons, Structure of Nuclei, Electrons and Atoms. J. of modern physics, 2010, vol.1, no.1, pp.175-184.
[6] Zubow K. V, Zubow A.V, Zubow V.A. Use of flicker noise spectroscopy for undestroed analysis of nano structures. Zavodskaja Laboratoria. Diagnostics of Material. 2008, vol. 74, no. 9, pp. 40-46 (in Russian).
[7] Zubow Kr. V., Zubow A.V., Zubow V.A. Cluster structure of liquid alcohols, water and n- Hexane. Journal of Appl. Spectr. 2005, vol.72, no. 3, pp. 321-328.

10.7. Water Cluster Structure in Mucosa of the Garden Slug and the Large Garden Snail

The development of bionics in the area of new artificial coating materials is not possible without understanding the skin structure of biological organisms. The best modern antifouling coatings cannot achieve the protection level of their natural analogs. For instance, spine-free animals belong to these analogs. The Garden slug and Large Garden snail are characterized to be coated by a unique gel skin, which protects them from enemies and drying out. Information on the skin biomatrix structure of these animals was not found in the literature by the authors; this refers particularly to the water state and water cluster structure in the skin hydrogel. Additionally, for example, it is unknown why hydrophilic sand does not stick to the "hydrophilic" surface of these animals. On the other side the Garden slug and Large Garden snail stick well to our hands and their slime is difficult to remove from our hands.

Using the Zubow gravitation mass spectroscopy (earlier referred to as "flicker noise spectroscopy,"FNS) [1, 2], a possible answers to these questions can be given.

The aim of the present chapter is to investigate the long-range order in the skin biomatrices of the Garden slug and Large Garden snail through GMS.

To get information on the LRO of water in the mucosa biomatrix, the Garden slug was analyzed by GMS (Figure 10.7.1). As is visible from Figure 10.7.1, the GMS-spectra for both the Garden slug spine and foot are nearly identical in the area of the small and middle clusters while there are some differences for the large oscillators ($\log m > 4.5$). Up to a scanning depth of 2-3 mm, the main LRO characteristics in the mucosa spine and foot are similar. The integral cluster distribution (W_x) has been described by:

For the spine:

$$W_x = 0.1062 \cdot \ln(m \cdot 10^{-5}) + 0.744$$

For the foot:

$$W_x = 0.1088 \cdot \ln(m \cdot 10^{-5}) + 0.733$$

Where $x = m \cdot 10^{-5}$, Da

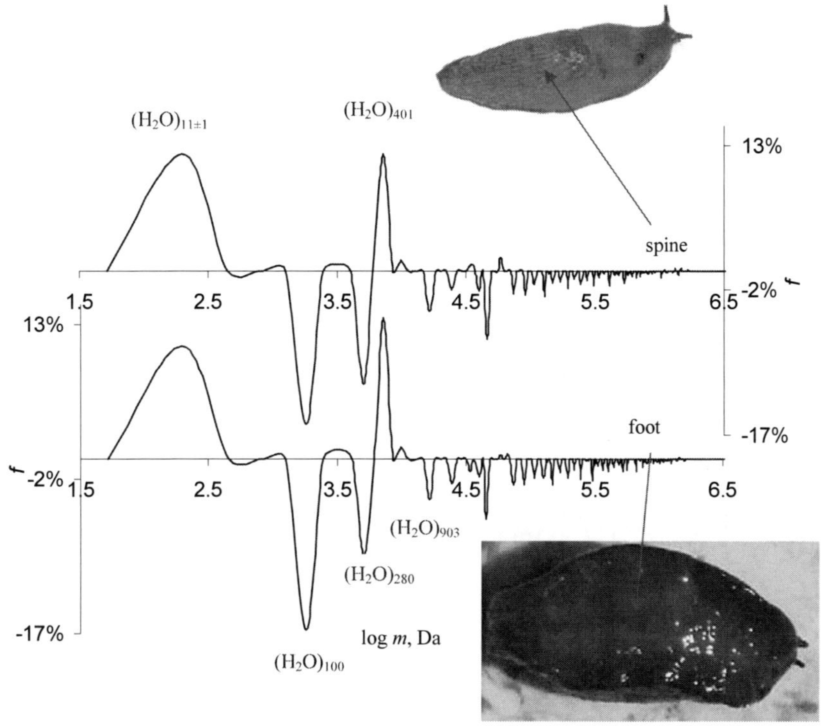

Figure 10.7.1. GMS-spectra of the Garden slug (*Arion vulgaris*) spine and foot at 293 K. Spine: M_{GMS} = 112,987 Da, part of collapsed clusters = 77%, and N = 52. Foot: M_{GMS} = 119,715 Da, part of collapsed clusters = 72%, and N = 54. July 22, 2007, atmospheric humidity = 85%, rain. The places where the GMS-sensor was installed are marked by arrows. A cluster ensemble with a mass up to 2.3 million Da has been analyzed, $p > 1$ N/m^2, the Zubow constant is equal to $6.4 \cdot 10^{-15}$ N/m. The cluster masses have been converted to the numbers of water molecules (shown against the signals). The water content in the slime is about 98%.

Additionally, the super cluster structure in the Garden slug mucosa has been analyzed using line spectra (Figure 10.7.2). In Figure 10.7.2, it is visible that there are some differences in the super cluster structure of the mucosa biomatrix of the Garden slug foot and spine. The cluster distribution (line density) in the mucosa of the foot differs from that of the spine (marked with an arrow and a horizontal bracket). The mucosa biomatrix of the Garden slug was concluded to consist mainly of water in the cluster form that is integrated in polysaccharide hydrogel. It has to be mentioned that at a polysaccharide concentration lower than 2 wt. %, the gel is characterized to be mechanically stable.

As is shown in Figure 10.7.1, the water clusters of 100 and 280 molecules are in the collapsed form while the base water cluster (12 molecules [3]) is in the expanded one. An optimal LRO for a water cluster ensemble in mucosa seems to be found through the unique selection instruments of nature. The harmony of collapsed and expanded clusters created here makes it possible for the Garden slug to not stick to the sand, to not dry out, and to always remain hydrophilic. The spectra of Figures 10.7.1 and 10.7.2 were obtained under strong shock waves, which must be felt by the animal with the consequence of defending itself by reacting on the structure of the mucosa hydrogel. Usually the Garden slug behaves in this way if it is attacked by its enemy, for example, at a manual collection during agricultural works. Furthermore, the GMS-spectra were influenced by changing atmospheric pressure or weather, e.g. hurricanes and droughts. A practical interest for this phenomenon could be found in industry, e.g. for the production of new coating materials.

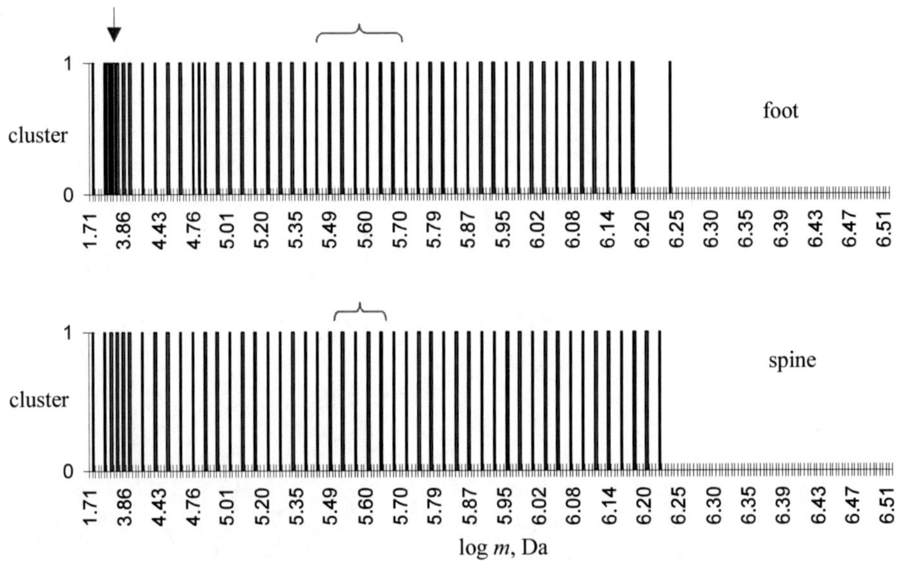

Figure 10.7.2. Line GMS-spectra of the Garden slug mucosa (see Figure 10.7.1).

The general GMS-spectrum of the foot from the Large Garden snail under weak shock waves is shown in Figure 10.7.3. As is visible, the spectrum of the Large garden snail greatly differs from that of the Garden slug (Figure 10.7.1), especially with regard to the content of the collapsed cluster signals with masses greater than log m = 5.5. These signals reflect the cluster structure of the foot biomatrices of the animal. Furthermore, all water clusters were analyzed to be in the expanded form (+*f*); here the animal was in a non-excited state.

It has to be mentioned that this spectrum was recorded under very weak shock waves that neither excite the animal nor destroy the LRO of the mucosa hydrogel. Similar spectra were found for liquids (solutions) and hydrogels (agarose) at long storage and under minimal influences on the side of outer energy fields (temperature changes, light, and mechanical and seismic waves) as well as for natural polymers (casein, amylopectin, collagen [4]). It stands out there is an area between 7.72 < log m > 7.47 (marked with a broad arrow) where there is not any signal at all; sometimes this "window" was found for other substances, too, e.g. rubber, synthetic materials, salt, and acid solutions. This fact we ascribed to an unstable behavior of super molecular structures at the level of cluster associates [4, 5].

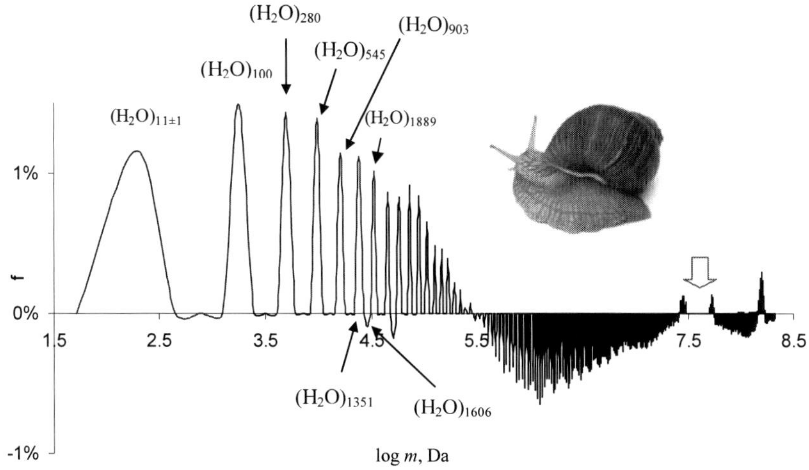

Figure 10.7.3. GMS-spectrum of the foot mucosa of the Large Garden snail (*Helix pomatia*). Cluster ensemble up to 200 million Da, $p < 1$ N/m^2, $M_{GMS} = 53{,}456{,}898$ Da, part of collapsed clusters = 73%, $N = 1{,}171$, Zubow constant amounts to $6.4 \cdot 10^{-15}$N/m at 291 K. May 3, 2008, 20:36 CET. Weak touch of the foot surface by the GMS-sensor.

As observed earlier, the LRO of water in the mucosa hydrogel reacts to the material of the surface on which the animal moves [4] as well as to changes of atmospheric pressure, lights, and temperature. Furthermore, mechanical excitations, chemical substances (water solutions of NaCl and acetic acid), and the planets [5] were found to influence the LRO of water.

The following scheme (Figure 10.7.4) gives an idea of how LRO has been influenced by the surface material on which the animal moves. According to this scheme, the cluster change between the expanded and collapsed forms permits the animal to monitor its surroundings and to find food. Expanded clusters actively interact with their surroundings with the help of hydrogen bonds. In contrast, clusters in the collapsed form are inactive because the hydrogen bonds are directed to the inside of the cluster.

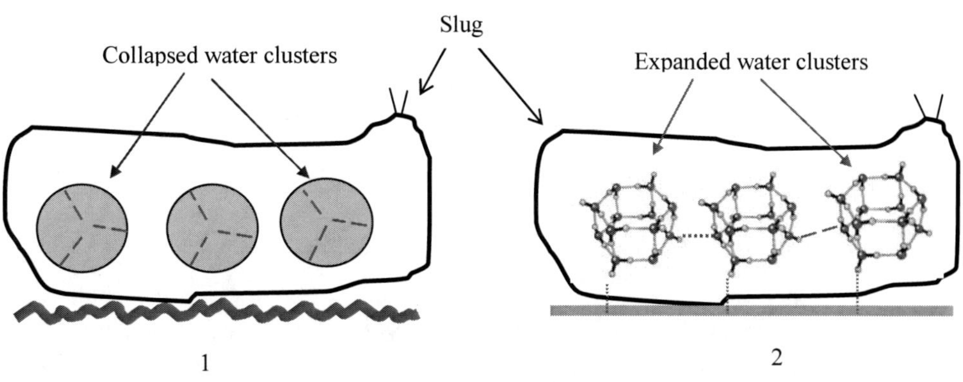

Figure 10.7.4. Model: biomatrix reaction to the excitement from surfaces. 1 – surface that excites the biomatrix (water clusters in hydrogel are in collapsed form) and 2 – surface that does not excite the biomatrix (water clusters in hydrogel are in expanded form, e.g. base water cluster). The hydrogen bonds are shown as dotted lines.

The change of the energy state in one water cluster ensemble seems to have an influence on the cluster ensembles in proteins and polysaccharides of the mucosa biomatrix, which leads to changed biopolymer conformations [6], §14. The nervous system uses this behavior for surface monitoring and administration of the biomatrix LRO.

Changed surroundings induce the animal to change its cluster structure, e.g. the cluster of 280 water molecules (scheme in Figure 2.1.1 that was made available by Professor M. Chaplin [7]), as well as to change the LRO in the biomatrix area that is in contact with the outside exciter.

Our experiments to model the LRO in the mucosa hydrogel of the Large garden snail with the help of agarose hydrogel (0.7 wt. % agarose) prepared with deionized, distilled rainwater yielded little differences in the GMS-spectra ($p < 1$ N/m^2). The spectra are only different from each other in the mass area > log $m = 5.5$ (Figure 10.7.3).

The correlation between the cluster oscillation frequency (ω, Hz) and cluster mass (*m*, Da) in the mucosa hydrogel of both animals can satisfactorily be described with the Zubow equation, §3.2.1.

Table 10.7.1. Mass clusters and their frequencies observed in the mucosa (comments in the text).

N	m, kDa	ω, kHz
1	0.2±0.02	22.2±0.1
2	1.8± 0.1	7.4±0.1
3	3.1± 0.1	5.5±0.1
4	4.9±0.5	4.4±0.1
5	7.0±0.5	3.7±0.1
6	9.5±0.5	3.2±0.1
7	12.5±0.5	2.8±0.1
8	15.7±0.5	2.5±0.1
9	26±3	2.0±0.1

$$m = 10^{11} \cdot \omega^{-2}$$

This equation is applicable for the complete molecular matter [5]. The masses and frequencies for the first nine clusters in the mucosa hydrogel are given in Table 10.7.1.

The relationship between cluster number and cluster mass was found to be well described by:

$$m = 268.19 \cdot N^{2.058}$$

where *N* is the order number of a cluster in the spectrum, beginning with the smallest cluster (0.2 kDa). This sequence builds in stationary gravitational waves of white noises that penetrate our Earth [5] and it shall be destroyed if the sample to be analyzed is near a wall (wall effect, §10.9) or if it is protected from white noise flow coming from clusters of different phases [8]. The sequence shall be destroyed, too, when the white noise in the sample is transformed into color noise, e.g. at changing temperature or pressure.

Conclusion

Water in the mucosa biomatrix of the Garden slug and the Large garden snail was observed to be in a cluster structure.

The animals steer the mucosa properties by influencing the long-range order of water in the hydrogel.

Thus, the water cluster structure in the biomatrix has to be understood as a unique sensor while analyzing outer events or while communicating with surroundings.

References to 10.7

[1] Zubow, K., Zubow, A., Zubow. V. Cluster structure of liquid alcohols, water and n-hexane. Journal of Applied. Spectroscopy, 2005, vol. 72, no.3, pp. 321-328.

[2] Zubow A.V, Zubow Kr. V, Zubow V.A. Investigation of the distribution of water clusters in vegetables, fruits and natural waters by flicker noise spectroscopy. Biofizika. 2007, vol. 52, pp. 585- 592 (in Russian).

[3] Bogdanov E. V, Mantrova G.M. Supermoleculare structures of water. Milimeter waves in biology and medicine, 2006, vol. 44, pp. 25-40 (in Russian).

[4] Zubow K.V., Zubow A.V, Zubow V.A. Principles of gravitation spectroscopy. New form of molecular matter. Processes. Fields". Aist Handels- und consulting GmbH, dep. R&D, Berlin, 2010, electronic book (www.zubow.de), p. 854. (in Russian).

[5] Zubow K.V., Zubow A.V., Zubow V.A. Ensemble of Clusters – New Form of Molecular Matter, Risks and Chances. Zubow Equations. In Advances in Chemistry Research, vol. 5, 2010, pp. 107- 145. Editor J.C. Taylor. Novapublisher. NY. USA.

[6] Zubow K., Zubow A.V., Zubow V.A. Water Clusters in Plants. Fast Channel Plant Communications. Planet Influence. Journal of Biophysics Chemistry. 2010, vol. 1, pp. 1-11.

[7] Chaplin M. (South Bank University of London, 2006) in http://www.lsbu.ac.uk/water/index.html

[8] Zubow K., Zubow A.V., Zubow V.A. Molecular Cluster Distribution in Sea Water Near the Wall. Antifouling aspect, Journal of Clusters, 2011, vol. 22, no. 4, pp. 603-619.

10.8. Water Clusters: The Role of the Surroundings

The mixing of different liquids should not lead to the formation of homogeneous systems—the so-called ideal solution—because of the cohesive and structural factors, the crucial role of which, unfortunately, is ignored by many researchers. If even simple liquids are considered to be inhomogeneous (long-range order at cluster level), then the mixtures of different miscible liquids are even more so. On the other hand, we can expect that heating will lead to the dispersion of the mixture of liquids and the cohesive and structural factors coupled with the positive influence of white noise will lead to the formation of a new long-range order. The mixtures are to be understood as nano colloidal systems where the nanoparticles of different liquids are not miscible with each other, although a phase boundary cannot be seen with the classical energy-rich methods. Some simple water clusters (up to hexamers, 3,329 cm^{-1}) in neon, however, were already recorded with IR spectroscopy by Hibarayashi and Yamada [1]. Nevertheless, a new non-destructive method for analyzing the long-range order in mixtures of liquids is necessary.

Understanding the liquids' states in mixtures will open new fields in nanotechnology, optimize the known processes in chemistry (drying, combustion, polymerization), model biological systems, and develop the remote communication of liquids.

In this section, we decided to study the long-range order in fuels as the most promising topic.

To get an idea of the long-range order in liquid fuels is of great economic importance in terms of optimizing combustion processes, dissolution, film formation, and even oil exploration. Scientific interest is dictated by the need to understand the formation mechanisms of clusters, their physical chemical properties as kinetically independent units, and the cluster distribution in a liquid. Furthermore, it is important to investigate cluster surfaces and their gravitational fields. The coming energy problems force us to search for new ways to use oil rationally, especially when it will be burned. Combustion engines have reached their limit and new engineering ideas cannot significantly reduce fuel consumption. In recent years, however, a number of patents have been applied concerning the possibility of reducing fuel consumption by adding water to it and dispersing the emulsion prior to injection into the cylinder combustion chamber [2, 3, 4, 5]. Literature published on this topic is being simultaneously researched. For instance, to Fu et al., [6] gasolines investigated with gas chromatography are characterized by a cluster structure (of up to 20 different clusters); different types of gasoline are even identified with the cluster structure. After passing through a porous ceramic membrane or with treatment from an electrical current, the water-gasoline mixture forms "ultra disperse" clusters.

Modeling the water cluster formation in n-hexane at 300 K, Mudzhikova et al. [7] found that three cluster types consisting of 27, 64, and 125 molecules are possible.

The cluster formation energy is found to be really low [8, 9, 10] where the cluster stability is ensured by the entropy factor.

Applying traditional methods of physical chemistry, it is difficult to detect and identify clusters [1, 9, 11] but computer simulations make it possible to calculate hypothetical structures with minimum energy.

For a better understanding of the material in this section, Figure 2.1.1 shows some modeled clusters of elementary liquids (see [12] ,too) and Figure 10.8.1 provides a review GMS-spectra of some liquid hydrocarbons. As investigative objects, *n*-hexane ($n^D{}_{20}$ = 1.3752), diesel (DIN EN 228), gasoline (ROZ DIN EN 98, FRG standards), and mixtures of liquids are used. The fuel mixtures are prepared at 293 K by simple mixing and the hydrocarbons' saturation with distilled water is prepared by shaking and following the decantation of the upper fraction. Samples are placed in a box that is maximally protected from outer energy flows [13], Figure 4.0.1. The measurement method is described in [8, 12, 13]. Before the measurement, the samples are stored for a week under the exclusion of light and mechanical fields. The GMS-sensor is directly placed inside the sample (10 ml). The negative signal in the GMS (*-f*) represents the energy release of a collapsed cluster leading to its higher density and crystallization. On the other side, the positive signal (*f*) indicates the energy absorption at the interaction of the shock wave with the expanded cluster and its melting. The values of the collapsed clusters are therefore described with a minus and that of expanded one with plus. For more information, see in [12].

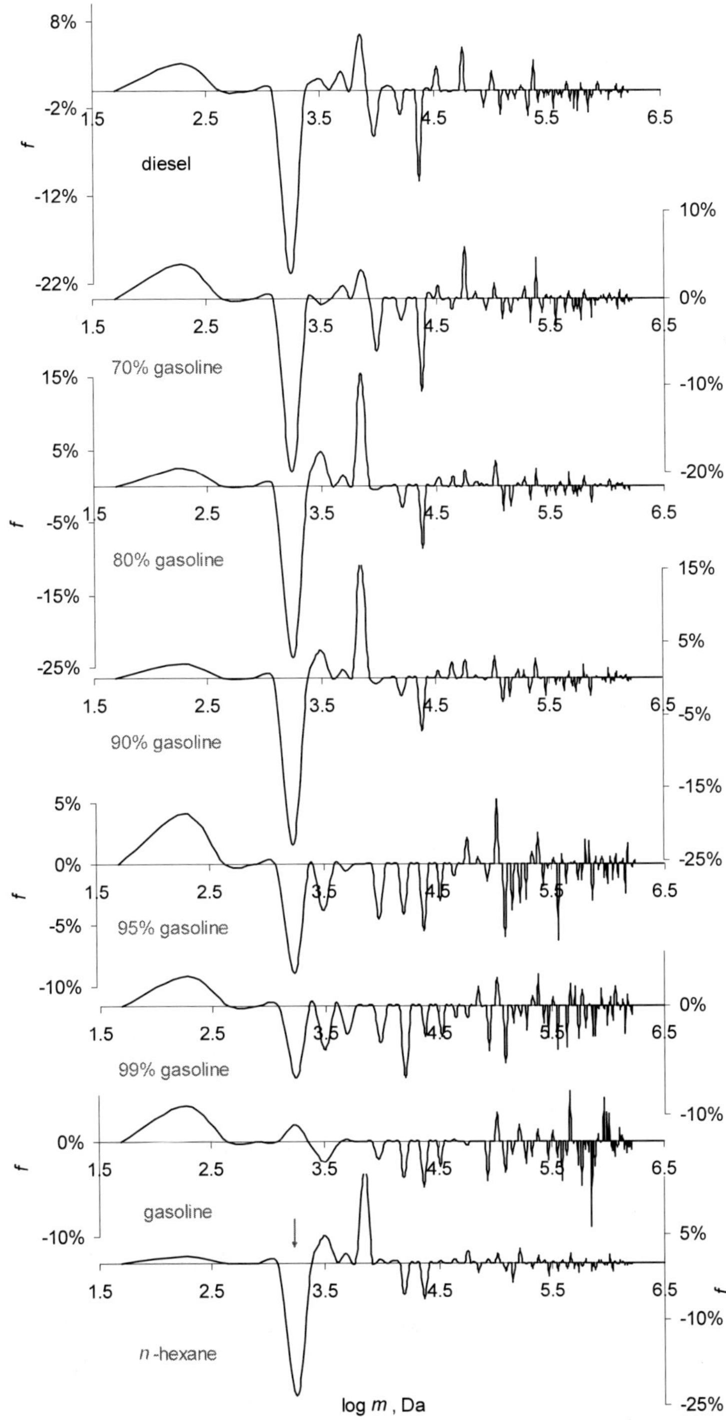

Figure 10.8.1. Review GMS-spectra of hydrocarbons at 286 K. Cluster ensembles up to 3 million Daltons, $p > 1$ N/m^2. Diesel and its mixtures with gasoline, vol. %.

According to Figure 10.8.1, the GMS-spectra of hydrocarbons and their mixtures are similar, however, there are some significant differences. For example, the cluster of 20 molecules oscillates in *n*-hexane [8] (labeled with an arrow) as a collapsed cluster while it is shown as expanded one in gasoline.

When 1 vol. % diesel is added to gasoline, then the GMS-spectrum changes dramatically. The strong change in the spectrum continues with the further addition of diesel. At a diesel content of 10 to 30 vol. %, the distribution of the clusters and their interaction with the surroundings approaches that of pure *n*-hexane.

There is not enough information about clusters formed by organic compounds [8, 9, 14] in the literature. Generally, the simplest molecule clusters, such as benzene dimers [15], complexes of water with acetone [15, 16], and other organic compounds with polar groups [17] were studied.

As shown in Figure 10.8.1, mixtures of organic compounds are characterized by their own long-range order at the level of cluster formation, i.e. they are also nano emulsions. Therefore, the entire range of clusters is of interest. If water is added to gasoline or *n*-hexane (Figure 10.8.2), then there is a harmonization of the cluster ensembles in the small and medium-sized clusters' range (up to $\log m < 3.7$). In this mass range, only the LRO in diesel changed but it did so marginally. Thus water affects the long-range order of light fuels (gasoline) where LRO of the mixture approaches that of medium fuels (diesel). Analyzing the signal intensities of water clusters in *n*-hexane gives an idea of where they are in the paraffin. As a result, in the mixture with water, the signal intensity is sharply increased at $\log m = 2.3$ in comparison with pure *n*-hexane (Figure 10.8.2, signal is indicated by arrow). This signal in pure water is defined as a signal of the base water cluster consisting of 11 ± 1 molecules [8, 18]. Its computer model is shown in Figure 2.1.1. The fact that this cluster appears in the expanded form in the hydrocarbon could be evidence for the destructive influence of *n*-hexane on it while the signal intensity increase indicates the individualization of the cluster as an oscillator leading to its weakened interaction with the surroundings. In gasoline and diesel, this effect is unpronounced. The signal intensity at $\log m = 2.3$ is very low and almost unchanged. Consequently, this cluster has lost its identity as a result of the interaction with the surroundings, and as a mass concentration it is almost nonexistent. The presence of, for example, aromatic structures and traces of alcohols, ketones, and other polar compounds in diesel fuel and gasoline contributes to the disappearance of the water cluster. The surface molecules of this cluster actively interact with the surroundings [17, 19, 20, 21, 22, 23] and thus it is exposed to the decomposition. If not, equilibrium between the collapsed and expanded forms will be reached, but this is rather unlikely.

LRO of water in hydrocarbons can be better understood by applying GMS-subtraction spectra. Here the signal intensities of hydrocarbons shall be subtracted from those of hydrocarbons saturated with water (Δf). This method is not uncontroversial, but it should be applicable to a series of similar substances. The GMS-subtraction spectra of water in hydrocarbons are shown in Figure 10.8.3. The figure shows different water cluster structures in hydrocarbons and bulk water. Note that the energy characteristics of individual clusters in different surroundings cannot be adequately compared. The GMS-spectrum of bulk water is to be considered as a comparison spectrum for this specific case of mixture preparation. In diesel and gasoline, the base water cluster [24] consisting of 11 ± 1 molecules is unpronounced and its signal intensity is close to zero, thus indicating a strong interaction with the surroundings.

The water cluster of 100 molecules in gasoline is highly individualized and presented in the collapsed form, while in diesel it is "smeared" as a result of the strong interaction with the surroundings. It still mainly remains, however, in the collapsed form. Otherwise, this cluster behaves in hexane, where it occurs as an expanded cluster.

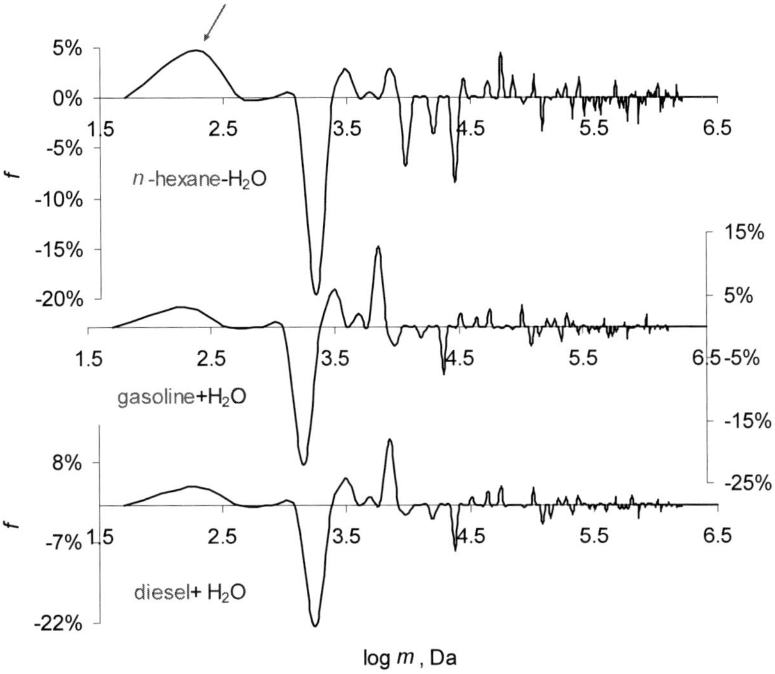

Figure 10.8.2. Review GMS-spectra of fuels saturated with water at 286 K. Mass ensembles up to 3 million Daltons, $p > 1$ N/m^2.

In bulk water, the signals of the clusters $(H_2O)_{100}$ and $(H_2O)_{178}$ are unavailable because of an equilibrium between the collapsed and expanded forms. As a result, the signals cancel each other (Figure 2.1.1). Consequently, for water clusters, the balance between the collapsed and expanded forms shall be actively influenced by hydrocarbons but the hydrocarbon LRO itself is altered, too. The signal of the Chaplin water cluster consisting of 280 molecules in hydrocarbons is expressed very poorly, while its expanded form in pure water is very strongly characterized. The hydrocarbon could affect the balance of expanded-collapsed so that the share of collapsed clusters approaches that of the expanded ones, leading to signal weakening. It is particularly striking that the signal intensities of the large clusters $(H_2O)_{402}$ and $(H_2O)_{545}$ have changed substantially. While they are described by weak signals in bulk water, they show very strong ones in hydrocarbons, especially in *n*-hexane and diesel. These clusters are present in expanded form in diesel but are in collapsed form in *n*-hexane in approximately equal parts. In gasoline, the cluster $(H_2O)_{545}$ shows a very weak signal, likely due to the equilibrium of different density forms. Table 10.8.1 summarizes the main parameters of the cluster formation of the investigated liquids.

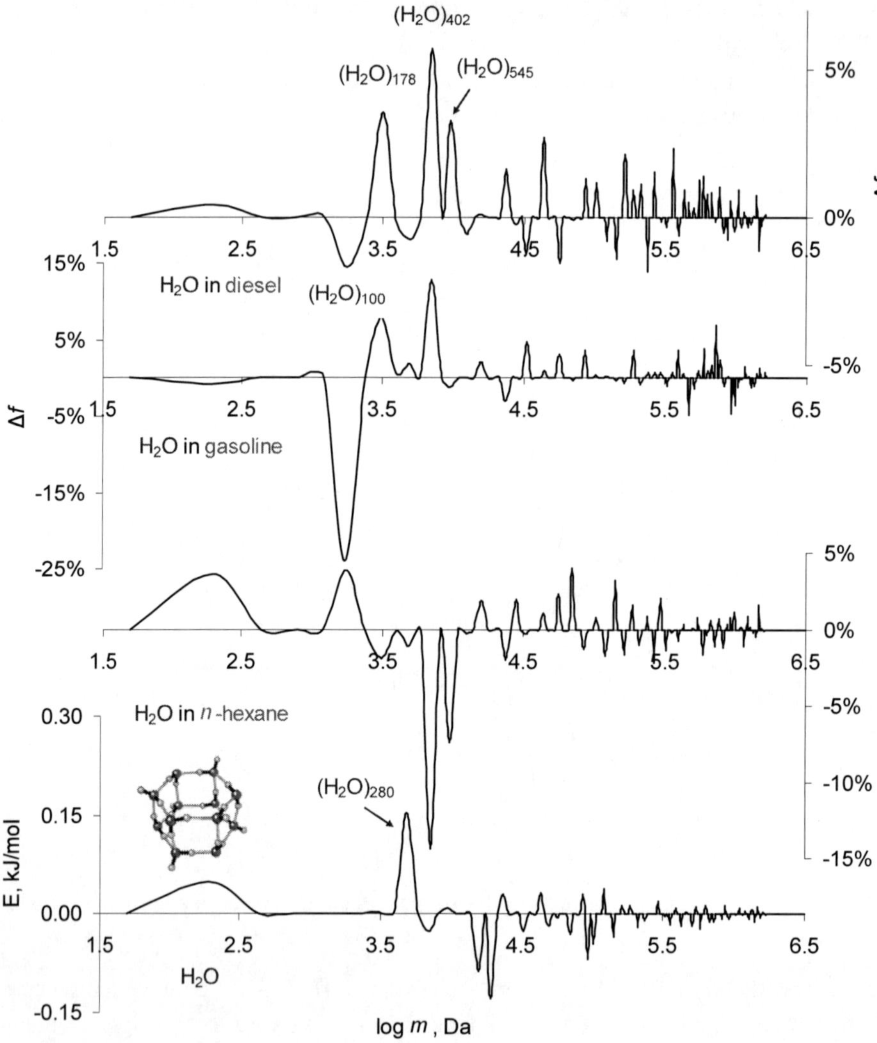

Figure 10.8.3. GMS-subtraction spectra for water in hydrocarbons (explanation in text). Mass ensembles up to 3 million Daltons, $p > 1$ N/m^2.

Table 10.8.1. Main characteristics of the long-range order in the analyzed liquids, where: D_c - part of collapsed clusters, M_{GMS} - average cluster mass, N - number of cluster kinds, and $F_{95\%}$ - number of cluster kinds with a signal intensity reproducibility of at least 95% (skeletal clusters)

Liquids	M_{GMS}, Da	D_c	N	$F_{95\%}$	f, 26±3 kDa
Diesel	190,241	69%	52	67%	-10.3%
Gasoline	465,554	66%	49	75%	-4.7%
n-Hexane	176,792	59%	52	62%	-6%
Gasoline-H_2O	168,254	57%	51	55%	-8%
Diesel-H_2O	167,966	58%	50	73%	-9%
n-Hexane -H_2O	236,972	64%	51	60%	-8%

As shown in Table 10.8.1, the addition of water to gasoline and diesel leads to a decreased average cluster mass, to reduced parts of collapsed clusters (nuclei of crystals), and only for the cluster kinds in diesel and gasoline, it was found that the part of the skeletal clusters had lessened. Attention is drawn to the very high M_{GMS} value in gasoline indicating the presence of large clusters. To prove this, the integral characteristics of the cluster fraction distribution need to be investigated.

The integral curves of the cluster fractions (W_x) in gasoline, *n*-hexane, gasoline with water, and *n*-hexane with water are described by the following relationships, respectively:

$$W_x = -0.0035(m \cdot 10^{-5})^2 + 0.108(m \cdot 10^{-5}) + 0.1417$$
$$W_x = 0.1106 \cdot \ln(m \cdot 10^{-5}) + 0.6945$$
$$W_x = 0.1122 \cdot \ln(m \cdot 10^{-5}) + 0.6968$$
$$W_x = 0.1162 \cdot \ln(m \cdot 10^{-5}) + 0.6406$$

According to the first equation, there are large fractions of big clusters in gasoline whose formation might be due to its long storage (2006) in the absence of mechanical and thermal influences being responsible for the long-range order destruction. The increased M_{GMS} could be explained by the long storing, too. The situation with long storing is known from the year 2006, when the gasoline demand in Germany fell and storages were overcrowded while the demand for diesel fuel had risen sharply.

The introduction of water into gasoline reduces M_{GMS} more than two times, and it is the same for gasoline and diesel fuel that is saturated with water. Hence water can be used to change the long-range order in hydrocarbons. Car enthusiasts have known for some time that the addition of water to liquid fuel affects engine performance.

Returning to Figure 10.8.3, the water cluster of 100 molecules in *n*-hexane and gasoline is characterized by different densities, which implies it has other surroundings. It can be assumed that in gasoline the surface of the water cluster is formed by water molecules so that the hydrogen atoms are directed inside the cluster while the oxygen atoms are mainly on the surface. In *n*-hexane, the surface water molecules are oriented differently; the oxygen atoms are directed towards the center but the hydrogen atoms are oriented to the surface where they interact with the hydrogen atoms of the *n*-hexane clusters through weak van der Waals forces. Here, this water cluster is partially loosened. In gasoline, the surface oxygens of the water cluster interact with their surroundings, too, but the interaction occurs because the water polarity is greatly weakened, finally leading to cluster individualization. Possible "partners" of such an interaction may be impurities of alcohols and ketones. In diesel, the number of possible partners is much higher and the water cluster loses its individuality, hence its signal is close to zero.

It should be mentioned that the lamellar reorientation of atoms on the surface of water clusters has already been observed by Yoshiteru [19].

The important role of water clusters in hydrocarbon nanoemulsions becomes clear when considering the scheme of how oxygen interacts with fuel at the level of the: (i) oxidant diffusion into the reductive cluster (fuel), (ii) chemical reaction itself, and the (iii) products' removal (combustion gases scheme).

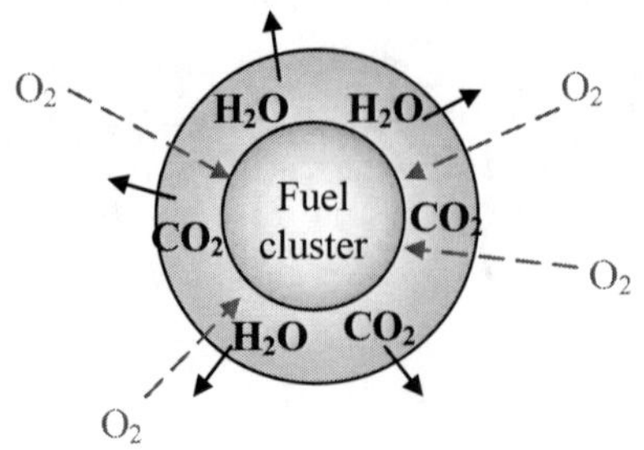

Without water clusters

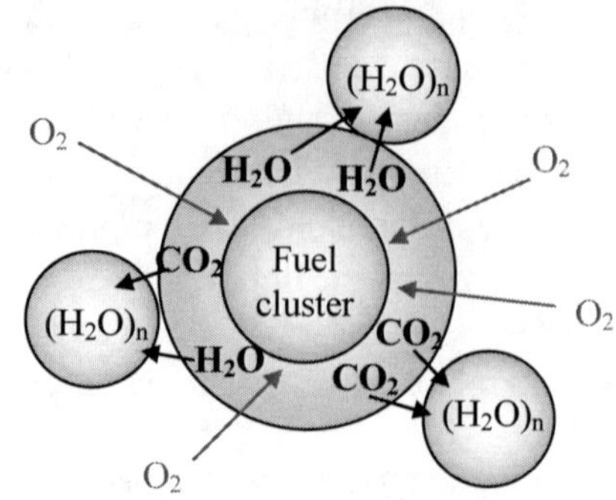

With water clusters

According to this scheme, the stream of reaction products prevent the penetration of oxygen into the fuel cluster but the addition of water provides the possibility for the fast removal of the reaction products. In the following, the combustion process will be accelerated, which leads to an increased fuel efficiency and higher engine power.

Table 10.8.2. A cluster with the mass of 26 kDa in different fuel mixtures (vol. %) and its interaction with the surroundings (*f* at 2 kHz, 286 K).

Fuel kinds	*f*
gasoline, 100%	**-4.7%**
99%	-3%
95%	-5%
90%	-7%
80%	-9%
70%	-11%
diesel, 100%	**-10%**

In conclusion, we considered the behavior of the unique cluster with a mass of 26±3 kDa as it was analyzed in both liquids and polymeric materials [25]. In the present study, this cluster is observed to be present in water, hydrocarbons, and their mixtures. For both the mixtures of water with hydrocarbons and water itself (GMS subtraction spectra for water in hydrocarbons according to the procedure given for Figure 10.8.3), the interaction of this cluster with the surroundings remained almost constant at -8%...-9% (see also Table 10.8.1). The cluster of 26 kDa is also present in mixtures of hydrocarbons but its oscillation and interaction with its surroundings depend on the type of fuel mixture. Table 10.8.2 presents the results obtained for this cluster.

As shown in Table 10.8.2, this cluster oscillates in all mixtures as a collapsed oscillator (-*f*). If the proportion of diesel increases in gasoline, then it also raises the individuality of this unique cluster (decreased interaction with surroundings). Based on the decrease of the *f* value, the presence of two cluster kinds in fuel mixtures can be assumed: gasoline and diesel clusters, both of which are immiscible. These are to be understood as an accumulation of similar clusters ("clouds") in a neighborhood of other clusters. In the saturated state, these mixtures of liquids can be imagined as systems consisting of these clouds, which are recognized by the different f values for the mixtures of 95% and 70% and for diesel. When water in the fuel or n-hexane is present, these hydrocarbon clusters seem to disappear and be replaced by a water cluster (stability of f amounts to 8%).

It has to be mentioned that all cluster sequences in hydrocarbons, water [8], polymers [25, 26], solutions [27], protein coils (lysozyme, L-arginine, Chymotrypsinogen-A [12]), and polysaccharides ([28] amylopectin, etc.) are characterized by similar masses. The cluster sequence consisting of the masses given in Table 10.8.3 (only the first nine sequence members) is common.

Table 10.8.3. Mass concentrations and their oscillation frequencies analyzed in liquids, polymers, and biomolecules (explanation in the text).

N	m, kDa	ω, kHz
1	0.2±0.02	22.2
2	1.8± 0.1	7.4
3	3.1± 0.1	5.5
4	4.9±0.5	4.4
5	7.0±0.5	3.7
6	9.5±0.5	3.2
7	12.5±0.5	2.8
8	15.7±0.5	2.5
9	26±3	2.0

The sequence is well described by the following relationship:

$$m = 268.19 \cdot N^{2.06}$$

CONCLUSIONS

At the level of molecular mass concentrations–clusters, liquid fuels are characterized by a long-range order; therefore, they are designated as nano-colloidal systems.

In fuel mixtures there proceeds a distribution of molecules according to their nature, and new cluster ensembles will be formed, which are stabilized by gravitational white noises.

The role of water in the formation of long-range order in liquid fuels is that it favors the formation of a thermodynamically stable nanoemulsion.

The hierarchy in the formation of nanoemulsions, represented by molecule clusters of the same nature, is a sequence of oscillating masses (clusters and their associates at the sub-micelle and micelle level).

The results obtained can be applied for the rapid and non-destructive monitoring of liquid fuels and other petroleum products.

REFERENCES TO 10.8

[1] Shinichi Hirabayashi, Koichi MT Yamada. The monocyclic water hexamer detected in neon matrices by infrared spectroscop. Chem. Phys. Letters 2007, vol. 435, pp.74-78.

[2] Urano Noritsugu. Functional water for improvement of hydrocarbon based liquid fuel JP 2001064662 A2 20010313. Application No. JP 1999-241435 1999 0827, IPC: C10L001-12. Jpn. Kokai Tokkyo Koho (2001), 5.

[3] Uzawa Masakazu. Method for treating for hydrocarbon fuel JP 2002256273 A2 20020911, Applic. No. JP 2001-59782 20010305. Jpn. Kokai Tokkyo Koho (2002), 11 pp. IPC: C10G032-00.

[4] McVitty Aaron. Improvement relating to fuel conditioning PCT Int. Appl. (2004) 19pp. WO 2004044101 A2 20040527. Appl. WO 2003- GB4955 20031114.

[5] Johnson Keith H, Zhang Bin. Stabilized water nanocluster-fuel emulsion designed through quantum chemistry PCT Int. Appl. (1998), 91 pp. WO 9821294 A1 19980522. Appl.: WO 97- US20779 1997114. IPC: C10L001-32.

[6] Fu da-you, Yuan Dong, Tan Wen-yuan, Xiang Shuang-guan. Cluster analysis method for gasoline. Jilin Huagong Xueyuan Xuebao. 2005, vol. 22, pp. 6-10.

[7] Mudzhikova GV, Brodskaya EN. Molecular dynamics simulation of small water clusters in nonpolar *n*-hexane. Vestnik Sankt Petersbugskogo universiteta, seria 4. Fisika, Chimija. 2004, vol.4, pp.71-77, in Russian.

[8] Zubow KV, Zubow AV, Zubow VA. Cluster structure of liquid alcohols, water and n-Hexane. J. of Appl. Spectr. 2005, vol. 72, pp. 321-328.

[9] Zakharov VV, Brodskaya EN, Laaksonen A. Molecular dynamics simulation of methanol clusters. J. Chem. Phys. 1998, vol.109, pp. 9487-9493.

[10] Shruti Maheshwary, Nitin Patel, Narayanasami Sathyamurthy. Structure and Stability of Water Clusters $(H_2O)_n$, n=8−20: An Ab Initio Investigation. J. Phys. Chem. A. 2001, vol.105, pp. 10525-10537.

[11] Fesenko E.E., Terpugov E.A. About unusual spectral properties of water in a thin layer. Biofisika. 1999, vol. 44, pp. 5-9, in Russian.

[12] Zubow KV, Zubow AV, Zubow VA. Principles of gravitation spectroscopy. New form of molecular matter. Processes. Fields. Aist Handels- und consulting gmbH, dep. R&D, Berlin, electronic book (www.zubow.de). 2010. (in Russian).

[13] Zubow KV, Zubow AV, Zubow VA. Ensemble of Clusters – New Form of Molecular Matter, Risks and Chances. Zubow Equations. In: Taylor JC, editor. Advances in Chemistry Research, New York: Novapublisher; 2010, vol. 5, pp. 107-145.

[14] Pagliai M., Cardini G., Righini R., Schettino V. Hydrogen bond dynamics in liquid methanol. J. Chem. Phys. 2003, vol.119, pp. 6655-6663.

[15] Tomas Rocha-Rinza, Luca De Vico, Valera Veryazov, Björn O. Roos. A theoretical study of singlet low-energy excited states of the benzene dimer. Chem. Phys. Letters. 2006, vol. 426, pp. 268-272.

[16] Herbert C. Georg, Kaline Continho, Sylvio Canuto. Converged electronic polarization of acetone in liquid water and the role in the n–π* transition. Chem. Phys. Letters. 2006, vol. 429, pp.119-123.

[17] Erin R. Johnson, Gino A. DiLabio. Structure and binding energies in van der Waals dimers: Comparison between density functional theory and correlated ab initio methods. Chem. Phys. Letters. 2006, vol. 419, pp. 333-339
[18] Lenz A, Ojamäe L. On the stability of dence versus cage-shaped water clusters: Quantum- chemical investigations of zero-point energies, free energies, basis-set effects and IR spectra of $(H_2O)_{12}$ and $(H_2O)_{20}$. Chem. Phys. Letters. 2006, vol. 418, pp. 361-367.
[19] Yoshiteru Yonetani. A severe artifact in simulation of liquid water using a long cut-off length: Appearance of a strange layer structureChem. Phys. Letters. 2005, vol. 406, pp. 49 - 53.
[20] Giuseppe Graziano. Benzene solubility in water: A reassessment. Chem.Phys. Letters. 2006, vol. 429, pp. 114 -118.
[21] Upadhyay D.M., Mishra P.C. Binding of benzene with water clusters $(H_2O)_n$, n=1–6, in the ground and lowest singlet excited states. THEOCHEM. 2002, vol. 584, pp.113 - 133.
[22] Keith M. Murdoch, Thomas D. Ferris, John C. Wright, Thomas C. Farrar. Infrared spectroscopy of ethanol clusters in ethanol–hexane binary solutions. J of Chem. Phys. 2002, vol. 116, pp. 5717-5725.
[23] Dolgopolova AV, Kushnarev DF, Qua KIm En, Kalabin GA, Schmidt FK. Proton and oxigen- 17 NMR spectroscopic study of hidrophilic-hydrophobic interactions in aqueaus solutions of petrolium products. Khimiya I technologiya vody. 1993, vol. 15, pp. 107-112, in Russian.
[24] Bogdanov EV, Manturova GM. Equiclaster model of water. Biomedicinskaja Radioelektronika. 2000, vol. 7, pp. 19-28, in Russian.
[25] Zysman V, Nguyen Tuan Q, Kausch, Henning H. Degradation on freezing dilute polystyrene solutions in *p*-xylene /J Polym. Sci., Part B: Polym. Phys. 1994, vol. 32, pp. 1257-1269.
[26] Zubow KV, Zubow AV, Zubow VA. Study of structural changes in a LDPE film by flicker- noise spectroscopy. Plasticheskie Massy. 2006, no. 3, pp. 32-34, in Russian.
[27] Zubow KV, Zubow AV, Zubow VA. Spectra of seed cristalls oscillation of NaCl in water solution. J. Appl. Spectr. 2005, vol. 72, pp. 766-772.
[28] Zubova KV, Zubov AV, Zubov VA. Low frequency movement of cluster-12 in potato amylopectin during growth. Influence of white noises. Khimiya Rastitel'nogo Syr'ya. 2009, no. 2, pp.77-87.

10.9. Weak Gravitational Radiation (GR) of Fast Rotating Bodies Resonating with the GR of Water

Applying the resonance caused by GR of water and rapidly rotating bodies provides further insights on the principles of gravitational mass spectroscopy.

A device for measuring the vectors of gravitational excitations is described in [1]; the base element of it is a fast rotating disk. A fundamentally new gravitational sensor for the analysis of the long-range order in liquids and polymers at the level of cluster ensembles is reported by the authors of the present monograph [2, 3, 4]. This GMS-sensor records the energy balance of a quasi-equilibrium process between the dissolution and condensation of protons in the physical vacuum and its histograms represent the accumulation of such nuclei in molecular clusters. The GMS-sensor, however, does not have its own vector.

As a device for the rotating iron rod (Ø 3 mm, 1 cm), an aluminum (Ø 12 cm) and rubber ring (Ø 0.7 m), magnet (magnetic stirrer 600 rpm), electro-magnet (electric centrifuge, 6000 rpm), and mechanical (manual, 180 rpm) drives are used. The water GMS-sensor is positioned 0.15 to 0.7 m from the point where the rotation axis and the surface in which the body rotates are crossing. To eliminate the influence of electromagnetic radiation and mechanical fields on the GMS-sensor (agarose hydrogel), it is placed in a metallic, grounded box, isolated from light and noises (4 cm polystyrene foam). The long-range order of water at cluster level is analyzed according to the methods described in [5, 6]. The scheme of the experiment is shown in Figure 10.9.1.

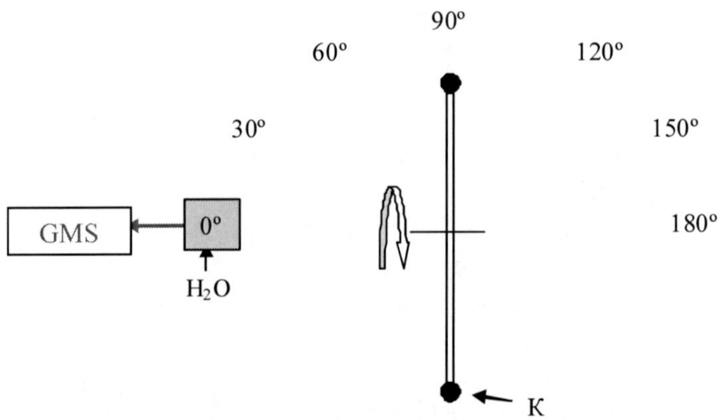

Figure 10.9.1. Scheme of the experiment. K – rotating rubber ring mounted on a bicycle wheel rim, H_2O – water sample (GMS-sensor) at position 0°, GMS–spectrometer. The axis of the ring is directed to the North. The GMS-sensor is moved in a clockwise direction. Prior to the measurement (30 s) the sample is kept for one minute in each position. In the experiments with metals, the bodies rotate horizontally.

Figure 10.9.2 shows how the water cluster ensemble with a mass up to 2.3 million Daltons changes in dependence based on the position of the GMS-sensor (Figure 10.9.1).

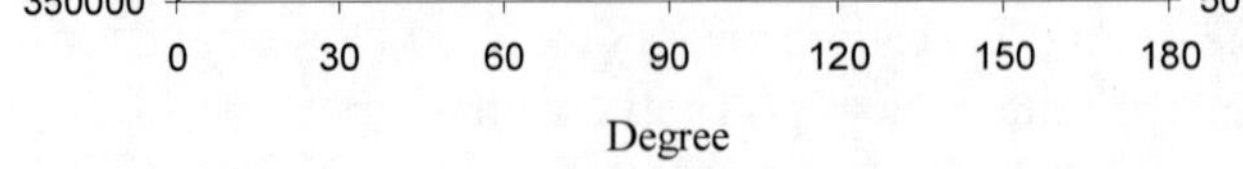

Figure 10.9.2. Average water cluster ensemble mass (M_{GMS}, curve 1) and number of water cluster kinds (N) in the ensemble (curve 2) versus the GMS-sensor position (Figure 10.9.1). The arrow denotes the maximum of curve 1.

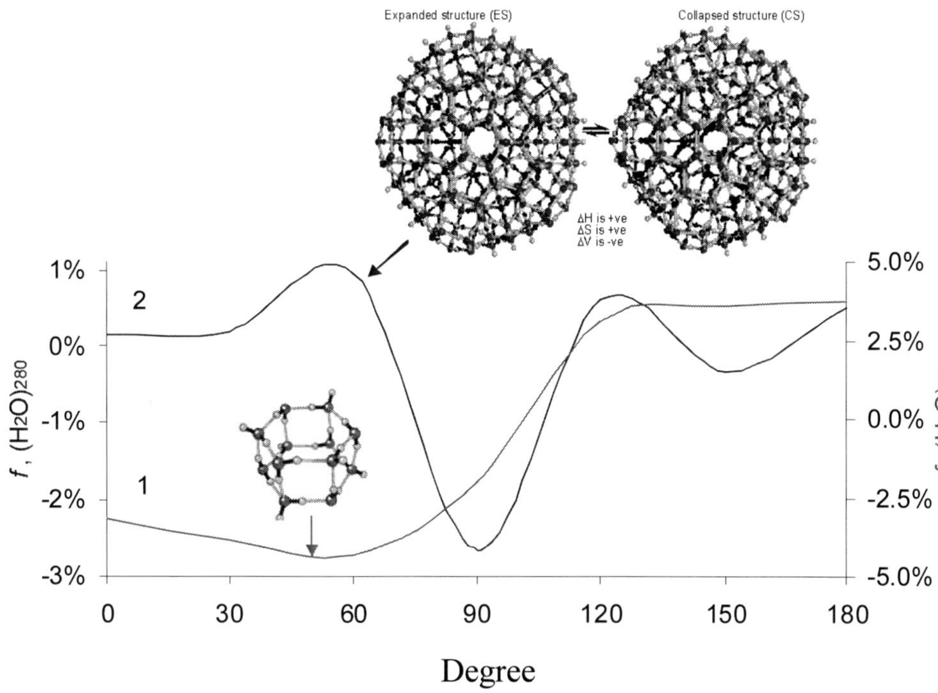

Figure 10.9.3. The behavior of the base (1) and Chaplin water cluster (2) in dependence on the spatial position of the GMS-sensor to the rotating O-ring (Figure 10.9.1). The above structures of the calculated water cluster models were kindly provided by Professors A. Lenz (Stockholm) and M. Chaplin (London).

As seen in Figure 10.9.2, the spatial position of water in relation to the axis of the rotating rubber ring strongly influences the basic characteristics of the cluster ensembles in the hydrogel of the GMS-sensor. While the number of cluster kinds does not change significantly, M_{GMS} increases by almost 47%. In the sequence of individual clusters the situation is even more pronounced. Thus, the signal intensities of the basic water cluster [7, 8] and Chaplin cluster of 280 molecules [9] indicate that clusters react selectively based on the gravitational radiation (GR) from a rotating body (Figure 10.9.3).

As seen in Figure 10.9.3, the water clusters react differently to the GR from a rotating body. Under GR influence, $(H_2O)_{12}$ changes from a collapsed state to an expanded one [10], whereas the much larger cluster of 280 molecules becomes highly thickened at 90°. Here the densification has to be understood as a higher ordering process inside the cluster proceeding, when the gravitational energy dominates over the thermal energy (destructive). Due to the fact that the GMS-sensor records the energy balance between the dissolution of protons in the physical vacuum and their condensation from it, $P \leftrightarrow E_g$, the radiation emitted by a rotating body can be called "gravitational" (GR) and its vector coincides with that of the centrifugal force. This serves as a reminder that this balance is realized with a super light velocity and its shift leads to the generation or absorption of GR.

Since the signals of water clusters (Figure 10.9.3) reflect the state of the gravitational balance in atomic nuclei, the reaction of these clusters on a rotating body's GR can be understood as resonance between the weak GR from water clusters and the GR of the rotating body. It follows that the maximum at 90° reflects the GR flow resulting from the shift of $P \leftrightarrow$

E_g in the direction of the centrifugal force. The shift is obviously minimal but the energy released is enormous and sufficient to be recorded by GMS.

Using other sources to drive the rotation of bodies, similar results are found. For the magnetic driver, they are presented in Figures 10.9.4 and 10.9.5.

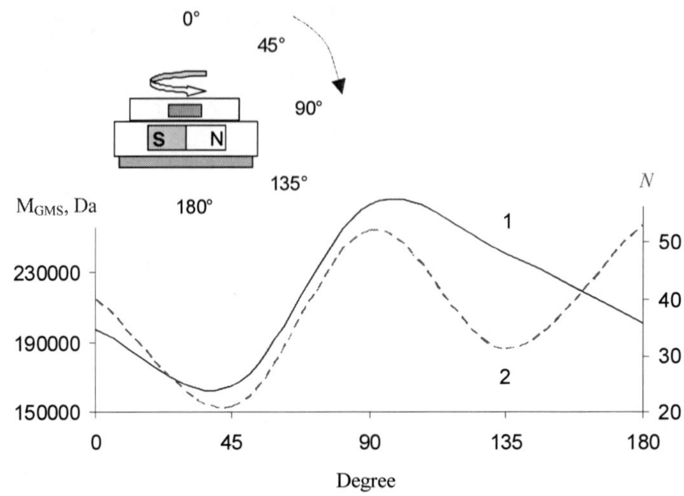

Figure 10.9.4. Average cluster mass in water sensor (1) and number of cluster kinds in the ensemble (2) versus GMS-sensor position. Rapidly rotating iron rod driven by magnetic stirrer (600 rpm, scheme in the upper left area).

As is visible from Figure 10.9.4, the water cluster ensemble reacts on GR released by the rapidly rotating iron rod. Unambiguous minima in the curves are found at 45° and 135°, indicating the ordering effect of GR on the water cluster ensemble in the GMS-sensor. Therefore, the magnetic field influences GR and its vector do not coincide with that of the centrifugal force.

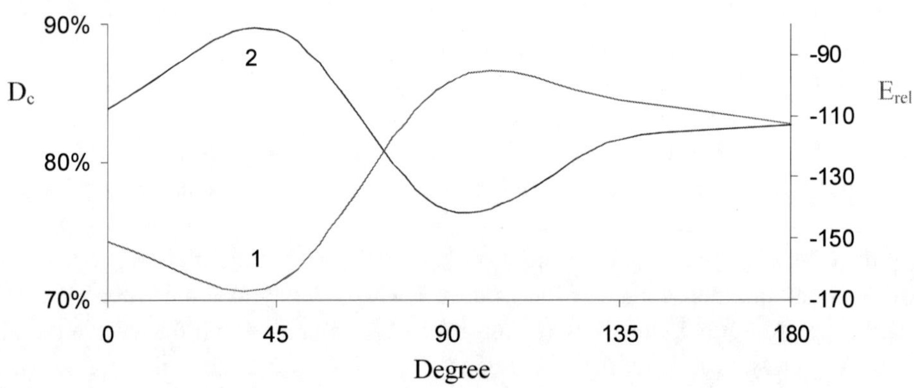

Figure 10.9.5. Cluster ensemble energy (E_{rel}) of water in GMS-sensor (1) and part of the collapsed water cluster (energy-rich, D_c, 2) dependent on the spatial position to the axis of the rotating body.

According to Figure 10.9.5, it follows that the direction of 45° is an especially strong GR vector that "pumps" energy into the collapsed clusters of the ensemble. In the same direction, however, the GR energy is observed to achieve a minimum, pointing to its drain, which is typical for GR [11].

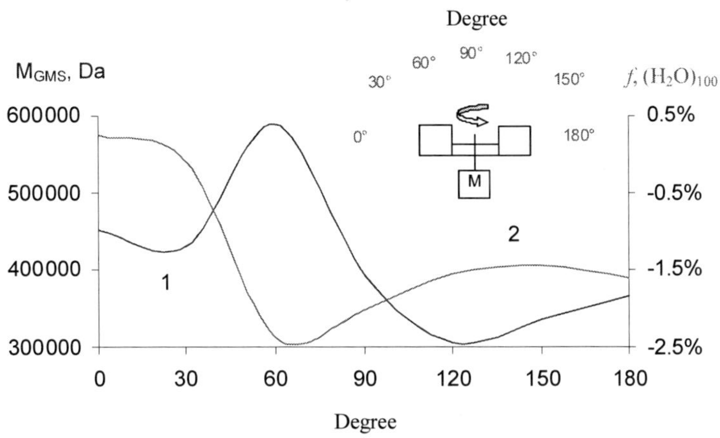

Figure 10.9.6. Remote effect of a rotating aluminum ring (6000 rpm) on the long-range order of water. 1 – average mass of water clusters in sensor and 2 – signal intensity of a water cluster consisting of 100 molecules.

In Figure 10.9.6, it is shown how a rotating aluminum ring installed in a laboratory centrifuge remotely influences the long-range order of water in a GMS-sensor that is placed at certain angles and at the distance of 25 cm from the axis of the rotating ring. The behavior of both curves at 60° is striking. While the average cluster mass in water achieves a maximum, the signal intensity of $(H_2O)_{100}$ is characterized to be the lowest (transformation of expanded form to collapsed form). Enhanced M_{GMS} values can be understood as increasing the GR flow in this direction, leading to a higher internal energy of individual clusters in the ensemble. In the present experiment (Figure 10.9.6), the number of clusters kinds remains almost the same, equal to 51±1.

Summarizing the experimental data and based upon previous observations on the interaction of weak gravitational radiation from water with the GR of celestial bodies [11], the following mechanism for the remote interaction between the GMS-sensor and the rotating body can be suggested. In reality, the balance between baryonic protons (P) and the physical vacuum $P \leftrightarrow E_g$ is slightly shifted toward the dissolution of the protons to observe them as GR. In this case, the energy will be lost and the molecular matter cooled down, e.g. as occurs with planets. In contrast, the condensation of protons from the vacuum is connected with energy release (E_g), for example, from the Sun.

In our experiments, the molecular clusters in the rotating body deformed under the centrifugal force should also consider a change in their atomic nuclei and their gravitational potential [12]. This process shifts the balance of the proton ↔ vacuum and thus reinforces GR. Then, the GR of the rotating body will resonate with the GR from the sensor. Here the stronger oscillator ensembles force the weak ones to subordinate them. According to this model, the nuclei in water clusters of the sensor will be forced to restructure, and this reorganization will be reflected immediately at the molecular level, e.g. in the water cluster ensemble energy, number of water cluster kinds, and part of collapsed clusters.

Therefore, rotating bodies generate gravitational radiation, which affects the structure of the local gravitational field and thus the surrounding molecular cluster ensembles. This effect is applied to create a fundamentally new gravitational telescope (§§19-20); it can also be used to monitor the GR and its influence on molecular matter, especially on water clusters in biomatrices [13].

To keep the monograph simple, the authors have therefore decided to discuss the GR effects only without explaining the reasons behind their arising (§§23-24). As we know, clusters reflect the energy state of stationary waves of gravitational radiation in white noise. The presence of molecular clusters, on the other hand, indicates that the proton jumps do not occur randomly but that they are subject to certain rules.

Thus there are two principal ways to analyze the long-range order in molecular matter:

1. Forced generation of shock waves in the material and analysis of the reflected ones from heterogeneities inside the sample
2. Analysis of white noises in the sample (sensor) and extraction of structurally stable waves that are generated by ensembles of jumping protons (H-solitons, signal cleaning)

The first direction is of interest for high-energy technological processes while the second one suits the analysis of the long-range order in solids and liquids, which are not exposed to colored noises. Depending on the aim of the work every investigator decides for himself which path should be chosen.

Conclusion

Weak shock waves can be used in molecular mass spectrometry to analyze molecular clusters; their sizes, shape, density, and structural heterogeneity; as well as to inform on the character of their interaction with their surroundings.

Using GR generated by rotating bodies creates a concrete molecular model and even an energy cluster can be influenced.

References to 10.9

[1] Smirnov V.N., Egorov N.V., Shchedrin I.S. A New Detector for Perturbations in Gravitational Field. Progress in Physics. April. 2008, vol. 2, pp.129-133.

[2] Zubow K.V., Zubow A.V., Zubow V.A. Ensemble of Clusters – New Form of Molecular Matter, Risks and Chances. Zubow Equations. In Advances in Chemistry Research, vol. 5, 2010, pp. 107- 45. Editor J.C. Taylor. Novapublisher. NY.

[3] Zubow K. V, Zubow A.V, Zubow V.A. Use of flicker noise spectroscopy for undestroed analysis of nano structures. Zavodskaja Laboratoria. Diagnostics of Material. 2008, vol. 74, no. 9, pp. 40- 46, in Russian.

[4] Zubow K.V., Zubow A.V., Zubow V.A. Study of structural changes in a LDPE film by flicker-noise spectroscopy. Plasticheskie Massy. 2006, no. 3, pp. 32-34, in Russian.
[5] Zubow A.V., Zubow K.V., Zubow V.A. The Investigation of the Salt Dissolving Mechanism and of the Ageing of the NaCl Solution in Water by Flicker Noise Spectroscopy. Journal of Chem. Prom. (S. Petersburg). 2007, vol. 84, no.1, pp. 40-48, in Russian.
[6] Zubow A.V., Zubow K.V., Zubow V.A. The Phenomenon of Water Clusters Distribution in Water and Solvated Clusters of Ion Pairs of NaCl in Solution Near the Wall. The Wall Effect. Journal of Chem. Prom. (S. Petersburg). 2006, vol. 83, no. 6, pp.300-306, in Russian.
[7] Bogdanov E.V., Manturova G.M. Equiclaster model of water. Biomedicinskaja Radioelektronika. 2000, no. 7, pp. 19-28, in Russian.
[8] Annika Lenz, Lars Ojamäe. On the stability of dence versus cage-shaped water clusters: Quantum-chemical investigations of zero-point energies, free energies, basis-set effects and IR spectra of $(H_2O)_{12}$ and $(H_2O)_{20}$. Chemical Physics Letters. 2006, vol. 418, pp. 361-367.
[9] M. Chaplin, SBU London. http://www.lsbu.ac.uk/water/index.html.
[10] Zubow K.V., Zubow A.V., Zubow V.A. Cluster structure of liquid alcohols, water and n-Hexane Journal of Appl. Spectr. 2005, vol.72, no. 3, pp. 321-328.
[11] K.V. Zubow, A.V. Zubow, V.A. Zubow. Principles of gravitation spectroscopy. New form of molecular matter. Processes. Fields. Aist Handels- und consulting GmbH, dep. R&D, Berlin, 2010, electronic book (www.zubow.de), p. 854 (in Russian).
[12] Oldershaw, Robert L. /arXiv.org, e-Print Archive, Physics (2008), 1-7, arXiv:0803.1197v1 [physics.gen-ph]. Publisher: Cornell University Library, CODEN: APRHCB http://aps.arxiv.org/ftp/arxiv/papers/0803/0803.1197.pdf.
[13] Zubow A.V., Zubow K.V., Zubow V.A. Investigation of the distribution of water clusters in vegetables, fruits and natural waters by flicker noise spectroscopy. Biofizika, 2007, vol. 52, no. 4, pp. 585-592, in Russian.

10.10. Water Clusters: The Influence of Seismic Activity

The phenomenon that the LRO in molecular matter reacts at a distance based upon processes in earthquake zones has already been observed by the authors [1]. The LRO in water at its cluster structure (Figure 10.10.1) was concluded to be a unique sensor for several events in the earth's core, especially for compression/extension processes, temperature jumps, energy impulses, and gravitational radiation from the epicenter [1]. The aforementioned factors destabilize cluster ensembles in the LRO.

In Figures 2.1.1 and 10.10.1, computer models for the simplest water clusters, consisting of the base water cluster $(H_2O)_{11\pm1}$ [2], are shown. The base water cluster belongs to an ensemble of dominating clusters that were formed and enforced in white noises: $(H_2O)_{11\pm1}$, $(H_2O)_{100}$, $(H_2O)_{178}$, $(H_2O)_{280}$, $(H_2O)_{545}$, $(H_2O)_{903}$, $(H_2O)_{1,351}$, $(H_2O)_{1,601}$, $(H_2O)_{1,889}$, etc. [3]. For a better understanding, the simplest solvated clusters of ion pairs are given; they can be applied to the remote monitoring of processes in the earthquake's epicenter, too [1].

It was the aim of the present section to find in a long-time test the interlinkage between the LRO in underground water and processes in earthquake zones.

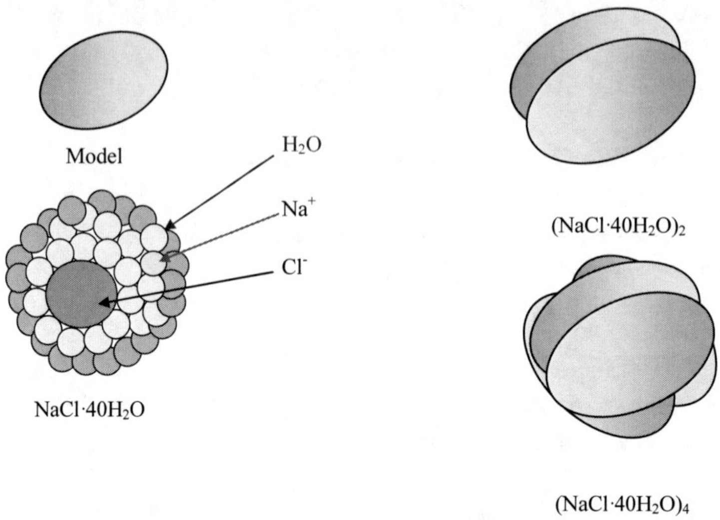

Figure 10.10.1. Solvated clusters of ion pairs (monomer, dimer, and tetramer) experimentally observed by the authors [4], §10.1.

The test set-up for the LRO in underground water as a scheme is shown in Figure 10.10.2.

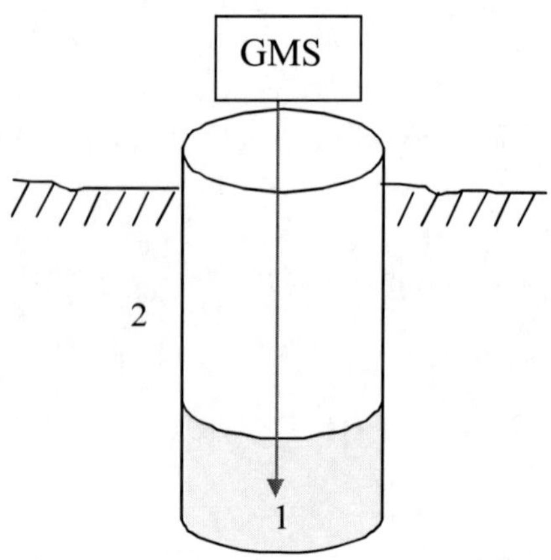

Figure 10.10.2. Gravitational noise sampling by the Zubow gravitation mass spectrometer. 1 – GMS-sensor in underground water and 2 – well of 3 m depth.

The LRO analysis was carried out according to the method described in [5, 6]. A cluster ensemble with a mass of up to 3.3 million Daltons was investigated at 279 K under constant atmospheric pressure at a place with the geographical coordinates 53°38′N, 12°35′E (North Germany). The water in the well was in a quiet condition. The principle of the gravitation mass spectroscopy (earlier called "flicker noise spectroscopy" [3]) is based on the fact that the gravitational energy of nuclei clusters in molecular matter shall be analyzed [7] (§24, by influencing the balance between proton-GR and the activated complex $\boldsymbol{p}$ + GR ↔ [$\boldsymbol{p}$-GR]$^{\neq}$). Additional information about the GMS-method is given in [8, 9] where the masses and oscillation frequencies of water clusters in underground water were calculated according to the Zubow equations using the Zubow constant $6.4 \cdot 10^{-15}$ N/m for water and salt solutions. Data about earthquakes was taken from the homepage, http://www.gfz-potsdam.de/news/recent/index.html.

Figure 10.10.3 provides information on how the energy of water cluster ensembles changed during the experiment. The changing energy of a complete water cluster ensemble as well as some extreme data is shown. The curve deviation from the base line should be explained by gravitational energy drain from the water cluster ensemble in the underground water to a stronger water cluster ensemble [10]. One of these extreme pieces of data was found to correlate with a signal from a seismograph that registered an earthquake in Puerto Rico (9:45, 19°25′N, 65°55′E, at a depth of 18 and 50 km). This correlation is explained by the fact that gravitational radiation spreads out with a super light velocity [3, 8, 10]. The other extreme signals cannot be ascribed to earthquakes but they also should not be ignored. They are an indication of the strong processes in the hydrosphere that are connected with the weakening of hydrogen bonds between water molecules. For example, a forced expansion of the hydrosphere under the influence of processes proceeding inside the Earth could belong to this categorization.

It has to be mentioned that the signals from the epicenter spread out as seismic [1] as well as gravitational waves [3, 8] where in the present paper only the latter ones shall be analyzed.

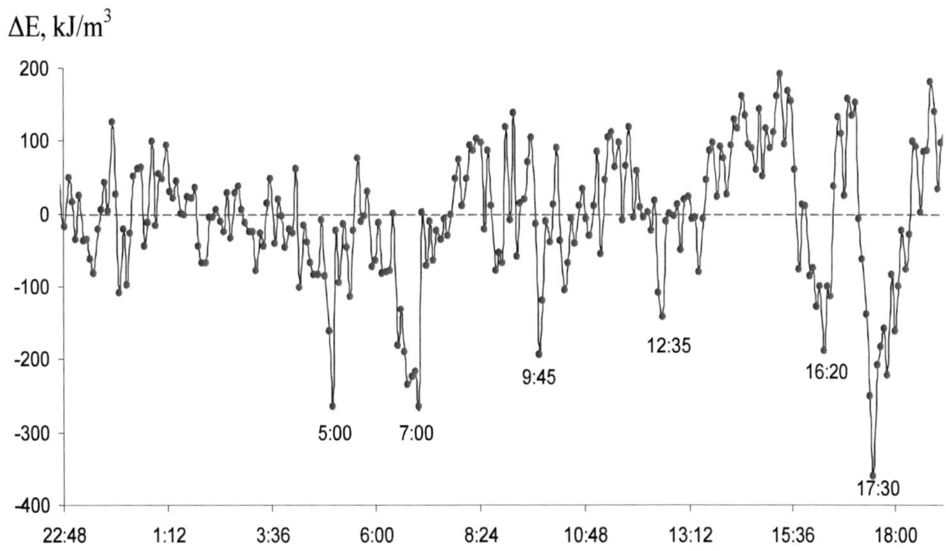

Figure 10.10.3. Energy changes of water cluster ensembles in underground water from August 13 to 14, 2009. The evaporation energy of water that is necessary for the complete destruction of all water clusters was used (zero line as dotted line) for the spectra calibration. Each point is the median of 4,000 measurements with a reproducibility of not less than 95%.

The deviation of the molecular mass of water cluster ensembles from the average value [6] seemed more sensitive to earthquake influence. Therefore, it shall be analyzed in detail (Figure 10.10.4). Only the signals that differ from around more than 30% of the average value are shown in this figure. As is visible, there are a number of signals that correlate with strong earthquakes whereas other ones correspond to weak earthquakes (Richter scale < 4.5). Furthermore, some signals did not indicate an earthquake. The group of signals that is marked with a horizontal bracket is connected to the development of events in the earthquake epicenter under the island of Java (8°73′S, 112°36′E) at a depth of 128 km at 13:05 (to the seismograph at 13:31). Remember that during the entire observation, the cluster types' number remained nearly unchanged (49…54). Therefore, the energy necessary for the cluster ensemble's destruction was not enough.

In the next section, the group of signals marked with a bracket in Figure 10.10.4 shall be discussed in detail. If the signals reflect the processes in the earthquake's epicenter, then a conclusion should be made that the dynamics of the cluster forms' changes depends on their interaction with the surroundings [1]. We assume that according to this work, the signals reflect real earthquake processes in the water containing layer of the Earth's core [10]. The appearance of expanded water clusters was ascribed to the destruction of water cluster ensembles in the earthquake epicenter whereas the appearance of collapsed ones characterized a decreased interaction with the surroundings mainly accompanied by the hydrogen bonds' formation inside the cluster. The water cluster formed in the underground water was therefore assumed to give an idea of the physical processes in the epicenter.

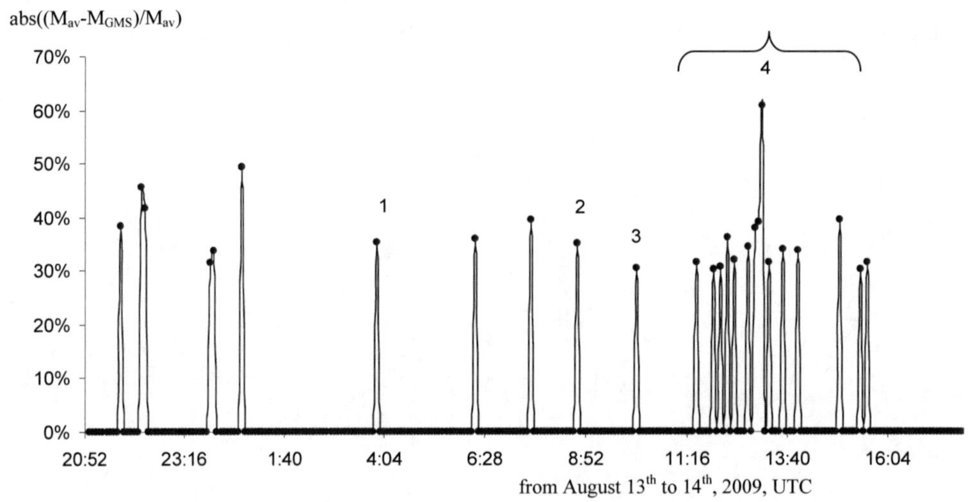

Figure 10.10.4. Selected signals (abs((M_{av}-M_{GMS})/M_{av})) that deviate more than 30% from the average mass of all water clusters (M_{av}) in an ensemble up to 2.3 million Daltons in the period from August 13 to 14, 2009 for all test points. M_{GMS} – average clusters' mass in underground water to every test point of one minute [13]. Earthquakes in: 1 – Japan (Honshu, 4.6 Richter scale, RS); 2 – Vanuatu, 5.4 RS; 3 – Puerto Rico, 4.9 RS; 4 – Java, 4.7 RS; 5 – Andaman, India, 5.3 RS; and 6 – the Philippines, 5.1 RS.

The cluster behavior of this group was compared with that of the base water cluster consisting of 12 molecules [2, 4] (Figure 10.10.5). As is visible, until the earthquake, the base water cluster was observed to mainly be in the expanded form, however, it swings alternately to the collapsed form. Once approaching the earthquake event (designated with a star), the frequencies of the pendulum (designated with broad arrows) increased. After the earthquake, the cluster is only in its collapsed form. This behavior of the clusters in the ensemble in underground water was characterized to reflect the remote influence of the clusters' analogs from the epicenter [1, 3, 10], which was provoked by gravitational resonance between similar clusters [3, 8, 10, 13]. Until the earthquake event, alternative compressing and expanding processes of rock formations in the epicenter were assumed to proceed in those interspaces that water infiltrated [11].

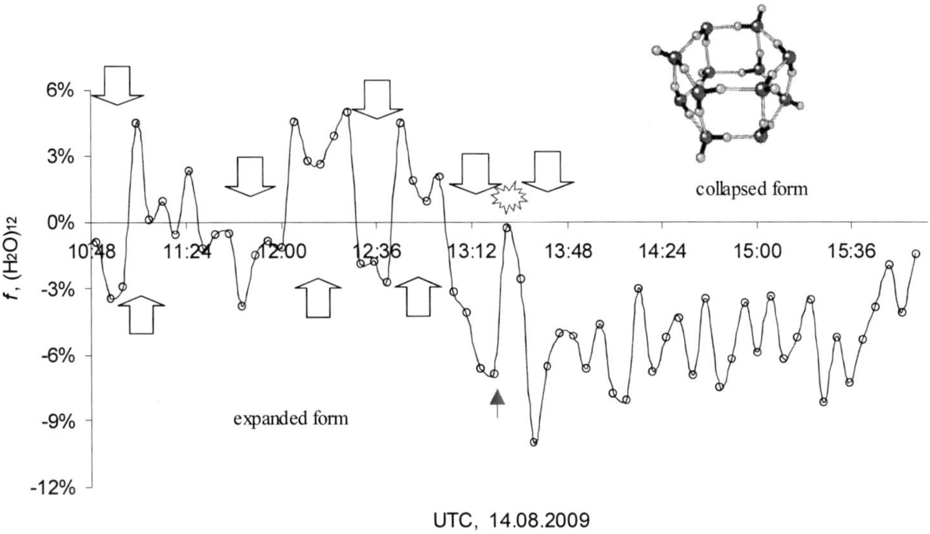

Figure 10.10.5. Signal intensities of the base water cluster during the earthquake on the island of Java (Figure 10.10.4, the points' group 4).

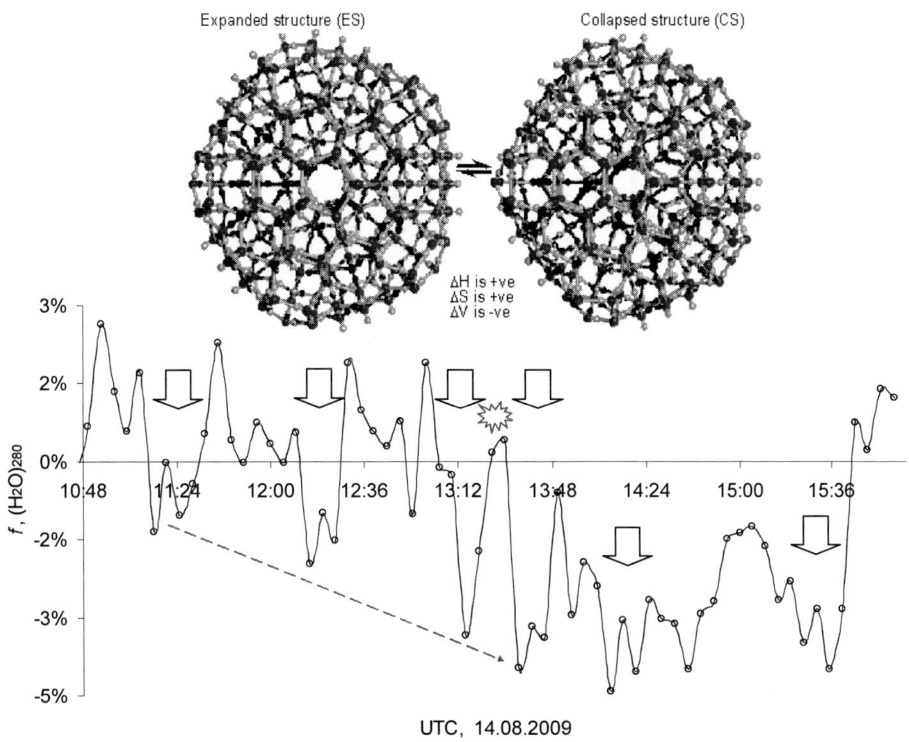

Figure 10.10.6. Computer model for the Chaplin water cluster $(H_2O)_{280}$ and its signal intensities during the earthquake (marked with star) on the island of Java.

As shown in Figure 10.10.6, until the earthquake event, the cluster $(H_2O)_{280}$ was mainly in its expanded form and after the earthquake, it changed to its collapsed form. After 2 h, however, the balance of collapsed–expanded shifted back to the expanded form. The behavior of this cluster was therefore found to be similar to that of the base water cluster it consists of [12].

In the next section, the curves of Figures 10.10.5 and 10.11.6 shall be compared. Until 13:00 the clusters changed differently. During the compression of $(H_2O)_{280}$ ($f < 0$) the base water cluster becomes more expanded ($f > 0$) and directly before the earthquake (13:12) both clusters are dominated by the collapsed forms that are enriched with potential energy. Furthermore, there it is visible that upon approaching the collapse (dotted line), the domination of the collapsed forms is amplified. In the rocks' pores, a mechanical compressing of these water clusters proceeds where this process is periodic (marked with arrows) and connected to the relaxation processes of similar times. These quasi-periodic processes correlate with mathematical models developed for a periodic shear-heating mechanism for intermediate depth earthquakes, according to which, temperatures can be increased up to 1,673 K [13]. These local temperature increases certainly influenced the water cluster ensembles.

As shown in Figure 10.10.7, the average molecular mass of clusters (M_{GMS}) in underground water was characterized by a distinct maximum before the earthquake. The ensembles of water clusters could run through several energy transformations inside the epicenter where M_{GMS} achieved the maximum at 13:05. To the authors, this maximum has the potential to be used for the short-term forecast of earthquakes.

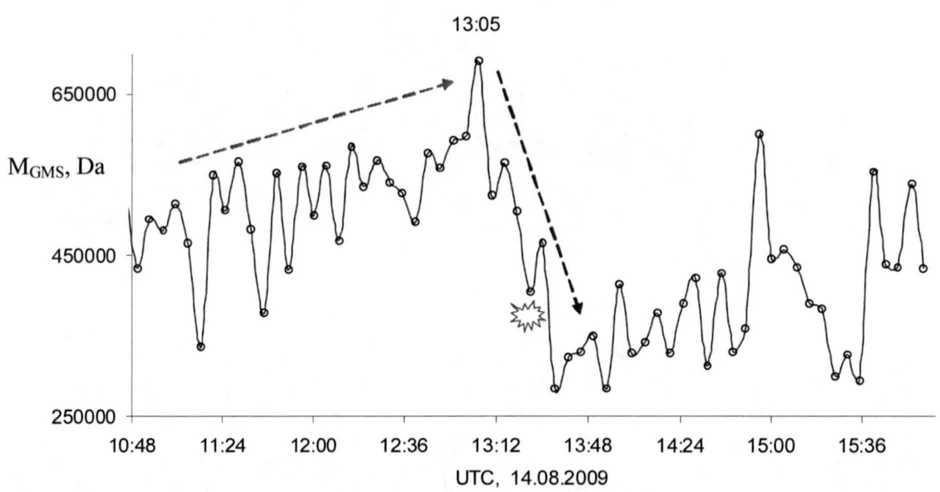

Figure 10.10.7. The average molecular mass in a water cluster ensemble up to 2.3 million Daltons during the earthquake on Java. The seismic maximum is ascribed with a star and the dotted lines show the trends of M_{GMS} increasing and strong M_{GMS} decline up until the earthquake.

Fifteen minutes until the earthquake, M_{GMS} almost doubled, which is explained by a fast gravitational radiation from the epicenter that excited the gravitational field of the water clusters in the well; this was a momentary occurrence.

The forced change of the water cluster forms in the epicenter can be understood from the position of varied hydrogen bonds' lengths (intercluster hydrogen bonds, Figure 10.10.8, zigzag lines) leading to another gravitational radiation [8]. The gravitational radiation changed enforced the potential energy of a cluster ensemble like an oscillator dimension. This leads to clusters' expansion or compression that is realized by the help of hydrogen bonds. Cluster compression has to be understood as the result of decreased hydrogen bonds' length through which the energy state of the protons in the hydrogen bonds shall be strongly influenced by oxygen atoms [8, 10] (Figure 10.10.8). Here the clusters were transferred into the collapsed forms that are enriched with potential energy. The compression was found to increase the dissolution velocities of the so-called "naked" protons [14] that were therefore (H) connected with their fast reverse condensation from the vacuum [8] (Figure 10.10.8). Through this, the gravitational field in the epicenter gets more strained. This observation was supported by the earlier information about the gravitational radiation from earthquake's epicenters [15]. The gravitational radiation of the compression zone causing green light (557.7 nm) in the atmosphere [16] was suggested to arise under the remote influence of the epicenter on the "naked" protons in the hydrogen bonds of the expanded water clusters in the atmosphere.

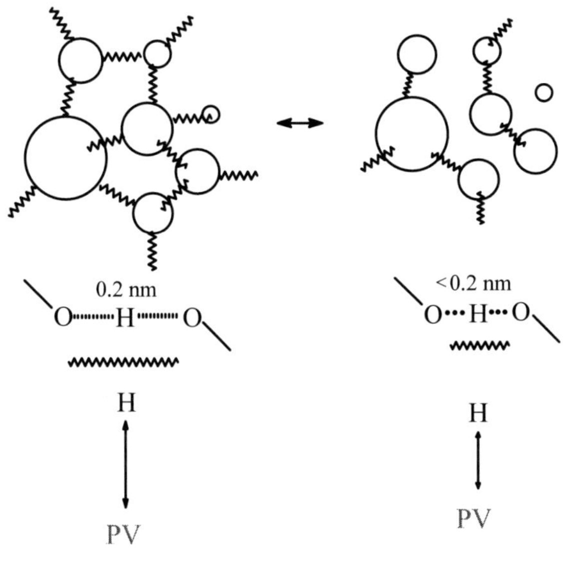

Figure 10.10.8. Two states of protons in hydrogen bonds in water that oscillate between baryonic matter (H) and the physical vacuum (PV) [8].

In the following, Figure 10.10.7 shall be discussed once again. At a constant number of cluster kinds in water and regarding the Zubow equation [3], the sum of a cluster of masses in the equation $M_{GMS} = \Sigma(|f|\cdot m)$ shall be a constant value, too. The great increase of M_{GMS} can therefore only be ascribed to f. As we already know, the rise of the absolute sum $\Sigma(|f|)$ is the result of clusters' individualization, reducing their interaction with the surroundings and changing their conformations. The hydrogen bonds' length changing is automatically connected with a changed water volume as well as with modified hydrodynamic properties of the liquids that can be recorded by dilatometers [17], viscosimeters [18, 29], or the flicker noise spectrometer [1].

To evaluate the physicochemical characteristics in the earthquake epicenter in Java the data from Figures 10.10.3, 10.10.5, 10.10.6, and 10.10.7 as well as the data of the remote cluster interaction [3, 10, 20] were used.

$$\Delta f = k \cdot \ln(M/R^2) + C$$

We start out from the above equation for the base water cluster $(H_2O)_{12}$, which connects the changed cluster signal intensities ($\Delta f = 7\%$, Figure 10.10.5, marked with thin arrow) at the observation place (North Germany) with the water mass in the epicenter (*M*, moles, 1 mol = 18 g) and the distance to the epicenter (*R*, m) [3, 10, 21]:

$$\Delta f = 1.162 \cdot M/R^2 - 0.0016$$

According to this equation, the excited water mass in the epicenter in Java ($R \approx 12.7 \cdot 10^6$, m, diameter of the Earth) was calculated to be about $179 \cdot 10^6$ t. To get an idea of the pressure developed in the epicenter, a simple variant of water heating in a hydro layer until it reaches 1,000 K [11] (shear-heating model, 1,673 K [14]) shall be discussed. The process developed here can be calculated with the help of the Mendeleev equation ($PV = n\mathrm{R}T$, where $n = 179 \cdot 10^{12}/18 = 9.94 \cdot 10^{12}$ moles of water, R = 8.31 J/mol·K, T = 1,000 K, and $V = 179 \cdot 10^{12}$ cm^3). The pressure is about 0.46 GPa, which has to be seen as an effective value of the pressure in hot water in the epicenter and which is lower than the pressure developed by the rocks' dehydration (2...9 GPa, [21]). As is visible from Figure 10.10.4, in the epicenter there is a variety of steam boilers (marked with a horizontal bracket). Concerning the earthquake on Java on August 14, 2009, it could have resulted from one of 14 boilers being overheated and explosively destroyed (Figure 10.10.8, if only 10% of the gravitational noises are filtered out, then 20…25 boilers can be observed). At the simultaneous heating-up of all the boilers in the epicenter until the critical steam state, the oldest and the most exhausted one shall be destroyed first.

The earthquake leads to a temporary post-seismic stabilization of the water cluster ensembles in the epicenter, which is at domination over the collapsed clusters. The collapsed clusters are enriched with potential energy [7, 10, 13] that they obtained from the boiler explosion (Figures 10.10.5 and 10.10.6). The stabilization process, however, is short and it is followed by the rocks' erosion processes through condensed water from layers above the epicenter. In the next step, this condensed water drips back into the boilers with regeneration of the beginning physicochemical state prior to the earthquake. The following boilers' aging evokes an explosion again and hence commences a new earthquake. This cycle permanently repeats (Figure 10.10.9, model of Soxhlet apparatus).

Let us return to Figures 10.10.4, 10.10.5, and 10.10.8. The periodic repeating of maximums and minimums until the collapse could be seen as support for the geyser model; namely, the periodic pores' infiltration with water [12] and then water heating up, upon which the water steam releases into the higher rocks' layers and finally the steam condenses with the water seeping back into the rock pores (boilers). Here the critical point is whether the rock walls of “boilers” can withstand the water steam pressure, and the earthquake breaks out if the walls do not withstand the pressure, leading to an explosion.

The Moon was observed to play an important role. It creates a high tide in liquid magma that causes an additional heating of water inside the boilers but it is difficult to find a direct correlation between the Moon's influence and an earthquake because of high magma viscosity and the time needed for water heating in the boilers.

Underground geysers in the epicenter are modeled as shown below (Figure 10.10.9).

A series of earthquakes that rocked the island of Java on October 30, 2010 confirmed the suggested model. At this time, the heating flow to the steam boilers was reinforced by the Moon, which seemed to cause magma high tide to the upper layers of the Earth's core. This led to an overheating of the underground water, which was connected with the destruction of the underground "Soxhlets' " stability.

The correlation between the earthquake in Java and the Moon's constellation can be proved with the help of the computer program ZET9 (www.astrozet.com). It only works, however, under the implication of the times needed for the arrival of the magma high tide at the steam boilers and for water heating.

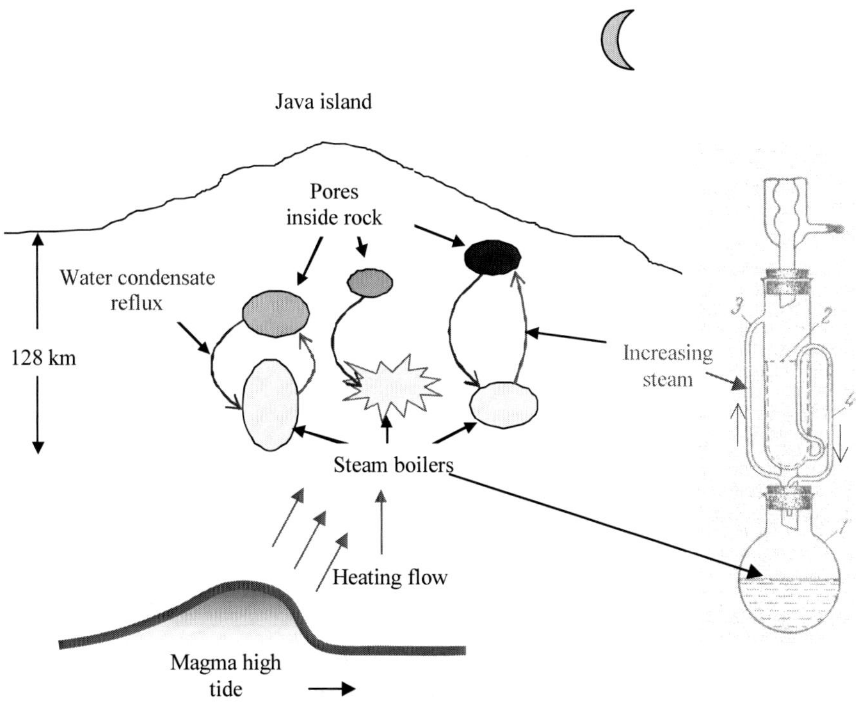

Figure 10.10.9. Model of underground periodic geysers consisting of three boilers that form the earthquake epicenters in Java. Water heating in boilers increased the pressure of steam accompanied with its release into the upper pores where the steam cooled down, condensed, and returned back into the boilers. On the right side, a scheme of the Soxhlet apparatus is shown, which is applied for the process modeling. 1 – liquid to be heated, 2 – condensate, 3 – steam flow tube, and 4 – water condensate reflux. In the upper part the reverse condenser is partially shown (http://en.wikipedia.org/wiki/Soxhlet_extractor).

CONCLUSION

Strong earthquakes in Java resulted from exploding underground "steam boilers" at the epicenters in a depth of 120…130 km.

The underground “steam boilers” were characterized by a specific distribution of at least 25 boilers.

The mechanism of water circulation under the island of Java is comparable to the Soxhlet principle.

The Moon could initiate strong explosions of steam boilers but this theory needs further investigation.

The Zubow gravitation spectrometer has the potential to be used for remote monitoring of earthquakes' epicenters to get an idea about their structure. Furthermore, it should be possible to investigate the earthquake’s evolution and to develop a strategy for early forecasting.

References to 10.10

[1] Zubov K.V., Zubov A.V., Zubov V.A. Phenomenon of Influences of natural disasters (hurricanes and earthquakes) on the long-range order in liquids. Risks and chances. Russian Journal of Applied Chemistry. 2007, vol. 80, №11, pp. 2001-2004.

[2] Bogdanov E.V., Manturova G.M. Equiclaster model of water. Biomedicinskaja Radioelektronika. 2000, no. 7, pp.19-28. (in Russian).

[3] Zubow K.V., Zubow A.V., Zubow V.A. Ensemble of Clusters – New Form of Molecular Matter, Risks and Chances. Zubow Equations. In Advances in Chemistry Research, 2010, vol. 5, editor J.C. Taylor. Novapublisher. NY, pp. 105-147.

[4] Zubowa, K. V.; Zubow, A. V.; Zubow, V. A. Oscillation spectra of seed NaCl crystals in aqueous solutions. Journal of Applied Spectroscopy. 2005, vol. 72, no. 6, pp. 840-847.

[5] Zubow K.V., Zubow A.V., Zubow V.A., Using of flicker noise spectroscopy for non-destroying analysis of nano structures. Zavodskaja Laboratorija. Diagnostics of materials. 2008, vol. 74, no. 9, pp.40-45 (in Russian).

[6] Zubov, A. V.; Zubov, K. V.; Zubov, V. A. Mechanism of sodium chloride dissolving in water and the process of the solution aging. Russian Journal of Applied Chemistry. 2007, vol. 80, no. 7, pp. 1249-1255.

[7] Zubow K., Zubow A.V., Zubow V.A. The Phenomenon of Proton Dissolving in Vacuum and of Proton Condensation from Vacuum. Two Forms of Protons, Structure of Nuclei, Electrons and Atoms. J. of modern physics. 2010, vol.1, no.1, pp. 33-43.

[8] Zubow K.V., Zubow A.V., Zubow V.A. The Dynamics of Low Frequency Movements of Molecular Clusters in the Hardening Process of Epoxide Resins. Chem Promislennost Segodnja. 2008, no. 9, pp. 12-21 (in Russian).

[9] K.V. Zubow, A.V. Zubow, V.A. Zubow. Principles of gravitation spectroscopy. New form of molecular matter. Processes. Fields. Aist Handels- und consulting GmbH, dep. R&D, Berlin, 2011, electronic book (www.zubow.de), p. 854. (in Russian).

[10] Shearing instabilities accompanying high-pressure phase transformations and the mechanics of deep earthquakes. Green, Harry W. Proceedings of the National Academy of Sciences of the United States of America. 2007, vol. 104, no. 22, pp. 9133-9138.

[11] Peacock, Simon M. Are the lower planes of double seismic zones caused by serpentine dehydration in subducting oceanic mantle? Geology. 2001, vol. 29, no. 4, pp. 299-302.

[12] Zubow A.V., Zubow K.V., Zubow V.A. Investigation of the distribution of water clusters in vegetables, fruits and natural waters by flicker noise spectroscopy. Biofizika. 2007, vol. 52, no. 4, pp. 585-592 (in Russian).

[13] Kelemen, Peter B.; Hirth, Greg. A periodic shear-heating mechanism for intermediate-depth earthquakes in the mantle. Nature (London, United Kingdom), 2007, vol. 446, no. 7137, pp. 787-790.

[14] L. Pauling, P. Pauling. Chemistry. H. Freeman and Company San Francisco, 1975.

[15] Hayes, T. J.; Valluri, S. R.; Mansinha, L. Gravitational effects from earthquakes. Canadian Journal of Physics. 2004, vol. 82, no.12, pp. 1027-1040.

[16] Kaladze, T. D.; Horton, W.; Garner, T. W.; Van Dam, J. W.; Mays, M. L. A method for the intensification of atomic oxygen green line emission by internal gravity waves. Journal of Geophysical Research, [Space Physics]. 2008, vol. 113, no. A12, pp. A12307/1-A12307/9.

[17] Roeloffs, Evelyn; Sneed, Michelle; Galloway, Devin L.; Sorey, Michael L.; Farrar,Christopher D.; Howle, James F.; Hughes, Jennifer. Water-level changes induced by local and distant earthquakes at Long Valley caldera, California. Journal of Volcanology and Geothermal Research. 2003, vol. 127, no. 3-4, pp. 269-303.

[18] Tameshige Tsukuda, Keinosuke Gotoh, Osamu Sato. Deep Groundwater Discharge and Ground Surface Phenomena. Bull. Earthq. Res. Inst. Univ. Tokyo. 2005, vol. 80, pp. 105-131.

[19] Omori Soichi, Komabayashi Tetsuya, Maruyama Shigenori. Dehydration event and earthquake in the subductig stab. Frontiers Science Series (2004), 44 (Physicochemistry of water in geological and biological systems), pp. 241-257. Publisher: Universal Academy Press, Inc.

[20] Zubow K., Zubow A.V., Zubow V.A. Water Clusters in Plants. Fast Channel Plant Communications. Planet Influence. Jour. of Biophysics Chemistry. 2010, vol.1, no.1, pp.1-11.

[21] Jung, Haemyeong; Green, Harry W., II; Dobrzhinetskaya, Larissa F. Intermediate-depth earthquake faulting by dehydration embrittlement with negative volume change. Nature (London, United Kingdom). 2004, vol. 428, no. 6982, pp. 545-549.

10.11. Phenomenon of the Influence of Natural Disasters on LRO in Liquids

For a long time, it was noted that natural disasters influence the state of health of animals, which have been used by people in the regions of frequent earthquakes for advance informing about their forthcoming. At the moment, there is no clear understanding of the reasons for these effects. It is only clear that animal behavior and forthcoming earthquake are connected. Japanese seismologists found that prior to an earthquake, liquids exert sharp changes in their properties. For example, the viscosity of olive oil grows anomalously [1]. The authors do not explain these phenomena, but intuitively point to certain connections between a natural disaster and the structure of a liquid. Another group of Japanese investigators [2] described the anomalous effects in the properties of water in the area of forthcoming earthquakes. These effects could not be omitted by specialists on chemical technology and the safety of chemical plants.

Let us attempt to find the sources of this interconnection basing it on the known low-energy changes in the structure of liquids that we described earlier [3, 4].

The aim of this chaptersection is finding a connection between long-range order in a liquid and natural disasters.

The objects of investigation were: distilled water, agarose hydrogel (0.7 wt. %), and solutions of salts and acid [NaCl (3.5 wt. %), HCl (2 wt. %), $AgNO_3$ (1.7 wt. %)]. The experimental procedure has been described in detail in [3-6]. The works on the study of long-range order in liquids in the North part of Germany were carried out in the period from February to May 2007. Comparison of the data on the long-range order (LRO) in liquids and natural disasters is made using the principle of "coincidence" in the timeframe of the events in the cell of measurement (53.08°N and 12.75°E) and the time of a disaster that has not been predicted to the moment (except the hurricane on February 17, 2007 over the North Sea and German territory). All event times are in Greenwich Mean Time.

Study of long-range order in the agarose hydrogel commonly used for modeling biological matrices showed the effects of the instability of long-range structure appearing as a sharp increase in the average value of the *f* parameter that characterizes the interaction of clusters with the environment [3, 4] (Figure 10.11.1).

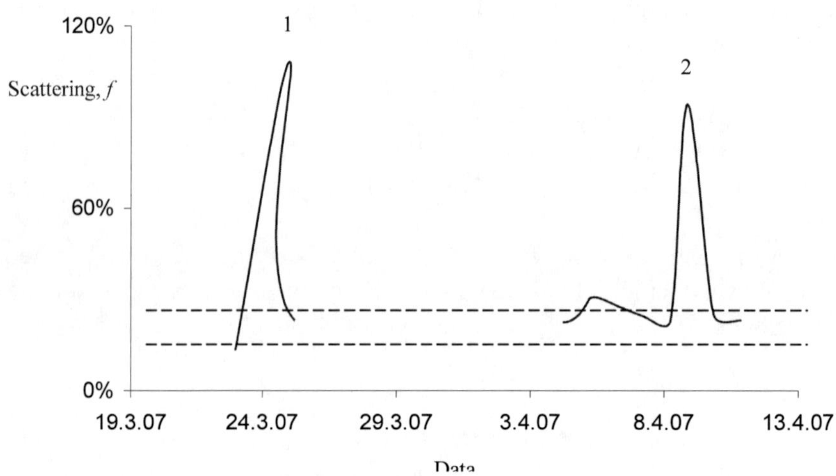

Figure 10.11.1. Interaction of all hydrogel water clusters with their surroundings (average *f* value) versus the observation time. The normal behavior of *f* is between the dotted lines. 1– earthquake in the Japanese Sea (March 25, 2007 at 00:42 (GMT), 7.2 magnitude on Richter Scale) and 2 – earthquake on the Kuril Islands (April 9, 2007 at 10:17 (GMT), 6.1 magnitude on Richter Scale). Earthquake Almanac, see http://www.gfz-potsdam.de/news/recent/index.html.

Average scattering of the *f* parameter thus reflects the "nervousness" of the system of super cluster structure in the hydrogel long-range order. The agarose hydrogel contains up to 99.3 wt. % of water, therefore, an increase in the cluster vibrations can be assigned with a great probability to the strengthened oscillation of the water clusters. The water clusters, as is known, affect conformations of biomolecules [7], hence, the influence direction and rate of certain biochemical processes in living organisms. Furthermore, they affect their biological reactions, that is, behavior. Thus we can trace the indirect influence of the earthquake's seismic wave impacts on the organism biomatrix and fix the connection between the behavior of people and animals and natural disasters. Damage to the LRO in the cluster structure of water in a biomatrix can influence the health state of the operators in chemical plants and this can introduce the factor of human risks, which should be taken into account. It should be noted that in the period of April 4- 9, 2007, there occurred many earthquakes. We succeeded in registering the events shown in Figure 10.11.1, proceeding from the time of achieving cells of GMS in Germany through seismic waves (4…7 km/s) and the time the experiment was carried out. At other events, the time of registering GMS-spectra and that of the coming earthquake shock wave did not coincide. We found that the coming shock wave activates the GMS-sensor differently and strongly depending on the chemical composition of the liquid. In some liquids, the LRO reaction on the shock wave impact can continue for several hours and is selective with respect to the mode of the molecular bunch. On the other hand, chemical processes in the earthquake epicenter can be understood by means of analysis of this selectivity, which reflects the resonance interaction of such oscillators in the measuring cell and in the earthquake epicenter. Therefore, chemical plants working in continuous regime can be used to possibly monitor events in the Earth's interior for a deeper understanding of the whole wide spectrum of the chemical processes of an earthquake.

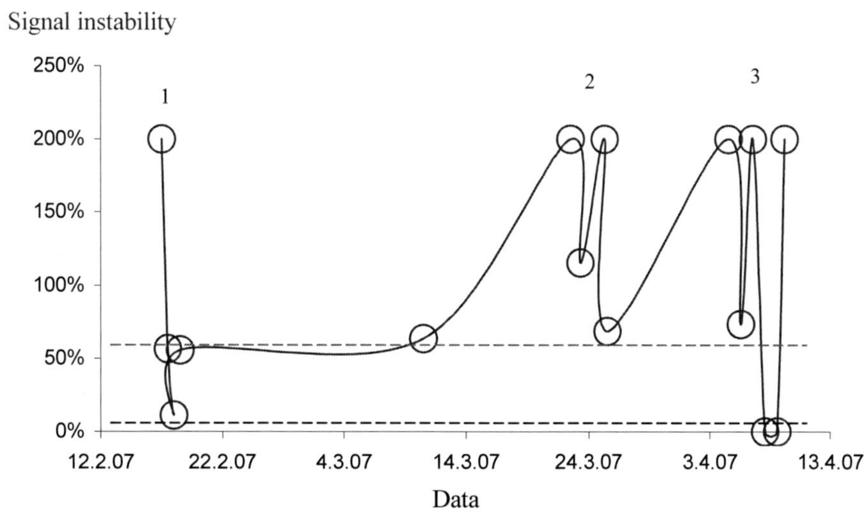

Figure 10.11.2. Interaction of all clusters in 3.5 wt. % NaCl with their surroundings (average *f* value) versus the observation time. The normal behavior of *f* is between the dotted lines. 1 – storm over the North Sea (February 16-17, 2007, wind gusts up to 200 km/h), 2 – earthquakes in the Pacific Ocean near Vanuatu (March 25, 2007 at 00:40 (GMT), depth of 93 km) followed by one in the Japanese Sea (March 25, 2007 at 00:42 (GMT), depth of 19 km), and 3 – a series of strong earthquakes in the region of South-West Pacific. Almanac of earthquakes, see http://www.gfz-potsdam.de/news/recent/index.html.

Figure 10.11.2 shows variations in the average value of the *f* parameter for ion pair solvated clusters of NaCl and water in 3.5 wt. % solution modeling sea water. From the figure, it is seen that these NaCl clusters respond to the damage of long-range order in sea water. Therewith, it is not significant whether this is damage due to hurricane (1) or due to compression in the epicenter of an earthquake in the ocean (2, 3). The long-range damage in the salt solution can be represented by Scheme I:

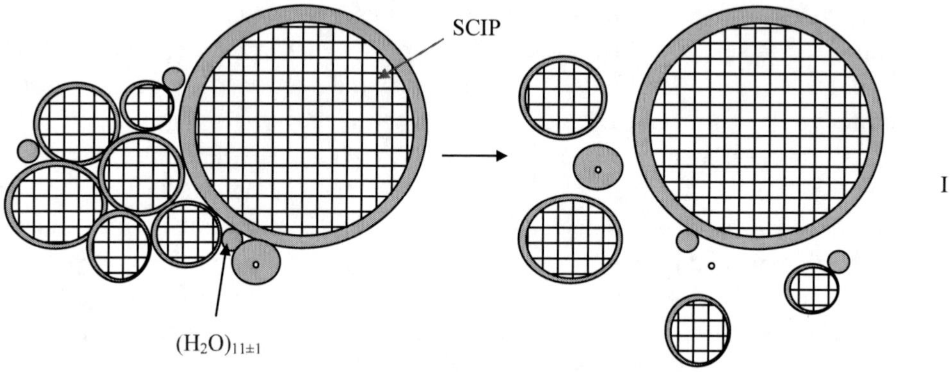

Such a scheme assumes the dispersion of the salt ion pair solvated cluster nanosuspension under the action of a shock wave from an earthquake or it assumes a resonance wave from the ion pair solvated cluster that is damaged due to a hurricane in the North Sea. This process is an unstable one, and the system quickly returns to a state close to its initial one even after a few hours. For some of the large NaCl ion pair solvated clusters, however, the process of achieving equilibrium of its interaction with the environment can occur in a few days.

We found that integral characteristics of solutions, including the average molecular weight of all clusters (M_{GMS}), "fills" the appearance of earthquake in two to three days, likely as a result of strong pressure in the earthquake zone prior to the collapse and the sea water LRO damage in the zone of epicenter. Understanding the developing process of the disintegration of SCIP in salt solutions within Scheme I should be taken into account in chemical technology, because this can lead to a sharp increase in the rate of chemical reaction and distortion of the stability of chemical plants. On the other hand, one can expect that the architecture of displacement of chemical reactors with forced dissolving and forced stirring near the reactors of fine synthesis can lead, according to Scheme I, to induced or continuous action on the LRO in the reactor for synthesis or to the psychological stability of the staff. Some differences occur in the response of hydrochloric acid solution to natural disasters.

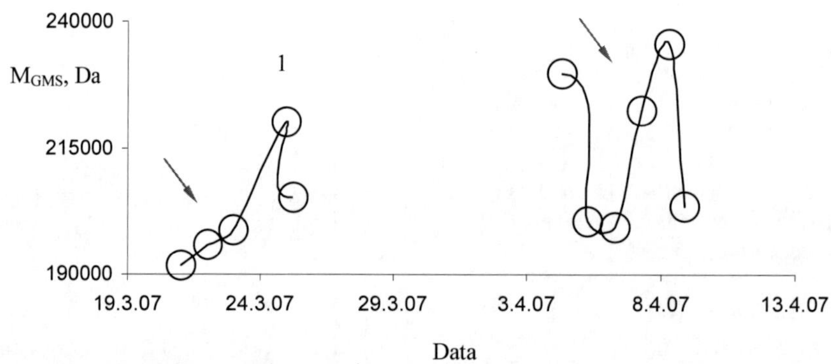

Figure 10.11.3. Average cluster mass M_{GMS} in an HCl solution versus the observation time in spring 2007. 1 – earthquakes in Japan and Vanuatu (see Figure 10.11.2).

Figure 10.11.3 presents a change in the average molecular weight of clusters in 0.6 M HCl solution. As seen from the figure, an increase in M_{GMS} upon the approaching earthquake in the ocean can be registered even in a few days (shown by arrows). Thus, the integral characteristics of long-range order can be used to predict earthquakes. According to the view of Japanese seismologists [1] that were in the region of the earthquake in the Sea of Japan, this occurs due to the dehydration of mineral pores under the action of very high pressure moving layers. We can assume that such dehydration strongly affects the structure of the solvated ion pairs in sea water due to a forced shift of equilibrium to the side of formation of HCl ion pair solvated clusters with a size less than the size of solvated clusters of NaCl ion pairs [5, 8, 9].

$$NaCl{\cdot}aq + H_2O \rightarrow HCl{\cdot}aq + NaOH{\cdot}aq$$

Thus, principally appears the possibility of obtaining distant information on the chemical processes in the zone of an earthquake and on their influence on technological processes with solutions.

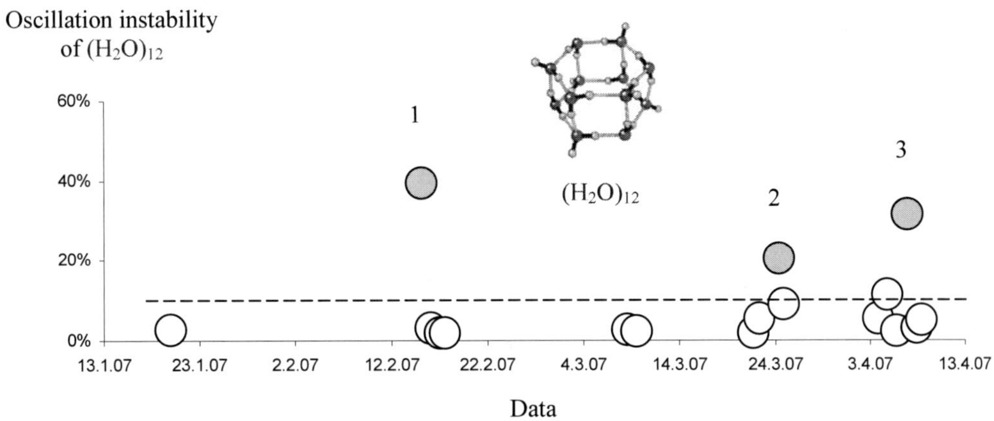

Figure 10.11.4. Oscillation instability of the base water cluster $(H_2O)_{12}$ in 0.1 N $AgNO_3$. Measuring cell at an earth depth of 1 m at 278 K. The calculated cluster structure was kindly provided by Professor Lenz [10]. 1– storm over the North Sea, 2 – earthquake in the Japanese Sea, and 3 – a series of earthquakes in the Pacific Ocean (see Figure 10.11.1).

Figure 10.11.4 presents the dynamic variations in the instability of the *f* signal of water cluster $(H_2O)_{12}$ [10] in 0.1 N $AgNO_3$ solution. This cluster is also found to be distantly sensitive toward weak energy streams of oscillation of its analogs in sea water affected by a hurricane or earthquake. In the series of earthquakes in the period April 4-9, 2007, it showed selectivity with respect to the shock waves attaining the place of measuring. Hence, seismic waves are the super positions of a series of waves, and the clusters respond actively to those defined by signals from their damaged analogs in the epicenter of a hurricane or earthquake. For chemical technology, this means that there is a necessity to select technology and equipment appropriate for the zone of frequent natural disasters.

Conclusion

Liquids are unique natural sensors of energetically weak processes in nature. Natural disasters distantly affect the long-range order in liquid. Changes in the long-range order in a liquid under the influence of seismic waves or resonance fields of molecular cluster oscillations can result in destabilization of technological processes and even lead to failures. These changes can be used for early information on forthcoming disasters through the continuous monitoring of the structure of the liquids applied in technological processes.

Risks of chemical processes due to this phenomenon lie in the quality of products and stability of the technological processes and in the area of the human factor: fatigue and overstrain. In our opinion, the situation can be improved by means of the inclusion into emerging technology of systems for monitoring distant influence on the long-range order in liquids and distribution of early information.

References to 10.11

[1] Omori Soichi, Komabayashi Tetsuya, Maruyama Shigenori. Dehydration event and earthquake in the subductig stab. Frontiers Science Series (2004), 44 (Physicochemistry of water in geological and biological systems), pp. 241-257. Publisher: Universal Academy Press, Inc.

[2] Tameshige Tsukuda, Keinosuke Gotoh, Osamu Sato. Deep Groundwater Discharge and Ground Surface Phenomena. Bull. Earthq. Res. Inst. Univ. Tokyo. 2005. vol. 80, pp. 105-131.

[3] Zubow A.C. Zubowa K.C., Zubow C.A. The Phenomenon of Water Clusters Distribution in Water and Solvated Clusters of Ion Pairs of NaCl in Solution Near the Wall. The Wall Effect Khimicheskaya Promyshlennost’, 2006, vol. 83, p. 300-306.

[4] Zubow, A.V., Zubow , K.V. and Zubow, V.A. The Investigation of the Salt Dissolving Mechanism and of the Ageing of the NaCl Solution in Water by Flicker Noise Spectroscopy. Khimicheskaya Promyshlennost’, 2007, vol. 84, no. 1, p. 40-48.

[5] Zubowa, K.V., Zubow , A.V. and Zubow, V.A. Spectra of seed crystals oscillation of NaCl in water solution. Journal of Applied. Spectroscopy, 2005, vol. 72, no. 6, p. 766-772.

[6] Zubowa, K.V., Zubow , A.V. and Zubow, V.A. Cluster structure of liquid alcohols, water and n-Hexane. Journal of Appl. Spectroscopy, 2005, vol.72, №3, pp. 321-328.

[7] Shiryayev A., Pagan D.L., Gunton J.D., Rhen D.S., Saxena A., Lookman T. Role of solvent for globular proteins in solution. J. Chem.Physics. 2005, vol.122, p.234911.

[8] Raji Heyrovska. Dependences of molar volumes in solids, partial molar and hydrated ionic volumes of alkali halides on covalent and ionic radii and the golden ratio. Chem. Phys. Letters. 2007, vol. 436, pp. 287-293.

[9] Daisuke Yokogawa, Hirofumi Sato, Shigeyoshi Sakaki. An integral equation theory for 3D solvatation structure: A new procedure free from 3D Fourier transform. Chem. Phys. Letters. 2006, vol. 432, pp. 595-599.

[10] Annika Lenz, Lars Ojamäe. On the stability of dence versus cage-shaped clusters: Quantum- chemical investigations of zero-point energies, free energies, basis-set effects and IR spectra of $(H_2O)_{12}$ and $(H_2O)_{20}$. Chemical Physics Letters. 2006, vol. 418, pp. 361-367.

Chapter 11

THE NATURE OF DOMAINS IN PROTEINS

It is known that macromolecule coils are usually inhomogeneous; this is explained by the hydrogen or donor-acceptor bonds between the functional groups of polymers that form clusters or domains. A clear understanding of the thermodynamics of such processes is still lacking in the literature. There is, however, the basic idea developed by the authors that the Earth's penetrating gravitational white noises stabilize organized structures [1]. According to this idea, the energy clusters (EC) in the gravitational noises promote the formation of molecular clusters. That means the last ones take the energy for their existence from EC [2, 3], Figure 3.2.1.

The dynamics of changes in the coils' heterogeneity in proteins is an important part to understand in the transformation of conformations under the influence of various environmental factors. For this understanding especially, a new non-destructive analysis method is needed in applied biochemistry. On the other hand, a serious problem in modern biochemistry is the lack of a fast and accessible method to investigate in vivo and in real time. This refers to the structures of proteins and their associates and to the cluster formation in biological liquids. It furthermore refers to solvated clusters of acids/bases, salt ion pairs, and water clusters as well as to their interaction with their surroundings. Understanding the full range of these processes as a unit will allow for a new level of problem solving in applied biophysics and biochemistry.

It is therefore the aim of the present chapter to use gravitational mass spectrometry for the analysis of mass heterogeneities in proteins and protein containing biomatrices.

In this chapter, on the example of simple proteins and their solutions in distilled water, we tried to show the possibilities of the GMS-method (earlier called the FNS-method) [2, 3, 4, 5] for the analysis of heterogeneity in proteins. The investigative objects were hen eggs, lysozyme solution (14 kDa, Sigma) in a sodium phosphate buffer (2 mg in 1.4 ml of buffer: pH = 7.5, 137 mmol NaCl, 2.7 mmol KCl, 12 mmol $Na_2HPO_4 \cdot 2H_2O$, 1.2 mmol KH_2PO_4), and chymotrypsinogen-A solution (Sigma, 25,651 Da) 12 mmol/l. The GMS-sensor was placed directly into the sample. The electrophoresis of proteins denatured at 368 K proceeded in the PAA-gel (polyacrylamide gel, 12 wt. % SDS buffer) at 20 μA and 290 K. The unreduced egg white was diluted in SDS (1:10, 5 min. 368 K) and 9 μl, which is used for the electrophoresis. The same procedure was made for the egg yolk, but with a buffer dilution of 1:50 and a sample volume of 4.5 μl. The molecular weights of denatured the proteins were analyzed in 12 wt. % PAA-gel.

The measurement method is described in [3] and in §3. Before the measurement, the samples were stored for a week under exclusion of light and mechanical fields. The GMS-sensor is directly placed inside the sample (10 ml). Here the sample was placed in a special earthed, metal box that was isolated from energy flows (heat, noise, mechanical and light) [2], Figure 4.0.1. The negative signal in the GMS (*-f*) represents the energy release of a collapsed cluster leading to its higher density and crystallization. On the other side, the positive signal (*f*) means energy absorption at the interaction of the shock wave with the expanded cluster and its melting. The values of the collapsed clusters are therefore described with a minus and that of the expanded one with a plus. For more information, see [2].

In Figure 11.1, the GMS-spectrum of lysozyme is given.

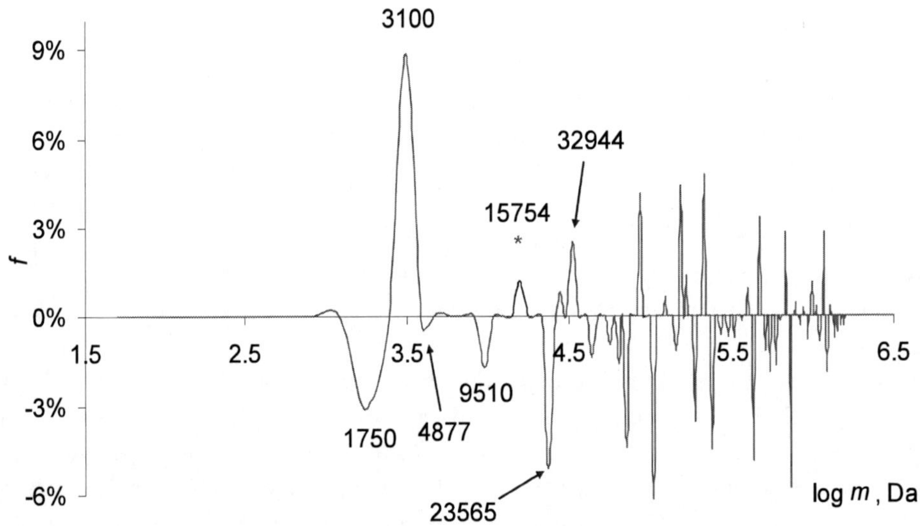

Figure 11.1. General GMS-spectrum of lysozyme. GMS is obtained by subtracting the spectra of the solvent clusters from those of the lysozyme solution clusters (mass ensembles up to 3.2 MDa). The Zubow constant amounts to $6.2 \cdot 10^{-15}$ N/m. 293 K. *Lysozyme coil signal. Scanning time was 30 s.

As is visible in Figure 11.1, the GMS subtraction spectrum reveals three main signals at 1,750; 3,100; and 9,510 Daltons (domains in the lysozyme coil) where the coil itself oscillates as a weak expanded cluster with a mass of 15,754 Da. The signals of larger masses than the mass of the coil are ascribed to coil associates and sub-micelle and micelle structures that are formed by coils in the solution. For example, the cluster with the mass of 32,944 Da belongs to the lysozyme dimer consisting of two coils and 4.4 wt. % water. Water seems to be there where the coils contact each other. The cluster signal with the mass of 23,565 Da being 1.5 times higher than the mass of the lysozyme coil itself represents a unique cluster that appears in all molecular systems [6, 7]. It is possible that this oscillator is formed compulsorily between one coil, the two domains of 3,100 and 4,877 Da, and nine water molecules (bound water). Such an oscillator is shown in the scheme (II) below as a part of an associated coil; it differs from the complete dimer only by the conformation.

According to a computer simulation on the basis of NMR data, lysozyme coils are found to consist of at least three areas of mass concentrations (60, 20, and 10%, Figure 11.2) [8]. This agrees well with GMS, which analyzed three types of large oscillating masses with weight fractions of 60, 21, and 12% (Table 11.1). The practical coincidence of these values has encouraged the authors to develop the method for use in biochemistry. With the GMS-method, a fast and direct determination of cluster/domain masses in Daltons is possible.

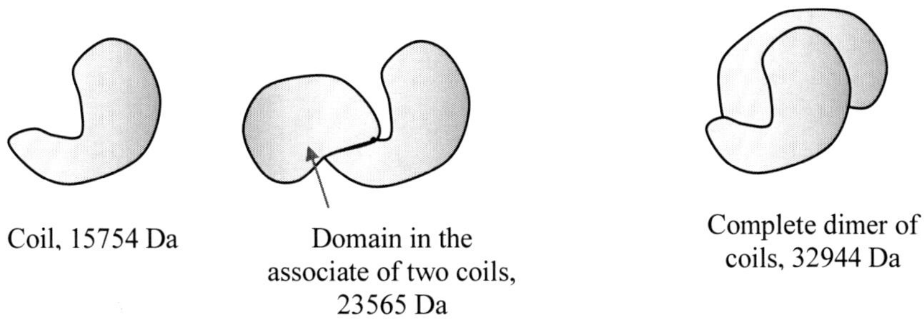

Scheme II.

Table 11.1. Cluster masses in the GMS-spectrum of a lysozyme solution. The spectra were obtained by subtracting the spectra of the clusters in the solvent from those of the clusters in the lysozyme solution (mass ensembles up to 16.3 kDa).

Oscillator kind at 293 K	Mass part of cluster %	*m*, Da	Cluster form
domain	up to 21	1,750	collapsed
–	up to 3	2,393	–
–	up to 60	3,100	expanded
–	up to 2	3,938	collapsed
–	up to 1	4,877	expanded
–	up to 12	9,510	collapsed
coil	–	15,754	expanded

The effect of pressure on the coil structure of lysozyme investigated by Rafaee [8] can be discussed in terms of a changed quality of white noises. Strengthening the noise background (translation into colored noise) as the result of increased pressure should lead to changes in organized systems such as protein coils, too.

The signal of the lysozyme coil in the GMS-spectrum is rather weak and is about three times less than that of the most common dimer form of this enzyme [9]. The lysozyme solution at 293 K reveals a strong signal (log m = 4.51) that is attributed to its dimer (expanded cluster, f = 3%). According to calculations based on GMS data, the dimer is analyzed to contain 8.9 wt. % water. The molecular mass of the oscillating coil amounts to 15,754 Da, which differs by 11% of the mass (14 kDa) that is determined by Sigma. This difference has been attributed to solvated water in the crystalline lysozyme (9 wt. %, [10]).

Hence, already in the solution, the dimer is to be regarded as a seed crystal. In the lysozyme coils themselves, no water clusters have been found; water therefore seems to be present in another form in the coils.

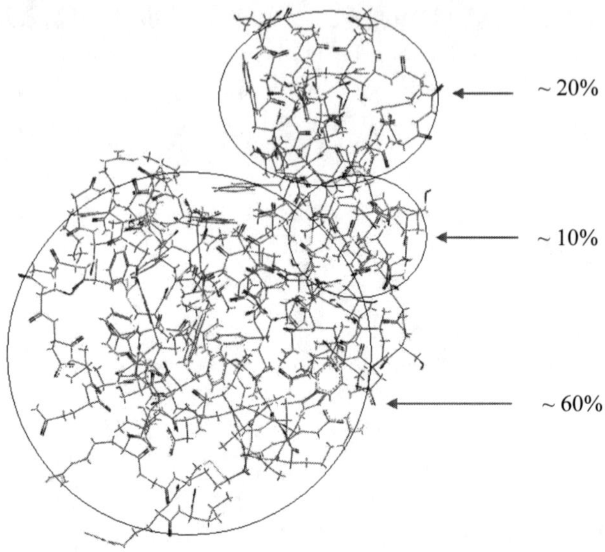

Figure 11.2. Computer model of a lysozyme coil on the base of NMR data [8], which is used for the comparison with the GMS results (Table 11.1).

With GMS, it is possible to get an idea of the heterogeneity inside the protein coils at the level of clusters/domains, and furthermore, to determine the cluster mass fractions as well as the masses of the individual clusters themselves.

GMS-spectra of this enzyme can be recorded both in the solid (powder) and dissolved state at all temperatures and pressures. These features facilitate the investigation of the mutual cluster transformations (expanded ↔ collapsed), cluster interactions with surroundings inside the coil, and the coil interactions with neighbors. Additionally, signals of coil associates in solutions (dimers, trimers, etc.) can be analyzed, which allows us to quickly determine the number of coils in these associates. Similar studies are carried out to analyze clusters in the coils of aldolase and chymotrypsinogen-A in PBS-buffer. Here oscillating clusters are found, too, and the cluster distribution at different temperatures can be calculated. Furthermore, using this method, it is possible to get some idea of the state of water included inside the coils, of the degree of the clusters' agglomeration in associates, and of their stability in solutions (Table 11.3).

Figure 11.3 shows the dynamics of the cluster signals in egg white in vivo at heating. Comparing the information provided in Figure 11.3 with the data obtained by electrophoresis separation of the components in the hens' eggs (Figure 11.4 and Table 11.2), the main egg components can be identified as well as its structure and density changes under increasing temperature. Concerning the molecular masses of the ingredients in egg yolk and egg white, there a good match between the GMS and electrophoresis data (Figure 11.4) has been found. This encouraged the authors to propose GMS as a new rapid analysis method (1…30 s) to determine the masses of domains and coils in proteins and to investigate their interaction with the surroundings. On the other hand, comparing the results from GMS and electrophoresis, it is possible to analyze the content of water and salts in the protein coils. In addition, super polymeric structures like dimeric, tetrameric, etc., protein coils that oscillate as a single unit (Table 11.2) can be recorded. This table shows a comparison of the results obtained by GMS and electrophoresis; it is based on the understanding that mass concentrations (clusters/domains, coils) are formed compulsorily under the influence of white noises [2]. To avoid anthropogenic noises (energy fields), the samples are kept in a box mostly isolated from these influences.

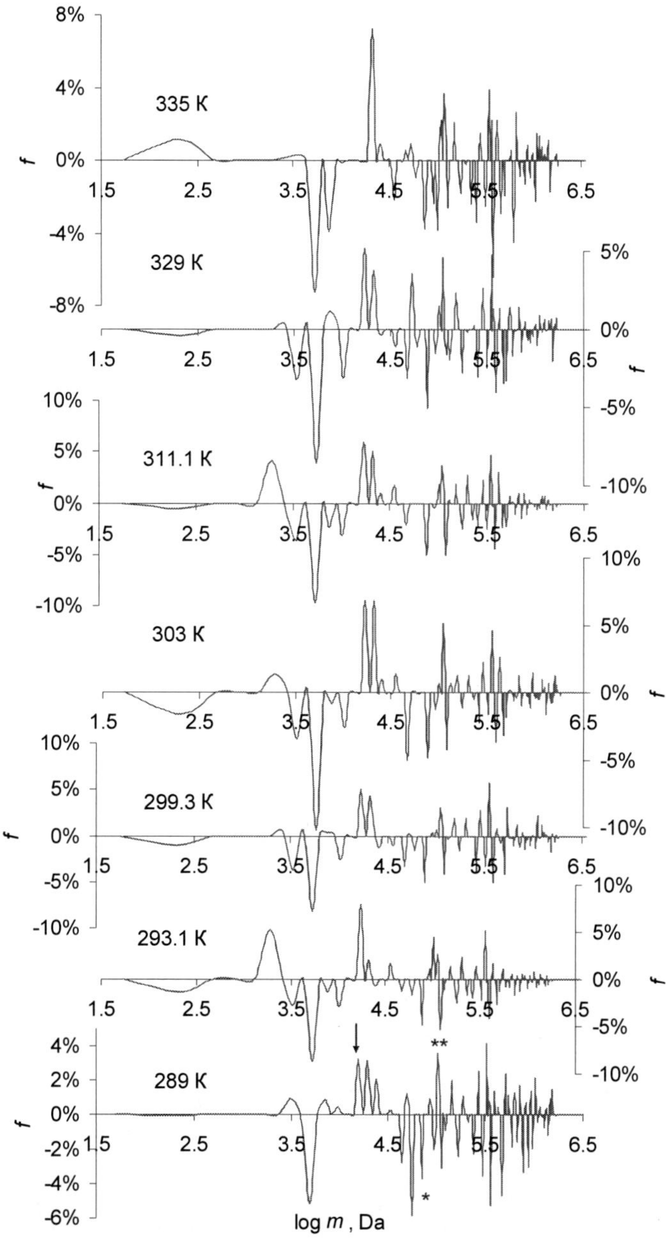

Figure 11.3. GMS-spectra of egg white biomatrices (in vivo) from free range hens. $P > 1$ N/m^2, Zubow constant = $6.2 \cdot 10^{-15}$ N/m, scanning time at each temperature is 30 s, arrow – lysozyme coil, *Ovalbumin coil, **Albumin coil.

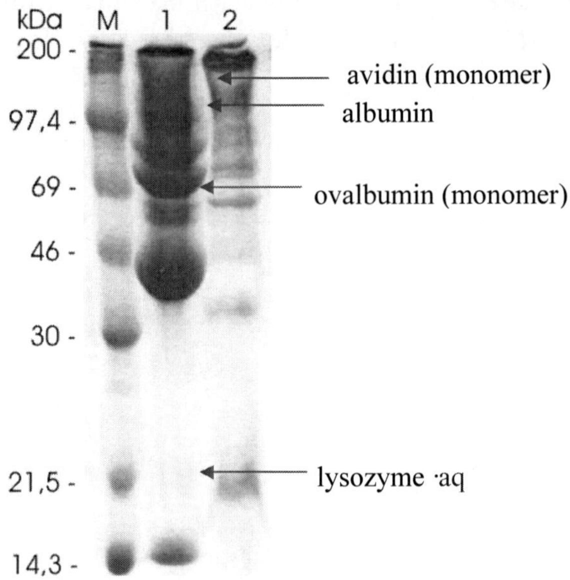

Figure 11.4. Protein electrophoresis of egg white (1) and egg yolk (2) at 293 K. M – reference substance being a mixture of proteins with known molecular weights ("Rainbow," Amersham Life Sci. Ltd).

Table 11.2. Comparison of GMS (30 s) and electrophoresis (Figure 11.4) on the example of hen eggs. Regarding GMS, the mass concentrations and their signal intensities are taken into account

Electrophoresis	335 K (GMS)		311 K (GMS)		293 K (GMS)		included H_2O	
log *m*	egg white	egg yolk	egg white	egg yolk	egg white	egg yolk	mass difference*	
lysozyme ·aq	log *m*	log *m*	log *m*	log *m*	log *m*	log *m*	log *m*	molecules
4.16	4.2	4.2	4.2	4.2	4.2	4.2	3.16	81
	-0.1%	0.7%	6%	2.9%	7.9%	-0.1%		
	ovalbumin·aq							
tetramer	5.27	5.27	5.27	5.27	5.27	5.27	4.21	897
	-0.2%	-1.2 %	2.7%	-0.1%	2.4%	0.2%		
monomer	4.64	4.64	4.64	4.64	4.64	4.64	3.01	57
4.63	0.6%	-2.1%	-2.1%	-0.6%	-1.2%	0.2%		
	avidin (sub. oxidase)							
dimer ·aq	5.21	5.21	5.21	5.21	5.21	5.21	3.56	203
5.20	-1.8%	+0.4 %	-2.5%	-1.5%	-2.5%	5.2%		
monomer ·aq	4.93	4.93	4.93	4.93	4.93	4.93	3.76	321
4.90	-2.3%	-0.4%	-0.5%	-2.5%	1.1%	0.5%		
	albumin							
4.84	4.85	4.85	4.85	4.85	4.85	4.85	3.12	73
[11]	-3.8%	-3.4%	-5.1%	-4.2%	-4.9%	0.5%		

*Difference of masses obtained by GMS and electrophoresis.

(-) – collapsed cluster and (+) – expanded cluster.

As is visible, the masses of GMS are slightly higher than that of the electrophoresis where these differences are ascribed to include water and possibly, traces of salt. The water concentration in coils is in the range from 2.2 to 10 wt. % (from 57 to 897 molecules); the smaller the mass of the coil, the higher the content of water.

The oscillation signals of the mass concentrations (domains) reflecting the albumin coils, show that in egg yolk at heating, the coils change their form from the expanded (293 K) to collapsed while they are present in the egg white only in the collapsed form in the whole temperature interval.

Regarding the lysozyme coils in the egg white, their expanded form reaches a maximum at 293 and 311 K where at 335 K the density of the coils even increases. Since the signal intensity (f) reflects the coils' interaction with the surroundings, two variants concerning how the temperature influences the density of the coils are possible:

1. Internal changes of the macromolecules' conformations leading to a loosening of the coils (weakening of the cohesive interactions)
2. Reinforcement of the adhesive interactions between the coil surfaces with their surroundings

If we assume that the conformation of macromolecules in the coil does not change significantly in a narrow temperature range, the large $|f|$ values suggest a decrease in the adhesive interactions of the cluster masses (coils) with the surroundings and an increase of the cohesive interactions in them. The more the GMS signal intensity, the less the interaction between the oscillator and the surroundings and the higher its individualization degree.

Collapsed clusters are observed to act as seed crystals of regular, crystalline structures; therefore, they could be interesting objects for studies in biochemistry. Figure 11.5 shows how the sum of the collapsed clusters in egg yolk changes at heating (12 K/h). The change does not proceed monotonously. The curve is characterized by some extremes, for example, and the maximum at 305 K may indicate a larger number of them at temperatures being close to the formation of the hen's embryos.

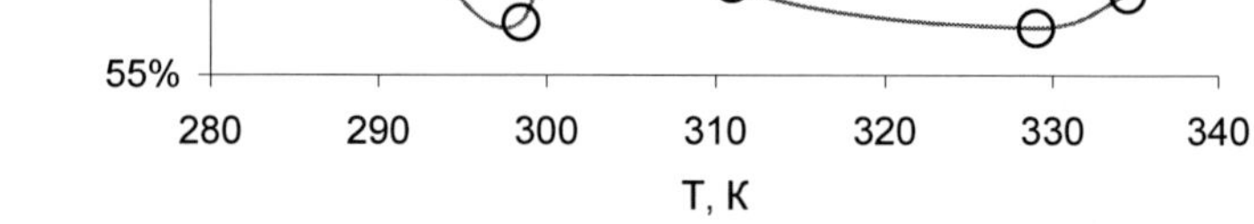

Figure 11.5. Temperature dependence of the sum of collapsed clusters in egg yolk.

Thus the transformation of "white noise" into "color noise" [1] by heat flow changes qualitatively in the cluster structure of egg yolk. The data presented in Figure 11.6 illustrates the nature of such changes.

Figure 11.6 presents the GMS-spectra of egg yolk clusters at heating in the dashed form. Visible egg yolk contains stable super cluster formations, marked by arrows but there are slight changes in the cluster distribution. With more prolonged thermostating, these differences become more pronounced and they can be applied to understand the beginning of embryo formation.

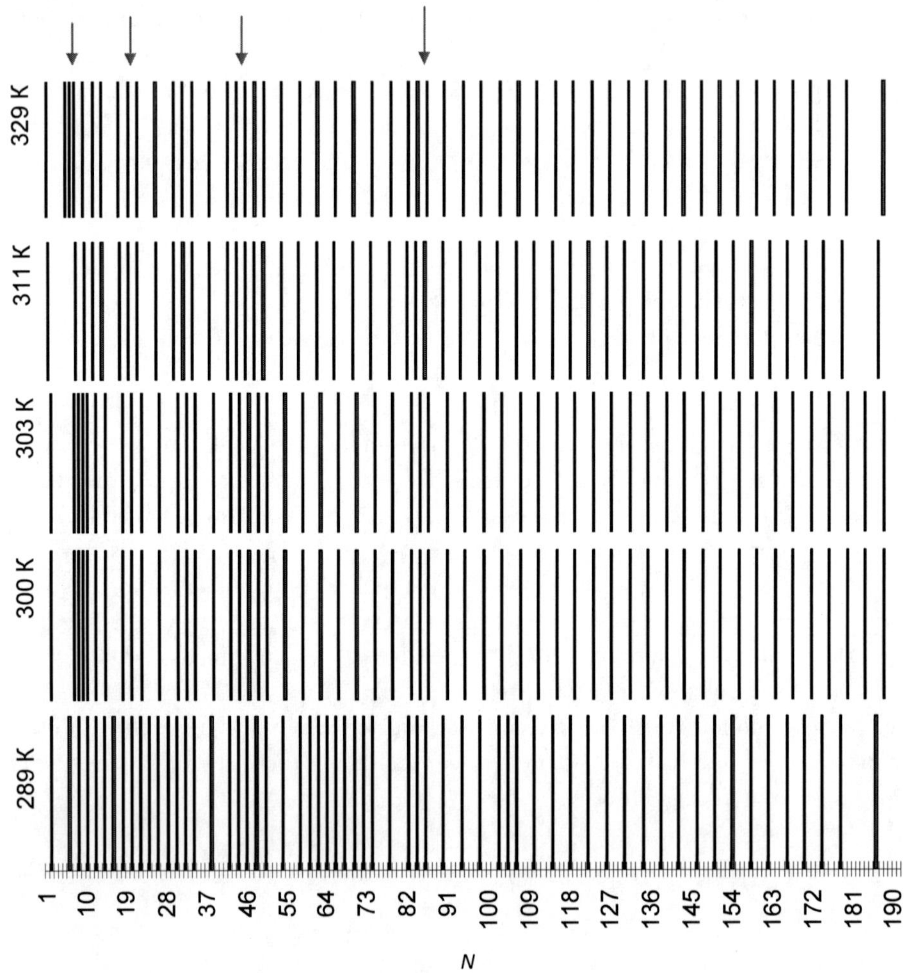

Figure 11.6. GMS-spectra (dashed form) of clusters in egg yolk at heating (12 K/h). Every line corresponds to one cluster regardless of its signal intensity; the distribution of the clusters is described by the following equation: log (m) = 0.9214 · ln (N) + 2.3216, where m is the mass of the cluster in Daltons and N is its serial number in the spectrum (see Table 11.3). Before each measurement there will be a 30 minute thermostating.

The presentation of the spectra in the dashed form promotes the automation of data administration. Strichcodes are a kind of "fingerprint" of a given biological system; they clearly show that the distribution of clusters in biomatrices is irregular.

Clusters themselves form super cluster structures. Such super cluster structures and their formation dynamics upon the heating/aging of biological systems as well as their interaction with physical fields and the effects of chemical substances on them may be an interesting research object for biochemists.

In Figure 11.7, the GMS-spectrum of chymotrypsinogen-A is given.

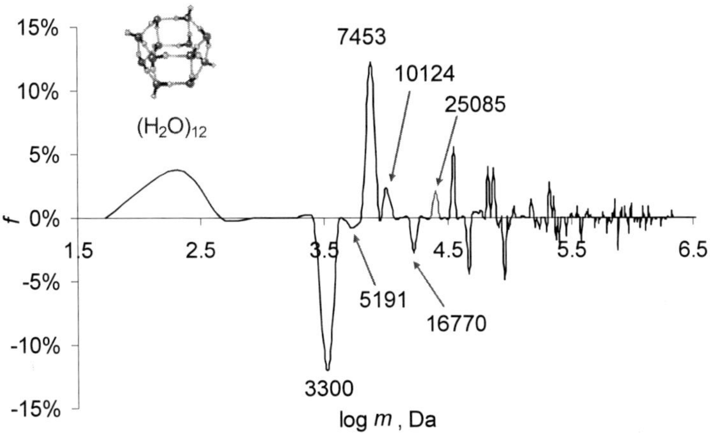

Figure 11.7. Overview GMS-spectrum of chymotrypsinogen-A. It is obtained by subtracting the water cluster spectrum from the chymotrypsinogen-A solution spectrum (mass ensembles up to 3.2 MDa). Zubow constant = $6.6 \cdot 10^{-15}$ N/m, 293 K. The numbers above the signals means the mass of the main domains and of the coil (25,085 Da). The model of the base water cluster has been kindly provided by Professor Lenz. Scanning time is 30 s.

The spectrum of chymotrypsinogen-A contains the signals of the base water clusters [12, 13], which are independent kinetic units that are included in the protein. The analysis of the main signals of this protein is given in Table 11.3. Table 11.3 presents the main characteristics of the domains in chymotrypsinogen-A coils, recorded with the highest GMS sensitivity. To obtain the real signals of the enzyme, the solvent signals are subtracted from those of the enzyme solution. As is visible from Table 11.3, the chymotrypsinogen-A coil is not homogeneous. It consists of at least nine domains that are characterized by different densities and formation energies. Investigating the temperature influence on intracoil processes and the coil's conformation, both are observed to be relatively stable at the domains' level and unstable at the micro domains' one (smallest clusters).

Comparing the data of this table with computer protein models (PDB), a satisfactory compliance has been taken but it should be noted that the data of PDB bases on crystalline enzyme structures do not really reflect their native state.

The successful application of the GMS-method for the rapid analysis of casein, collagen, and proteins in muscle tissue is described in [4, 6].

The sequences of investigated clusters in natural waters [14], distilled water [5], polymers [7, 15, 16], solutions [17, 18], protein coils, and polysaccharides (lysozyme, L-arginine, chymotrypsinogen-A, amylopectin, etc. [6]) are characterized by similar masses. Common to all is a series consisting of the cluster masses shown in Table 11.4 (only 31 members are listed in an ordered sequence). These sequences, including lysozyme (Table 11.1), are satisfactorily described by the following Zubow equations:

$$\log m = (0.80 \pm 0.13) \cdot \ln N + (2.34 \pm 0.12)$$

where N is the order cluster number in the spectrum beginning with the smallest mass of 0.2 kDa (N = 1).

The Zubow equation relating the cluster mass (*m*) with its oscillation frequency (ω) is the following:

$$m = 10^{11} \cdot \omega^{-2}$$

Table 11.3. Real signals of domains in chymotrypsinogen-A coils (Sigma, 25,651 Da). Aqueous solution (0.012 mmol/l) at 310 K. For the spectra calibration regarding the energy, the formation/destruction energies of known water clusters [5] are used, which have been determined in the same measuring cell. Comments are in the text

Part of mass %	*m*, Da	ω, kHz	E, kJ/mol	Oscillators
0.2	2,547	6.3	0.004	expanded domain
16	3,300	5.5	0.616	–
1	4,193	4.9	0.019	–
15	5,191	4.4	-0.396	collapsed domain
0.3	6,246	4.0	0.011	expanded domain
47	7,453	3.7	-1.626	collapsed domain
0.01	8,767	3.4	-0.002	–
0.2	10,124	3.2	-0.216	–
20	16,770	2.5	1.687	expanded domain
–	25,085	2.0	-1.804	collapsed coil

The applicability of the Zubow equations for all molecular matter compounds points to a common mechanism for the formation of clusters/domains, which get the energy for their organization from energy clusters [2, 6].

CONCLUSION

The GMS method allows for more accurate, reliable, and faster determination of the mass of protein domains in coils as well as the mass of coils and coil associates. Additionally, it allows us to get an idea of the sub-micelle and micelle coil structures of proteins in their solutions and biomatrices.

Under anthropogenic influence the white noise is transformed into a color one, which changes the stabilizing effect of the noise on organized systems. Hence the long-term effects of anthropogenic noises in urban environments, factories, and transport are assumed to affect at least the super cluster structure of biological systems. In our opinion, there should be a limit of such an overload, above which the irreversible changes in biological systems take place, e.g. in protein conformations, water clusters, and solvated clusters of ion pairs, which refers to the whole biomatrix and occurs as a consequence to accelerate the mutagenesis processes.

At the domain formations in proteins and biomatrices, hydrogen and donor-acceptor bonds are observed to play a secondary role only while the energy clusters in stationary gravitational waves play the "first violin." They serve as an energy source for the formation of domains and clusters.

Table 11.4. Clusters/domains (log *m*) and oscillation frequencies under the influence of shock waves observed in liquids, polymers, and biomolecules (explanation in the text)

№, N	Water 293 K	Methanol 293 K	*p*-Xylene 293 K	*n*-Hexane 293 K	Agarose gel 3 wt. %	Egg white 293 K	Egg yolk 293 K	Chymotrypsinogen-A 293 K	Amylopectin 295 K	ω, Hz
1	2.30	2.28	2.28	2.28	2.30	2.30	2.30	2.31	2.32	22,222
2	–	–	3.24	3.24	3.26	3.26	–	–	3.28	7,353
3	3.39	3.38	–	–	3.39	–	–	3.41	–	6,289
4	3.51	3.49	–	–	3.51	3.51	3.51	3.52	3.53	5,525
5	3.61	3.60	–	–	3.61	–	–	3.62	–	4,902
6	3.70	3.69	–	–	3.70	3.70	3.70	3.72	3.73	4,405
7	3.78	3.77	–	–	–	–	–	3.80	–	4,016
8	3.86	3.85	–	–	3.86	3.86	3.86	3.87	3.89	3,676
9	3.93	3.92	–	–	–	–	–	3.94	–	3,390
10	3.99	3.98	–	–	3.99	3.99	3.99	4.01	4.02	3,155
11	–	–	–	–	4.11	–	–	–	4.14	2,755
12	4.21	4.20	–	–	4.21	4.21	4.21	4.22	4.24	2,451
13	4.30	–	4.29	4.29	–	4.30	4.30	–	–	2,203
14	4.39	4.37	4.37	4.37	4.39	4.39	4.39	4.40	4.41	2,004
15	–	–	4.45	4.45	–	–	–	–	–	1,838
16	4.53	4.52	4.52	4.52	4.53	4.53	4.53	4.54	4.56	1,695
17	–	–	4.58	4.58	4.60	–	–	–	–	1,575
18	4.65	4.64	4.64	4.64	4.65	4.65	4.65	4.67	4.68	1,471
19	4.71	4.70	4.70	4.70	–	4.71	4.71	4.70	–	1,377
20	4.76	4.75	–	–	4.76	4.76	4.76	4.75	4.74	1,297
21	–	4.80	–	–	4.81	–	–	4.78	4.79	1,225
22	–	–	–	–	–	–	–	4.83	–	1,192
23	4.86	4.85	4.85	4.85	4.86	4.86	4.86	4.87	4.89	1,160
24	–	4.89	4.89	4.89	4.91	–	–	4.92	–	1,103
25	4.95	4.93	4.93	4.93	4.95	4.95	4.95	4.96	4.97	1,050
26	4.99	4.97	–	–	4.99	4.99	4.99	5.00	–	1,002
27	5.03	5.01	5.01	5.01	5.03	5.03	5.03	5.04	5.05	959
28	5.06	–	–	–	–	5.06	5.06	–	–	919
29	5.10	5.09	5.09	–	5.10	5.10	5.10	5.11	5.13	882
30	–	5.15	5.15	5.15	–	–	–	–	–	832
31	5.17	5.18	–	–	5.17	5.17	5.17	5.18	5.19	817

Other gravitational fluctuations (energy cluster state in stationary gravitational waves), e.g. outside of our galaxy, change the domain distribution in proteins, which should bring about new properties of proteins and automatically new life forms.

References to 11

[1] Horsthemke W.; Lefever R. Noise-Induced Transitions. Theory and Applications in Physics, Chemistry and Biology. Edit. Hermann Haken, vol. 15; Springer-Verlag: Berlin, Heidelberg, NY, Tokio, 1984.

[2] Zubow K.V., Zubow A.V., Zubow V.A. Ensemble of Clusters – New Form of Molecular Matter, Risks and Chances. Zubow Equations. In Advances in Chemistry Research, vol. 5, 2010, pp. 107-145. Editor J.C. Taylor. Novapublisher. NY.

[3] Zubow K., Zubow A.V., Zubow V. A. Water Clusters in Plants. Fast Channel Plant Communications. Planet Influence. Journal of Biophysics Chemistry.2010, vol.1, no.1, pp.1-11.

[4] Zubow K. V, Zubow A.V, Zubow V.A. Use of flicker noise spectroscopy for undestroed analysis of nano structures. Zavodskaja Laboratoria. Diagnostics of Material. 2008, vol.74, no. 9, pp. 40-46. (in Russian).

[5] Zubow Kr. V., Zubow A.V., Zubow V.A. Cluster structure of liquid alcohols, water and n-Hexane. Journal of Appl. Spectr. 2005, vol.72, no.3, pp. 321-328.

[6] K.V. Zubow, A.V. Zubow, V.A. Zubow. Principles of gravitation spectroscopy. New form of molecular matter. Processes. Fields. "Aist Handels- und consulting GmbH", dep. R&D, Berlin, 2010, electronic book (www.zubow.de), p. 854. (in Russian).

[7] Zysman V, Nguyen Tuan Q, Kausch, Henning H. Degradation on freezing dilute polystyrene solutions in *p*-xylene. Journal of Polym. Sci., Part B: Polym. Phys. 1994, vol. 32, pp. 1257-1269.

[8] Rafaee M., Tezuka T., Akasaka K., Williamson M. Pressure-dependent Changes in the Solution Structure of Hen Egg-white Lysozyme. J. Mol. Biol. 2003, vol.327, pp. 857-865.

[9] Wilson, Lori J.; Kim, Yeong Woo; Baird, James K. Lysozyme Self-Association in Aqueous NaCl at pH 4.0 and 20 °C. Crystal Growth & Design. 2002, vol.2, no.1, pp. 41- 43.

[10] Wunderlich B. Macromolecular physics. Crystal structure, morphology, defects. Vol. 1. Academic Press, NY 1973.

[11] Harvey Lodish, David Baltimore, Arnold Berk, S. Lawrence Zipursky, Paul Matsudaira, James Darnell. Molekulare Zellbiologie 2 Auflage. 1996. Berlin, NY, Walter de Gruyter, S. 92.

[12] Bogdanov EV, Manturova GM. Equiclaster model of water. Biomedicinskaja Radioelektronika. 2000, no.7, pp. 19-28, in Russian.

[13] Lenz A, Ojamäe L. On the stability of dence versus cage-shaped water clusters: Quantum-chemical investigations of zero-point energies, free energies, basis-set effects and IR spectra of $(H_2O)_{12}$ and $(H_2O)_{20}$. Chem. Phys. Letters. 2006, vol. 418, pp. 361-367.

[14] Zubow Kr., Zubow A., Zubow V.A. Investigation of the distribution of water clusters in vegetables, fruits and natural waters by flicker noise spectroscopy. Biofizika, 2007, vol. 52, no. 4, 585-592, in Russian.
[15] Zubow K.V, Zubow A.V, Zubow V.A. Study of structural changes in a LDPE film by flicker-noise spectroscopy. Plasticheskie Massy. 2006, no.3, pp. 32-34, in Russian.
[16] Zubova K.V., Zubov A.V., Zubov V.A. Cluster-like supramolecular structure and is changes at low pressure of LDPE film. Plasticheskie Massy. 2005, no.12, pp. 36-39, in Russian.
[17] Zubow Kr. V., Zubow A.V., Zubow V.A. Oscillation spectra of seed NaCl crystals in aqueous solution. Journal of Appl. Spectr. 2005, vol.72, no. 6, pp. 840-847.
[18] Zubow K., Zubow A.V., Zubow V.A. Molecular Cluster Distribution in Sea Water Near the Wall.Antifouling aspect. Journal of Clusters. 2011, vol.22, no. 4, pp. 603-619.

Chapter 12

DOMAINS IN BIOMATRICES

The structure of egg white and egg yolk at the level of cluster ensembles and micro domains is not completely clear, hence, there is no method for a fast and safe determination of statistical ensembles. Using the gravitation mass spectroscopy, it is possible to investigate the long-range order in liquids, protein coils, and protein coil associates in solutions [1, 2, 3]. The German firm, "AIST Handels und Consulting GmbH," suggested this method be applied for the LRO analysis of liquids, clusters in proteins, and for the analysis of their interaction with the surroundings [4, 5]. Statistical ensembles in liquids, seed crystals, clusters, and protein coil associates play an important role in the early stage of embryo origin.

The aim of the present chapter is to apply the multidimensional digital signal processing [6, 7] to the investigation of the biomatrices in hens' eggs as well as to find out whether there are differences between the eggs from free-range hens and battery hens (bird flu, quarantine 2005 in Germany).

To analyze egg yolk and egg white, fresh eggs from free-range hens (10 pieces, June 2005) and battery hens (10 pieces, 45 days after the beginning of the quarantine, October…December 2005) were chosen. The egg yolk was separated from the egg white and placed in the measurement cell of the GMS of the German firm, "Aist Handels und Consulting GmbH," where the spectra were recorded with a scanning time of 30 s.

Every measured signal was cleaned from gravitational noises raised by proton dissolving/condensing in a physical vacuum [8] and analyzed. The experiments were repeated three times. The measuring error was 10±5%, the temperature stability ±0.1 K, and the GMS-sensor was installed directly in the egg yolk (in vivo, Figure 12.1).

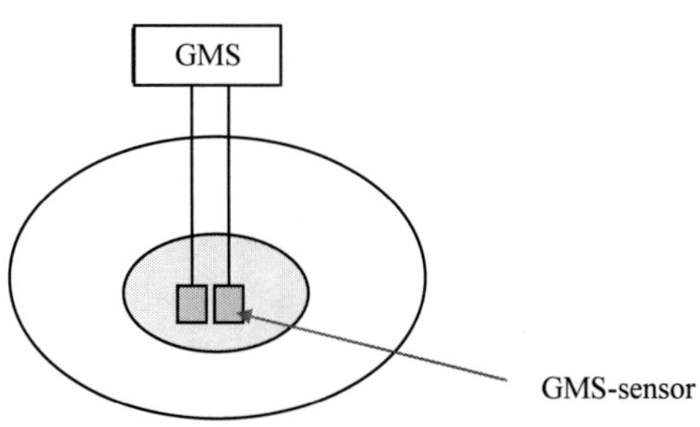

Figure 12.1. Measurement scheme.

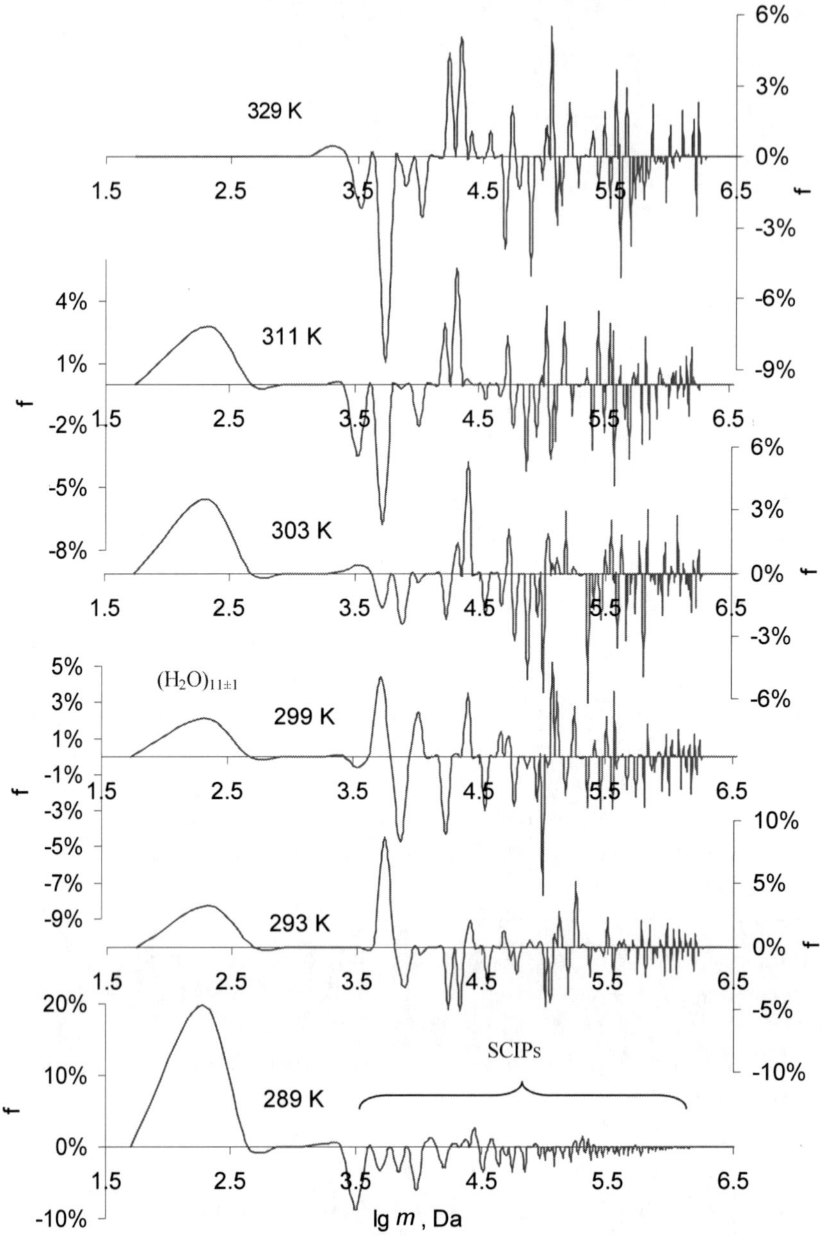

Figure 12.2. GMS-spectra of egg yolk at different temperatures. Free-range eggs with strong shock waves and a Zubow constant = $6.4 \cdot 10^{-15}$ N/m [11]. The value *f* reflects the energy characteristics of an oscillator.

Device calibration and measurement procedures were carried out as described in [6]. The GMS-spectroscopy is based on known methods of the statistical analysis of very weak gravitational signals [1]. Yamauchi [9] suggested a new analogous analysis method for molecule oscillations with intramolecular and/or intermolecular energy transfer that is given in time-frequency characteristics. Using this method, which is based on short-time Fourier transformation, energy density fluctuations can be detected, too, just like with GMS. In the following, it shall be described how real cluster signals were cleaned from noises of the measurement cell [5].

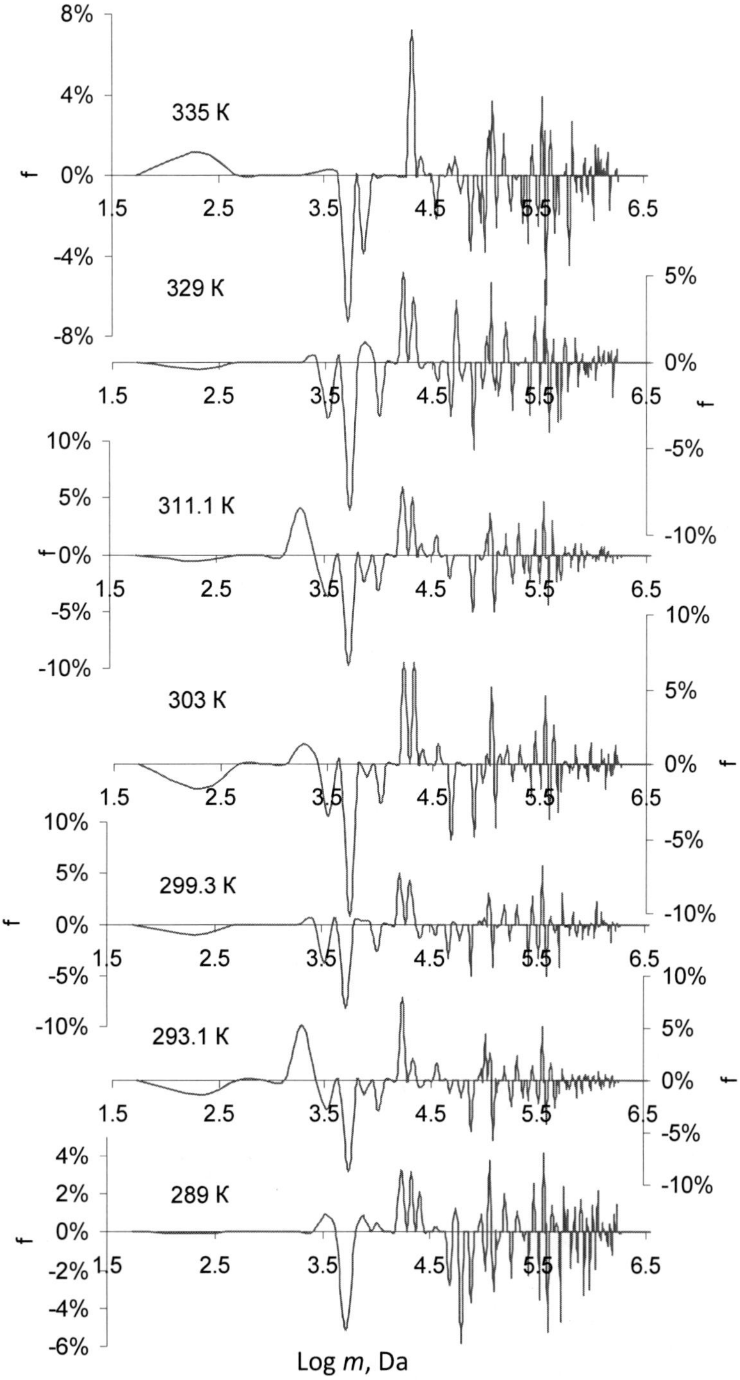

Figure 12.3. GMS-spectra of egg white at different temperatures. Free-range eggs. Strong shock waves. Zubow constant $6.4 \cdot 10^{-15}$ N/m.

As shown in Figure 12.2, the main clusters are represented as solvated clusters of ion pairs in egg yolk at 289 K [6]. Since the SCIP signals are considerably stronger than those of the other oscillators they overlap these. The presence of high molecular SCIP could be an indication that these clusters are confined in limited spaces. Upon heating the egg yolk, the SCIP signals disappeared but new signals, typically for organic clusters, appeared [7]. Most likely cations (metals, ammonium) in SCIP participate in the formation of new oscillators on the base of proteins, and they could act as a catalyst in the beginning biochemical reactions of embryo growth. The oscillator signal of the small expanded water cluster $(H_2O)_{11\pm1}$ is observed to be reduced by almost six fold (http://www.lsbu.ac.uk/water/index.htm, [10]). Before heating, water clusters and SCIP are in an intermolecular hydrophilic space of the biomatrix. With heating, the surroundings have been changed from hydrophilic molecules to hydrophobic ones. The contact of water clusters with the hydrophobic lipids leads to a weakened interaction with the surroundings and the water clusters implode in the collapsed form. Now, with increasing temperature, the structured water and SCIP start to participate actively in biochemical reactions of the embryo growth. Therefore, at low temperatures, the salts of the egg yolk are deposited in SCIP and water appears mainly as small clusters. The nano emulsion in egg yolk seems to be destroyed upon temperature increase and its destructive products take part in the biochemical processes of embryo growth.

As shown in the Figures 12.2 and 12.3 the GMS-spectra of egg white and egg yolk are very different. One sees this particularly at the GMS-spectrum of egg yolk at 289 K, which is typical for SCIP [6]. According to Belitz [12], the salt content in egg yolk and egg white amounts to 0.6 and 1.1 wt. %, respectively. If salts are available as oscillating SCIP, then the SCIP signals could influence the complete spectrum and overlap the other signals. With temperature rise these signals disappeared and consequently so did the SCIP. The signal of $(H_2O)_{11\pm1}$ is recorded as an expanded oscillator in egg yolk (Figure 12.2) and it only disappeared at 329 K, which is explained either by the small water clusters that were included in the biochemical processes or by the fact that a new thermodynamic balance with the collapsed clusters was achieved [6]. In contrast, in egg white, these water clusters were found to be in a collapsed form that was transformed to the expanded one at 335 K (Figure 12.3). The concentration of the base water cluster $(H_2O)_{11\pm1}$ is very low at 289 K in egg white; water could be in non-oscillating structures, for example, as adsorption water on protein molecules. Water desorption from protein molecules at low heating (299.3 K) leads to the formation of collapsed water clusters near proteins (scheme, Figure 12.3). At 335 K, the hydrogen bonds between the nitrogen bases were destroyed because of the beginning of protein denaturation. This process gives new possibilities for new hydrogen bonds with these collapsed water clusters, accompanied with their immediate transition to the expanded state (scheme I, the model of the base water cluster was given with kind permission of Professor A. Lenz [13]). Thus, the evolution of the water cluster structure and its important role in biological systems can be studied.

As we found out earlier, the cluster structure depends on temperature, pressure, mechanical fields [4], chemical/biochemical substances, and weak gravitational fields [5]. Furthermore, ensembles of clusters in liquids were observed to influence biological processes [14], particularly under space conditions (space station) and under the origin of life on other planets.

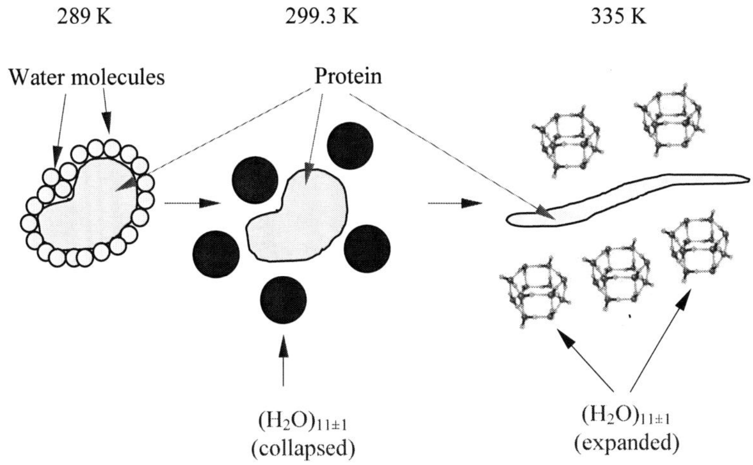

Scheme I.

In Figure 12.4, the sum of collapsed clusters in free-range eggs versus temperature is given. The concentration of the collapsed clusters in both egg components changes differently upon heating. At 300 K there is a maximum for egg white since the maximum for egg yolk is shifted to 305 K, which could be caused by a large number of collapsed clusters at the temperature near the embryo formation (310 K). This high sum of collapsed clusters can be seen as a preliminary stage for the egg yolk's protection from temperature fluctuations.

The other GMS-signals in Figures 12.2 and 12.3 reflect the oscillations of domains, macromolecule coils, and coil associates.

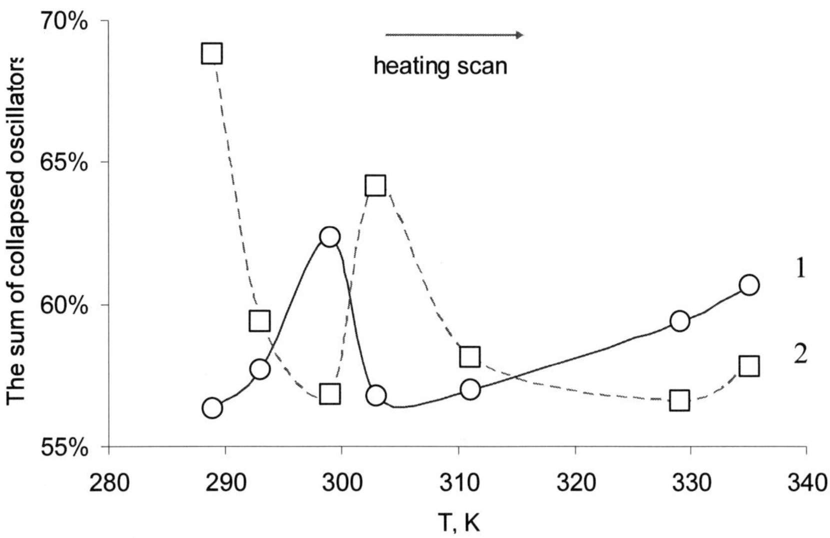

Figure 12.4. Temperature dependence of the sum of collapsed oscillators in eggs (0.2 K/min.). 1 – egg white and 2 – egg yolk.

It would be of interest to investigate how the clusters behave in the eggs of battery hens. This possibility was presented during the worldwide bird flu epidemic in 2005 when all hens were locked in (quarantine in North Germany). To compare the egg quality of free-range hens (June 2005) and battery hens (November 2005, after 45 quarantine days), a small group of isolated hens from a private hens grower (1 hen per m^2) was chosen. During the quarantine fewer eggs had been laid, and at the same time, feeding was around 30-fold less. After the quarantine, egg production was normal again. As shown in Figures 12.5 and 12.6 the GMS-spectra of egg yolk and egg white are generally similar, however, they greatly differ from those of the free-range eggs (Figures 12.2 and 12.3). Furthermore, they are very close to the spectrum of 0.9 wt. % NaCl solution [6] containing SCIP and to that of egg yolk at 289 K (Figure 12.2). The SCIP signals are much stronger than those of the other oscillators and therefore they overlap them. The super molecular structure in the egg yolk and egg white has been greatly changed. This state was observed for free-range eggs at low temperatures only (Figure 12.2, 289 K) whereas it appears at all temperatures for the embryo originating from battery eggs. This can be explained by the insufficient movement and stress of the hens. It stands out that for battery eggs there are signals in the area of high oscillating masses (Figures 12.5 and 12.6), which are unavailable for free-range eggs. This could be an indication of the fact that the battery eggs are structured not only in the area of small seed crystals but also in that of large polymers, probably at the level of protein associates and of large SCIP [15]. Some differences are observed for expanded clusters; in the spectra, they are marked with an arrow. In our opinion, these signals can only partially be assigned to SCIP. They rather characterize protein oscillators. Unlike free-range eggs the SCIP oscillations do not disappear at heating (Figures 12.5 and 12.6). Therefore, the embryo growth after the nano emulsion destruction in egg yolk starts according to different scenarios of protein-SCIP interaction, which influences the eggs' quality. The stable presence of SCIP in eggs could be caused by stress and hypo dynamics during the quarantine that led to a salt enrichment in the eggs although there are not any differences between free-range and battery eggs visually.

As shown in Figure 12.6, upon heating the battery egg white, changes in the GMS-spectra are little where the signals of the small water clusters $(H_2O)_{11\pm1}$ and $(H_2O)_{100}$ remain stable. In contrast to free-range eggs the super molecular structure in egg white has lost its temperature dependence; it is dominated by SCIP signals. The "island" of cluster signals (indicated with arrow) probably moves to high masses as a result of its association and coagulation processes at warming.

The GMS-spectra of egg white from free-range and battery eggs given in Figures 12.3 and 12.6 are very different from each other. The detailed understanding of these differences could be the object of a special investigation. The stress factors of the quarantine therefore influenced the super molecular structure formation in egg white and finally the egg quality in general. Psychological pressure and lack of light, room, movement, and quiet are the main stress factors that led to an increased production of stress hormones in the hens′ body and accelerated the aging and destruction processes of biomatrices. These hormones possibly prevent the participation of SCIP in biochemical processes of the egg white synthesis and embryo growth.

The change of collapsed clusters in eggs after 1.5 month quarantine is shown in Figure 12.7. At warming, the concentration of collapsed clusters in egg white reduced monotonously, but it did so with a little increase in the temperature interval, 305…315 K. This behavior is strongly different from that of free-range eggs (Figure 12.4). The super molecular structures in the biomatrix were therefore concluded to be strongly different in free-range and battery eggs. These differences should be explained with the fact that hens in small closed rooms are permanently under psychological stress.

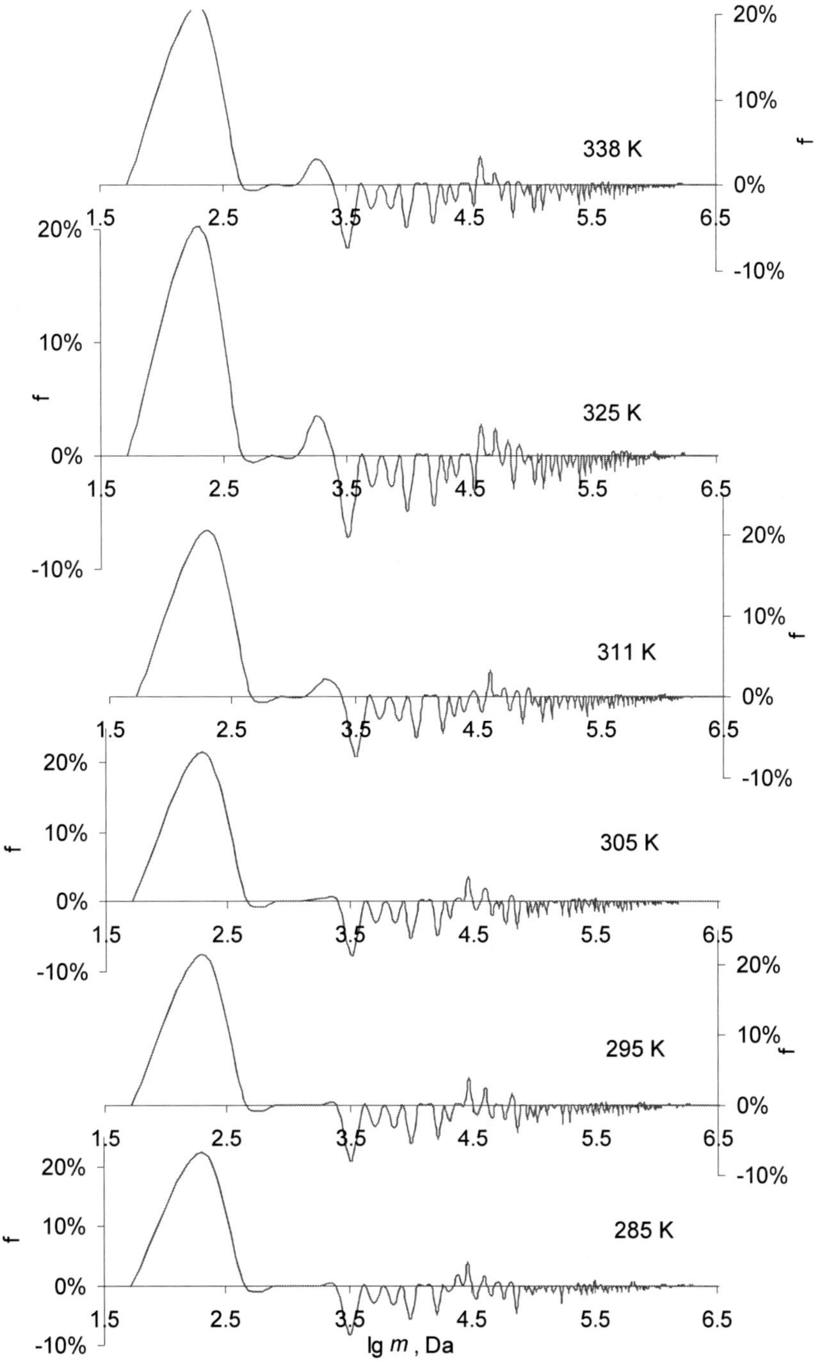

Figure 12.5. GMS-spectra of egg yolk at different temperatures after 45±3 quarantine days.

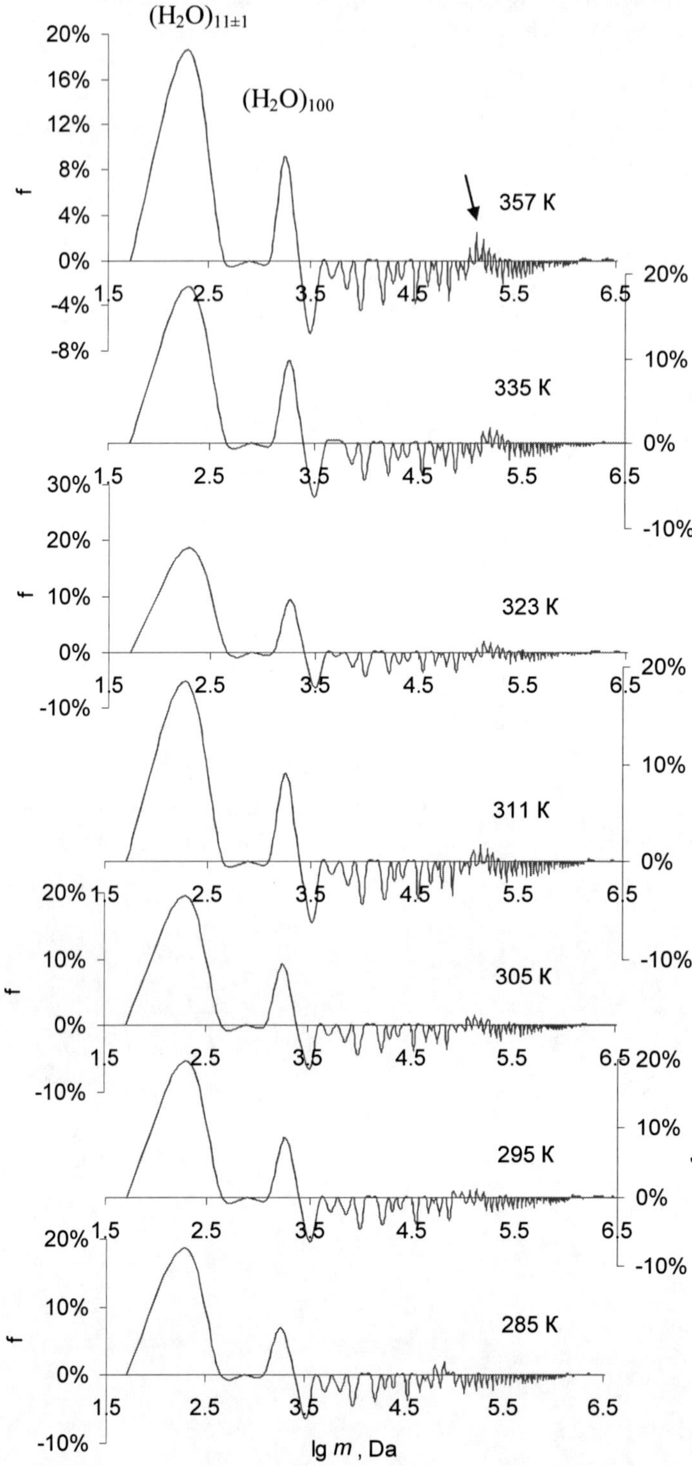

Figure 12.6. GMS-spectra of egg white at different temperatures after 45 quarantine days.

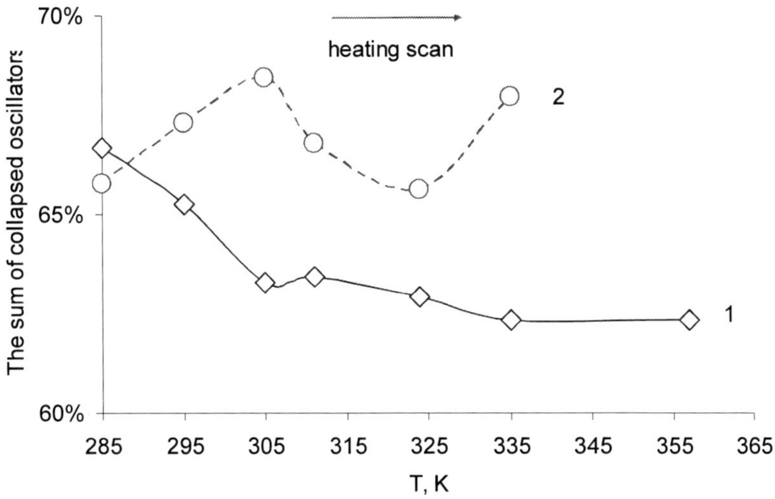

Figure 12.7. Temperature dependence of the sum of collapsed oscillators in egg white (1) and egg yolk (2) during the quarantine in North Germany in 2005 at 0.2 K/min.

In Figures 12.8 and 12.9, the GMS-spectra of the clusters' signal ratios in egg yolk and egg white of free-range and battery eggs are given. The absolute signal intensities of Figure 12.3 were divided by those from Figure 12.6 at the same temperature. The same procedure was made for the signal intensities of Figures 12.2 and 12.5. By these measures, it was simpler to analyze the changes in the liquid structure at the level of oscillating clusters, domains, protein coils, and their associates. In addition, it is possible to see how these signals developed at heating. Large oscillators changed the most with regard to both coils and their associates. Comparing the signal intensities of free-range and battery eggs, it was found that they are for considerably higher for the first ones. From the other side, it is known that the signal intensity primarily depends on the interaction of an oscillator with the surroundings and then on the oscillator concentration [16]. The high signal intensity in free-range eggs is therefore ascribed to a weaker interaction of the coils and their associates with the surroundings, which is only possible when protein coils are ordered in a denser structure. With a temperature increase from 287 to 311 K (Figure 12.8), the signal ratio is reduced from 40…70 to 10…15, however, it rises up to 100 above 311 K. At 311 K, the highest number of signals was measured. They describe the differences between the two egg types. The oscillator signal with a mass of 5.3 kDa is ascribed to the main domain in lysozyme [17], but those with 153 and 371 kDa are ascribed to the conalbumin dimer (76 kDa [12, p. 440]) and ovalbumin octamer (44.5 kDa, [12, p. 440]), accordingly. The oscillator with a mass of 1,520 kDa also seems to be part of the associate of protein coils, which decomposes at heating.

The strongest changes take place in egg yolk (Figure 12.9) where the signal ratios can achieve 200. The oscillations with masses of 390 kDa should belong to lipovitelline dimers (400 kDa, [12, p. 444]). They disappear as oscillators at 287 and 295 K and appear at 311 K.

In the following, we want to analyze the oscillations of ovalbumin in egg white (see table), which appears as a monomer with different water contents: 6.3 wt. % for free-range eggs and 1.5 wt. % for battery eggs (Figure 12.9). Additionally, ovalbumin was observed as octamers, too. The first octamer type, containing 4.3 wt. % water (expanded oscillator) was found for both egg types whereas the latter one with 9.2 wt. % water (collapsed oscillator) was discovered for free-range eggs only (Figure 12.10). The water concentration in ovalbumin was determined by subtracting the mass for ovalbumin (44.5 kDa, [12]) from the experimentally observed oscillating mass.

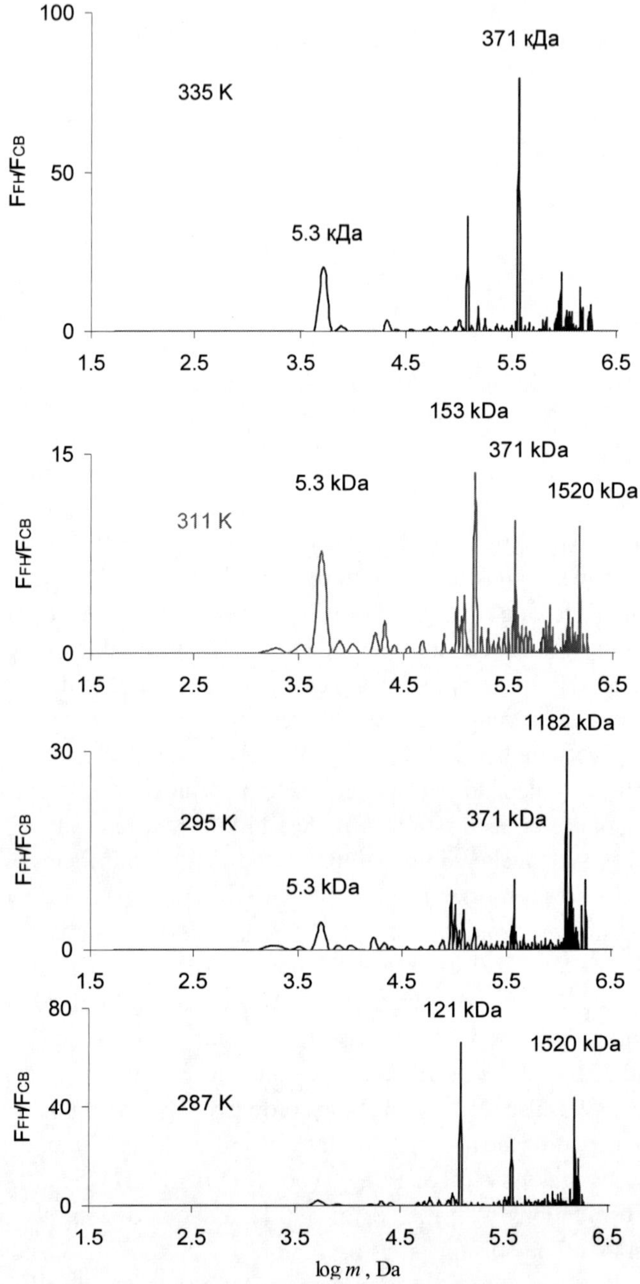

Figure 12.8. GMS-spectra comparing the oscillators in egg white (free-range eggs) at different temperatures. F_{FH}/F_{CB} – the ratio of oscillator signals in free-range eggs (F_{FH} = abs(f_{FH})) to those in battery eggs (F_{CB} = abs(f_{CB})). Oscillators with the same mass were compared.

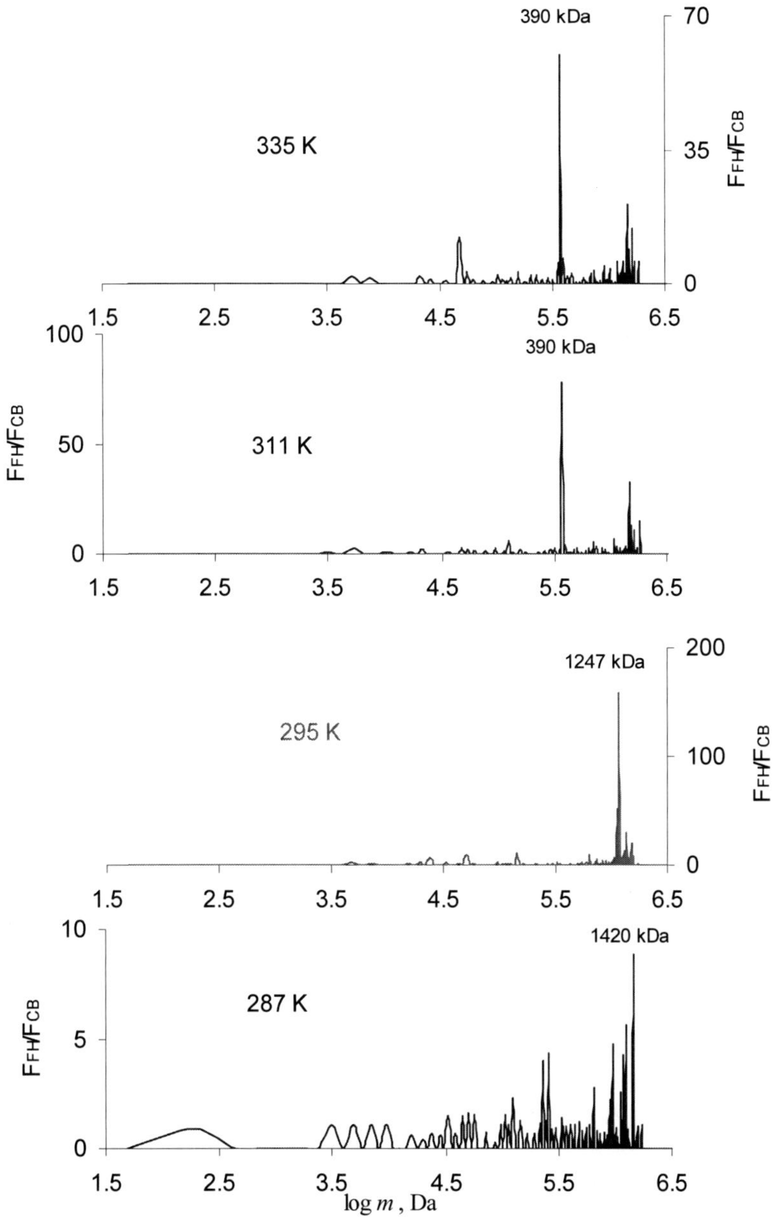

Figure 12.9. GMS-spectra comparing the oscillators in egg yolk (free-range eggs) at different temperatures. F_{FH}/F_{CB} – the ratio of oscillator signals in free-range eggs (F_{FH} = abs(f_{FH})) to those in battery eggs (F_{CB} = abs(f_{CB})). Oscillators with the same mass were compared.

As is visible in Figure 12.10, the oscillating ovalbumin monomers in free-range eggs actively interact with the surroundings whereas those in battery eggs are in a "frozen" state. The loss of elasticity in the interaction with the surroundings could be an indication of the permanent stress the hens were exposed to. At 305 K the ovalbumin monomer is the most mobile in its vibrations and at this temperature a minimum is available in the curve. The monomer coils are in a collapsed form with minimal interaction with their surroundings.

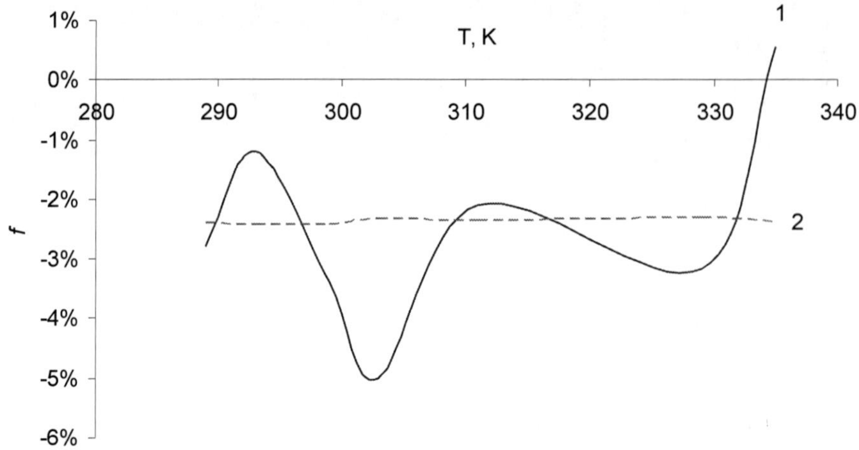

Figure 12.10. Temperature dependence of the oscillations caused by ovalbumin monomers (f) in egg white. 1 – free-range eggs (6.3 wt. % water) and 2 – battery eggs (1.5 wt. % water).

Differing from free-range eggs, only one octamer type was found in battery eggs, which could be caused by the hen's exposure to permanent stress (Figure 12.11). The octamers in battery eggs are similar to the monomers in that they are in a "frozen" state (ordered SCIP shell like water on the protein surface [18], scheme); those interactions with the surroundings are constant. The octamer's surroundings prevent its normal interaction with its neighbors or the octamer could perhaps now be surrounded by other neighbors. The lower water content in the ovalbumin coils of battery eggs in comparison with that of free-range eggs (Table 12.1) and the large number of SCIP (Figure 12.6) could give an answer to the monomer's surroundings. Thus, the ovalbumin coils in egg white (battery eggs) are in a cage that is formed by SCIP. SCIP strongly limit the oscillation possibilities of coils and their associates and consequently their interaction with the surroundings.

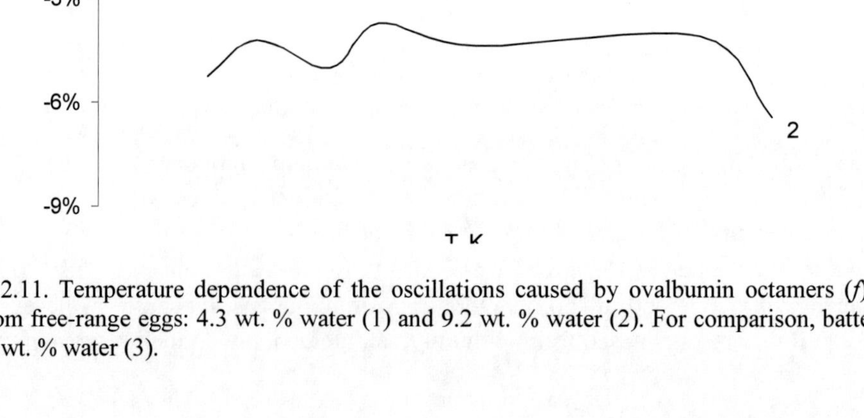

Figure 12.11. Temperature dependence of the oscillations caused by ovalbumin octamers (*f*) in egg white from free-range eggs: 4.3 wt. % water (1) and 9.2 wt. % water (2). For comparison, battery eggs with 1.5 wt. % water (3).

Oscillations of ovalbumin dimers and tetramers were also found in the GMS-spectra (Table 12.1). Possible structures of ovalbumin with the surroundings for both egg types shall be given in Figure 12.12. According to this model, the ovalbumin coils from battery eggs are a little compressed and smaller compared with those of free-range eggs.

The oscillations of monomers and their associates are limited by the surface of the surroundings and by an interaction with surfaces fragments (Figures 12.1 and 12.12) and they can be identified with GMS-spectroscopy, too. For both egg types, the following models concerning the state of the coils and their interaction with the surroundings have been suggested (Figure 12.12).

Table 12.1. The content of included water (wt. %) in ovalbumin coils

Oscillator (ovalbumin)	Free-range eggs	Battery eggs
–	289 K	285 K
monomer	6.3	8.5
dimer	4.1	4.3*
tetramer	0	1.7
octamer I	4.3	4.3
octamer II	9.2	–

*For the dimers the signals lie thick beside each other.

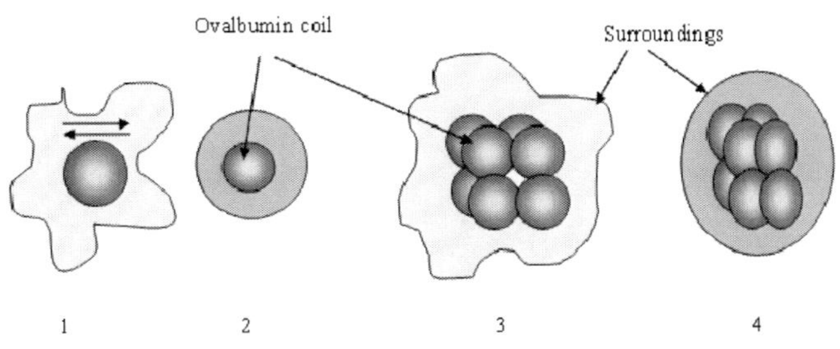

Figure 12.12. Models describing ovalbumin coils and their interaction with the surroundings in egg white. 1, 2 monomers and 3, 4 octamers. 1, 3 – free-range eggs and 2, 4 - battery eggs. The ordered shell of SCIP is shown in grey.

Differing from the ovalbumin dimers of free-range eggs, those of battery eggs show a wide spectrum of dimer signals with almost the same masses.

Based on the data from Table 12.1 and Figure 12.12, the following models for the formation of protein associates can be represented.

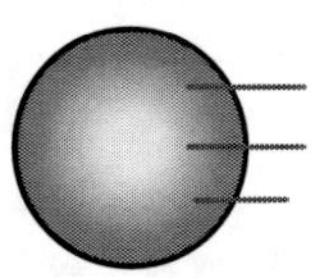

Monomer in free-range eggs
(6.3 wt. % included water)

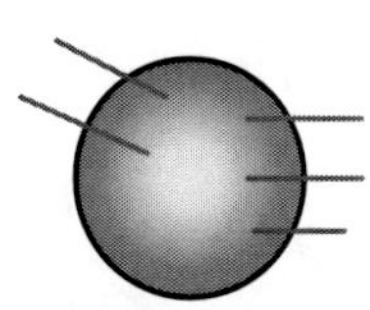

Monomer in battery eggs
(8.5 wt. % included water)

The ovalbumin coils with asymmetrical hydrophilic groups are solvated with water molecules (drawn as lines) where the hydrophilic groups are localized on the protein surface (ordered shell [18]).

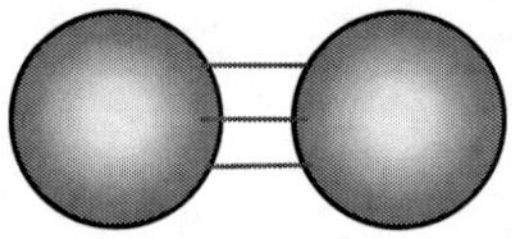

Model of an ovalbumin dimer from free-range eggs
(4.1 wt. % included water)

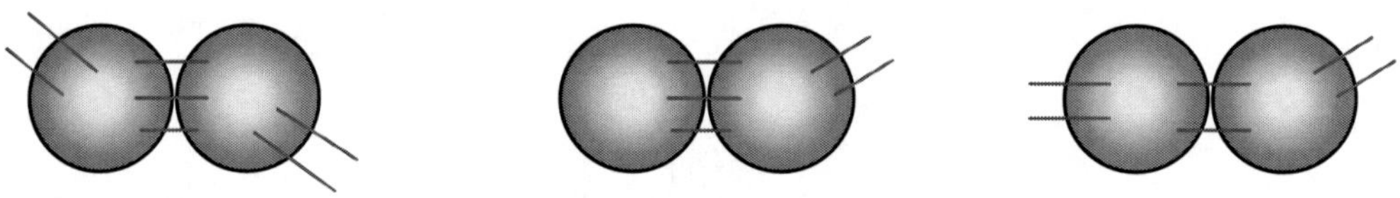

Models of ovalbumin dimers from battery eggs
(4.3 wt. % included water)

Both the water content and the number of possible combinations for ovalbumin dimers from battery eggs should be higher according to this model.

Ovalbumin tetramer model from free-range eggs
(0 wt. % included water)

To this model, the surface hydrophilic groups of the coils only interact with themselves and the tetramer is therefore "dry."

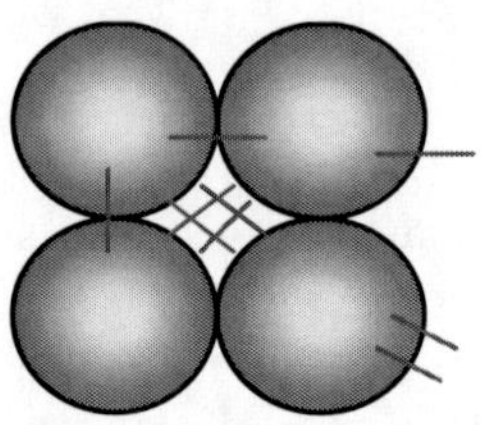

Ovalbumin tetramer model from battery eggs
(1.7 wt. % included water)

For this ovalbumin tetramer from battery eggs, the hydrophilic groups are partially unordered and the tetramer is no longer "dry."

Stress for hens was therefore found to change the coils' conformation and super molecular structure in the biomatrices as well as change the surroundings of the protein coils and their interaction with neighboring molecules.

CONCLUSION

The GMS method can be applied to quickly control (0.5…30 s) the eggs' quality at the level of the super molecular structure in egg yolk and egg white biomatrices. Stress and captive breeding destroyed the biomatrices of eggs at the level of cluster ensembles and changed the surroundings of proteins and protein conformations. The obtained results would also be interesting to analyze for the solution of other problems, for example, for the development of bio-cosmic technologies for space flights and space stations.

REFERENCES TO 12

[1] Zubow, V.A. Statistische Ensemble in Bieren. Neue Qualitätskriterien. Brauwelt. 2005, no.27, pp. 795-796.

[2] Zubow V.A. Flicker Noise Spectroscopy für die Lebensmittelanalytik. Neue Qualitätskriterien. Deutsche Milchwirtschaft. 2005, no. 21, pp. 926-927.

[3] Dan E. Dudgeon, Russel M. Mersereau, 1984. Multidimensional digital signal processing. Prentice-Hall, Inc., Englewood Cliffs.

[4] Zubow K.V., Zubow A.V., Zubow V.A. Ensemble of Clusters – New Form of Molecular Matter, Risks and Chances. Zubow Equations. In Advances in Chemistry Research, vol. 5, 2010, pp. 107-145. Editor J.C. Taylor. Novapublisher. NY.

[5] Zubow K., Zubow A.V., Zubow V. A. Water Clusters in Plants. Fast Channel Plant Communications. Planet Influence. Journal of Biophysics Chemistry.2010, vol.1, no.1, pp.1-11.

[6] Zubow Kr. V., Zubow A.V., Zubow V.A. Oscillation spectra of seed NaCl crystals in aqueous solution. Journal of Appl. Spectr. 2005, vol.72, no. 6, pp. 840-847.

[7] Zubow Kr. V., Zubow A.V., Zubow V.A. Cluster structure of liquid alcohols, water and n-Hexane. Journal of Appl. Spectr. 2005, vol.72, no.3, pp. 321-328.

[8] Zubow K., Zubow A.V., Zubow V.A. The Phenomenon of Proton Dissolving in Vacuum and of Proton Condensation from Vacuum. Two Forms of Protons, Structure of Nuclei, Electrons and Atoms. J. of modern physics. 2010, vol.1, no.1, pp.175-184.

[9] Yusuke Yamauchi, Hiromi Nakai.Hybrid approach for ab initio molecular dynamics simulation combining energy density analysis and short-time Fourier transform: Energy transfer spectrogram. J. Chem. Phys. 2005, vol. 123, pp. 034101-1- 034101-9 .

[10] Maheshwary, S.; Patel, N.; Sathyamurthy, N.; Kulkarni, A. D.; Gadre, S. R. Structure and Stability of Water Clusters $(H_2O)_n$, n = 8-20. The Journal of Physics Chemistry A. 2001, vol. 105, pp.10525-10537.

[11] Zubow K.V., Zubow A.V., Zubow V.A. Using of flicker noise spectroscopy for non destroyng analysis of nano structures, Zavodskaja Laboratorija. Diagnostics of materials. 2008, vol. 74, no. 9, pp.40-45. (in Russian).

[12] H.-D. Belitz, W. Grosch., 1987. Lehrbuch der Lebensmittel. Dritte, überarbeitete Auflage. S.439. Springer-Verlag, Berlin, Heidelberg, NY, Paris, Tokio.

[13] Lenz A, Ojamäe L. On the stability of dence versus cage-shaped water clusters: Quantum- chemical investigations of zero-point energies, free energies, basis-set effects

and IR spectra of $(H_2O)_{12}$ and $(H_2O)_{20}$. Chemical Physics Letters. 2006, vol. 418, pp. 361-367.

[14] K.V. Zubow, A.V. Zubow, V.A. Zubow. Principles of gravitation spectroscopy. New form of molecular matter. Processes. Fields. Aist Handels- und consulting GmbH, dep. R&D, Berlin, 2010, electronic book (www.zubow.de), p. 854. (in Russian).

[15] Yannis; Kierzek, Andrzej M.; Saenger, Wolfram. Cluster formation in aqueous electrolyte solution observed by dynamic light scattering. The Journal of Physical Chemistry. 2000, vol. 104, no.15, pp. 3405- 3406.

[16] Drago R.S. Physical Methods in Chemistry. Philadelphia, London, Toronto. W.B. Saunders Company. 1977.

[17] Rafaee M., Tezuka T., Akasaka K., Williamson M. Solution structure of lysozyme at low and high pressure. Journal of Molecular Biology. 2003, vol. 327, pp. 857-861.

[18] Shiryaev A., Pagan D. L., Gunton J. D. Role of solvent for globular proteins in solution. The Journal of Chemical Physics. 2005, vol. 122, pp. 234911 (0-10).

Chapter 13

CLUSTER-12 IN POTATO AMYLOPECTIN DURING GROWTH

Investigating the low frequency movement of chain fragments of biopolymers could be helpful in understanding the formation mechanism of biopolymers as well as the spectra of their conformation transformation. This could be fundamental for studying the biopolymer surroundings and their interactions with them [1]. To the authors, the investigation of biopolymers on the level of clusters, domains, and super cluster structures is one of the main problems in modern biochemistry, for example, viral disease as a result of the conformation change of biomolecules. Therefore, the structure of the natural polymer starch consisting of amylopectin and amylose is not yet completely clear, especially in the amorphous area. On the other side, the mechanism concerning the formation, development, and aging is unknown, too. Therefore, the amylopectin structure on the level of branching (BD) and polymerization degree (PD) is worth examining as well as the dynamics of coil conformation and its interaction with the surroundings. The main publications [2-4] are based on the work of Banks and Mair [5] that were made in the seventies of the 20th century.

It is known that amylopectin in granule form is partially in a crystal state [2, p. 314]. Although according to the physical chemistry of polymers, branched macromolecules crystallize badly or not at all. This also should apply to the highly branched amylopectin macromolecules. The crystal structures of these polymers possess many defects and branching of the crystals will be repelled [6].

The rolled up conformation of macromolecular branches is thermodynamically stabler than the open one. To model the crystallization process of a macromolecule in which to every 20…30 units one branch point with 20…30 units belongs is extremely difficult. Therefore it can be only the talk of a certain order but not of a crystalline structure. The problem is really complicated and the above-mentioned authors did not discuss it, probably due to the missing methods. To Stryer [3], branched biopolymers (amylopectin and glycogen) were chosen by nature because of their fast synthesis and destruction. To destroy a crystalline structure is much more complicated than to destroy an amorphous one. Another open question is how such short chains in real biomatrices can form double spirals [2, 4]. Unfortunately, almost all experiments were not carried out with pure starch or amylopectin but with their solutions or gels where the role of water was neglected. It is obviously the amylopectin and starch structures in granule that are incomparable with those in gel or solution, which means that the latter one cannot be applied as a base for understanding the native biopolymer structure.

There is not any fast and non-destructive method for the registration of the movements of large mass concentrations in solid polymers at the moment. Kortstee and Satoh [7, 8] tried to get some information on these parameters using an enzyme that split amylopectin. They found that fermentative amylopectin splitting leads to the formation of small chains with PD equal to12 units and that these chains, according to XRD-analysis, behave like the spirals of two meanders where each meander consists of six α-D-glycopyranose units [9].

By examining the low frequency movement of biopolymer chains under mechanical influence and heating, Vallés [10] studied the thermal decomposition of starch. The observed activation energies were assigned to destructive processes of branched points. At 383 and 433 K, some unexplainable effects were detected.

The investigation of the starch structure is not only of theoretical interest but it is of practical, too; in the near future, starch will be the base raw material for new food products [11] as well as the new source for biodegradable packing materials [9, 10]. Its fine structure could be an indicator of changes in plants caused by anthropogenic influences (climate warming and environmental pollution) and the intensification of agriculture. The quality of starch as a technological raw material depends on the concentration and quality of amylose and amylopectin, which are the main components in starch.

There are many papers [7-13] that describe the main amylopectin characteristics PD and BD. The synthesis mechanism of starch is well studied [14], although on the level of super molecular formation of clusters and super cluster formation and their interaction each with another in a biomatrix of potato tubers, there are some uncertainties.

Kuakpetoon [16] applied oxidation (2% NaOCl) to determine the PD of macromolecule branches in amylopectin where he found a branch length between 6 and 50 α-D-glycopyranose units. This method, however, seems to be a little inaccurate because the oxidation is unselective, especially for polymers.

After the ultra sedimentation of starch, which was previously crushed in a ball mill, Tester [17] found amylose chains with PD of 12, 14, 24, 50-51, and 79 in the decomposition products. This method also delivers only rough values. Experiments using NMR, optical spectroscopy, and chromatography are complicated; they need a lot of time and are hard to reproduce.

Macromolecule fragments in branches of amylopectin probably form sub coils, which are caused by cohesion-cooperation forces of α-D-glycopyranose units and hydrogen bonds. The rolling-up conformation of branches is thermodynamically more favorable than the open structure because of increased entropy [6]. The variety of different coils in such a model (grape-like clustering [18]) can reflect both PD and molecular masses of branches. The heating of such a super coil is connected with its disintegration, leading to single sub coils (Scheme I).

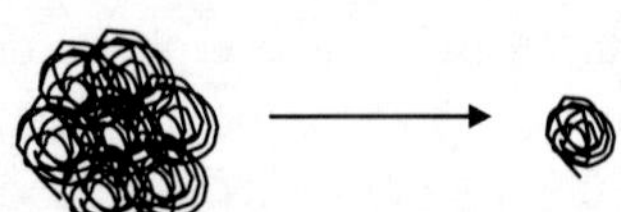

(I)

This behavior can be explained by the fact that at heating, the vibration of sub coils increases, leading to the splitting-off of sub coils from the main amylopectin chain. The masses of sub coils are different. Additionally, the sub coils differ in their structure position in the super coil as well as in their interaction with each another or with their surroundings. Thus, the stepwise thermal decomposition of amylopectin can characterize both the super coil structure and the forces that stabilize the sub coils. If this model is right, then the oscillations of both the super coils and the sub coils can be proved by GMS. Furthermore, the GMS method is able to detect single chain units in starch at heating.

The aim of the present chapter is to investigate the energy characteristics of the low frequency movement of cluster-12 (12 α-D-glycopyranose units) in potato starch and pure potato amylopectin during potato growth.

Starches from different types of potatoes ("Agria-N" from Kartoffelzucht Böhm KG and "Kalena" from NORIKA Nordring-Kartoffelzucht-und Vermehrungs-GmbH, both from Germany) prepared by a standard method were chosen as investigative objects. Pure amylopectin was delivered from the Bavaria State Research Center LfL [19]. The spectra were measured with the spectrometer of the Aist Handels und Consulting GmbH, Germany (www.zubow.de), where the energy (*f*) caused by the interaction of the clusters with a shock wave is a function of the molecular mass of the oscillating clusters. The potato samples (tubers) were taken at the very same place (planted on 27.04.2006) in summer time at the period when the absorption of sun energy by the green part of the plants was the highest [20]. Starch was extracted from cleaned potato tubers through a standard method (cutting the tuber in water into very small pieces by disintegrator, decantation, and filtration of the solid potato part; starch precipitation; starch separation by decantation and filtration; washing the solid starch with distilled water; drying on air at a temperature not higher than 308 K up to a constant weight). The air dried starch represented a white powder with a water content of no more than 11…13 wt. % whereas pure amylopectin contained ca. 1 wt. %. A 7…10 mg starch sample (low pressed powder) was placed in the GMS measuring cell ([21], Figure 13.1), and then it was heated up to 338 K with a heating rate of 6…10 K/min under air free atmosphere. The same procedure was continued for the following temperatures: 373 K, 393 K, 433 K, 473 K, and 523 K. The temperature precision was ±3 K and the measurement time was about 30 s.

For calibration of the energy GMS-spectra, an inner standard was used for which the enthalpy of water adsorption on pure cellulose (dried in vacuum at 383 K) was taken (49.3 J/g [22, p.121]). The energy GMS-spectra were calculated using the sum of the cleaned signals ($\Sigma|S|$) in the temperature interval 363…393 K (effect of water desorption from starch surface, [12, 13]) and sample weight. The formation energy of $(H_2O)_{12}$ that is present in starch amounted to 65±10 J/mol (oscillation signal at $\log m = 2.28$ in GMS-spectrum of starch extracted from potato on July 23, 2006, Figure 13.2). This value sufficiently agrees with the formation energy in bulk water for this cluster ~75±10 J/mol [23]. The measuring error for the cluster mass is up to 15% in the interval 0….$\log m = 2.5$, up to 6% in the interval $\log m = 2.5$…$\log m = 4.5$, and up to 20% in the interval $\log m = 5.5$...$\log m = 6.5$. For the energy calculation, the error is ±15%. The method is based on the multidimensional digital signal processing of weak interactions of spectrometer shock waves with clusters in the studying substances [24].

An equation connecting masses of clusters (m, Dalton) with the frequencies of their oscillations (ω, Hz) was found by the author of this present chapter [25, 26].

$$m = (Z \cdot N_A \cdot 1000)/(4\pi^2 \cdot \omega^2)$$

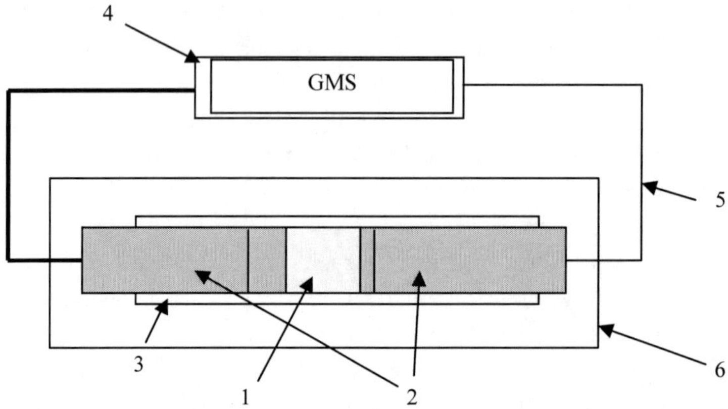

Figure 13.1. 1 – powder or polymer film; 2 – sensor system; 3 – Teflon pipe; 4 – GMS-spectrometer, including a microprocessor, program service for signal clearing and signal extraction, and shock waves and control system; 5 – interface; and 6 – thermostat.

where Z is the Zubow constant $(7\pm2)\cdot10^{-15}$ N/m) and N_A is the Avogadro number ($6.02\cdot10^{23}$). The equation can be simplified to:

$$m = 10^{11}\cdot \omega^{-2}$$

Furthermore, the author developed a spectrometer that was calibrated by Chaplin water clusters (http://www.lsbu.ac.uk/water/index.html), e.g. $(H_2O)_{280}$, m = 5,040 Dalton in water bi-distillate at 4.4±0.1 kHz, lysozyme (m = 15,754 Dalton, 2.5 kHz), α- and β-casein (25±3 kDa, 2 kHz, $Z = 6.4\cdot10^{-15}$, N/m), and styrene clusters of 241 molecules being in liquid styrene (25±3 kDa, 2 kHz, $Z = 6.6\cdot10^{-15}$, N/m) and in polystyrene ([27], 25±3 kDa, 2 kHz, $Z = 6.6\cdot10^{-15}$, N/m). The energy needed for their stabilization was taken from noises [15], because their destruction was caused by heat fluctuation (E = kT). The dependence of the cluster number (K = 1, 2, 3, 4, 5....) in the molecular matter on its oscillation frequency was described as follows:

$$m = (250\pm20)\cdot K^{(2.2\pm 0.2)}$$

Due to the fact that amylopectin is a branched molecule, its crystallization proceeds with difficulty. As shown with scanning electron microscopy, in amorphous lamellae of starch granules it is present in the form of coils [28]. The amylopectin chains build spirals where their meanders consist of six units. The spirals possess a low pronounced regularity [9] and they are in a revolved conformation, which is caused by hydrogen bonds and the cohesion interaction between the units. According to the oscillation model of coils (Scheme I), consisting of 2, 4, 6, and 8 spirals with 12, 24, 36, and 48 units, correspondingly, the oscillations must be detected by GMS-like oscillation of molecular mass concentrations in a spherical form.

Starch is thermally stable up to 523 K. TG/DTA measurements show neither weight losses nor any thermal effects until this temperature [12, 13]. The number of possible side reactions can therefore be decreased and it is possible to study them through GMS according to the low frequency movement of large molecular masses before the polymer destruction takes place. The thermal cluster movements can be regarded from the point of view of the cluster's interaction with its surroundings, and through the description of the energetic characteristics (activation energy of polymer flow) of such movement a known mathematical method can be used [29].

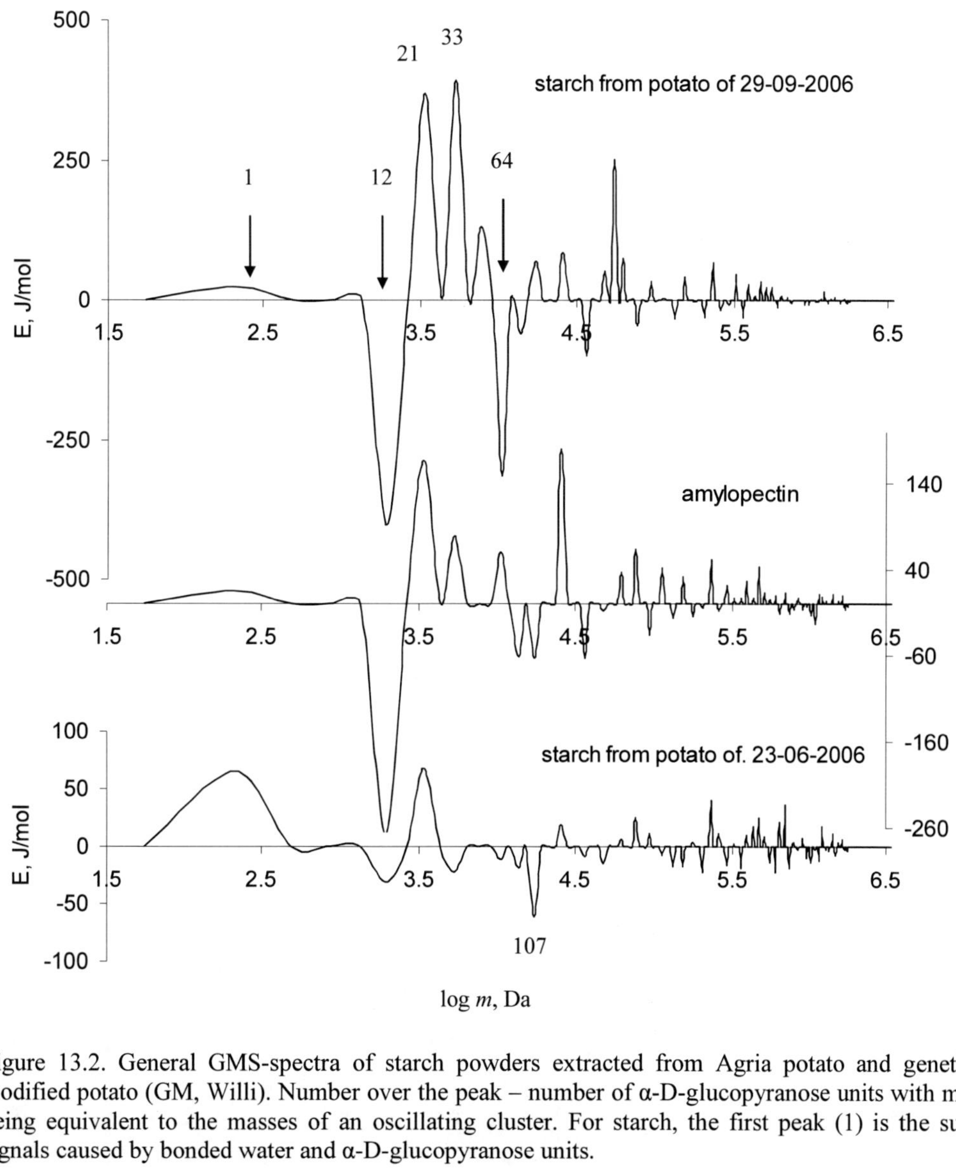

Figure 13.2. General GMS-spectra of starch powders extracted from Agria potato and genetically modified potato (GM, Willi). Number over the peak – number of α-D-glucopyranose units with masses being equivalent to the masses of an oscillating cluster. For starch, the first peak (1) is the sum of signals caused by bonded water and α-D-glucopyranose units.

In Figure 13.2, some GMS-spectra of starch and pure amylopectin are given. Number 12 marked the signals of molecular mass concentrations that are equivalent to 12 α-D-glycopyranose units. Modeling this mass concentration with Cochrane, it can be concluded that it is a spiral of two meanders. The model of this cluster (cluster-12) is given in Scheme II.

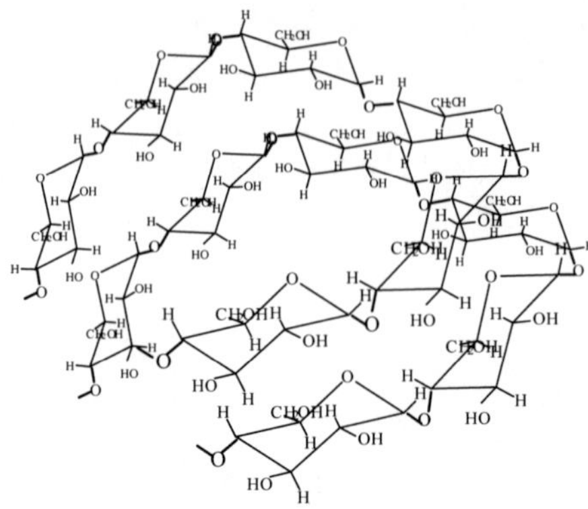

(II)

This spiral is stabilized by hydrogen bonds between units that are in the inner of spiral. The stability will be supported by constant signals in the whole investigated temperature interval. Cluster-12 is a collapsed structure that becomes more condensed under the influence of shock waves. Its structure as an oscillator can be schematically represented as shown below (Scheme III):

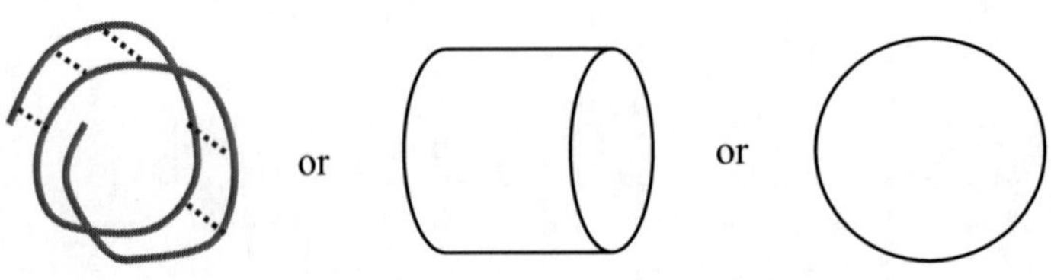

(III)

Possible variants of how cluster-12 in starch is linked with the main chain (dotted line, Scheme IV):

IV

a *b* *c* *d*

The stability of the structures could be explained by the fact that the hydrogen bonds (α-1,5) between the meanders in cluster-12 are stronger than the links between the meanders and the main chain. Through Cochrane modeling, it was shown that the form of this oscillator approaches a sphere and that its interaction with the surroundings does not mainly proceed through hydrogen bonds.

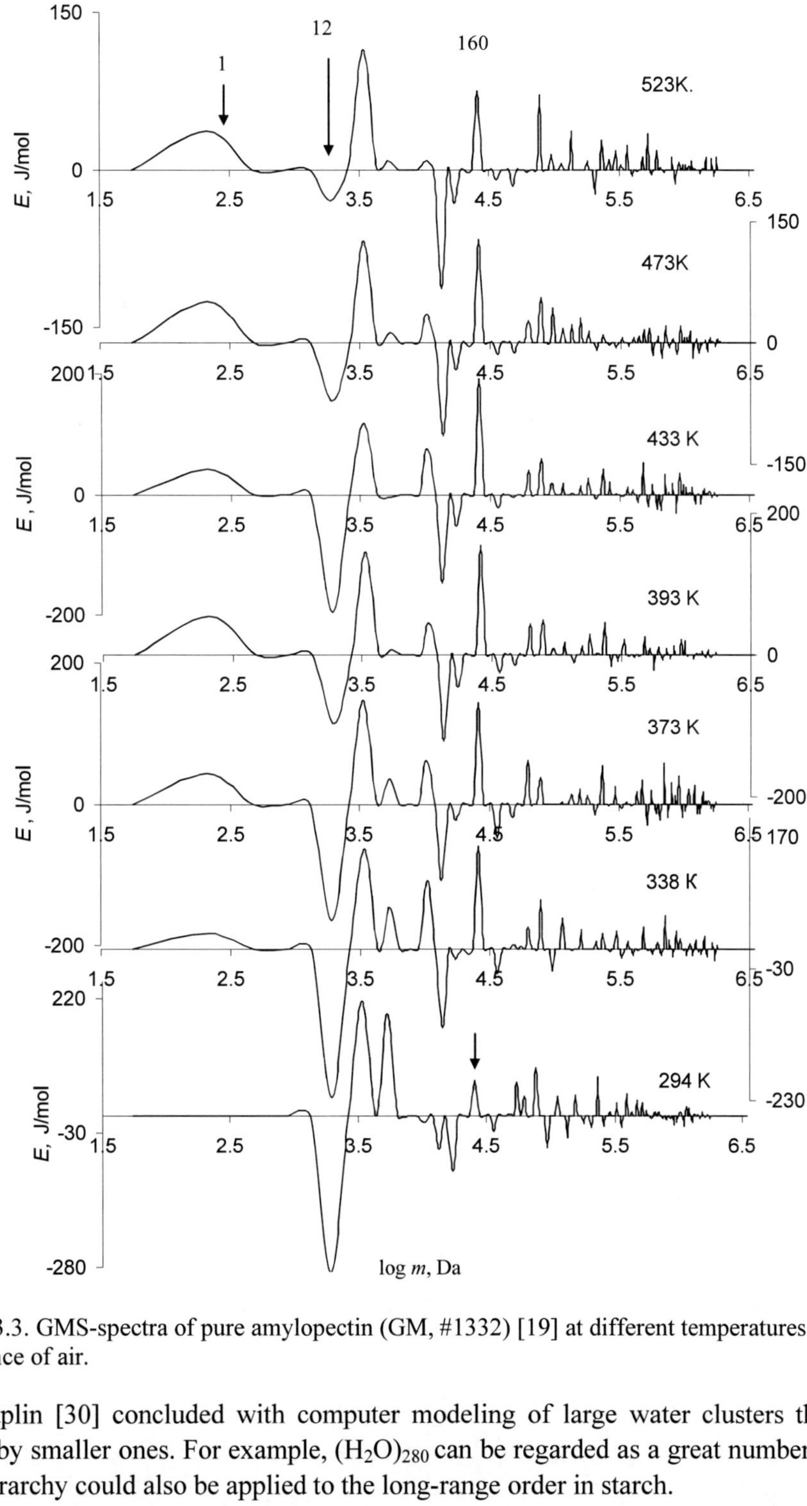

Figure 13.3. GMS-spectra of pure amylopectin (GM, #1332) [19] at different temperatures. Heating in the absence of air.

Chaplin [30] concluded with computer modeling of large water clusters that they are formed by smaller ones. For example, $(H_2O)_{280}$ can be regarded as a great number of $(H_2O)_{14}$. This hierarchy could also be applied to the long-range order in starch.

As shown in Figure 13.3, the formation energies of amylopectin clusters in a GM potato are in the interval between –280...+220 J/mol. At increasing temperatures, cluster-12 becomes unstable and its formation energy moves monotonously against zero. The large cluster-160 behaves quite differently and has a maximum stability at 393 K. At 294 K (Figure 13.3) in the region of mass oscillations in amylopectin up to log $m < 2.8$ there are not any signals. Although beginning with 338 K, a signal of mass oscillation appears that is equivalent to the mass of the α-D-glucopyranose unit.

There could be two types of α-D-glucopyranose unit oscillations in an amylopectin chain (cluster-1, Scheme V):

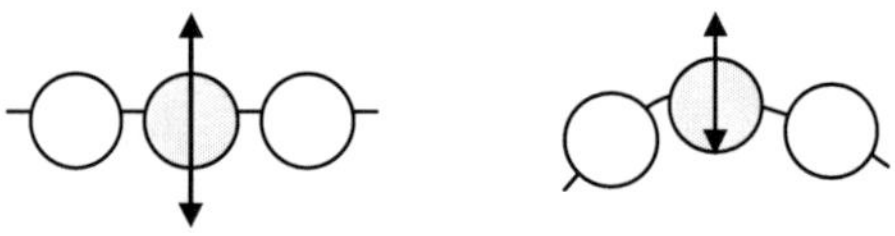

expanded collapsed

(V)

This signal (log $m = 2.8$) in the GMS-spectra is the sum of signals caused both by the collapsed and expanded cluster-1. With the help of Scheme V, the behavior of cluster-1 can be satisfactorily explained. Temperature increase leads to straight polymer chains (denaturizing) and the balance between expanded and collapsed clusters is shifted to the expanded cluster-1. It is emphasized that in this spectrum region (log $m < 2.8$), the oscillations of the α-D-glucopyranose unit is actively overlapped by water cluster oscillations, which means the sample has to be prepared carefully [26].

The small cluster-12 seems to be the base unit in the super cluster organization of starch, which means it has a periodically repeating structure unit (secondary building unit, SBU). Regarding PD (Figures 13.2, 13.3), the following row of clusters can be suggested: 12, 21, 33, 47, 64, 84, 107, 160, etc. Considering the error in the mass measurement and carrying out the calculations for one SBU, this row is as shown: 1, 2, 3, 4, 5, 7, 9, 13, etc. In these rows, signals of clusters, which only sometimes appear in the spectrum, are marked by an underlined number, for example, 4·SBU (Figure 13.3, Scheme VI). On the other side, larger oscillators behave differently. Spirals of 5 SBU were completely destroyed at 338 K whereas spirals consisting of 13 SBU (Figure 13.2, marked with an arrow, Scheme VII) at 294 K are probably in a balance between the collapsed and expanded form. Upon heating, they first shift to expanded clusters (T~338 K) but those are then thermally destroyed.

Cluster-160 of 13 SBU is thermally stable where its signals as oscillator are present in all spectra (Figure 13.2). At heating, its signal intensity increased, indicating that its interaction with its surroundings has got weaker. By modeling these clusters one can understand that they possess a nearly spherical form that for its part consists of SBU. This sphere is comparable with a forced wounded spiral in which one meander is one SBU (Scheme VI). Unexpectedly, its cluster mass agrees more or less with the mass of the polystyrene cluster that was extracted by Zysman from a diluted solution [27]—for which an explanation is given later.

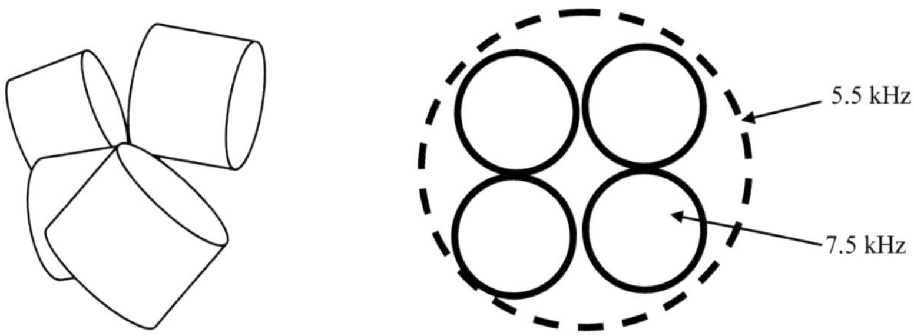

(VI)

Grape-like model of the cluster (4 SBU)

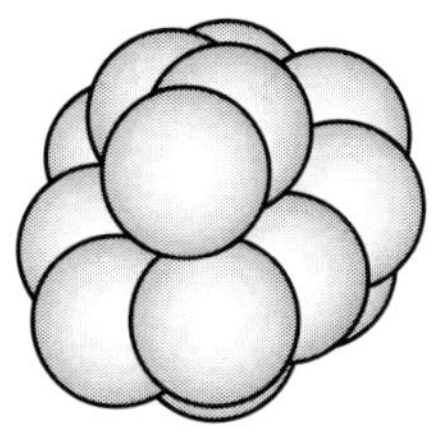

(VII)
Grape-like model of cluster-160 (13 SBU)

The integral distribution of clusters in starch and in the two pure amylopectin types is different.

Potato starch (Agria-N, 23.06.2006), approximation with high degree polynom:

$$W_y = -2\cdot10^{-6}\cdot x^6+9\cdot10^{-5}\cdot x^5-0.0021\cdot x^4+0.0219\cdot x^3-0.1157\cdot x^2+0.3371\cdot x+0.2027$$

where $x = m\cdot10^{-5}$

Potato starch (Agria-N, 6.07.2006):

$$W_y = -2\cdot10^{-6}\cdot x^6+9\cdot10^{-5}\cdot x^5-0.002\cdot x^4+0.022\cdot x^3-0.1302\cdot x^2+0.3947\cdot x+0.22$$

Potato starch (Agria-N, 01.09.2006):

$$W_y = 0.1058\cdot\ln(m\cdot10^{-5})+0.7074, \text{ satisfactory}$$

Potato starch (Agria-N, 29.09.2006):

$$W_y = 0.1092\cdot\ln(m\cdot10^{-5})+0.7533, \text{ satisfactory}$$

For amylopectin #1332, the integral cluster distribution is well described by:

$$W_y = 0.1199\cdot\ln(m\cdot10^{-5})+0.6771$$

For the amylopectin Willi:

$$W_y = 0.1173\cdot\ln(m\cdot10^{-5})+0.681$$

The temperature-forced movement of clusters can be described with the parameter of the cluster interaction with its surroundings (*f*). Therefore its change can be used as a characteristic for the temperature-forced movement, for instance, for the Arrhenius description of the activation energy of the cluster movement in starch ($f = A \cdot e^{-E/RT}$) [10].

In Figure 13.4, the temperature dependence of *f* for cluster-12 at different potato ripening stages is shown.

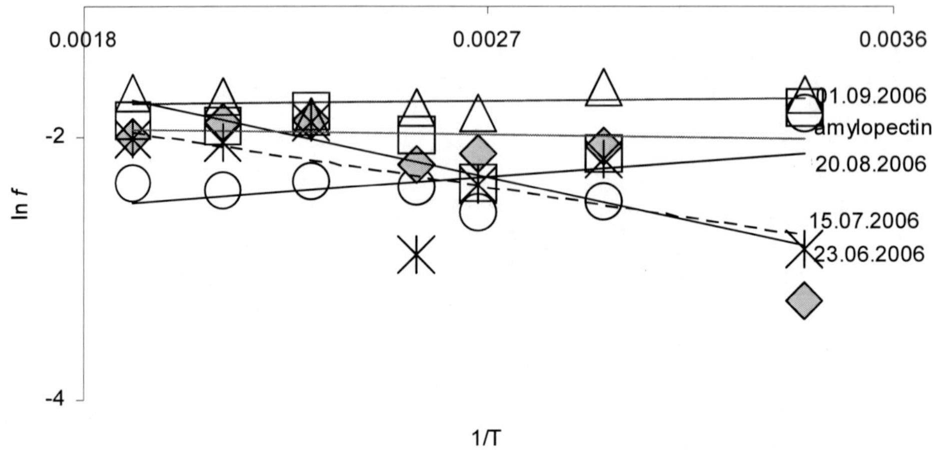

Figure 13.4. Temperature dependence of *f* for cluster-12 at different potato ripening stages.

As is visible, the thermal activation of the movement (E_{act}) of cluster-12 in starch and amylopectin can be calculated by this method. At low temperatures, the test points were strongly strewed, which is explained by the presence of structured water in starch (10...13 wt. %). A thermal elimination of water in native starch, however, could lead to the destruction of the long-range order in the amorphous part. This means that studying the forced thermal movement of cluster-12 in the absence of structured water cannot provide any real idea on the movement and the influence of the surroundings on cluster movement. Additionally, the lines' behavior was influenced by the ripening time. With an increasing ripening grade, the lines get horizontal and the test points are less strewed.

Using the data from Figure 13.4, the activation energy for the forced, thermal movement of cluster-12 was calculated (Figure 13.5).

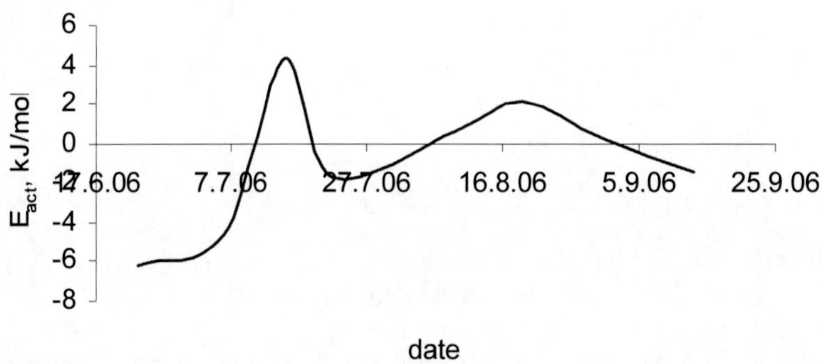

Figure 13.5. The change of the activation movement energy of cluster-12 in starch during potato growth (Agria).

As shown in Figure 13.5, the activation movement energy for cluster-12 interacting with its surroundings changed during potato growth. Until July 27, 2006 there was a sufficient correlation between increasing E_{act} and increasing molecular mass of macromolecular chains in the amorphous part of starch (Figure 13.6). The first peak position and the curve character agreed, regarding the time, with those in the dependence of the molecular masses of amylopectin super coils on the potato growth time. The negative E_{act} values could be an indication of an unstable (metastable) cluster-12 state in young potatoes (young starch) and for that there are not any outer barriers for the movement activation. On the contrary, there is even an outer stimulation, e.g. caused by an intensive influence of the main chain. The structures ***a*** and ***d*** (Scheme IV) matched these spatial conditions. In the early stage of the potato growth, the negative E_{act} values for cluster-12 reflected forced movements and they showed some similarities with the top part of climbing plants when they were watched with a camera in slow motion. On July 15, 2006, E_{act} got positive, which means that there is a real energy barrier that disturbs the movement of cluster-12 or that there is no more forced influence on the part of the main chain (structures ***a*** and ***d***, Scheme IV). It gets narrow in the amorphous part of macromolecule starch and therefore, cluster-12 movement becomes difficult. This state can be ascribed to structures ***b*** and ***c*** (Scheme IV). The greatly decreased E_{act} on July 25, 2006 correlated well with the highly decreased molecular masses (MM) of amylopectin super coils, which resulted from their enzymatic destruction [18]. By debranching [18] the surroundings of cluster-12, it received a little relief leading to its freer movement. The destructive process of amylopectin was finished after the temperature was decreased to lower 298 K and the debranching enzyme was inactivated. As a result, the MM of super coils rose again on from August 10-20, 2006 (Figure 13.6), which in turn caused a rise of E_{act}. In the ripened potato tuber, E_{act} became negative again because of the potato's aging and because of an increasing densification/crystallization degree of amylopectin as well as the pushing out of cluster-12 on the coil surface.

In aged starch, the movement of cluster-12 was stimulated by the main chain once again. Since the enzymatic destruction of amylopectin super coils proceeded on the surface, the clusters-12 were also mainly there. This was proven by a model stimulating this cluster movement by the main chain (Scheme IV). Unlike this, the inactive clusters-12 remained inside the coils and a differentiation between them and active oscillators is difficult. Temperature decrease and tuber aging (Figure 13.6) broke the rise of MM in amorphous starch that led to increased amylopectin crystallization. During crystallization, cluster-12 will be pushed out from the crystal visible at the monotonous decrease of the activation barrier after August 20, 2006 (Figure 13.5).

Processes characterized with negative activation energy are well known in polymer chemistry [29] and they are used to understand the mechanism of reactions including synthesis, structure formation, and crystallization (aging) of amylopectin.

It will be emphasized that the rows of above investigated clusters and of clusters in natural [31] and distilled water [23, 25, 32], in polymers [21, 27, 31, 33] as well as in solutions [34], in alcohols and hydrocarbons [23, 26, 35], and in protein and polysaccharide coils (lysozyme, L-arginin, chimotrypsinogen-A, amylopectin, etc.) [25] have similar masses. The row consisting of mass concentrations is general for all investigated clusters (see Table 13.1 below for the first nine members).

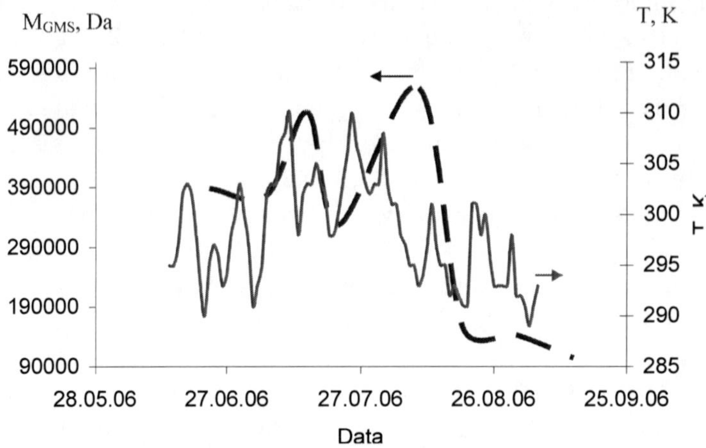

Figure 13.6. The average molecular masses of amorphous amylopectin super coils isolated from potatoes (dotted line) type Agria, red line – the average day temperature line.

Table 13.1. Mass concentrations against their oscillation frequencies observed in liquids, polymers, and biomolecules

n	*m*, kDa	kHz
1	0.2±0.02	22.2±0.1
2	1.8± 0.1	7.4±0.1
3	3.1± 0.1	5.5±0.1
4	4.9±0.5	4.4±0.1
5	7.0±0.5	3.7±0.1
6	9.5±0.5	3.2±0.1
7	12.5±0.5	2.8±0.1
8	15.7±0.5	2.5±0.1
9	26±3	2.0±0.1

The above row was satisfactorily described by Zubow equations.

The existence of such a stable row for quite different substances and for their states cannot be accidental. A convincing explanation could be that it arises in stationary waves of white noises that penetrate the Earth [25, 36]. The stability of this row can be destroyed, e.g. under temperature influence that charges white noise with thermal components and changes it to color noise. For example, at the heating of a water NaCl solution up to 368 K, frequencies and masses of oscillating solvated clusters of ion pairs [34] were shifted to higher masses and at the quick quenching of SiC nanoparticles (~10^6 K/s) from the plasma of adiabatic compression [37], the harmony of this row was destroyed [25].

Conclusion

The molecular mass concentration, equal to 12 α-D-glycopyranose units, was the most stable building unit (cluster-12) in the amorphous part of starch.

Cluster-12 represented a spiral of two meanders, the thermal stability of which was determined by hydrogen bonds between the meanders.

The formation energy of this cluster was in the interval from -35 to -220 J/mol and changed in dependence on the cluster-12 surroundings and ripened potato state.

The activation energy for cluster-12 movement in starch and amylopectin was from -6 to 5 kJ/mol and changed synchronously with the changing molecular masses of super coils during potato growth.

It was assumed that clusters were formed in stationary waves of white noises penetrating the Earth.

Even with the clusters in starch and amylopectin (grape-like model), the row containing the following cluster-12 units as secondary building units can be derived: 1, 2, 3, 4, 5, 7, 9, 13, etc.

REFERENCES TO 13

[1] Mazul V.M., Zaizewa E.M.Tsherbin D.G. Intramolecular dynamics and functional activity of proteins. Biophisika. 2000, vol. 45, no. 6, pp. 965-989, in Russian.

[2] Belitz H.-D., Grosch W., Schieberle P. Lehrbuch der Lebensmittelchemie. 5. Auflage, Berlin, Heidelberg, NY, Barcelona, London, Mailand, Paris, Singapur, Tokio, Springer Verlag 2001, p. 863.

[3] Lubert Stryer. Biochemie. 4 - Aufl. Spectrum Akademischer Verlag, Heidelberg, Berlin, Oxford, 1996, p. 690.

[4] Linewich L.I. Starch. Chemical encyclopedia. Moscow, «Soviet encyclopedia». 1990, vol. 2, p. 498.

[5] W. Banks, D.D. Muir, "Structure and chemistry of the starch granule," In: P.K. Strumpf, E.E. Conn, Eds., The biochemistry of plants, Academic Press, New York, 1980, Vol. 3, 321-330.

[6] Tuedze R., Kawai T. Physical chemistry of the polymers. Russian edit. M, Chimija, 1977, p.167.

[7] Anne J. Kortstee,Angela M.S. Vermeesch, Beja J.de Vries, Evert Jacobsen, Richard G.F. Visser. Expression of Escherichia coli branching enzyme in tubers of amylose-free transgenic potato leads to an increased branching degree of the amylopectin The Plant Journal. 1996, vol. 10. no.1, pp. 83-90.

[8] Hikara Satoh, Aiko Nishi, Kayuhiro Yamashita, Yoko Takemoto, Yasumasa Tanaka, Yuko Hosaka, Aya Sakurai, Naoko Fujita, Yasunori Nakamura. Starch-Branching Enzyme I- Deficient Mutation Specifically Affects the Structure and Properties of Starch in Rice Endosperm Plant Physiology. 2003, vol. 133, no 3, pp.1111-1121.

[9] Mia Sjöqvist, P. Gatenholm. The Effect of Starch Composition on Structure of Foams Prepared by Microwave Treatment Journal of Polymers and the Environment. 2005, vol. 13, no. 1, pp. 29-37.

[10] Vallés A. Lluch., Felippe A.M., Greus A.R., Cadenato A., Ramis X., Salla J.M., Morancho J.M. Thermal Analysis Characterisation of the Degradation of Biodegradabile Starch Blends in Soil. J. Appl. Polym. Sci. 2005, vol. 96, pp. 358-371.

[11] Beatrice Conde-Petit, Jeannette Nuessli, Eva Arrigoni, Felix Escher, Renato Amado. Perspective of Starci in Food Science. Chimia. 2001, vol. 55, no. 3, pp. 201-205.

[12] Aggarwal P., Dollimore D. The combustion of starch, cellulose and cationically modified products of these compounds investigated using thermal analysis./Thermochimica Acta. 1997, vol. 291, pp. 65-72.

[13] Aggarwal P., Dollimore D., Heon K. Thermal Analysis Study of Two Biopolymers, Starch and Cellulose. J. of Thermal Analysis. 1997, vol. 50, pp. 7-17.

[14] Lehrbuch für Botanik für Hochschulen Gustav Fischer Verlag 1993. 33. Auflage neubearbeitet von Peter Sitte, Hubert Ziegler, Friedrich Ehrendorfer, Andreas Bresinsky. S. 276.

[15] Pyda M. Conformational Heat Capacity of Interacting Systems of Polymer and Water. Macromolecules. 2002, vol.35, pp. 4009-4016.

[16] Kuakpetoon D., Ya-Jane Wang. Structural characteristics and physicochemical properties of oxidized corn starch varying in amylase content. Carbohydrate Research. 2006, vol. 341, pp. 1896-1915.

[17] Tester R.F., Patel T., Harding S.E. Damaged starch characterisation by ultracentrifugation. Carbohydrate Research, 2006, vol. 341, pp. 130-137.

[18] Batlle N, Carbonell J.V, Sendra J.M. Determination of Depolymerization Kinetics of Amylose, Amylopectin, and Soluble Starch by Aspergillus Oryzae –Amylase Using a Fluorimetric 2-p-Toluidinylnaphthalene-6-Sulfonate/Flow-Injection Analysis System. Biotechnology and Bioengineering, vol. 70, no. 5, December 5th 2000, pp. 544-552

[19] Reichmann M., Schwaryfischer A. Reine Amylopektin-Stärke aus Kartoffeln. Kartoffelbau. 2006, no. 4, pp. 170-173.

[20] Schittenhelm S. Kartoffelkraut und Trockenheit. Kartoffelbau. 2006, no. pp. 164-165.

[21] Zubow K. V., Zubow A.V., Zubow V.A. Study of structural changes in a LDPE film by flicker-noise spectroscopy. Plasticheskie Massy. 2006, no. 3, pp. 32-34.

[22] N.I. Klenkova. Structura i reaktionnaja sposobnost cellulosy. Nauka. Leningr. Branch. L., 1976.

[23] Zubow Kr. V., Zubow A.V., Zubow V.A. Cluster structure of liquid alcohols, water and n- Hexane. Journal of Appl. Spectr.2005, vol.72, no.3, pp. 321-328.

[24] Dan E. Dudgeon, Russel M. Mersereau. Multidimensional Digital Signal Processing. Prentice-Hall, Inc., Engelewood Cliffs. 1984.

[25] K.V. Zubow, A.V. Zubow, V.A. Zubow. Principles of gravitation spectroscopy. New form of molecular matter. Processes. Fields. "Aist Handels- und consulting gmbH", dep. R&D, Berlin, 2011, electronic book (www.zubow.de), p. 854 (in Russian).

[26] Zubow K. V., Zubow A. V., Zubow V.A. Use of flicker noise spectroscopy for analysis of nano structures. Zavodskaja Laboratorija. Diagnostika Materialow. 2008, vol. 74, no. 9, pp. 45-49, in Russian.

[27] Zysman, Viviane; Nguyen, Tuan Q.; Kausch, Henning H. Degradation on freezing dilute polystyrene solutions in p-xylene. J. of Polym. Sci., Part B: Polym. Phys. 1994, vol. 32, no.7, pp. 1257-1269.

[28] Tang H., Mitsunaga T., Kawamura Y. Molecular arrangement in blocklets and starch granule architecture. Carbohydrate Polymers. 2006, vol. 63, pp. 555-560.

[29] George Oudian. Principles of polymerization. McGraw-Hill Book Company, NY 1970.

[30] Chaplin M. (South Bank University of London, 2006) in http://www.lsbu.ac.uk/ water/index.html

[31] Zubow K. V., Zubow A. V., Zubow V.A. Investigation of the distribution of water clusters in vegetables, fruits and natural waters by flicker noise spectroscopy. Biofizika, 2007, vol. 52, no. 4, pp. 585-592.

[32] Andersson P., Steinbach C., Buck U. Vibrational spactroscopy of large water clusters of known size. Eur. Phys. J. D. 2003, vol. 24, pp. 53-64.

[33] Zubova K.V., Zubov A.V., Zubov V.A. Cluster-like supramolecular structure and is changes at low pressure of LDPE film. Plasticheskie Massy. 2005, no.12, pp. 36-39.

[34] Zubow Kr. V., Zubow A.V., Zubow V.A. Oscillation spectra of seed NaCl crystals in aqueous solution. Journal of Appl. Spectr.2005, vol.72, no. 6, pp. 840-847.

[35] Zakharov V.V., Brodskaya E.N., Laaksonen A. Molecular dynamics simulation of methanol clusters. J. Chem. Phys. 1998, vol. 109, pp. 9487-9493.

[36] Halberg Franz; Cornelissen Germaine; Regal Philip et al. Chronoastrobiology: proposal, nine conferences, heliogeomagnetics, transyears, near-weeks, near-decades, phylogenetic and ontogenetic memories /Biomedicine & pharmacotherapy. 2004, vol. 58 Suppl 1, pp. 150-187

[37] Zubov V.A, Scholtz R., Braun M., Zubowa H.-L., Kolbanovskii Yu.A., Ovsyannikov A. A., Shchipachev V.S. Structure and Properties of Inorganic Polymer on the Base of Silicon Carbide from High Temperature Gas Phase Russian Journal of Inorganic Chemistry. 1994, vol. 39, no. 11, pp. 1708-1711.

Chapter 14

WATER CLUSTERS IN PLANTS, FAST CHANNEL PLANT COMMUNICATIONS, AND THE PLANET'S INFLUENCE

It is well known that for the most part, water molecules form clusters [1]. Though clusters were discovered in biological matrices [2], up until now there is still no method by which clusters in plants can be identified during growth in vivo. Okonchi Shoichi [3] was suggested that water molecules are in the form of clusters in living organisms. In Figure 2.1.1, computer models of some water clusters are given. In our laboratory, we developed a gravitation spectrometer for water cluster identification in the biomatrices of plants [4, 5, 6]. Knowing the state of water clusters in plants could be helpful in understanding the relation between biochemical processes at the nanoscale level during growth and its qualitative yields.

The aim of the present chapter was to investigate water cluster ensembles in potato tubers and apples during their growth period in summer 2006. Furthermore, we were aiming to understand whether cluster communication with each other or with their surroundings is possible.

Potato starch from different cultivars ("Agria-N" from Kartoffelzucht Böhm KG and "Kalena" from NORIKA Nordring-Kartoffelzucht-und Vermehrungs-GmbH, both of Germany) was prepared by a standard method and were chosen as investigative objects. Pure amylopectin was provided by Bavaria State Research Center LfL. The mass spectra were measured with the gravitation spectrometer of Aist Handels und Consulting GmbH, Germany (www.zubow.de), where the energy (*f*) caused by the interaction of the clusters with a shock wave is a function of the molecular masses of the oscillating clusters. The molecular masses of the clusters were calculated by Zubow equations (§3.2.1) [5]. The calibration of the device was carried out with the clusters $(H2O)_{11\text{-}12}$ [7, 8, 9], $(H2O)_{100}$, and $(H2O)_{280}$ both through Chaplin [10] in bi-distilled water at 295 K [1]. The potato samples (tubers) were taken from the very same place (planted on April 27, 2006) whereas the apple sample of "alstar" cultivar was withdrawn from the very same tree and from the very same branch. Pure agarose of Aldrich (Tgel = 315 K, Tfl = 533 K) was used for preparing a model biomatrix. Agarose hydrogels were prepared by dissolving agarose (0.3, 0.8, and 3 wt. %) in boiling distilled water $(15\pm5)\cdot10^{-8,}$ om^{-1} . cm^{-1}) with follow up cooling to room temperature. Below, the general Scheme (I) of the GMS-measuring cell for the investigation of clusters in samples is shown [6]. The average day temperature (T_d, from 09:00 until 18:00) between June 8th (x_0 =

1987) and September 1st (x_i = 2070) can be described with $T_d = 3E\text{-}05 \cdot x^3 - 0.1889 \cdot x^2 + 393.73 \cdot x - 273034$.

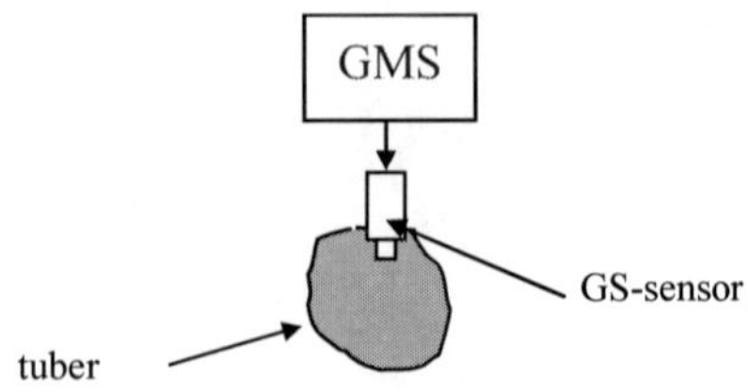

(I)

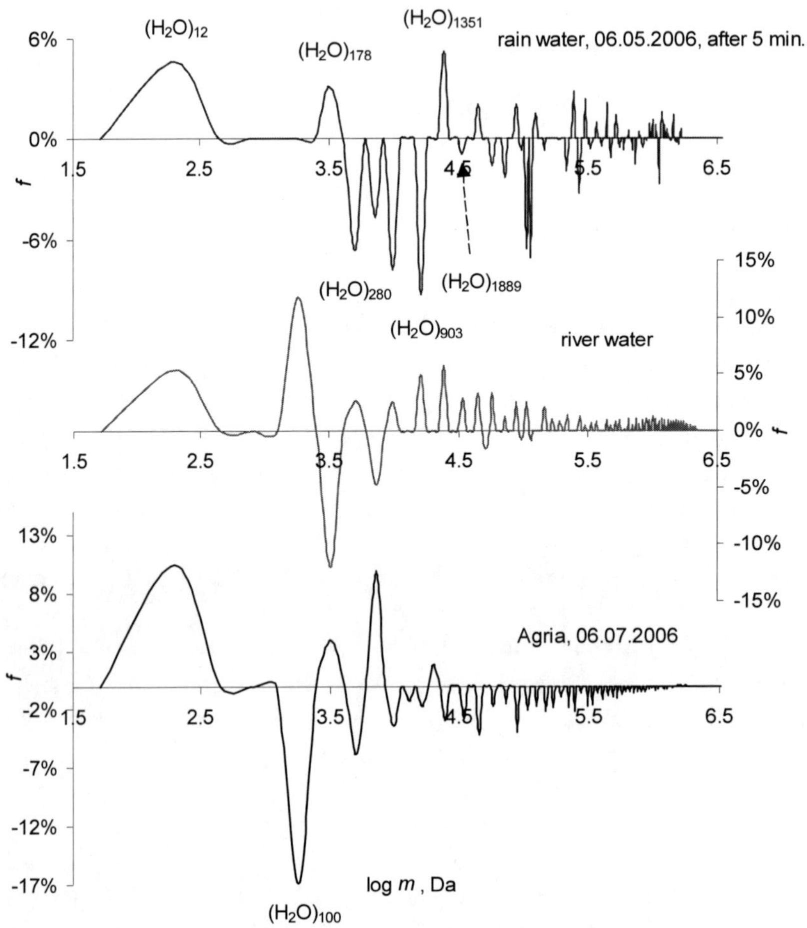

Figure 14.1. GMS-spectra of water in different substances, $p > 1$ N/m2.

The GMS-method, called in earlier works by the authors the FNS-method, is described in detail in [1, 5] (for algorithms see in Figure 4.2.1). The negative value of a signal in the GMS-spectrum means energy absorption, which is necessary for melting the collapsed cluster. On the other side, the positive signal is an energy release at the point of interaction of the shock wave with the expanded cluster, leading to its destruction. The signals of the collapsed clusters were therefore described with minus and those of the expanded one with a plus. The

Zayzev program, Astronet, version ZET 9 (www.astrozet.net), was used to get information on celestial body positions.

Typical GMS-spectra for rain and river water and water in potato tuber tissue are shown in Figure 14.1. As is visible, all clusters modeled in Figure 2.1.1 are present here. Water clusters in materials were found to be in different forms (collapsed or expanded), which depends on the interaction of the clusters with their surroundings [1]. In this study, we report the analysis data of water clusters (12, 100, 178, 280, and 1889) in potato tubers and apples during their growth in North Germany in 2006. The signal of clusters with 11±1 water molecules is stable in all the investigated samples and the expanded form (+*f*) of clusters dominates over the collapsed one (-*f*). Remarkably, the collapsed form is more enriched with energy than the expanded one.

The cluster of 100 water molecules in the examined potatoes is in its collapsed form but in river water this cluster is expanded and in rainwater it is often missing. This means that the plants need energy to adapt the irrigation water cluster structure (river, lake, rain, etc.) to the water clusters in biomatrices. Irrigation water must logically have similar water cluster (long-range order) forms as those occurring in plants. It is therefore assumed that the irrigation water weakens the immune system of the plants and indirectly reduces their energy potential.

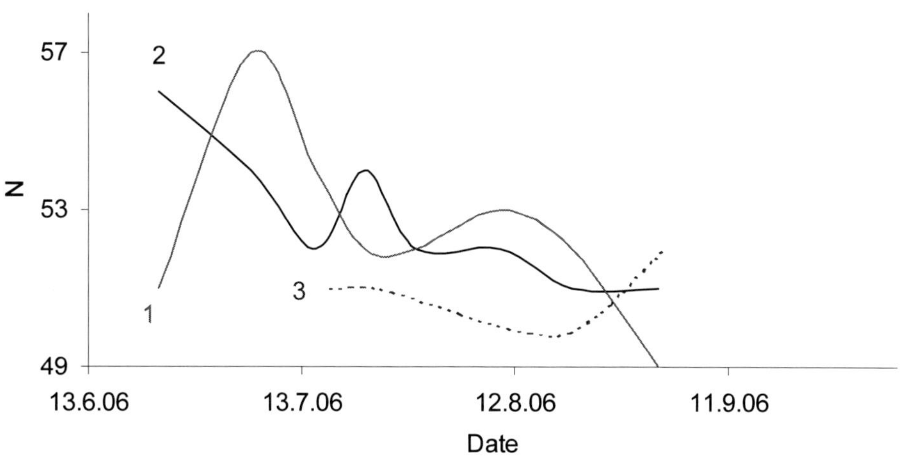

Figure 14.2. Number of cluster kinds in cluster ensembles of: 1 – tubers of Agria, 2 – tubers of Kalena, and 3 – apple.

In Figure 14.2, the development of all clusters (water, biomolecules, solvated clusters of ion pairs [1, 11], and bio polymers [2]) in plants during growth in summer 2006 is shown. There are strong changes in the total number of clusters, especially in the middle of summer and generally it trends toward decreasing. Differing from potatoes, in apples, the total number of clusters is less strongly dependent on growth time.

For a better understanding, we analyzed some water clusters in detail; we started with $(H_2O)_{12}$. In Figure 14.3, the degree of individuality or how this cluster interacts with its surroundings (*f*) is dependent on the given season. In both potato cultivars, the $(H_2O)_{12}$ dynamics are nearly synchronous during growth. As is visible, on July 1, 2006 the cluster almost completely lost its individuality, which can be explained by a strong interaction with the biomatrix of the potato tissue. After that the interaction weakens and the cluster's individuality increases up to 12%. Beginning with August 1, 2006 there are not any important changes. In apples, cluster $(H_2O)_{12}$ possesses a much higher individuality than in potatoes. This is visible as a maximum on August 12, 2006 (up to 15%). In potatoes, this cluster interacts more strongly with surroundings in the biomatrix than that of the apples. Possible cluster interaction with surroundings is shown in Scheme II.

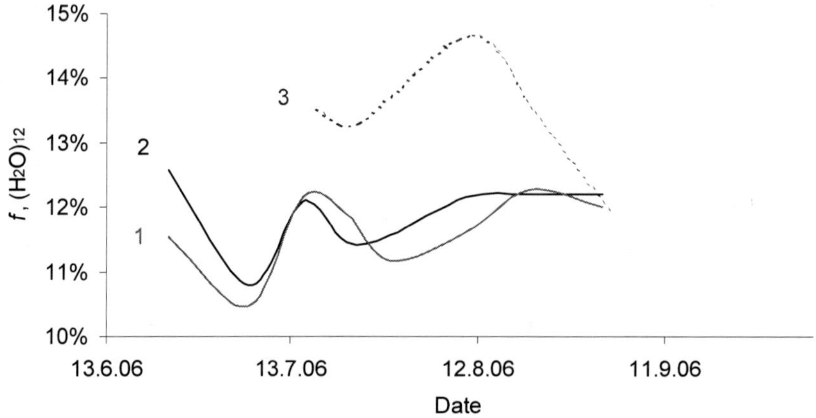

Figure 14.3. Relative rate of the water cluster $(H_2O)_{12}$ in cluster ensembles of: 1 – tubers of Agria, 2 – tubers of Kalena, and 3 – apple.

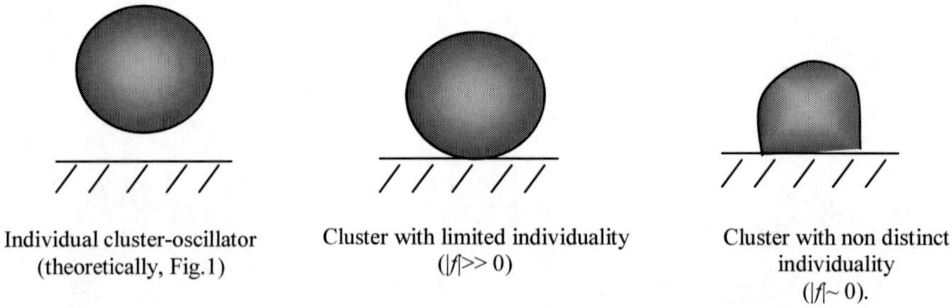

(II)

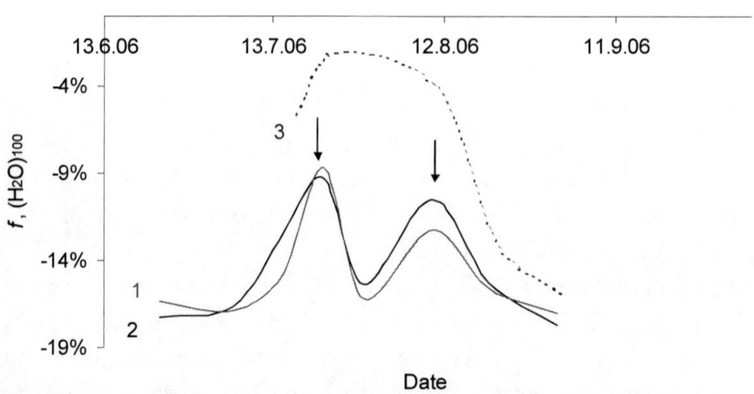

Figure 14.4. Relative rate of the water cluster $(H_2O)_{100}$ in cluster ensembles of: 1 – tubers of Agria, 2 – tubers of Kalena, and 3 – apple.

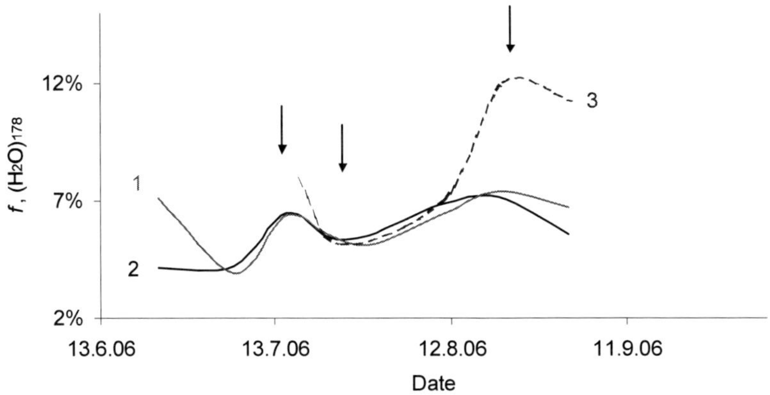

Figure 14.5. Relative rate of the water cluster $(H_2O)_{178}$ in cluster ensembles of: 1 – tubers of Agria, 2 – tubers of Kalena, and 3 – apple.

The interaction of the cluster consisting of 100 water molecules with its surroundings is given in Figure 14.4. This water cluster is in the collapsed form only during the indicated growth period. The interaction with its surroundings is very weak ($|f| >> 0$). The curves for both potato cultivars have two peak areas where the cluster is active at the point of interaction with its surroundings. The peak maxima are on July 15th and August 10th (arrows). Both peaks can be compared with those in Figure 14.4, but they are shifted for one month to a later time. For the $(H_2O)_{100}$ cluster in apples, there was found a broad peak between July 15th and August 10th. In this time range, the individuality reaches nearly zero, indicating an unknown influence on the cluster formation on the part of the surroundings in the biomatrix (see Scheme II). Thus, similar processes involving cluster $(H_2O)_{100}$ proceed both in potatoes and in apples.

In the following these investigations were applied to the water cluster $(H_2O)_{178}$ (Figure 14.5) that exists in the expanded form only. The curves are similar with two maxima on July 15th and August 20th. Compared with the two potatoes cultivars, the curve for apples shows higher *f*-values due to a stronger individuality of the clusters, which means these clusters' interactions with the surroundings is lower. On the other side, the behavior of all three curves is similar and possibly reflects general biochemical processes in plants (marked by arrows). Computer animation, however, did not yield this cluster. According to Chaplin [10], the clusters consisting of 160 and 184 water molecules were only computed. To the authors, cluster $(H_2O)_{178}$ is caused by biomatrix influence. The mass analog of this cluster was found for polysaccharides extracted from plants during the time from July 15th to August 20th, too.

The next discussion concerns the water cluster of 280 molecules (Figure 14.6) that is present in the collapsed form only in all investigated plants. Independently, on plant type for this cluster, we recognized clear individuality between July 15th and August 20th. Outside this time interval that means until July 15th and after August 15th, all curves approach the zero line (x-axe). The cluster seems to react on changes in the biomatrices, too. The influence of the biomatrix leads to a dominating collapsed form. Later, the August 15th aging processes in potatoes and apples causes biomatrix compression. The biopolymers contract and press out the included water molecules comparably with the gel-xerogel transformation. $(H_2O)_{280}$ clusters shall be destroyed through the densification of the biomatrix in ripe plants.

Figure 14.7 shows how the formation energy of this cluster changes during the plant growth. E_{form} varies in the interval between 0.4 and 1.3 kJ/mol and achieves its highest value in the end of July. This result agrees satisfactorily with the formation energy for the cluster in fresh distilled water (0.4 kJ/mol, [1] and with the experimentally observed value of ~1 kJ/mol [10, 13]).

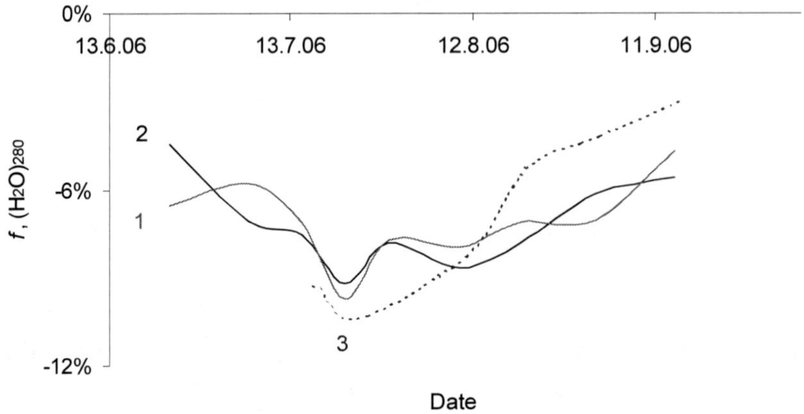

Figure 14.6. Relative rate of the water cluster $(H_2O)_{280}$ in cluster ensembles of: 1 – tubers of Agria, 2 – tubers of Kalena, and 3 – apple.

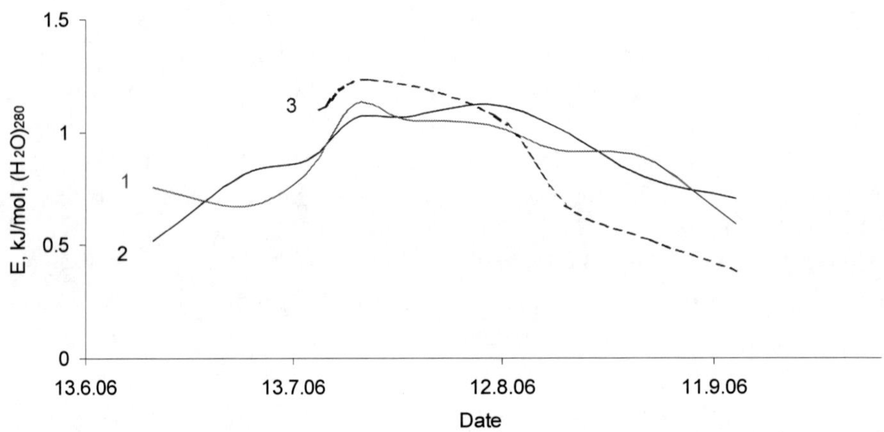

Figure 14.7. Changing formation energy of $(H_2O)_{280}$ in plants during growth. 1 – tubers of Agria, 2 – tubers of Kalena, and 3 – apple.

Finally, we shall study the water cluster $(H_2O)_{1889}$, which was calculated by Chaplin [10], too (Figure 14.8). There are strong differences between the oscillations in the two potato cultivars on the one hand and in the apples on the other hand but generally, the *f* value changes were found to be lower than for the other smaller clusters (-4%). In potatoes, the interaction of this cluster with the surroundings during growth seems to be very sensitive and in September the signal disappeared completely, which applies for the apples, too.

The average molecular mass of all clusters in plants during growth is represented in Figure 14.9. By comparing it with Figures 14.4 and 14.6, it can be concluded that the behavior of the average molecular mass of clusters is dependent on the time; it shows some similarities with the signals of $(H_2O)_{100}$ and $(H_2O)_{280}$. Remarkably, the absolute maxima for all curves are at the same time. This could be an indication for the fact that similar biochemical processes on the level of super molecular structures proceed in the biomatrix of all investigated plants. Comparing the results of Figures 14.4, 14.5, and 14.9 at least two to three transformation processes in plants during growth can be concluded.

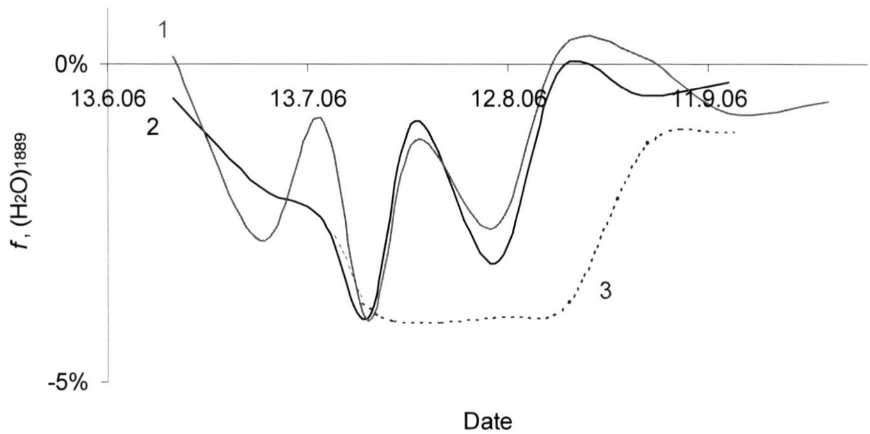

Figure 14.8. Relative rate of the water cluster $(H_2O)_{1889}$ in cluster ensembles of: 1 – tubers of Agria, 2 – tubers of Kalena, and 3 – apple.

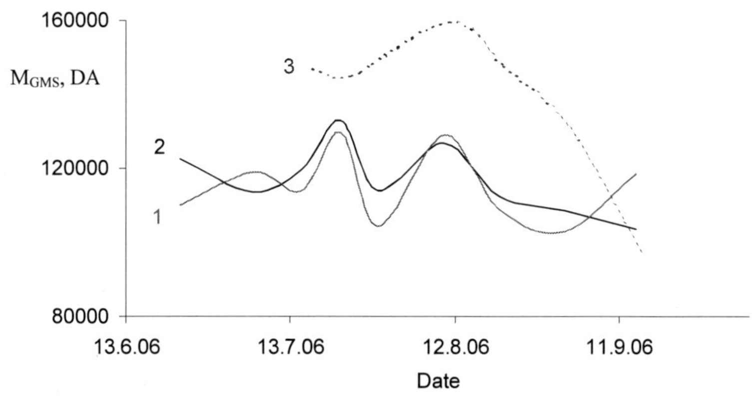

Figure 14.9. M_{GMS} of all clusters in vivo versus the growth time: 1 – tubers of Agria, 2 – tubers of Kalena, and 3 – apple.

Based on the results discussed above, the following model for water cluster forms in the biomatrices of potato tubers has been suggested. In this model (Figure 14.10), the biomatrix is represented as a line net and water clusters as balls. It is visible how the changing biomatrix influences the water cluster form. In a young potato, $(H_2O)_{100}$ and $(H_2O)_{178}$ are in collapsed form whereas $(H_2O)_{280}$ and $(H_2O)_{1889}$ are in an expanded one (dotted line). In the middle of summer, the biomatrix growth is very fast, which leads to large pores and the pressure on the part of the biomatrix on the water clusters becomes weaker. In young plants, the individuality of clusters is more developed. With plant ripening, the biomatrix loses water due to the fact that all super molecular structures contract and release water in the plant tissue like a gel-xerogel transformation (syneresis).

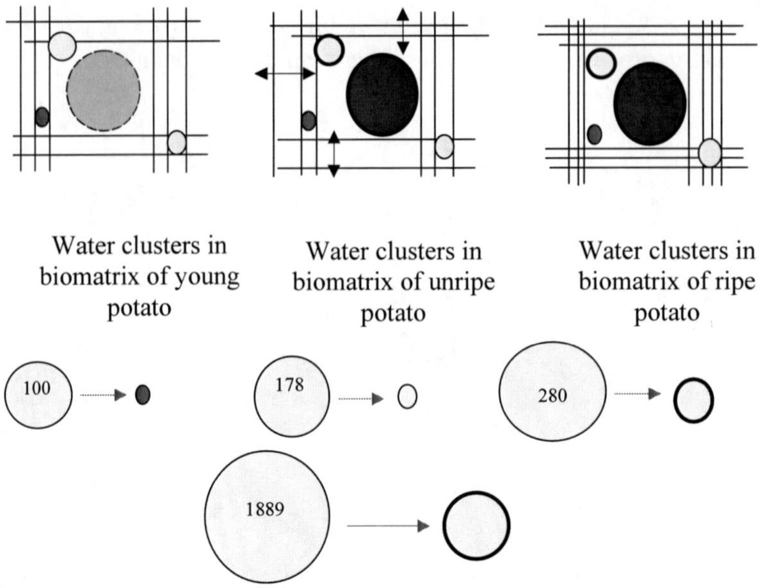

Figure 14.10. Model of water cluster interaction with biomatrix with a dependence on growth time.

For a better understanding of the biomatrix influence on the water clusters, amylopectin was extracted during the potato growth and the M_{GMS} values were measured and compared with the M_{GMS} values of all the clusters in the potato tissue (Figure 14.11). The M_{GMS} of amylopectin (grape-like clustering) is much higher than the corresponding value for all the clusters in the potato tissue. Curve 1 as well as curve 2 has two maxima, though concerning their position, there are some differences. The first maximum (marked with a star) in curve 1 on July 6th corresponds with the minimum in curve 2. On the other side, the second maximum of curve 1 (August 10th) clearly correlates with the second maximum in curve 2. The minimum (July 22nd) in curve 1 shall be explained with a debranching process of amylopectin at temperatures higher than 298 K [14]. The similarities between both curves assume that the two events, which take place in amylopectin (biomatrix) and in water clusters in the potato tissue, are interconnected.

In the next section, some small water clusters shall be briefly compared with M_{GMS} of amylopectin (amorphous area) as a possible representative of the macromolecular structure in the biomatrix (Figure 14.11). Both in young and ripe potatoes, the water cluster $(H_2O)_{100}$ is in a free undisturbed state (Figure 14.4) whereas during the growth (unripe state, Figure 14.10), when the biomatrix changes, a new interaction between this cluster and the biomatrix leads to a new cluster form. Both sides seem to influence each other (see the correlation between curves 1 and 2 in Figure 14.4 and curve 1 in Figure 14.11). In contrast, $(H_2O)_{178}$ does not participate in active biochemical processes in the biomatrix (bad correlation between curves 1 and 2 in Figure 14.5 and curve 1 in Figure 14.11) due to that, it is actively involved in the super molecular structure net of the biomatrix, which retains this cluster. Only July 22nd shows collapsed cluster $(H_2O)_{280}$ with a strong individual behavior in potatoes and apples

(Figure 14.6), but in young and old plants, the biomatrix destroys the cluster with the *f* value approaching zero (Figure 14.10).

Concerning the biomatrix influence on large clusters, $(H_2O)_{1889}$ is the most sensitive one (Figure 14.8). During growth it is permanently forced to change its form (Figure 14.10).

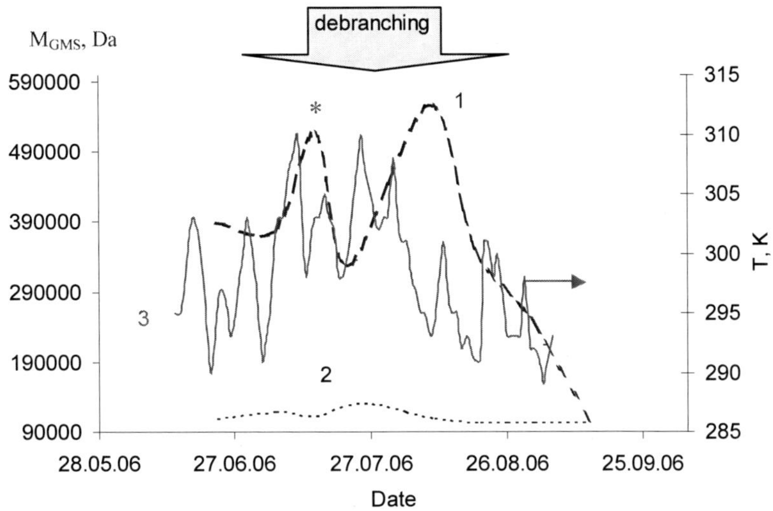

Figure 14.11. Average molecular masses (M_{GMS}) of amylopectin super coils isolated from the potato cultivar Agria (1) and of all the clusters in potato tuber tissue of Agria in vivo (2), the average day temperature line (3).

For the integral distribution of clusters in plants we found the following equation:

$$W_x = a \cdot \ln (m \cdot 10^{-5}) + b$$

where W_x is the integral rate of cluster fractions, *m* is the mass of clusters in Daltons, and the coefficients *a* and *b* change depending on the ripe degree. We describe these changes with the following equations for the apple ("elstar," 83±2 wt. % water in ripe state):

$$a = -5E\text{-}09 \cdot t^4 + 5E\text{-}05 \cdot t^3 - 0.1816 \cdot t^2 + 295.49 \cdot t - 180235$$
$$b = 6E\text{-}09 \cdot t^4 - 6E\text{-}05 \cdot t^3 + 0.2311 \cdot t^2 - 378.76 \cdot t + 232739$$

where *t* is the date of the test.

For potato tubers, the character of *a* and *b* changes and cannot be described with these equations. The values of *a* and *b* strongly depend on the ripening time and they reflect some changing processes in the super molecular structures in tuber tissue during potato growth.

The signal frequencies (kHz) we detected directly for the water clusters in the tubers of two potato cultivars and one apple cultivar consisting of 11±1; 100; 178; 280; 402; 545; 715; 903; 1,119; 1,351; 1,606; and 1,889 molecules were: 22.2, 7.3, 5.5, 4.4, 3.7, 3.1, 2.8, 2.4, 2.2, 2.0, 1.8, and 1.7, accordingly.

14.1. Water Cluster Communication Field

Now we shall discuss a hypothetical oscillation model of one element being part of a multi element system (cluster like dimension). The oscillation of one molecule (P) can be envisaged as a pendulum and the oscillation of the dimension of P like a dimension of pendulums (PD), oscillating in one rhythm (minimization of the total energy and harmonization of the model, scheme in §3.2.1). In these systems, the pendulum oscillations depend both on the nature of the pendulum and its interaction with its surroundings. The driving force (DF) of the association of some P in PD is the resonance that harmonizes the whole system, leading to a higher stability and to a minimum of the total energy in the new ensemble. Furthermore, PD oscillations influence other oscillation systems in their surroundings by resonance, where a stronger PD forces weak PD to oscillate in the rhythm of the strong PD. How strongly the surroundings shall be influenced by PD depends on the number of P in the ensemble. The information transfer from PD to P and backwards proceeds through the medium of the pendulums.

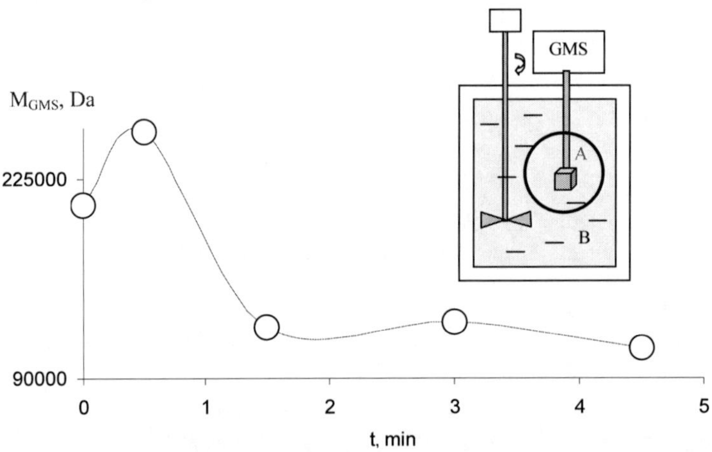

Figure 14.1.1.Change of M_{GMS} in vessel A by stirring water in vessel B. Both vessels are filled with distilled water at 286 K. The stirring engine at GMS-measuring was switched off. The equipment was isolated from light and mechanical vibrations.

It would be interesting to investigate the information transfer between clusters of water (PD) being in different vessels and of water in a model biological tissue (hydrogel). Agarose hydrogels are often used in biochemistry and microbiology where their chemical structure is well known and they can be considered biological tissue [15, 16]. If this information transfer channel really exists, then it should be observed by analyzing the influence of one (PD_A, vessel A) on the other one (PD_B, vessel B). For that the water cluster ensembles in vessel A were stimulated by a forced stirring of water (destruction of long-range order). Contemporaneously, the water in vessel B at the distance *x* was in a calm state. The interaction of the cluster ensembles belonging to water in different vessels was measured with the help of the Zubow gravitation spectrometer given in Figures 14.1.1 and 14.1.2 [6].

Water clusters like oscillators or pendulum ensembles have to communicate with each other through a resonance field. Here, one dimension of a water cluster-pendulum in vessel A (Figure 14.1.1) interacts with another one of the same water cluster-pendulum in vessel B through the resonance between these pendulums (scheme in §3.2.1) where the resonance field is not homogenous, since it is influenced by a number of other surrounding resonators (water clusters of the neighboring river, and water in the floor, wood, clouds, etc.). To check it, a simple experiment was carried out.

Before beginning the experiment and in order to achieve a thermal and mechanical balance, the system was stored for 6 h excluding the influence of outer thermal flow, light, and mechanical fields. At this time the characteristics of LRO in water (Figure 14.1.1), agarose hydrogel, and potato tissue (Figure 14.1.2), in which the GMS-sensor was installed, were measured three times. Then, water in vessel B was stirred for 1 min by a Teflon stirrer, the electrical engine was switched off, and after 1 min, the GMS-spectra were taken in vessel A at an interval of 5 min. As is visible in Figure 14.1.1, forced destruction of the LRO in water in vessel B changes the LRO in the water of vessel A (strong increase of M_{GMS}). This reaction supports the correctness of Scheme II. According to Scheme II, water clusters as oscillator ensembles interact with each other by resonance. We found that water clusters react differently upon the forced activation of their analogs in vessel B. Shortly after, at 30 s, a maximum M_{GMS} (250 kDa) was achieved; M_{GMS} decreased to 120 kDa and approximated then to a stable level. The rare cluster $(H_2O)_{137}$ is the most sensitive one to the resonance field influence of water that is placed in another vessel; it appears and disappears. The vibration process lasts 5 minutes and is comparable to a decreasing oscillation. Hence by the integral characteristics of the long-range order in liquids, e.g. M_{GMS}, changes in the fine cluster structure cannot be analyzed by the observer for the process was finished after 1 minute. At the level of individual clusters, to get an answer on the excitation from analogs in the other vessel, more time may pass in the form of some minutes or even hours.

The question raised is whether water clusters, for instance in agarose hydrogel or in plant tissue (receiver), can communicate with cluster ensembles in stirred water (transmitter) that is at a distance of 18 cm and 31 cm. To find an answer for this question, the next equipment was developed (Figure 14.1.2).

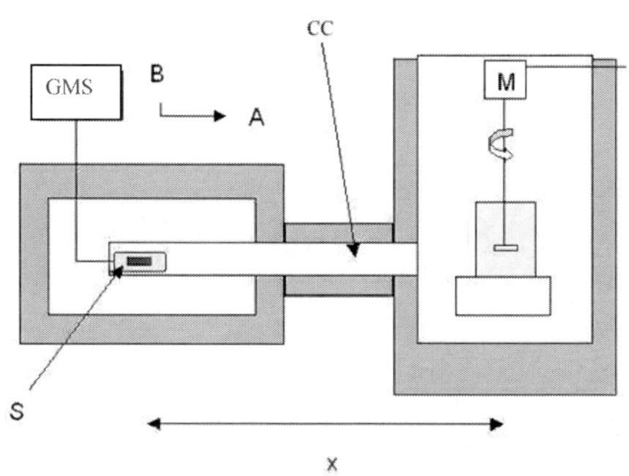

Figure 14.1.2. Scheme for the determination of the information field caused by the interaction of water clusters that are distant from each other. 286 K, x = 18 and 31 cm. GMS-sensor and water-stirring equipment were placed in a sound isolated polystyrene foam containing box. Before each GMS-measurement (1 min) the stirring engine (M) was switched off. A – direct vessel communication by the communication channel (CC, pipe), B – contact-free variant (CC is directed to the line of the shortest interaction with an angle of 90°), and S – sample (GMS-sensor is in 3 wt. % agarose gel or directly in tuber tissue).

These communication signals (change of the average mass of the secondary activated water clusters in the receiver cell) are shown in Figure 14.1.3. In the first 10 minutes when the stirrer is switched off, M_{GMS} is nearly unchanged. Then one minute of stirring causes (transmitter) a strong increase of M_{GMS} in the agarose hydrogel (receiver cell). This signal is unsymmetrical—while the increase is very steep, the decrease proceeds more slowly. Based on this result, wireless communication between these two systems, between the transmitter (activated water) and receiver (water clusters in agarose hydrogel), was concluded. The same stirring procedure was repeated 10 minutes and as is visible, the answer is similar but of a lower intensity.

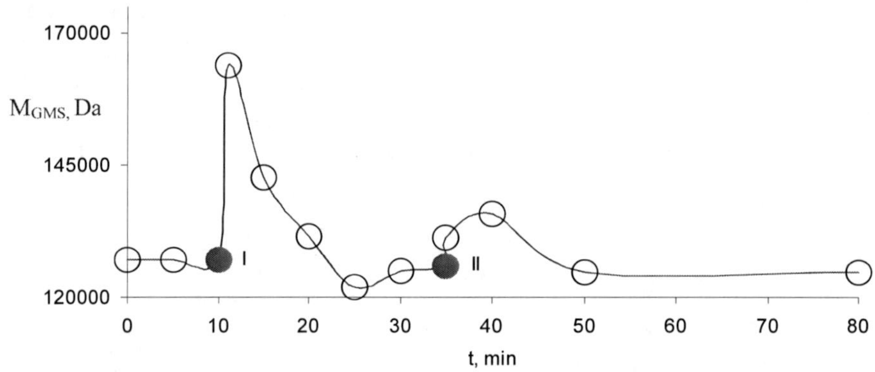

Figure 14.1.3. Effect of directed resonance interaction of water clusters in the measuring box with water clusters in the stirring vessel (Figure 14.1.2). I – measuring after 1 minute stirring and 1 minute break, directed position (Figure 14.1.2 in the direction of A) and II - measuring after 1 minute stirring and 1 minute break, non-directed position (Figure 14.1.2 in the direction of B).

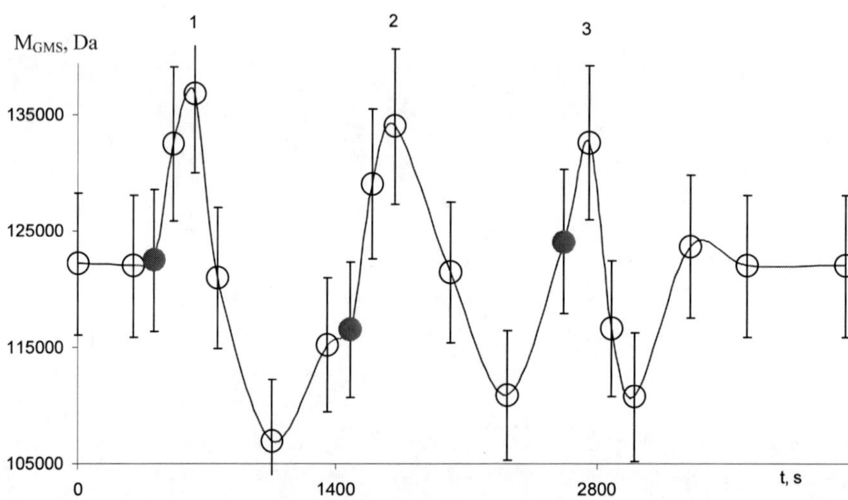

Figure 14.1.4. Change of the average molecular cluster mass in a potato tuber (in vivo) under the resonance field influence of oscillating water clusters in a 2 L water vessel that is at a distance of 18 cm (Figure 14.1.2), 286 K. Communication medium – air. The black circle means the switching on of the stirrer for 1 minute. The spectra were taken when the engine was turned off.

Analogous experiments were carried out with potato tubers in which the GMS-sensor was installed (Figure 14.1.2). As for agarose hydrogel for the potato biomatrix (in vivo), an answering reaction on the forced destruction of long-range order in water in a vessel that is at a distance of 18 cm was observed, too (Figure 14.1.4). The test was made three times. The first peak is symmetrical, although compared with agarose hydrogel the decrease of M_{GMS} is deeper. The second and third effects are similar. The triple repeated experiments clearly support the fact that there exists a communication channel between water clusters in tubers and the water clusters of the surroundings.

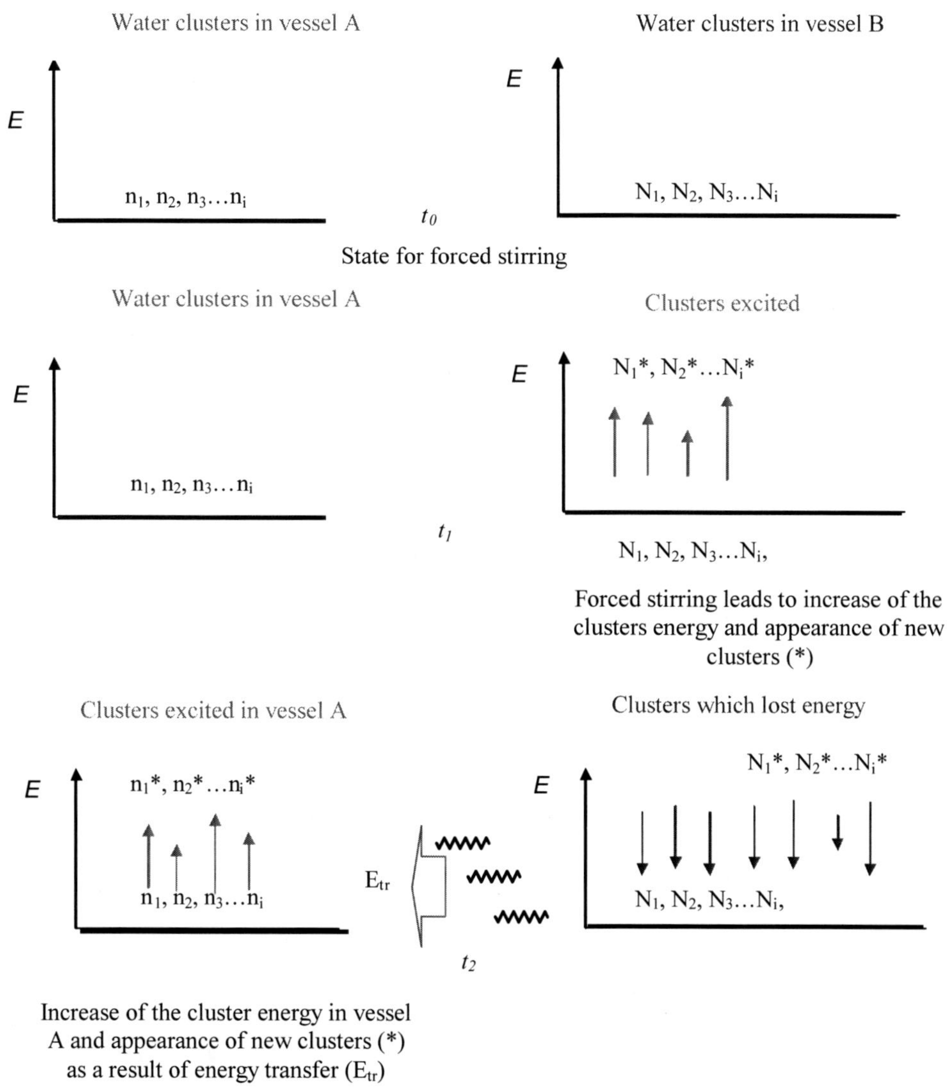

Figure 14.1.5. Scheme illustrating how the energy levels of water clusters in hydrogel were filled by energy transfer. Each oscillator is marked by n_1, n_2, n_3…n_i (number of water cluster kinds in hydrogel) and by N_1, N_2, N_3…N_i (number of water cluster kinds in forced stirred water). The energy levels of water clusters in hydrogel will be filled as soon as the energy (E_{tr}) is delivered from stirred water.

Several water clusters in potato tubers react differently upon impact of a resonance field caused by the transmitter (forced stirred water in vessel); some clusters react quickly, within some seconds, whereas others need more time (range of some minutes). The great increase of M_{GMS} and then its fast decrease to lower than the M_{GMS} level in a tuber without field influence indicates that this influence can be regarded as a shock wave, which consists of a falling and a reflective wave. It is comparable with a low tide and high tide. The energy flow from activated clusters (see Figure 14.1.5) achieves the level of water clusters in the potato tuber after the average cluster mass in the biomatrix increased. This process is characterized by the condensation of small clusters and domination of a large one. In the following 15-20 min after the energy impact, energy dissipation in the biomatrix thermodynamic balance has been recovered (Figure 14.1.4). The elasticity of this interaction can serve as a characteristic for the stability of the biomatrix against the influence of fields caused by other molecular oscillators in the hydrosphere of the earth (natural and artificial irrigation, influence of large water masses, e.g. clouds).

The detailed study of resonance processes of some clusters has shown that new clusters appear whereas old ones disappear, finally, leading to changed surroundings by different clusters. The rare water cluster containing 226 molecules is formed upon the forced stirring of water [17]. In the potato tuber, it is present, too, although the signal is of low intensity (f = 0.1%) and it appears only after the third impact at 2,750 seconds (Figure 14.1.4). Its oscillation intensity first increases eight times, then after 125 seconds (2,875 s, Figure 14.1.4), it decreases by 38 times. Later after 245 s (2,995 s) it increases again by 27 times and after 1,385 s (4,136 s), it calms down while the other clusters react five times weaker. Remarkably, this cluster is present in the communication net of agarose hydrogel-water, too, (Figure 14.1.2) but it is not in the water-water system (Figure 14.1.1). In the water-water system, the most active role is played by another cluster, namely $(H_2O)_{137}$.

The activation of $(H_2O)_{226}$ seems to be caused by the presence of polysaccharides in the biomatrix. We assume that this cluster is integrated in the biomatrix pores of the tuber tissue and that it leaves these pores under the impact of shock waves where this process is reversible (Scheme III).

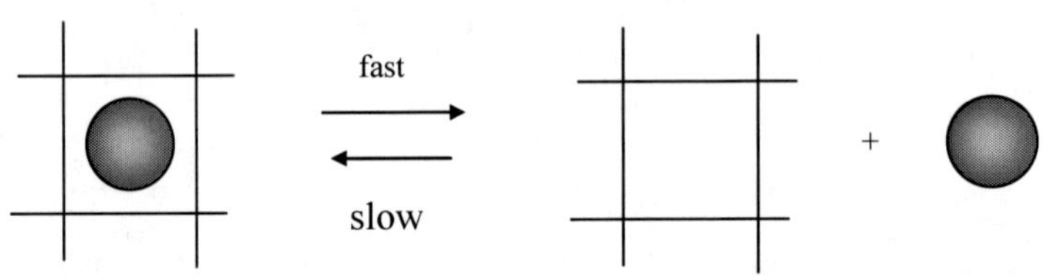

(III)

The appearance of the new unstable state in the tuber biomatrix (Scheme III, right part of equation) gives a signal to the plant, which can be interpreted as the beginning of the irrigation, the approaching of rain clouds, or the rain beginning. As a consequence, the plant switches on a preparation mechanism for irrigation. Analogous communication between plants and their surroundings at approaching drought, frost, winds, etc., is possible under the participation of other cluster signals.

Additional studies showed there is a fast communication channel in the tuber-tuber system, namely, a mechanically destroyed tuber influences its neighboring tuber in which a GMS-sensor is installed (Figure 14.1.2).

The information field of water clusters' interaction in tubers is sensitive to the water quality, namely, to rain, river, or distilled water (Figure 14.2) as well as to neighbors containing a large amount of water and possessing a cluster field like a dimension of resonators (scheme in §3.2.1). This information channel diversity permits us to use it for plant communication. The amount of communication channels can be estimated for the first nine members of the cluster sequence by the next equation:

$$C = \sum_{i=1}^{9} \binom{9}{i}$$

The equation gives 509 communication channels that can be multiplied by the number of kinds of super molecular structures (n), for instance, water clusters surrounding uniform biomolecules in the biomatrix, which appears like a unique oscillator block. The entire information field capacity is the following:

$$C_o = C \cdot n$$

The transformation velocity is in the sound frequency area and depends on the traffic kind (cluster transmitter and cluster receiver), too, where the transformation quality can be influenced by external factors, e.g. resonance fields of water clusters belonging to other systems.

For all clusters, the oscillation frequencies are known. Therefore, it is possible to develop a program and equipment for an early warning system for plants about weather changes and parasite attacks. Additionally, it is possible for this system to monitor and stimulate the ripening process (Scheme IV).

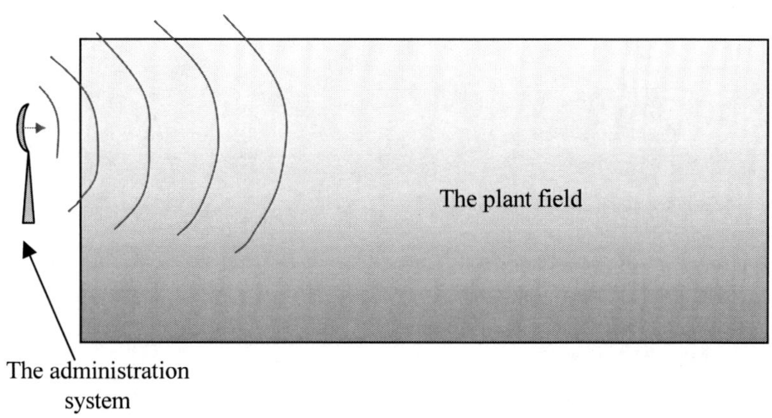

(IV)

In Figures 14.1.5 and 14.1.6, we shall represent the mechanism of possible communication between biological systems realized by the resonance fields of water clusters.

We see two dimensions of equal oscillators (state t_0, Figure 14.1.5) that are divided from each other by a molecular space, for instance, air. The forced activation of the oscillators of one dimension (in vessel B, Figure 14.1.5) by, e.g. pumping with potential energy, causes a new energy set of clusters (state t_1) that release this energy (E_{tr}) after the forced energy pumping is finished and the system achieves the starting thermodynamic state (t_2). The released energy will be transferred as a gravitational energy wave to other dimensions of oscillators where it causes an answer in the form of the oscillation of the whole system as well as of some individual clusters of the same type. Figure 14.1.5 can be applied for cluster communication in liquids as well as in biomatrices.

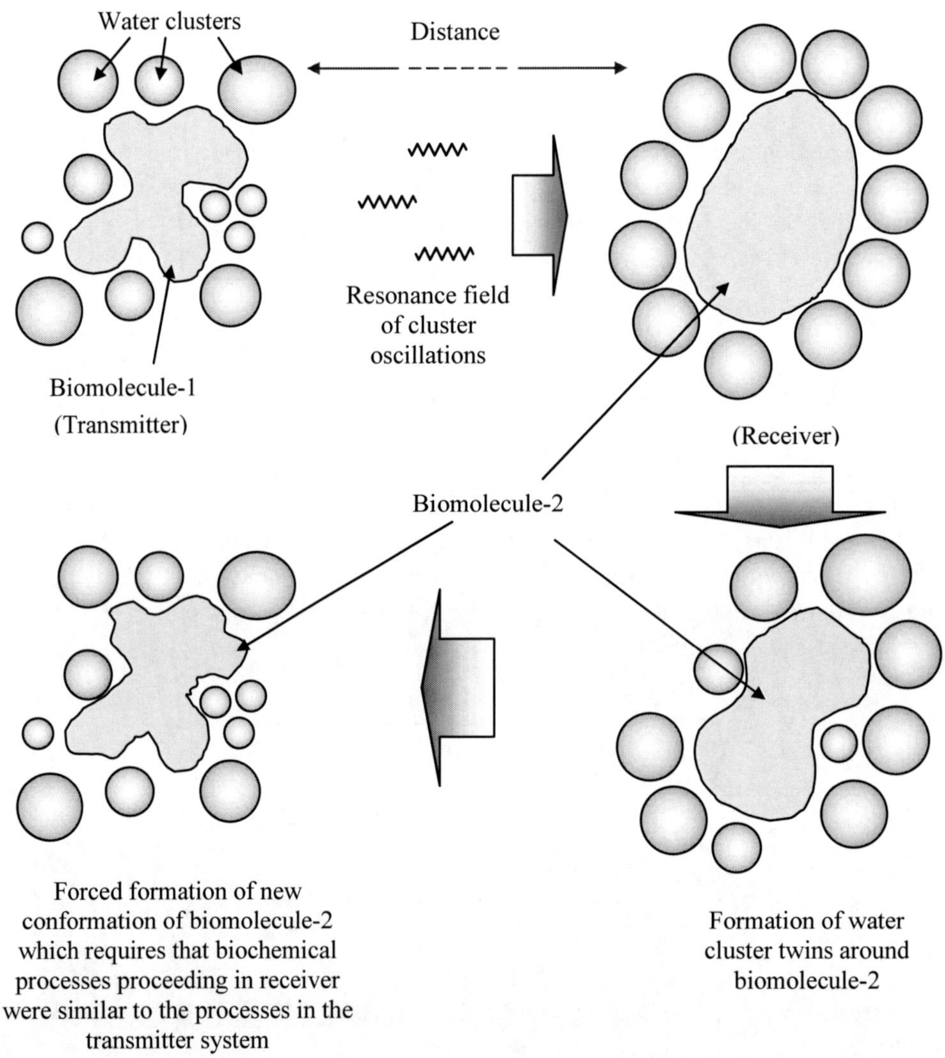

Figure 14.1.6. Model for information transfer from one biomolecule to another through the forced change of the biomolecule conformation under the influence of water clusters.

The influence of water clusters on the conformation of a biomolecule surrounded by different water clusters is shown in Figure 14.1.6. Furthermore, it is shown how water clusters change the communication of the first biomolecule with a second one that is distant and that can belong to another plant. It was shown by [18] that water around a biomolecule changes its conformation. Additionally, we found that at resonance, the interaction of water cluster ensembles with other analogous water cluster ensembles will cause the weak ensembles to adapt to the dominant ones. This adaptation process or harmonization of water cluster oscillations forces biomolecules in the receiver system to accept the conformation of the dominating transmitter system. The consequence is that the receiver biomolecules have to participate in the same biochemical processes like the transmitter biomolecules. These processes change the biological and physiological behavior of plants.

Practically, we observed this effect in summer 2006 for potatoes produced by "Kloke" GmbH in North Germany. The influence of the water quality on the green plant part was visible. For the potato cultivar Agria, irrigation with underground water led to color loss followed by immunity weakening. On locations of the same potato field where the irrigation machine did not work, the green plant part had a dark green color. According to Figure 14.2 the cluster structure of water used for the irrigation is not suitable to the water cluster structure in the plants. $(H_2O)_{12}$ (22.2 kHz), $(H_2O)_{178}$ (5.5 kHz), and $(H_2O)_{280}$ (4.4 kHz) are similar for the potato tissue and rainwater. In contrast, $(H_2O)_{100}$ (7.3 kHz), $(H_2O)_{178}$, and $(H_2O)_{280}$ in the river and underground water are different and oscillate in a form opposite to water in the potato. This means that the plants need more energy for the adaptation of irrigation water to its own water clusters.

Based on the oscillator interaction (scheme in §3.2.1) and the experimental proof for water, agarose hydrogel, and potato tissue, it was concluded that this effect could be applied to other plants. Therefore, water clusters play an important role as a cost factor for water choice. On the other side, this damage can be reduced through the corresponding preparation of irrigation water, for instance, by influencing the artificial water cluster communication field or with help of physical measures, or better, through correctly choosing irrigation water.

Figure 14.1.7 shows how the planets develop the relative molecular cluster energy in the interval ***C*** [6, 12] in which masses and sizes of clusters are comparable with those of starch globula. In this one time only, nearly all the planets were together on one side of the sun. The surface of the 3D graphic is characterized by some regular maxima and minima that correlate with planet influences (from left to right marked with red arrows): sunset at 19:19, Moon-Uranus opposition (180° at 21:28:25) and Mars-Pluto opposition (175°) in the slice plane of proton resonance [4, 6]. All of these events led to strong changes in the energy characteristics of all the cluster ensembles in the potato biomatrix (marked with a broad arrow). Furthermore, the other arrows shall be designated to the appearance in this slice plane of: Saturn at 23:38, the Moon at 00:39, second gravitational focal point of the Moon's orbital at 04:53, center of our galaxy at 07:00, sunrise at 07:17, Pluto at 07:18, Jupiter at 08:35, Neptune at 10:56 and nearly simultaneously Mercury and Venus at 12:03, Uranus at 12:31, and the Sun at 13:17. Then, the signal intensities were highly decreased while preserving their regularity. In [4, 6] (§10.6) we found out that the planets influence the molecular matter by their gravitational radiation, which leads to a changed balance between the proton dissolving in a physical vacuum and proton condensation from the physical vacuum. Thus, gravitational fields of planets were observed to greatly impact the cluster structure of molecules in biomatrices (long-range order) as well as consequently the biochemical processes. These phenomena are signals for the turning on/off of biological cycles in living matter.

CONCLUSION

Water in plants is present as clusters.

These water clusters oscillate and communicate through oscillation with other water clusters of the surroundings.

Water clusters and their distribution (collapsed or expanded) play an important role during plant growth. Their behavior in plants could be used as an indicator for biomatrix development during plant growth.

Plants communicate with each other and with their surroundings through the resonance field of oscillating water clusters.

The right choice of irrigation water concerning the long-range order of water clusters could promote plant growth and reduce costs.

The influence on plant communication by the resonance field of water cluster oscillation could be applied for early warning systems on weather changes (drought, rain, frost, sun radiation, etc.) or for the regulation of biochemical processes during ripening and storage (inhibition of gel-xerogel transformation), as well as for manufacturing processes.

Water clusters in plants were found to be unique sensors for the gravitational influence of celestial bodies.

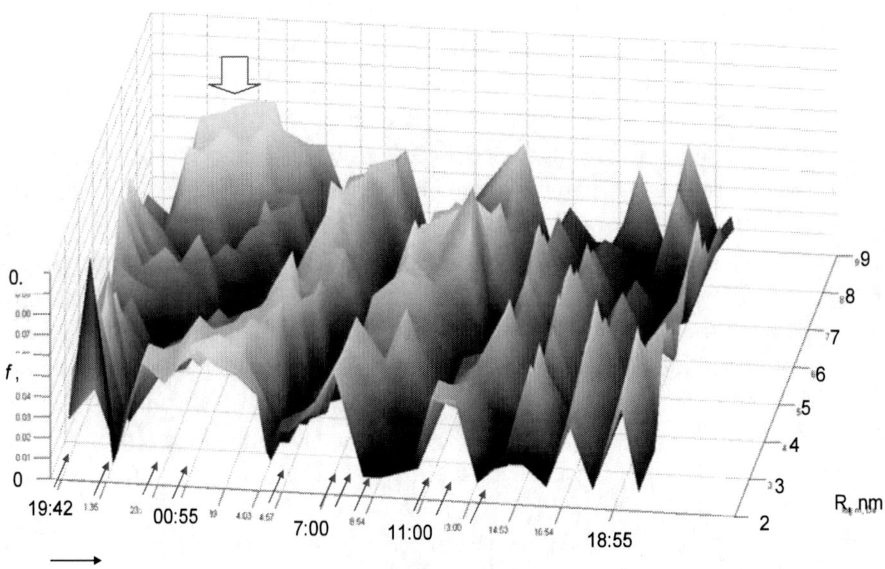

Figure 14.1.7. 3D graphic of GMS-spectra of potato biomatrix (interval C – for clusters of middle sizes) during full moon from March 20, 2008 (Berlin, summer time), *f* – energy rate of a cluster in a cluster ensemble.

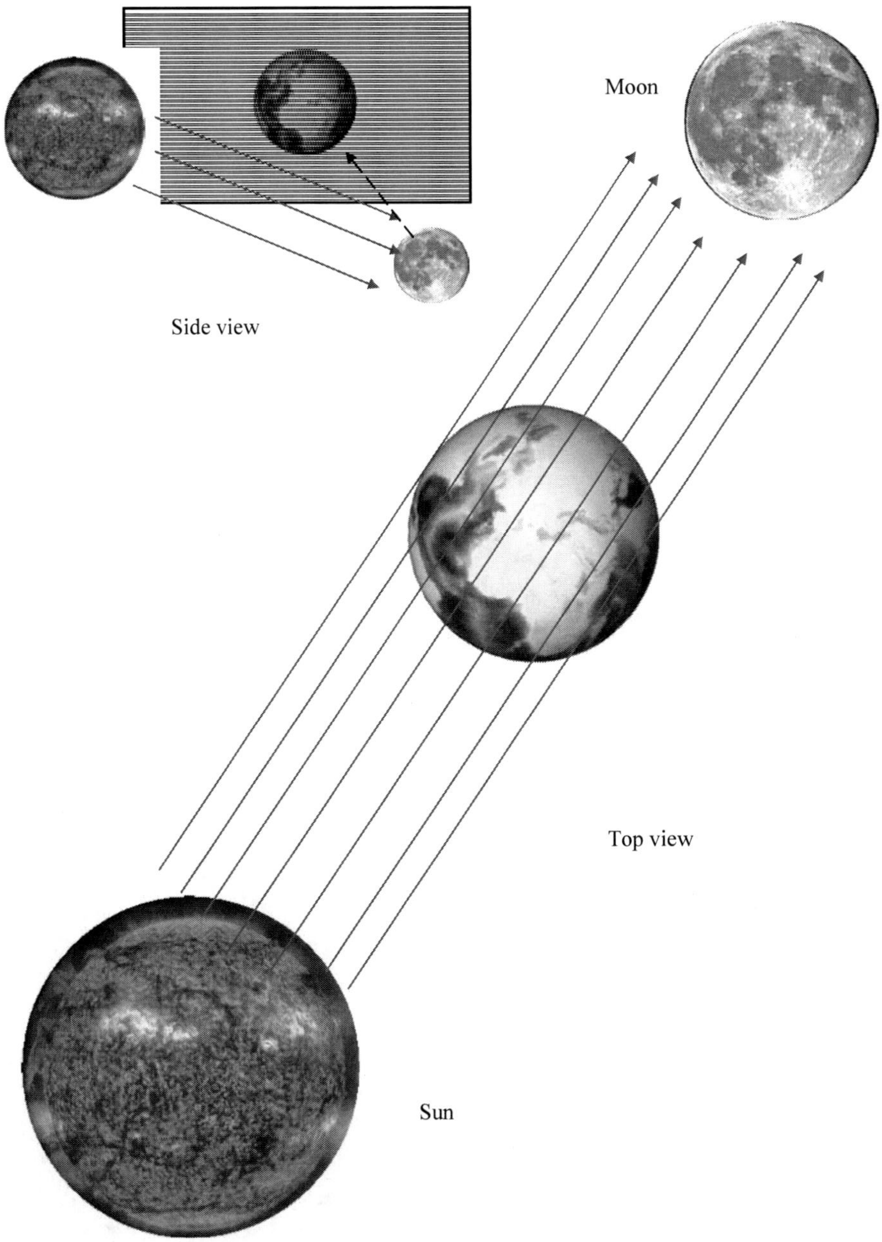

Figure 14.1.8. Full moon constellation. The scale has not been respected.

REFERENCES TO 14

[1] Zubow K. V, Zubow A.V, Zubow V.A. Cluster structure of liquid alcohols, water and n-hexane. Journal of Appl. Spectr. 2005, vol. 72, no.3, pp. 321-328.

[2] Zubow A.V, Zubow Kr. V, Zubow V.A. Investigation of the distribution of water clusters in vegetables, fruits and natural waters by flicker noise spectroscopy. Biofizika, 2007, vol. 52, no. 4, pp.585-592 (in Russian).

[3] Okonchi Shoichi. Water structure in living organism. Mizu no Tokusei to Atarashii Riyo Gijutsu.: 2004, pp.317-326.

[4] K.V. Zubow, A.V. Zubow, V.A. Zubow. Principles of gravitation spectroscopy. New form of molecular matter. Processes. Fields. Aist Handels- und consulting GmbH, dep. R&D, Berlin, 2010, electronic book (www.zubow.de), p. 854. (in Russian)

[5] Zubow K. V., Zubow A.V., Zubow V.A. Low Frequency Movement of Cluster-12 in Potato Amylopectin During Growth. Influence of white Noises. Journal Chemistry of raw plant material. 2009, no.2, pp. 81-88 (in Russian).

[6] Zubow K.V., Zubow A.V., Zubow V.A. Ensemble of Clusters – New Form of Molecular Matter, Risks and Chances. Zubow Equations. In *Advances in Chemistry Research*, vol. 5, 2010, pp. 107-145. Editor J.C. Taylor. Novapublisher. NY. USA.

[7] Bogdanov E. V, Mantrova G.M. Equicluster water model. In Biomedical Radioelectronics. 2000, no. 7, pp.19-28. (in Russian).

[8] Lee H.M, Suh S.B, Kim K.S. Structures, energies and vebrational spectra of water undecamer and dodecamer: An ab initio study. J. Chem. Phys. 2001, vol. 117, no. 24, pp 10749-10756.

[9] Lenz A, Ojamäe L. On the stability of dence versus cage-shaped water clusters: Quantum-chemical investigations of zero-point energies, free energies, basis-set effects and IR spectra of $(H_2O)_{12}$ and $(H_2O)_{20}$. Chemical Physics Letters. 2006, vol. 418, pp. 361-367.

[10] Chaplin M. Structural forms in the icosahedral water cluster, aviable via South Bank University of London, 2009, in http://www.lsbu.ac.uk/water/index.html

[11] Zubov A.V, Zubov K.V, Zubov V.A. Mechanism of sodium chloride diossolving in water and the process of the solution aging. Russian Journal of Applied Chemistry. 2007, vol. 80, no. 7, pp.1249- 1255.

[12] Zubow K. V, Zubow A.V, Zubow V.A. Use of flicker noise spectroscopy for undestroed analysis of nano structures. Zavodskaja Laboratoria. Diagnostics of Material. 2008, vol. 74, no. 9, pp. 40-46. (in Russian).

[13] Lee H.M, Suh S.B, Kim K.S. Erratum: " Structures, energies and vibrational spectra of water undecamer and dodecamer: An ab initio study" [J. Chem. Phys. 114, 10749 (2001)] J. Chem. Phys. 2001, vol. 115, pp. 7331-7334.

[14] Batlle N, Carbonell J.V, Sendra J.M. Determination of Depolymerization Kinetics of Amylose, Amylopectin, and Soluble Starch by Aspergillus Oryzae –Amylase Using a Fluorimetric 2-p- Toluidinylnaphthalene-6-Sulfonate/Flow-Injection Analysis System. Biotechnology and Bioengineering. 2000, vol. 70, no. 5, December 5th 2000, pp. 544-552.

[15] Kadri A; Lorber B; Charron C; Robert M. C; Capelle B; Damak M; Jenner G; Giege R. Crystal quality and differential crystal-growth behaviour of three proteins crystallized in gel at high hydrostatic pressure. Acta Crystallographica, Section D: Biological Crystallography. 2005, vol. D61, no. 6, pp. 784 -788.

[16] Zhang, Jianling; Takayama, Hisao; Matsuba, Takashi; Jiang, Ru; Tanaka, Yoshinori. Induction of apoptosis in macrophage cell line, J774, by the cell-free supernatant from Pseudomonas aeruginosa. Microbiology and Immunology. 2003, vol. 47, no. 3, pp. 199-206.

[17] Zubow V.A., Zubow K.V., Zubow A.V. The Phenomenon of Water Clusters Distribution in Water and Solvated Clusters of Ion Pairs of NaCl in Solution Near the

Wall. The Wall Effect. Journal Chemical Industry (Sankt Petersburg, Ru). 2006, vol. 83, no. 6, pp. 300-306.

[18] Shirayev A, Pagan D.L, Gunton J.D, Rhen D.S, Saxena A, Lookman T. Role of solvent for globular proteins in solution. J. Chem. Physics. 2005, vol.122, p. 234911.

Chapter 15

FORCED FORMATION CLUSTERS IN WHEAT GRAIN

If we understand how the LRO of water in biomatrices and starch nanostructures in grain shall be formed then it should be possible to get an idea of the mechanism of the wheat ripening process and then to develop a plan of how the process can be steered to look for possible starch synthesis. At this time, there is no information concerning the dynamics of the structure change in the grain biomatrix and starch during the wheat ripening process.

Structural differences in starches of similar wheat types were previously investigated by calorimetry [1] where the thermal characteristics of the melting of single starch molecular fragments were determined and their structures given. These structures are little different from the classic suggestion of Banks and Mair [2].

Amylopectin granula are assumed to be in the crystal state but XRD patterns are not distinctive and show a high amorphous halo that is an indication for a low structured polymer [3, p. 314, 4]. On the other side, from the physical chemistry of polymers, it is known that branched polymers do not crystallize or only do so badly. Then if it is a polymer with very hard chains, it may additionally have branches to which belong polysaccharides. Crystal structures of these polymers are characterized by many defects where the branches are thrown out from the crystal lattice [5, p. 167].

It is well known that the folded conformation of macromolecule fragments in branches is thermodynamically more favorable than the unfolded one. The macromolecules' folding and the folding of their branches leads to a cluster structure formation [6], which is determined by gravitational fields, and the molecular clusters (domains) shall be built at the expense of energy clusters in physical vacuum [7]. The last one is caused by the heterogeneity of the physical vacuum at a stationary gravitational wave level [8]. Additionally, as was shown earlier by the authors, molecular clusters in biomatrices and starch actively interact with gravitational radiation from celestial bodies [9]; therefore, this fact should not be ignored.

The aim of the present chapter was to understand the formation process of the LRO in biomatrices and starches of wheat grain during ripening.

The wheat type B growing in North Germany (Mecklenburg/Vorpommern) was used as the investigative object where grains were collected from the lower ear part (Figure 15.1, 20 pieces). The sensor of the Zubow gravitation mass spectrometer [15] was placed directly inside the grain and the LRO scanning in the biomatrix was carried out according to the method described in [8, 9]. Remember that the division of clusters into collapsed and expanded forms was interpreted using the energy state of the cluster's interaction with its surroundings. Starch was obtained from 20 wheat ears. First, the grains were ground in distilled water, and then the suspension was squeezed out. The liquid part was taken for sedimentation where the precipitate of the second decantation was washed with distilled water. The obtained white powder was dried at 298 K on air and finally over silicagel. The grain samples were collected from 20 random ears every second day (afternoon) on the very same agricultural area. LRO analysis was recorded no later than 15 minutes after sampling. The air temperature (T, K) in the agricultural area at the sampling time from July 11th to August 22, 2010 was described by the approximation $T = 10^{-5}t^4 - 1.5454t^3 + 93584t^2 - 3\cdot10^9t + 3\cdot10^{13}$, where t is the calendar data from July 11th through August 22, 2010. The summer in 2010 was characterized by high temperatures up to 310 K (July 11th) and dryness. The thermal properties of starch were analyzed using sample heating under an air-free atmosphere (6 K/min) and the pressure influence was recorded at 298 K.

The base characteristics for the development of the wheat grain biomatrix are shown in Figure 15.1. In the area of grain growth, there are two events in the LRO that leads to a decreased cluster kinds' number (marked with horizontal bracket). Let us discuss these events in detail (Figure 15.2).

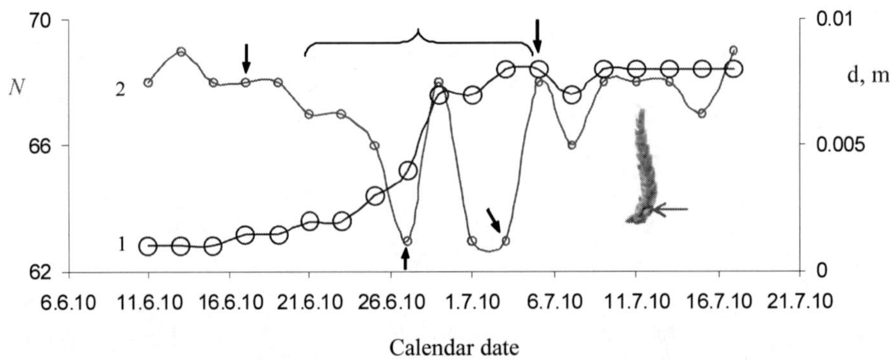

Figure 15.1. Change of the maximal wheat grain size (curve 1, d, meter) and number of cluster kinds in grain biomatrix (curve 2, *N*) during the ripening process. Diapason of cluster ensemble masses up to 3.2 million Daltons. Active events of grain growth are described by horizontal bracket. The water content in the biomatrix after July 17, 2010 was lower than 50 wt. %. The grain to be analyzed is marked with a horizontal bracket.

As shown in Figure 15.2, the GMS-spectra greatly differ from each other. While the spectrum of June 17, 2010 shows a distribution of collapsed water clusters, sub-micelle, and micelle structures [6, 9] (-*f*) that are typical for a stable biomatrix, the other spectra cannot be characterized in such a way. The grain size enlargement was therefore concluded to influence the LRO in biomatrix, which at this time greatly changed. The clusters mainly consisted of expanded structures (+*f*). After forced ripening and grain size stabilization, due to strong weather dryness on July 5, 2010, the LRO in the biomatrix approached the LRO of June 17, 2010 and almost matched it on July 13, 2010.

To understand the processes in the biomatrix during the ripening period, the data shown in Figure 15.3 shall be discussed. Upon intensive grain growth (Figure 15.1), a strong decrease of the density of all clusters in an ensemble, including the base water cluster, was observed. $(H_2O)_{12}$ [10] gets a positive value f, being an indication for its forced expansion.

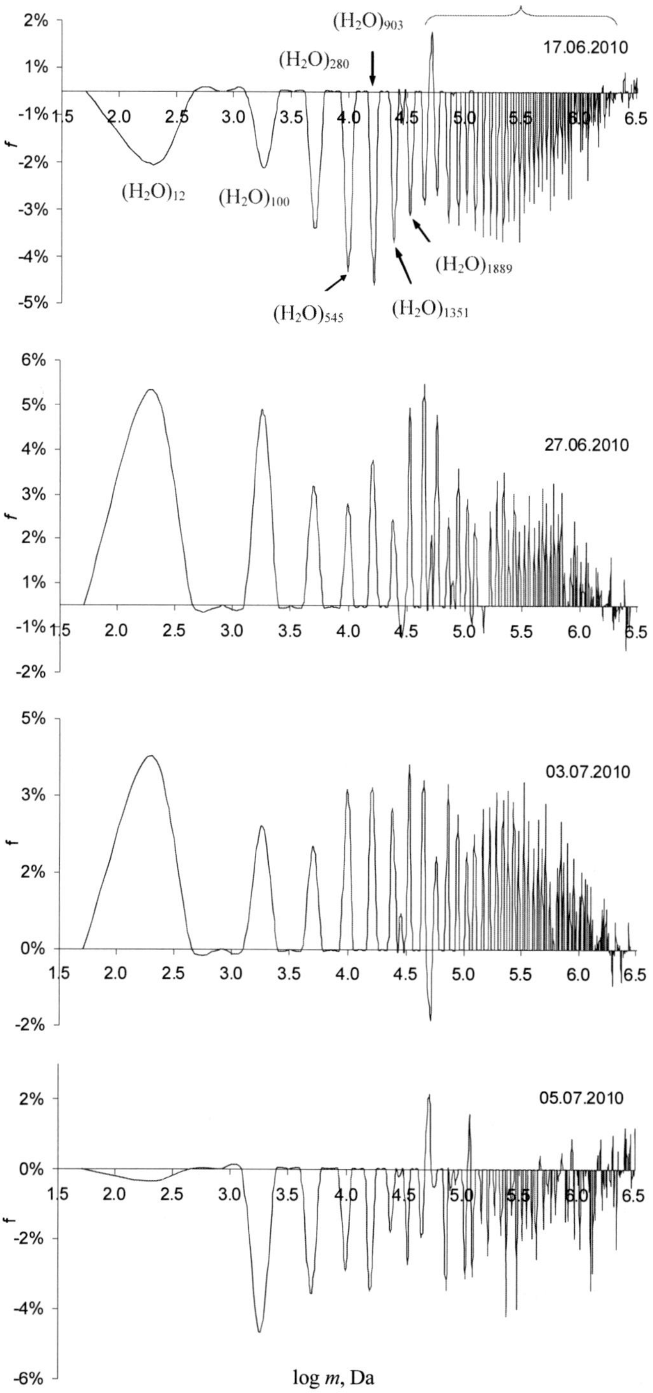

Figure 15.2. GMS-spectra of wheat grain biomatrix at the time that it is marked by arrows in Figure 15.1. Diapason of cluster ensemble masses up to 3.2 million Daltons. Weak shock waves. The Zubow constant amounts to $6.4 \cdot 10^{-15}$ N/m. The interval of clusters, sub-micelle, and micelle structures built by polysaccharides and proteins are marked by a horizontal bracket (***C***).

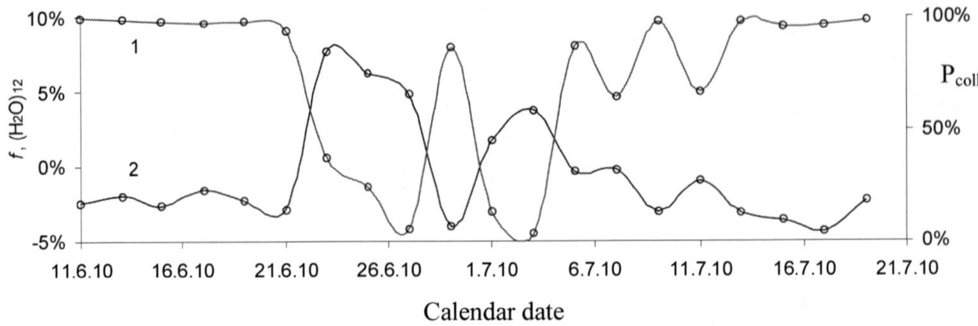

Figure 15.3. The rate of collapsed clusters (1, P_{coll}) and signal intensities (2, f) of the base water cluster in the wheat grain biomatrix during ripening.

As is visible from Figure 15.3, there are two events on June 23rd and June 27, 2010 that are seen as one large overlapped signal. These events were identified by some water clusters, too. They were assumed to concern not only the base water cluster but the complete water cluster ensemble in the biomatrix. These two earlier events are the beginning of the LRO changing in the biomatrix as well as the start of the ripening.

To get an initial idea of the mechanism of processes proceeding in the biomatrices, the results from Figure 15.4 shall be discussed. As is shown, the average molecular cluster mass ($M_{GMS}=\Sigma(|f|\cdot m)$) achieved its highest value on June 23, 2010 (filled point), which permits its distinguishment from earlier events by using the mechanism. Due to the fact that the cluster masses remained unchanged at this time, the rise of M_{GMS} has to be ascribed to an increased absolute sum $\Sigma(|f|)$. This effect could only be explained by a reinforced interaction of the clusters with their surroundings [8]. The average cluster mass remains constant even when the part of the collapsed clusters (Figure 15.3) and the number of cluster kinds (Figure 15.1) greatly changes—as was the case on June 27, 2010 and July 3, 2010. The events of this time should be connected less with the surroundings' interaction and more with the fact that they were likely caused by some decreasing of the cluster kinds' number (Figure 15.1).

Thus, the hierarchy of the events taking place in the LRO of biomatrix could be represented as follows: strengthening of clusters' interaction with surroundings, destruction of some cluster kinds, rearrangement of clusters in ensemble, and recovery of some earlier destroyed clusters. The processes in the LRO of the growing grain biomatrix were considered from the position of the micelle structure in which the role of surfactants (SAS) was played by simple proteins and lipids (Figure 15.5). In the first event (Figure 15.4, I) in the biomatrix, the cluster kinds' numbers and the part of the collapsed clusters decreased while the interaction of clusters with their surroundings rose. Here the water clusters were assumed to be destroyed under the influence of SAS, which led to the formation of the simplest sub-micelle structures [6] in the grain seed that then coagulated to a micelle structure on June 27, 2010 (Event II, Figure 15.4). At this time, the N value declined a little (Figure 15.1, marked with arrow) [11] whereas the cluster interaction with its surroundings remained unchanged.

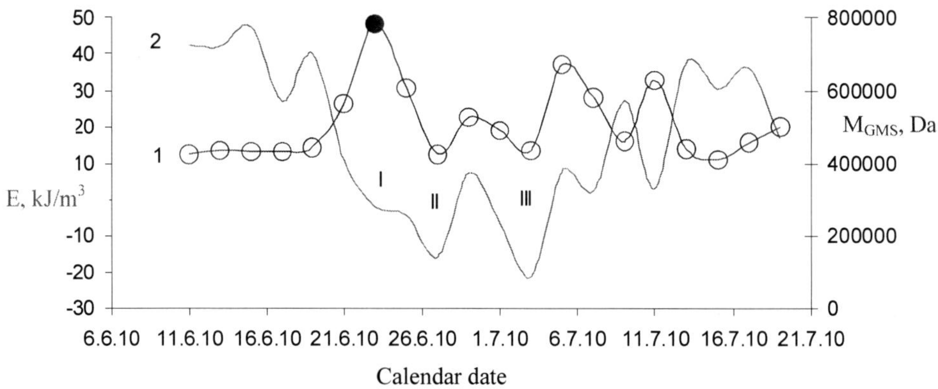

Figure 15.4. Average cluster molecular mass (1, M_{GMS}) and energy (2, E) of the cluster ensemble (up to 3.2 million Daltons) in the biomatrix. The device calibration was carried out according to the evaporation energy of water (44.1 kJ/mol). ($M_{GMS} = \Sigma(|f|\cdot m)$), where Σ is the mathematical symbol for sum of all the clusters (from $N = 1$ up to N in this investigated ensemble) and m is the cluster mass in Daltons.

Event III (Figure 15.4) was ascribed to the condensation processes of glucose to starch macromolecules inside of the sub-micelle and micelle structures (Figure 15.5) [9]. Water remove from the biomatrix during ripening changed the LRO in which amylopectin, amylose, proteins, lipids, and solvated clusters of ion pairs now started to dominate. Thus, the starch synthesis was concluded to be provoked by glucose accumulation in sub-micelle structures that were then, depending on their filling degree with glucose, transformed to micelle and later to super micelle structures. Condensation of glucose into ordered micelles led to starch formation (granula) where proteins, lipids, SAS, and salts were concentrated on the micelles' surface; from that, later a shell between granula forms.

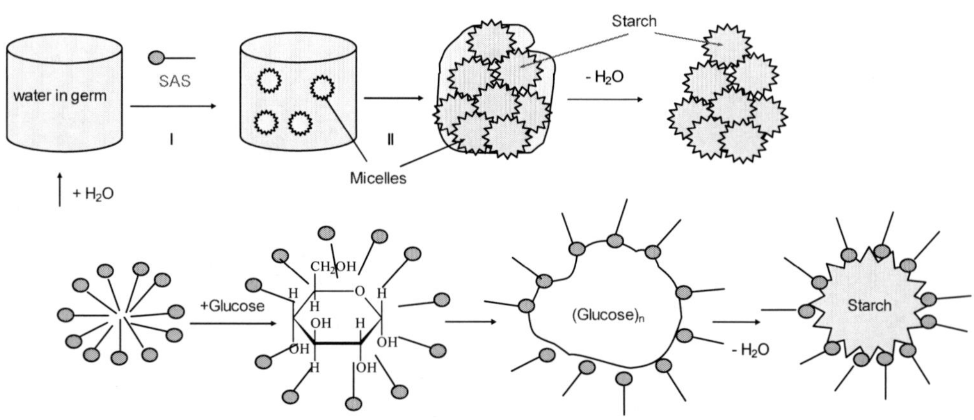

Figure 15.5. Models for the formation mechanism of micelle structures in wheat grain biomatrix during ripening and for starch synthesis.

To understand structural changes in starch during wheat ripening, the LRO of polysaccharides was investigated under heating [9] and pressure [11]. Both by heating and pressure, the movement of individual clusters as well as the thermodynamic stability of all cluster ensembles as a unit can be influenced. First it shall be discussed how the integral characteristics of the cluster distribution in starch for different ripening stages changed. The integral curves for the cluster fraction distribution (W_x) in starch extracted from wheat grain on July 4th and August 14, 2010 appear to be identical, hence supporting the cluster formation mechanism in starch according to the grape bunch principle [12]. There are, however, some differences in the cluster mass ranges of 500,000 and 1,500,000 Daltons becoming clearer at starch heating (Figure 15.6).

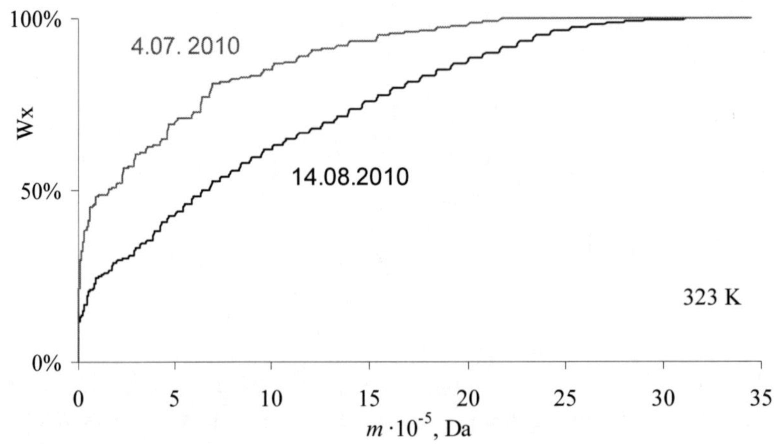

Figure 15.6. Integral curves for cluster fraction distributions (W_x) in starch extracted from wheat grain on July 4th and August 14, 2010. Heating up to 323 K where $W_x = |f{\cdot}m|_i + |f{\cdot}m|_{i+1} +_{...}+ |f{\cdot}m|_N$, m cluster mass in Daltons, and i – cluster number from 1 to N.

As is visible from Figure 15.6, the starch cluster fractions of various ripening grades are characterized by different thermal stability. In the fully ripe grain, starch clusters were more expanded but upon heating up to 323 K they were transformed to collapsed ones. The part of the collapsed clusters in starch at the ripening stage on July 23, 2010 shall be discussed (Figure 15.7).

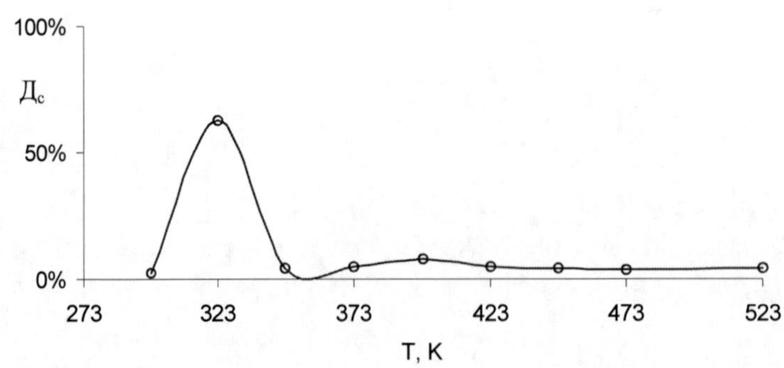

Figure 15.7. Part of collapsed clusters ($Д_c$, the part of signals with negative f value) in starch extracted from wheat grain on July 23, 2010.

At 325 K, the part of the collapsed clusters in starch rises and then quickly falls again (Figure 15.7). After cooling down the sample from 523 to 298 K, P_{coll} agrees with P_{coll} at the beginning of the heating process. This behavior could indicate the reversibly destroyed cluster's interaction with its surroundings proceeding at the melting of specific fractions of the hard polymer chains. The lower the cluster interaction with its surroundings, the more dense it is, and conversely that means the interaction with the surroundings causes more expanded structures. The GMS-spectra, however, are not completely identical before and after heating; some differences are detectable only with integral characteristics.

On the other side, the unknown melting processes in starch were described at 325 K [1]. The nature of these processes is not completely cleared up yet, but the endothermic reconstruction processes in the tertiary structures of starch at the LRO level could proceed at these low temperatures. The melting enthalpy of the starch chains' fragments (2.6 kJ/mol) from Kozlov [1] were used for the calibration of the GMS spectrometer and related to the highest point in the curve (Figure 15.7). Several heating effects of the curves for starch at different ripening stages are shown in Figure 15.8. During ripening, a LRO reconstruction was observed to proceed in the starch of the grain biomatrix, which leads to the domination of some melting/crystallizing structures at different temperatures.

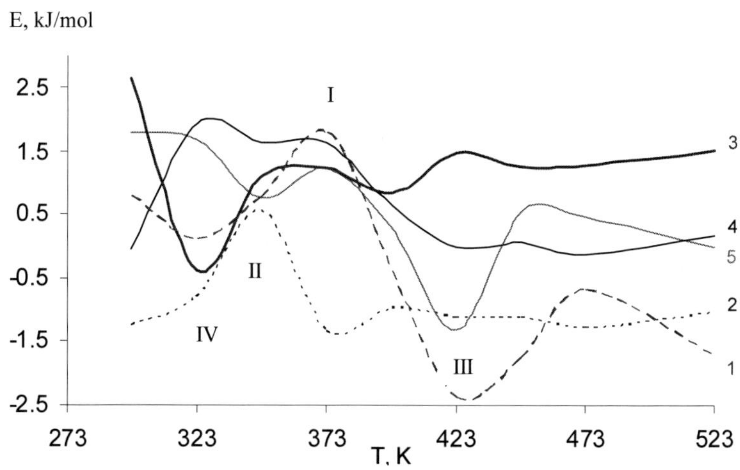

Figure 15.8. Thermal effects of different starch types extracted from wheat grain at different ripening stages at heating. 1 – July 4, 2010; 2 – July 13, 2010; 3 – July 23, 2010; 4 – August 4, 2010; and 5 – August 14, 2010.

"Young" starch (curves 1 and 2) is characterized by an exothermic crystallization process (clusters' densification and ordering) at 373 K (I, curve 1), which has been shifted to 348 K (II) on July 13, 2010. Besides this, there is an endothermic effect at 425 K (III) that has to be seen only as a melting of the polymer chains' fragments, because up to 523 K, polymer destructions were not observed by TG/DTA [13, 14]. On July 23, 2010 extracted starch is described by one endo effect only at 325 K (IV, curve 3). In starch from old wheat grain (August 4, 2010), there are only low exothermic effects of crystallization at 325 and 373 K, which were observed in the starch of much older grains (curve 5), too, although the latter shows a strong endothermic peak at 423 K, which is ascribed to its melting.

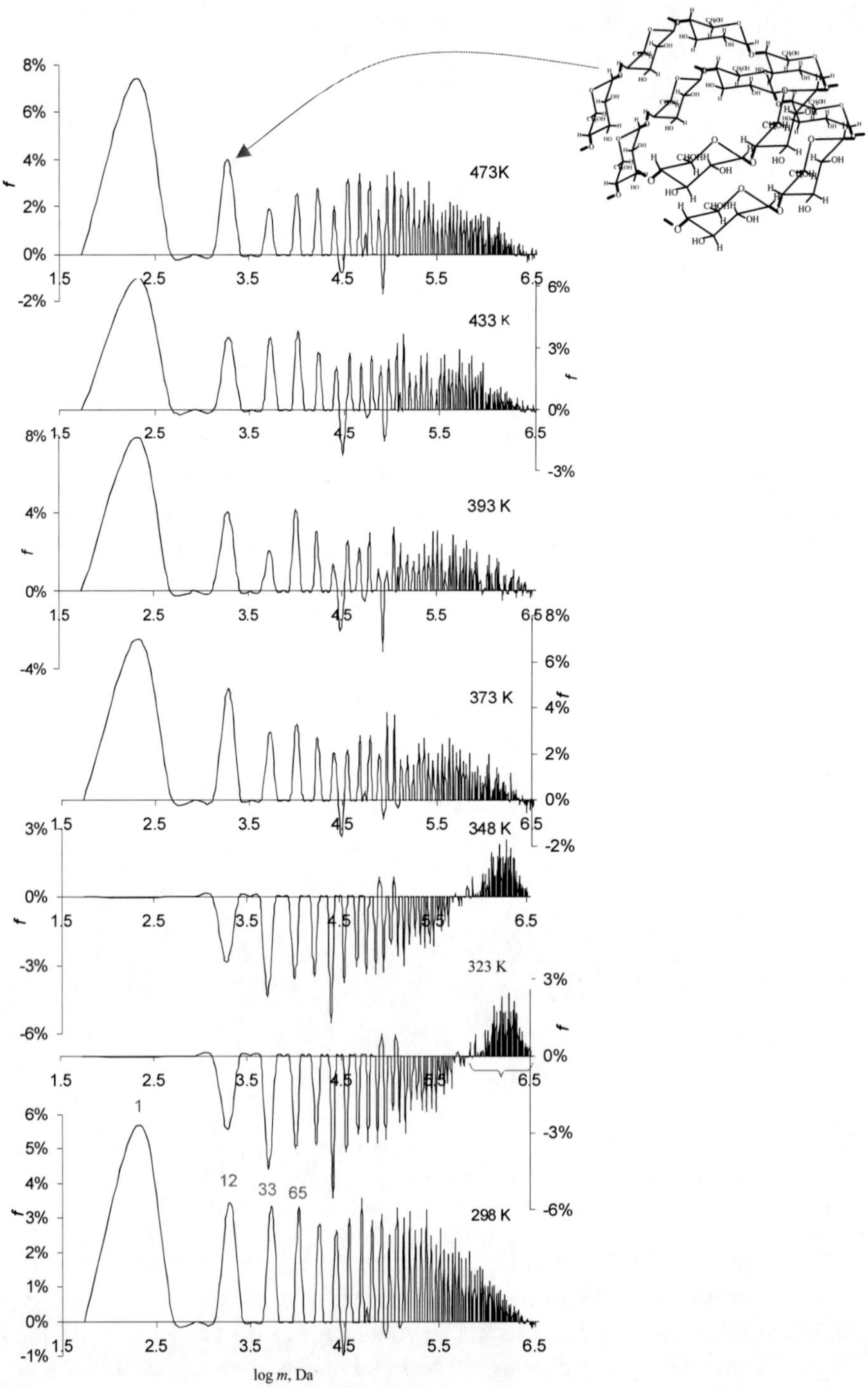

Figure 15.9. Change of cluster signals in starch extracted from wheat grain on July 23, 2010. On the right side the model of the base starch cluster (12 units' spiral of α-D-glycopyranose, 2 meanders) is shown [11, 15]. Cluster ensemble up to 3.2 million Daltons, 6 K/min. Zubow constant is equal to $6.8{\cdot}10^{-15}$ N/m. Weak shock wave ($p < 1$ N/m^2). The number of α-D-glycopyranose units in a cluster is given over the signal. Low pressured powder in air-free atmosphere.

The GMS-spectra of starch extracted from wheat grain on July 23, 2010 are given in Figure 15.9. Already a temperature increase by 25 K leads to the radical rearrangement of the LRO in the polysaccharide. The main part of the simplest clusters and some sub-micelle clusters (up to log m = 5.5) became free of their interaction with the surroundings and collapsed while the micelle structures (log m > 5.8) remained unchanged. The integral cluster parts' distribution (Figure 15.10) is characterized by the domination of super micelle structures in the mass interval marked by a horizontal bracket in Figure 15.9, which is caused by the melting of sub-micelles and micelles. The state remained unchanged up to 373 K. After that and because of an increased cluster interaction with the surroundings, the collapsed clusters were destroyed to expanded forms.

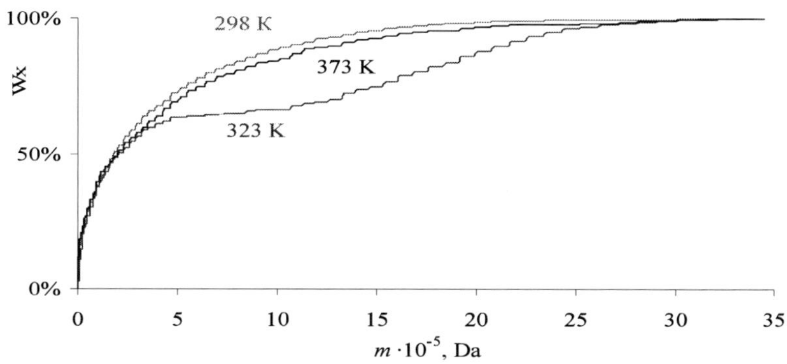

Figure 15.10. Integral cluster parts' distribution in starch extracted from wheat grain on July 23, 2010 at 298, 323, and 373 K.

Sub-micelle and micelle structures were found to partially disappear upon heating the polysaccharide up to 323 K (mass interval from 0.5 million to 2.5 million Dalton) while further heating up to 373 K recovered the distribution state at 298 K. The melting of clusters seems to be accompanied with their partial destruction where some clusters should be rebuilt. This is the base for the formation of new LRO at higher temperatures. The super micelle structures that are marked with a horizontal bracket in Figure 15.9 are characterized to be unstable, too, and they were destroyed to create new sub-micelle and micelle structures at temperatures higher than 373 K. Here the signals of α-D-glycopyranose members showed an interesting behavior (1, Figure 15.9); they are represented in the whole temperature range as expanded structures except at 323 and 348 K. At these temperatures, the signals seem to be a super position of signals reflecting the units' state both in small and super molecular clusters.

In the next section, the GMS-spectra of "young" starch in a cluster ensemble up to 4 billion Daltons shall be discussed (Figure 15.11). As is visible, huge clusters of defined molecular masses were cut from the mass ensemble after heating. In the base polysaccharide, at 298 K in the area of super micelle structures > log m = 6.5, the signals are randomly, at first glance, indicating the poorly organized clusters as expanded and collapsed. They were, however, destroyed at heating and recovered at cooling. Here a melting process of nanostructures proceeds that has to be understood as a breakup of clusters independently on their form or interaction with the surroundings. What will these regularities be for starch from wheat grain of different ripening times?

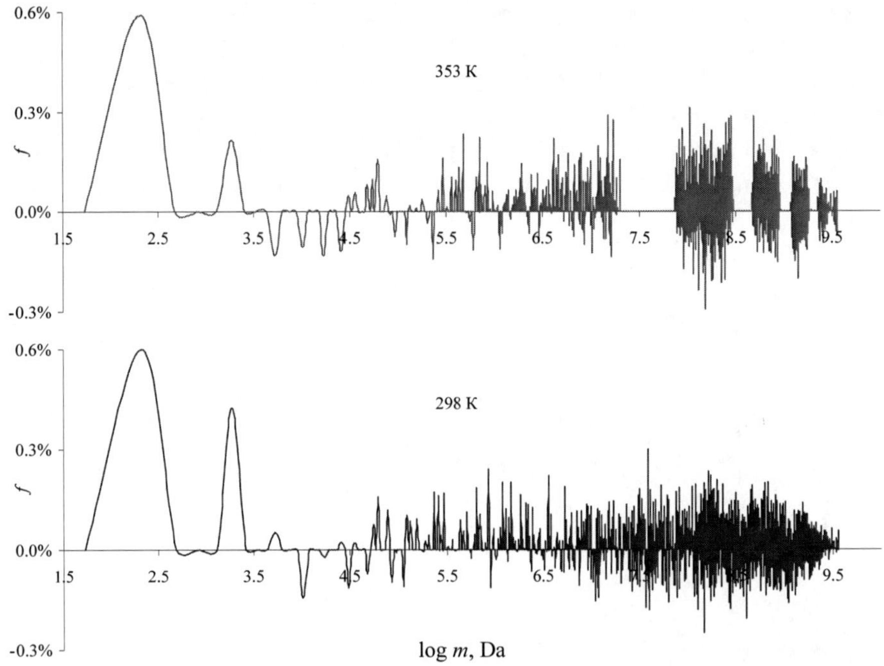

Figure 15.11. Change of cluster signals in "young" starch extracted from wheat grain on July 4, 2010. Cluster ensemble up to 4 billion Daltons, 6 K/min. Zubow constant is equal to $6.8 \cdot 10^{-15}$ N/m. Low pressured powder in air-free atmosphere. Weak shock wave ($p < 1$ N/m^2).

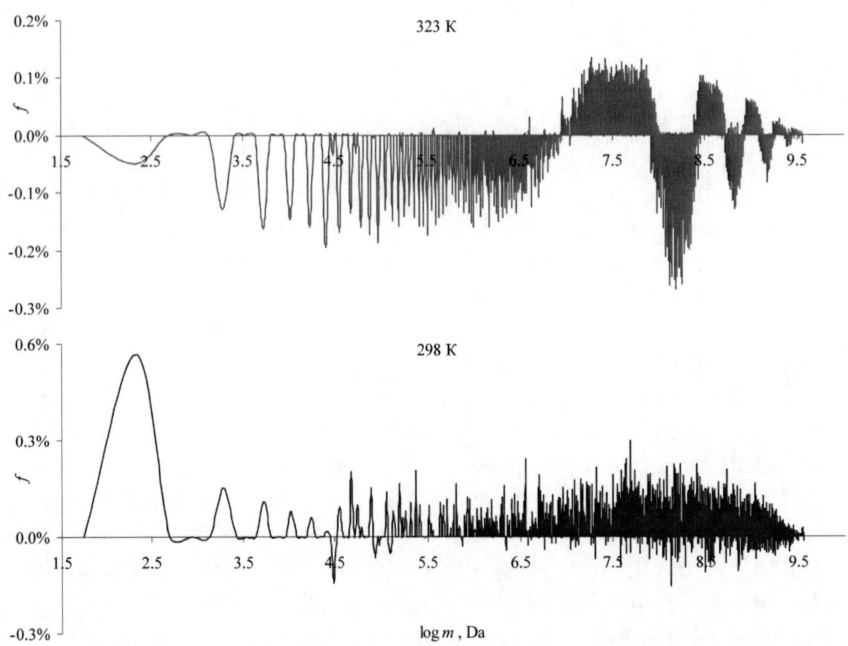

Figure 15.12. Change of cluster signals in "young" starch extracted from wheat grain on July 23, 2010. Cluster ensemble up to 4 billion Daltons, 6 K/min. Zubow constant is equal to $6.8 \cdot 10^{-15}$ N/m. Low pressured powder in air-free atmosphere. Weak shock wave ($p < 1$ N/m^2).

Aside from the spectra of July 4, 2010 (Figure 15.11), the analog spectra on July 23, 2010 are given in Figure 15.12. Melting processes in the polymer proceed in the area of super micelle structures but the sample heating did not lead to cluster destruction in the range of large masses as was found for "young" starch (Figure 15.11). On the other side, the cluster distribution in the area of super micelle structures was found to be more uniform than in "young" starch.

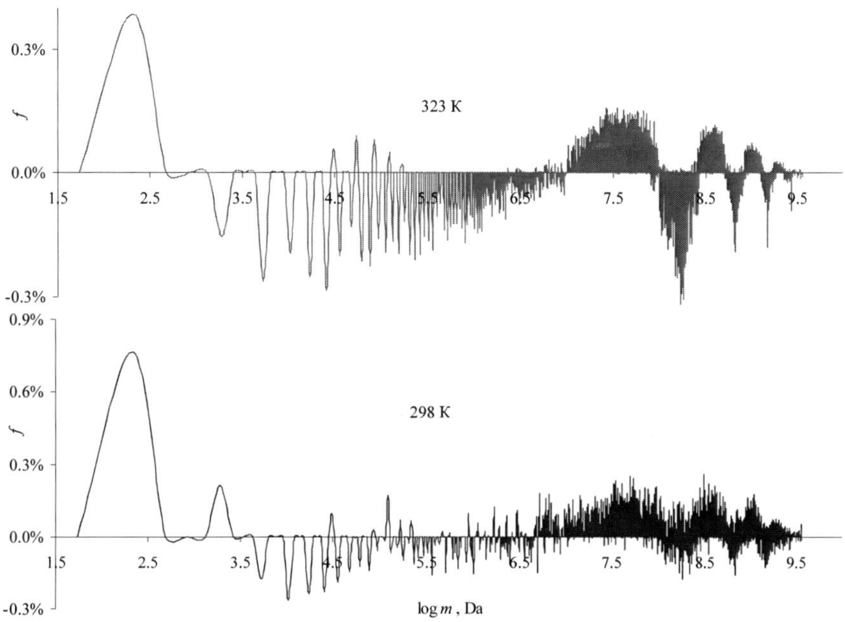

Figure 15.13. Change of cluster signals in "young" starch extracted from wheat grain on August 14, 2010. Cluster ensemble up to 4 billion Daltons, 6 K/min. Zubow constant is equal to $6.8 \cdot 10^{-15}$ N/m. Low pressured powder in air-free atmosphere. Weak shock wave ($p < 1$ N/m^2).

The final starch analysis was made on August 14, 2010 (Figure 15.13). As is shown, the heating process is accompanied with strong LRO reconstructions, especially in the super micelle region even though the processes in "old" starch strongly differ from those of the "young" one. This difference can be applied for a qualitative polysaccharide analysis and furthermore, for understanding the LRO formation mechanism.

At the starch samples' heating up to 523 K, and in order to adapt to the new thermodynamic conditions, some reorganization processes in the LRO take place. Due to the unstable LRO, however, the processes cannot always be definitively interpreted. For example, the GMS-spectra sequence of starch on July 13, 2010 (period of intensive grain growth, Figure 15.1) shows strong changes in the cluster ensemble for the heating of temperatures from 298 to 523 K, but it did so without selective cluster reaction on temperature for "young" and "old" starches. This could be a special case of a structure that is not completely organized here.

The polymerization degree (PD) of starch branches (amylopectin) can be approximately estimated by using the average cluster molecular mass and the number of cluster kinds the clusters decomposes into at heating. The polymerization degree for starch extracted from wheat grain of different ripening times is given in Figure 15.14. A correlation between the PD and temperature of wheat grain sampling was observed only after July 13, 2010. Before this time, the PD is smaller, which is explained with a beginning growth of amylopectin branches. Here, in young grain, the branches' growth in amylopectin seems to dominate over the falling temperature. The PD declining after July 13, 2010 was explained to be connected with a temperature debranching proceeding at temperatures higher than 298 K [9] and leading to rapid polymer aging.

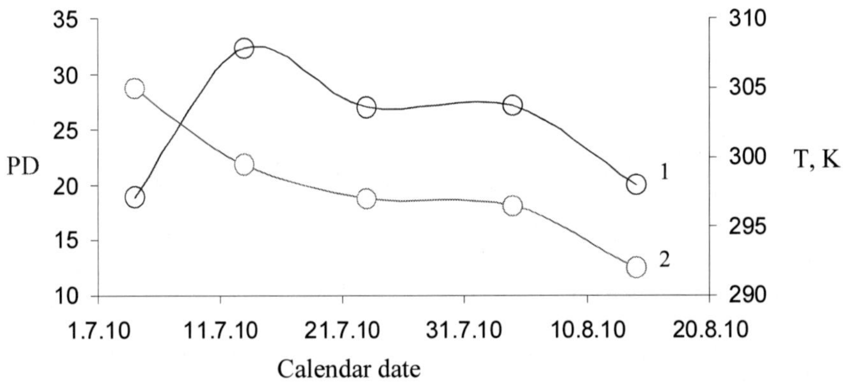

Figure 15.14. Changes in the polymerization degree in branches of starch (1) extracted from wheat grain of different ripening stages and change of the air temperature (2) at grain sampling from the agricultural field.

Upon heating starch extracted from grains of different ripening stages, it was observed that both some old clusters disappeared and new ones arose. The dynamics of these processes at starch heating from 298 to 323 K is shown in Figure 15.15.

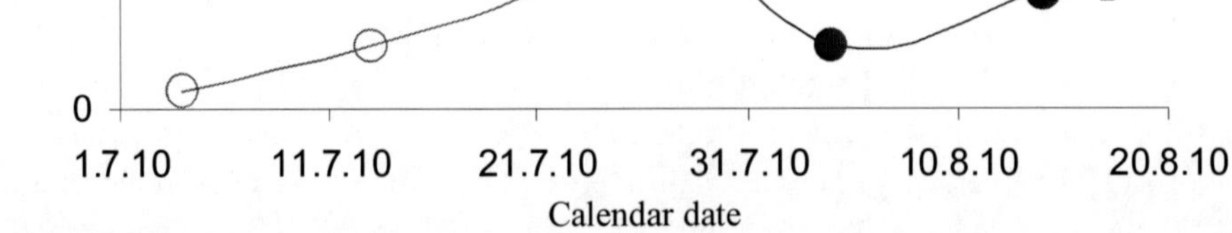

Figure 15.15. Cluster kinds' dynamics in starch extracted from wheat grain of different ripening stages. Heating from 298 to 323 K. 1 – arising of new cluster kinds (n_1) and 2 – disappearance of old ones (n_2).

The new clusters' arising and the old ones' disappearance have to be understood as a result of a destroyed clusters' interaction with its surroundings, which is comparable with coagulation/melting. As is visible from Figure 15.15, the arising of new clusters and disappearance of old ones does not precede synchronously or especially explicitly for "young" starch. For ripe wheat grain (August 14, 2010), however, the number of disappearing clusters is nearly equal to that of the arising clusters in starch. This case also occurs when curves 1 and 2 cross each other on July 26, 2010. The highest number of new cluster kinds on August 4, 2010 was assumed to be the LRO reconstruction. After heating up to 523 K, the differences between curves 1 and 2 became marginal for all starch types.

It would be of interest to see the changes in the dynamics of the whole number ($S_{(d+a)}$) of mobile cluster kinds (sum of appearing (S_a) and disappearing (S_d) cluster kinds in a narrow temperature interval, 25 K) in starch from differently ripened wheat grain (Figure 15.16).

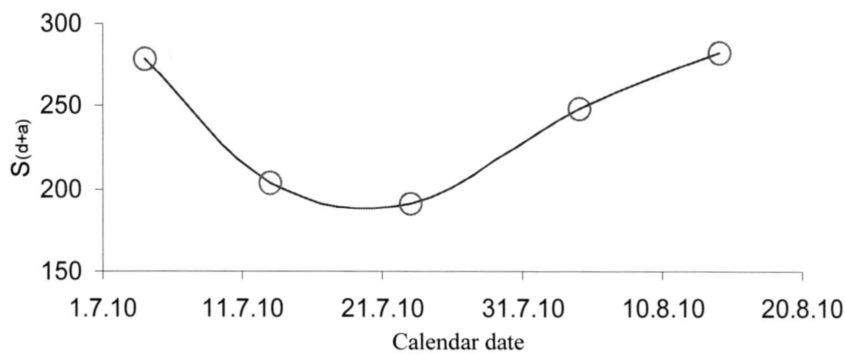

Figure 15.16. Whole number of mobile cluster kinds ($S_{(d+a)}$) in starches of differently ripened wheat grain. Polymer heating from 298 to 523 K. Cluster ensemble up to 3.2 million Daltons.

The whole number of mobile cluster kinds in "young" and "old" starch seems to be nearly equal even though, for the remaining starch types, it reduced by 40%. These events were assumed to be the dynamics of two processes: intensely growing polymer chains in "young" starch and the chains' debranching in "old" starch. Here the debranching in solid granula is less likely caused by enzymes but is rather the result of mechanical disintegration because of rising internal tensions in the branched polymer. The mobility of enzymes in a solid and dry granule is highly limited. Many unlinked clusters, or at least clusters that are weakly linked with surroundings, seem to be in "young" starch, however, they are in "old" starch, too, as a result of mechanical decomposition of amylopectin branches (curves 1 and 5 in Figure 15.8). This conclusion proves the model for the starch nanostructure as a grapes' bunch.

Thus, the aging of starch in ripe wheat grain was assumed as a potential energy that minimizes polysaccharide chains and as formation of compact, ordered nanostructures in the LRO where the polymer chains shall partially be released from the crystallizing starch structures [5]. Logically, from July 11th to July 31, 2010, the most compact LRO structures must be built in starch granula.

Let us return to the results of Figure 15.11. The inversion of the expanded clusters (+*f*) into collapsed ones (-*f*) in the whole mass interval of the cluster ensemble indicates the clusters' dependency on each other in this ensemble. Additionally, there is the formation of a unique energy field in which every cluster must subordinate itself to the other ones as a consistent energy pendulum [9]. If an ensemble is forced to change into another energy state, e.g. at heating or under pressure, then a corresponding reaction in the form of the LRO reconstruction has to be expected. As is visible in Figure 15.11, after heating up to 323 K, the cluster ensemble changed precisely with the small clusters and large ones partially becoming more individual with fewer interactions from the surroundings whereas another group of very large clusters (super micelles in expanded form) kept hold of their interaction with their surroundings. The "melting" in starch as ascribed by Kozlov [1] is less the result of the double spiral decomposition and more that of an endothermic cluster reconstruction in the ensemble. Processes of forced LRO reconstruction in starch at heating from 298 to 523 K were observed to be accompanied by at least two endothermic effects of "pseudo melting" at 323...348 K and 448±10 K, which strongly depends on the ripening grade of wheat grain. These properties of starch types can probably be applied for their identification, however, it needs further investigations.

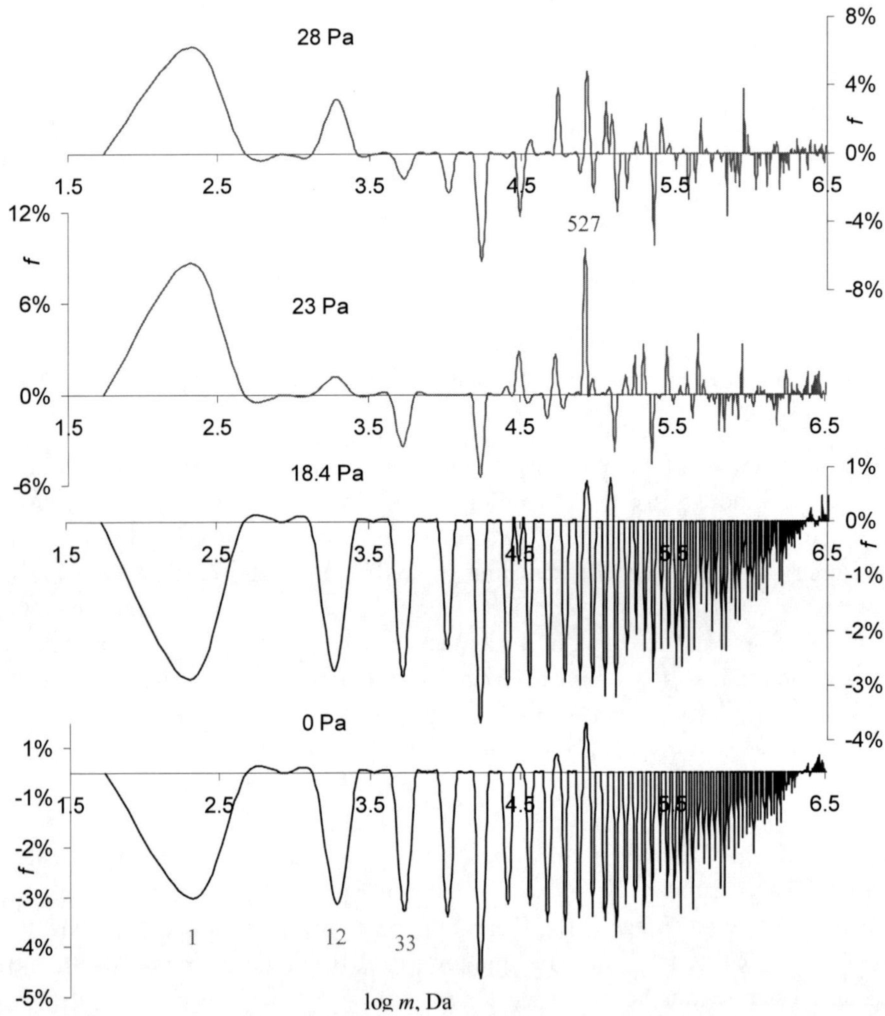

Figure 15.17. Influence of pressure on the LRO in starch extracted from wheat grain on August 14, 2010. Zubow constant is equal to $6.8 \cdot 10^{-15}$ N/m. Weak shock wave ($p < 1$ N/m^2) at 298 K.

Starch was found to be sensitive to low pressure, e.g. compression [11]. Shear forces were observed to destroy the polymer clusters' interaction with their surroundings, which influences the whole oscillator ensemble. In order to understand this phenomenon, the GMS-spectra of starch at different pressures are given in Figure 15.17. As is visible, the starch cluster ensemble remains stable to 18.4 Pa. The following pressure increase greatly changes the LRO structure, though. Here the changes are stronger than those found for heating. Strikingly, at 23 Pa the α-D-glycopyranose units (1) and the spiral of two meanders (12 polymer units) started to actively interact with the surroundings. This event was interpreted as the expanded state of molecular clusters. On the other side, the cluster of 33 units remained in a collapsed state and the larger clusters either disappeared or changed to a new thermodynamic balance (expanded-collapsed). The expanded cluster with a mass equal to 527 α-D-glycopyranose units was characterized by only one high signal at 23 Pa that was interpreted as relatively stability at this pressure. It seems to belong to skeletal clusters of this starch cluster ensemble [11, 15].

CONCLUSION

The LRO in the biomatrix of growing wheat grain is represented as molecular clusters that are in a dynamic balance with the gravitational fields of their surroundings, growing/aging processes, and weather conditions.

The polysaccharide synthesis in the wheat grain biomatrix was found to proceed in micelle structures that were transformed into large starch-containing super micelle structures during ripening and aging.

The LRO in starch was characterized to change during wheat ripening.

Starch samples extracted from wheat grain of different ripening grades are different with regards to their physicochemical properties.

It should be possible in principle to synthesize starch during one to two weeks under laboratory conditions.

REFERENCES TO 15

[1] Kozlov S.S., Krivandin A.V., Shatalova O.V., Noda T., Bertoft E., Fornal J., Yuryev V.P. Structure of starches extracted from near-isogenic wheat lines. Journal of Thermal Analysis and Calorimetry. 2007, vol. 87, no. 2, pp. 575-584.

[2] Banks W., Muir D.D. Structure and chemistry of the starch granule. In: P.K. Strumpf, E.E. Conn, Eds., The biochemistry of plants, Academic Press, New York, 1980, vol. 3, pp. 321-330.

[3] H.-D. Belitz, W. Grosch, P. Schieberle. Lehrbuch der Lebensmittelchemie. 5. Auflage, Berlin, Heidelberg, NY, Barcelona, London, Mailand, Paris, Singapur, Tokio, Springer Verlag 2001.

[4] Mia Sjöqvist, P. Gatenholm. The Effect of Starch Composition on Structure of Foams Prepared by Microwave Treatment. Journal of Polymers and the Environment. 2005, vol. 13, no. 1, pp. 29-37.

[5] R. Tuedze, T. Kawai. Physical chemistry of the polymers. Russian edit. M, Chimija, 1977.

[6] Zubow K.V., Zubow A.V., Zubow V.A. Using of flicker noise spectroscopy for non destroyng analysis of nano structures. Zavodskaja Laboratorija. Diagnostics of materials. 2008, vol. 74, no. 9, pp. 40-45, in Russian.

[7] Zubow K.V., Zubow A.V., Zubow V.A. Experimental Platform for the Investigation of the Structural Heterogeneity of the Physical Vacuum. Vacuum Energy Risks and Chances. In: Horizons of World Physics. Editor A. Reimer. Novapublisher, New York. 2012, vol. 217.

[8] Zubow K.V., Zubow A.V., Zubow V.A. Ensemble of Clusters – New Form of Molecular Matter, Risks and Chances. Zubow Equations". In: Advances in Chemistry Research, Editor J.C. Taylor, Novapublisher, New York. 2010, pp. 107-145.

[9] Zubow K.V., Zubow A.V., Zubow V.A. Water Clusters in Plants. Fast Channel Plant Communications. Planet Influence. Journal of Biophysics Chemistry. 2010, vol.1, no.1, pp. 1- 11.

[10] Bogdanov E.V., Mantrova G.M. Equicluster water model. Biomedical Radioelectronics. 2000, no. 7, pp. 19-28, in Russian.

[11] Zubow K.V., Zubow A.V., Zubow V.A. Principles of gravitation spectroscopy. New form of molecular matter. Processes. Fields. Aist Handels- und Consulting GmbH, dep. R&D, Germany, 2010, electronic book (www.zubow.de), p. 854. (in Russian).

[12] Batle N., Carbonell J. V., Sendra J. M. Determination of Depolymerisation Kinetics of Amylose, Amylopectin, and Soluble Starch by Aspergillus Oryzae – Amylose Using a Fluorimetric 2-p-Toluidinylnaphthalene-6-Sulfonate / Flow-Injection Analysis System. Biotechnology and Bioengineering. 2000, vol. 70, no. 5, pp. 544-552.

[13] Aggarwal P., Dollimore D. The combustion of starch, cellulose and cationically modified products of these compounds investigated using thermal analysis. Thermochimica Acta. 1997, vol. 291, pp. 65-72.

[14] Aggarwal P., Dollimore D., Heon K. Comparative Thermal Analysis Study of Two Biopolymers, Starch and Cellulose. J. of Thermal Analysis. 1997, vol. 50, pp. 7-17.

[15] Zubow K.V., Zubow A.V., Zubow V.A. Low frequency movement of cluster-12 in potato amylopectin during growth. Influence of white noises. Khimiya Rastitel'nogo Syr'ya. 2009, no. 2, pp. 77-87.

Chapter 16

THE DIRECTION OF GRAVITATIONAL RADIATION

Macromolecule chains in liquid polyethylene glycol (PEG) were shown by Cojocario and Natansohn [1] to be in a twisted conformation where the chains' ends are near each other. Furthermore, Wang and Krenzer [2] found out that the PEG chains are of spiral conformation, which shall be favored by water. The distance between meanders in the spiral they evaluated to be 0.5 nm. On the other side, as shown by statistical mechanics, the distance between the ends of 12 members' chains is ~2 nm [3] and the entropy and energy of conformation is about $8.3 \cdot 10^{-23}$ J/K and 1.5 kJ/mol, correspondingly. This data also underlines the twisted character of PEG chains. It is informed by a special dependence of chemical reactions' velocities of the terminal OH groups on the polymer chains' length, although there is no explanation for it [4]. The authors suggest the idea that only the inter-chain interactions in PEG influence chemical reactions.

The spiral conformation of PEG in non-water solutions has been described by the Alessi group, as well as the fact that coil conformations shall be formed by PEG in water solutions [5].

All this took us to the thought that there is a well organized long-range structure in liquid PEG. To understand it permits us to explain much of its properties, for example, pore formation and surfactant and plasticizer properties.

In our earlier works we described the phenomenon [6, 7, 8] that protons of hydrogen bonds (water, starch, agarose hydrogel) shall be dissolved in a physical vacuum (PV) under the influence of the planets' gravitational radiation. The resonance between the weak gravitational radiation coming from molecular matter and the gravitational radiation (GR) of celestial bodies was shown to cause this phenomenon [8]. This proton dissolving and condensation in a PV should change the macro characteristics of the sample, e.g. viscosity, surface tension, or adsorption; therefore, it is of interest to examine this phenomenon by analyzing, for example, the viscosity of these synthetic polymers containing hydrogen bridges.

The aim of the present chapter was to investigate the long-range order in liquid PEG and in its water solutions (> 0.00001 wt. %) through the gravitation mass-spectroscopy as well as how it shall be influenced by celestial bodies' GR.

Liquid polyethyleneglycol-400, the dynamic viscosity of which can be easily measured using a rotating metal cylinder (isolated and earthed box) with minimal measuring error, was chosen as the investigation object. 11,0 ml of PEG-400 (Merck, 1,150 kg/m^3, 7.3 sSt) was placed in the measuring cell of "Reotest-2" (measuring error ≤ 4 %), which is supported by computer equipment that makes measuring possible within the first 0.16 s of the cylinder rotation. The rotation time was equal to 3 s with a relaxation interval for the PEG sample of 60 s. The device was placed in a temperature-constant room (±0.2°) and it was additionally protected from seismic waves [6]. The determination of the planets' constellation proceeded according to the Zaytzev computer program, ZET 9 (www.astrozet.net). The coordinates of the measurement place were equal to 53°38′N, 12°35′E. SPGPR has to be understood as the plane going through the Earth's rotating axis and the sample place on the Earth's surface (slide plane of gravitational proton resonance, SPGPR). The computers were calibrated with the help of radio signals. CET is Central European Time. The PEG gravitation mass spectra, which were earlier called "FNS" (flicker noise spectroscopy), were recorded with the Zubow gravitation spectrometer [7] according to the method given in works [9, 10]. Here the PEG sample was placed in a special earthed, metal box that was isolated from energy flows (heat, noise, mechanical, and light, Figure 4.0.1) [6]. The box was 10 m away from the viscosimeter.

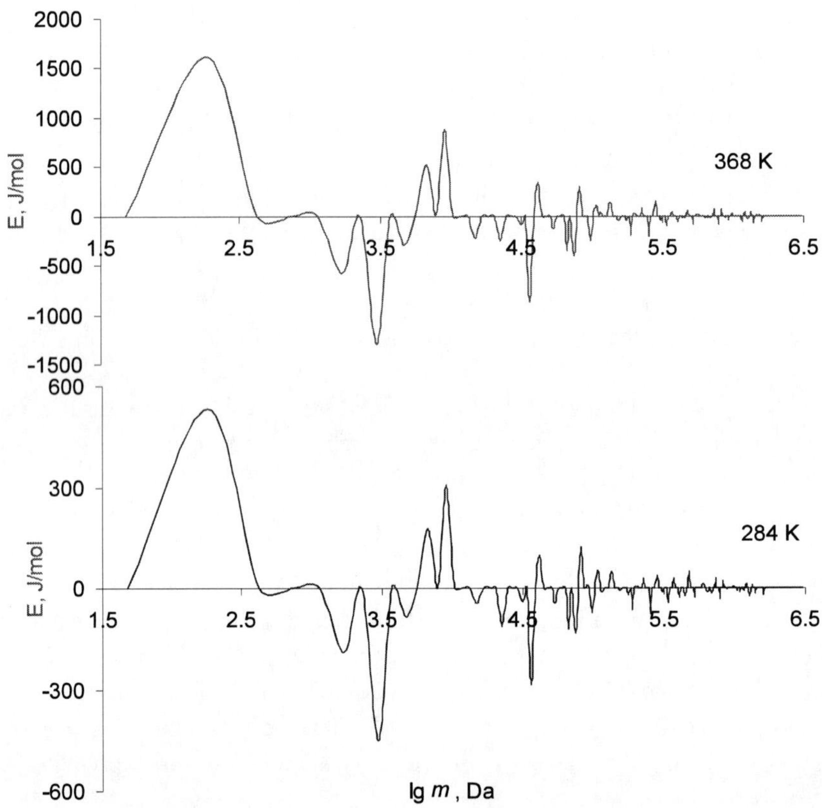

Figure 16.1. Energy GMS-spectra of PEG. The amount of PEG molecules in clusters is marked with a number opposite the signals. Zubow constant is 5.9 E-15 N/m, with a strong shock wave [9].

In Figure 16.1, the GMS-spectra of clusters in liquid PEG shall be shown. Differing from water [11], LRO in PEG has been changed more slowly. The spectra were found to be reproduced with high stability. As shown in Figure 16.1, PEG possesses LRO at the level of clusters (oscillators with masses up to log $m < 3.5$) and super clusters (log $m > 3.5$). Clusters were divided into collapsed and expanded ones where their distribution is not uniform. Their formation/destruction energies were observed to be around –1.5…2 kJ/mol. Therefore, PEG is less a homogeneous liquid but rather a heterogeneous nano-emulsion consisting of molecular clusters in the liquid polymer.

In liquid PEG, the following signals for small clusters were observed: $(PEG)_4$ (7.3 kHz), $(PEG)_7$ (5.5 kHz), $(PEG)_{11}$ (4.4 kHz), $(PEG)_{16}$ (3.7 kHz), $(PEG)_{22}$ (3.2 kHz), $(PEG)_{36}$ (2.5 kHz), and $(PEG)_{54}$ (2.0 kHz). The strong signal at log $m = 2.6$ belongs to the oscillation of small spirals that are built by 5-membered PEG chains, as shall be shown below (Figure 16.2).

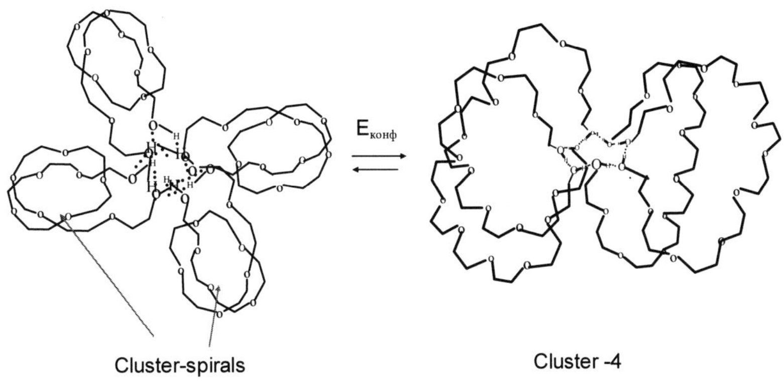

Figure 16.2. Model for the oscillation mechanism of small spirals (expanded clusters, 22.2 kHz) into cluster-4 (7.3 kHz). Spirals before the shock wave influences are marked by arrows; the nucleus of the cluster-4 shall be shown in Scheme I, see below.

At the forced destruction of LRO, e.g. at heating of 283 K to 303 K, a new expanded oscillator with a mass of nine PEG molecules and a signal intensity that is 120 times higher than that at 283 K was observed. After further warming to 323 K, it disappears again. Additional warming led to changed signal intensities in the area of the large clusters (> log m = 5.5) only.

In the GMS-spectra of PEG that was subjected to influence by a light shake or dilution with water, the signal of cluster-4 disappeared. Furthermore, in PEG and in its water solutions, there were no signals found for cluster-2 or cluster-3 as well as for those coils consisting of single molecules. The oscillations of other clusters and the cluster associates remain almost untouched by this influence. We believe that as for salt solutions [12], time is needed to achieve the thermodynamic balance in the liquid PEG solutions and in liquid PEG; this time is remarkably higher than the time of the technological application of this pore-building substance (template).

The absolute formation energy of cluster-4 was found to be three times lower than the conformation energy of the polymer chain (Figures 16.1 and 16.3), which is proof for its structure that was given in Figure 16.2.

Therefore, cluster-4 $\{HO\text{-}[CH_2CH_2O]_{10}\text{-}H\}_4$ was concluded to consist of a core and a shell of polymer spirals. Its core is built by eight terminal hydroxyl groups (Scheme I) even though it does not show any signals in the GMS-spectra.

The spirals in liquid PEG belong to expanded oscillators (+*f*); they melt and get expanded under the shock wave influence. These signals are the strongest ones in the GMS-spectra (Figure 16.1). The signal intensity ratio of spirals to cluster-4 cannot be higher than 4:1 (Figure 16.2).

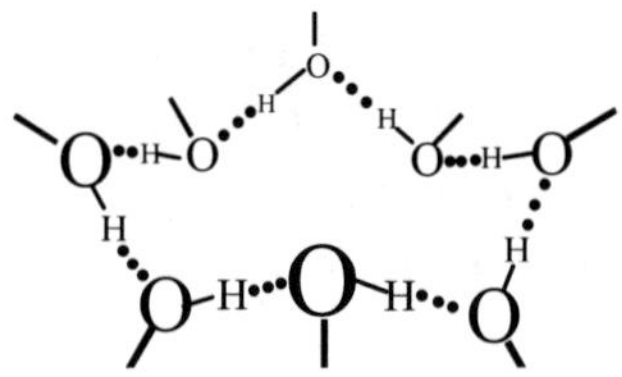

(I)

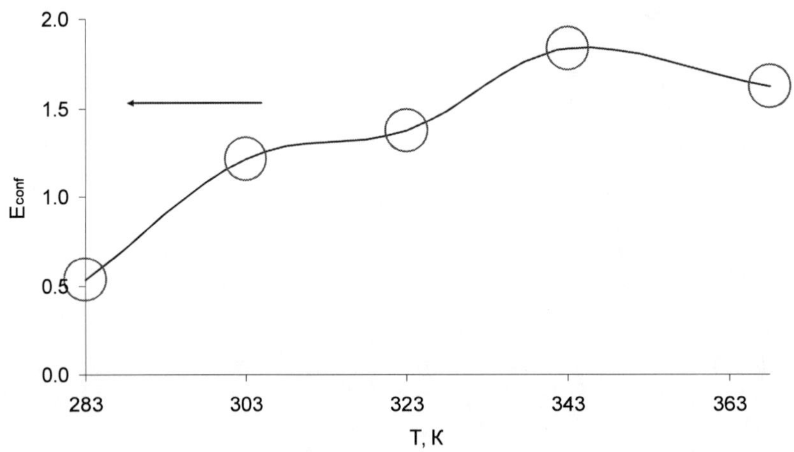

Figure 16.3. The oscillation energy of spirals (conformation energy, kJ/mol) in PEG chains versus temperature. Results were obtained after device calibration with distilled water [10]. The values of Livadara L [3], marked with arrow, agree well with our results (1.2...1.8 kJ/mol) in the temperature interval of 303 to 373 K.

On the other side, it is known that polyethylene oxide, an analog to PEG, crystallizes in a monoclinic primary building unit consisting of four polymer chains' fragments in spiral conformation [17]. One can therefore imagine the oscillation of cluster-4 as an oscillating sphere, resulting as the product of cluster rotation (Scheme II).

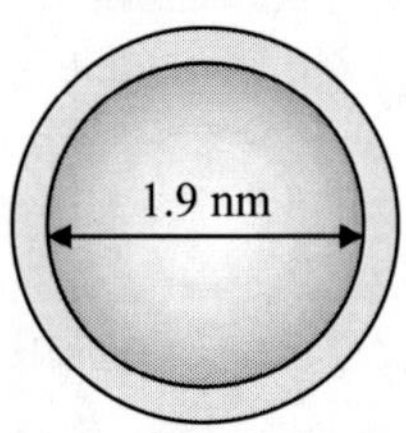

(II)

The sphere diameter is ~1.9 nm. Under the consideration that the distance to the surrounding molecules is ~0.2 nm, a porous material with mesopores around 2.3 nm can be expected. Such mesopores were described in the works [13, 14, 15, 16] and they could serve as proof for the models of cluster-4 and cluster-7 (Scheme III).

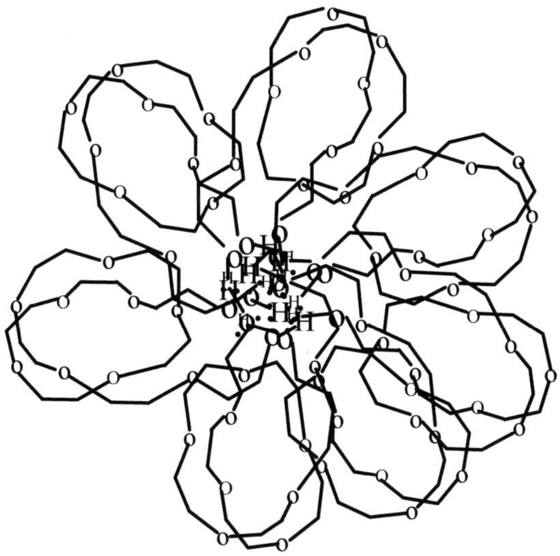

(III)

Cluster-7 oscillates as a collapsed one and its structure can be seen as a cluster-4 that has three additional PEG molecules, although it differs from cluster-4 in that there are 14 terminal OH groups in its core that connected to each other by hydrogen bonds. These hydrogen bonds stabilize the cluster-7 structure. The oscillation intensity of cluster-7 was observed to be two times higher than that of cluster-4 (Figure 16.1). It has to be mentioned that Chaplin (http://www.lsbu.ac.uk/water/index.html) described the formation of large water clusters ($(H_2O)_{280}$) in relation to the small ones consisting of 14 molecules.

Thus, cluster-4 and cluster-7 are the main kinetic units in LRO of PEG.

Clusters with masses of 11 and 22 PEG molecules oscillate as expanded structures. They are probably dimer associates of cluster-4 and cluster-7 (cluster-11) and of those of two clusters-11 (cluster-22). Their GMS signals are only weakly shown and they are typically for associated clusters (Figure 16.1). The clusters with masses of 36 and 54 PEG molecules are represented in the collapsed form; those structures are different from those of cluster-4 and cluster-7. The Cochrane modeling of these large clusters did not yield any clear answer.

Cluster-4 and cluster-7 associates' oscillation as collapsed structures (cluster-11) could be an indication that these clusters flow together in the liquid PEG even on a core level. Furthermore, at 347 K, abnormal deviations in the distribution of the clusters in their energy formations, in the number of cluster kinds, and in the average molecular masses and content of collapsed clusters were observed in liquid PEG. At this temperature, there seems to precede a structural redesign in LRO, the analogous effects of which we found for other organic liquids and water [11]. As was shown earlier [17], polyethylene oxide, a PEG analog, produces clusters as a reaction with water, too, although the authors did not give any explanation for it. Using GMS-spectroscopy the authors of the present work found out that PEG form some clusters in water solutions (Figure 16.4). The average cluster mass M_{GMS} in PEG water solutions was found to have a maximum at 12±3 wt. % (284 K). Both decreasing and increasing of the PEG concentration, either lower or higher than 12±3 wt. %, led to a smaller M_{GMS}, whereas in diluted solutions it decreased particularly quickly. This concentration maximum agreed well with the value of 14 wt. % for modified PEG water solutions found by Cojocario [1]. They are probably an indication for the LRO reconstruction in liquid water surfactant solutions.

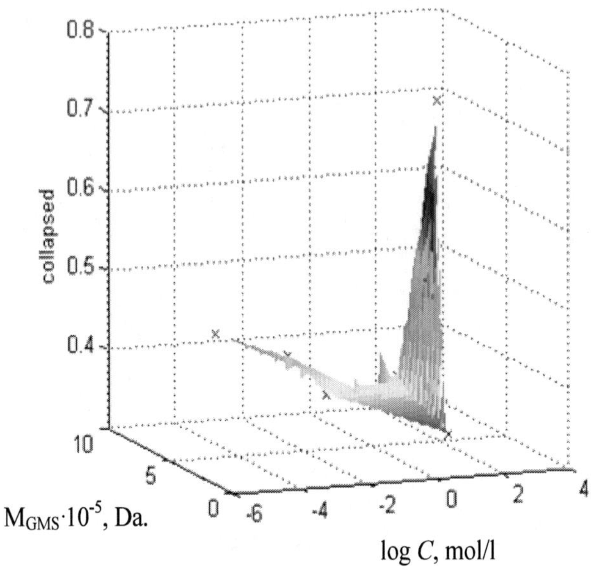

Figure 16.4. Influence of the PEG water solution concentration (*C*) on the part of the collapsed clusters and on the average cluster mass (M_{GMS}, Dalton) at 283 K.

In PEG and in its water solutions, a stable cluster with a mass of 1,602 kDa was analyzed; the cluster is present at all investigated temperatures. It is possibly a broader cluster in the statistical cluster ensemble, hence larger clusters are destroyed by heat (k*T*). We believe that these cluster giants do not have sufficient cohesion energy to hold the cluster together under strong heat influence. The concentration of collapsed clusters in solutions was found to be in the interval of 56 to 62 wt. % of all clusters where the maximum at 62 wt. % was ascribed to the PEG concentration of 1 wt. %. With rising temperature, the part of collapsed clusters in PEG increases from 56 to 58 wt. %. If the water concentration in the solution rises, then the LRO destruction of the polymer proceeds, which leads to reduced amounts of PEG cluster associates (increasing M_{GMS}). This effect is the strongest at a PEG concentration < 10 wt. %. PEG cluster-4 has not been observed in solutions. Analyzing the PEG GMS-spectra, which were obtained by subtracting the water GMS-spectra from those of the PEG water solutions (LRO of PEG only [7]), it shall be shown how strongly the spiral oscillations (Figure 16.2) depend on the solution concentration. The PEG clusters associates are in their expanded form if the PEG concentration is higher than 10 wt.%, whereas cluster-7, cluster-11, and cluster-16 were analyzed to be mainly in their collapsed form. At this concentration, the solvent does not seem to influence the cohesion interaction inside the polymer cluster. At PEG concentrations lower than 10 wt. % (better of 1 wt. %), however, the situation changes dramatically; in the polymer chains, the concentration of the expanded clusters is equal to that of the collapsed one. After further dilution, the collapsed clusters dominate over the expanded ones. This is possible only when one water molecule is installed into the spiral structure and when it connects the chain's segments that are lying opposite each other by hydrogen bonds (Scheme IV).

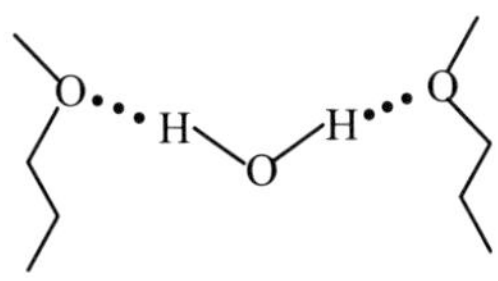

(IV)

In [18, 19, 20], there was information about the intermolecular interaction of PEG chains with water molecules. We found the cross linking of chains by water, according to Scheme IV, to transform the expanded oscillators in the collapsed ones. To Oesterhelt [20], the installation of water in PEG is connected with the formation of a non-planar super molecular structure that shall be stabilized by water molecules and characterized by a free formation energy of $1.2 \cdot 10^{-20}$ J [20].

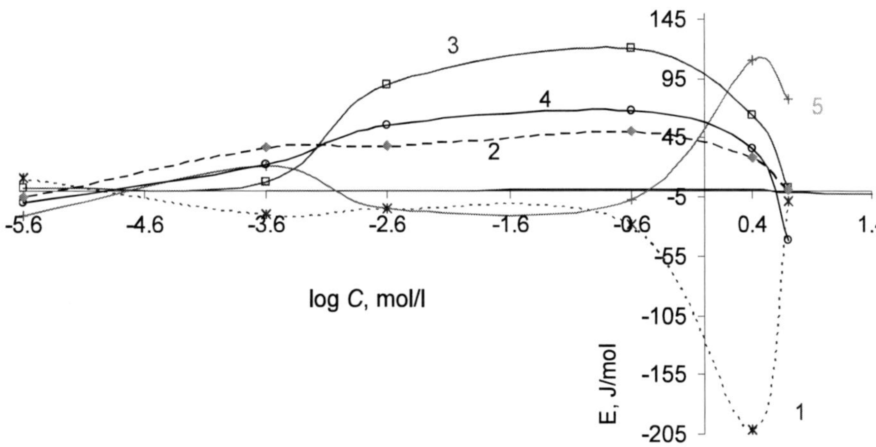

Figure 16.5. The oscillation energy of small PEG clusters and their associates in water solutions. The value E was obtained by using an internal standard, spiral oscillation in PEG chains (Figure 16.3), and the difference between the cluster signal intensities in solutions and those in the solvent. 1 – cluster-7, 2 – cluster-11, 3 – cluster-16, 4 – cluster-22, and 5 – PEG spiral.

As is visible in Figure 16.5, with decreasing PEG concentration, the energy of the cluster oscillation decreases indicating the PEG cluster destruction in a "good" solvent—water. The curse of curves 1 (cluster-7) and 5 (spiral oscillation) seems to be interesting because they change oppositely; there could be a connection between the two events. The spiral expansion in the PEG chains caused by water leads to the collapsed cluster-7, which is still further condensed under the shock wave influence of the device. The relation of the other clusters to the PEG spiral is comparable with cluster-7 but it is shown less strongly. At the PEG concentration in water of ~12±3 wt. %, the density of all the clusters has been strongly increased with the exception of cluster-7, whose density on the contrary is highly reduced. It becomes expanded under the influence of weak shock waves from the side of the spectrometer. It has been mentioned that at this concentration, the average molecular cluster mass has a maximum. This anomaly was explained to be caused by the fact that the macromolecule ends are strongly approached [1]. At this concentration, a structure transformation on the LRO level takes place that leads to the appearance of new clusters, to the changed clusters' interaction with the surroundings, and to a new thermodynamic balance,probably.

In Figure 16.6, the change of the dynamic viscosity of PEG at the time when the Earth lies on the line connecting to the Sun and giant mass clusters of dark matter (VIRGOHI 21) in the constellation Virgo shall be given. At the opposition of the giant mass of concentrations to PEG, the polymer viscosity has been greatly decreased and reversibly leads to negative values. The negative viscosity can be understood only as the result of highly weakened hydrogen bonds. The interaction of the polymer with the metal surface of the measurement cylinder seems to be reduced so strongly that its viscosity decreases more than at the cylinder rotation in air (control). Thus, the giant mass concentrations of the Sun and VIRGOHI 21 shifted the proton–vacuum balance to the side of the proton dissolving into the physical vacuum [8]. A sucking out of the boson protons in PEG in the physical vacuum proceeds and the Earth is transparent for such a form of gravitation. In this case, we have to do it with the secondary gravitation, namely, with the “ripple” of the main gravitation waves [21] that spread with light speed [6].

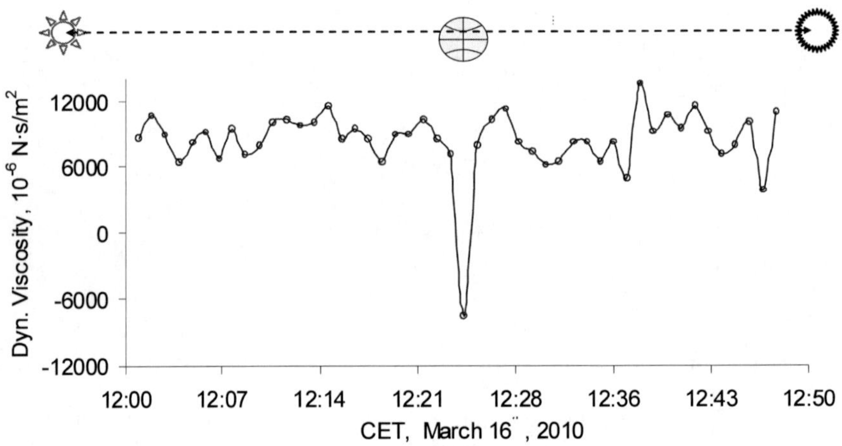

Figure 16.6. Dynamic viscosity of PEG when VIRGOHI 21 and the Sun are in an opposite constellation and they are in SPGPR.

As is visible in Figure 16.7, the weak gravitational radiation (WGR) from PEG is in resonance with GR of planets where all the planets shift the proton–PV balance to the dissolution side, accompanied with viscosity decreasing. Due to the fact that the reverse viscosity reductions were much higher than the measuring error of the device (4%), these signals were ascribed to the planets' constellation. We believe that they were caused by GR of mass concentrations in space and/or the GR waves' interference inside the sample [6, 8]. On December 6, 2009 the planets were beside each other and on the same side facing the sun and their influence on the polymer viscosity was investigated at the time when they appeared in

SPGPR. Even such little mass concentrations, like the dwarfs Mercury and Pluto, influenced the viscosity. The Sun's influence is much lower here than in the case of its opposition with the dark matter (Figure 16.6).

Furthermore, it could be interesting to analyze whether the Earth's satellite, the Moon, changes the PEG viscosity.

In Figure 16.8, there are the observation results from when the Moon appears in SPGPR. The viscosity here behaves quite differently; it is reversible and it rises spontaneously. To understand this, the fact that the PEG adheres to the measurement cylinder surface more strongly due to its reinforcement of the hydrogen bonds must be understood. In this case, when the Moon appears in SPGPR, the proton–vacuum balance shall be shifted to the side of the proton's condensation from vacuum.

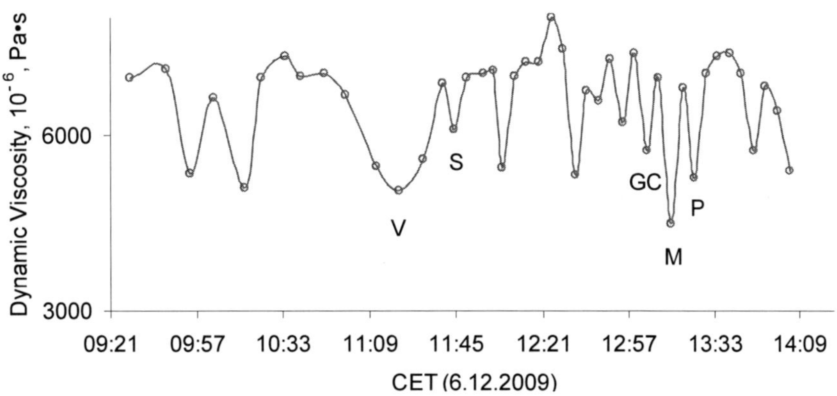

Figure 16.7. Change of the dynamic viscosity of PEG at 289 K on December 6, 2009. V – Venus in SPGPR, S – the Sun in SPGPR, GC – the Milky Way's center in SPGPR, M – Mercury in SPGPR, and P – Pluto in SPGPR.

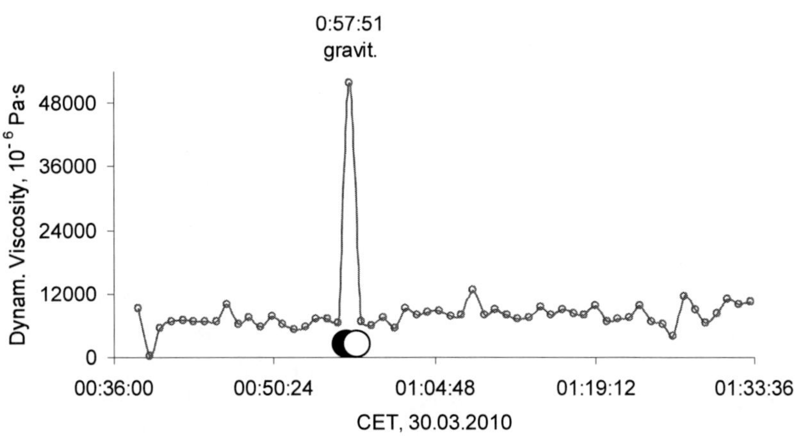

Figure 16.8. Change of the dynamic viscosity of PEG at 290 K at full moon and the Moon's appearance in SPGPR (Moon–Earth–the Sun, 178°).

These effects can only be explained from the position of a newly developed understanding of the molecular matter, namely, it is built by mass cluster ensembles [6, 8, 9].

The potential energy of cluster ensembles of the Moon is smaller than those of the Earth [7]. This leads to the conclusion that the Earth removes the gravitation energy from the Moon through the reinforcement of proton condensation from the vacuum in hydrogen bonds in the sample.

Under the same conditions, the Moon is in SPGPR but it is also together with the Sun in opposition to the Earth (4:25:28, 180°, Figure 16.9); in this case, the effect is repeated but it is done so with reduced viscosity. The Moon can cause decreased PEG viscosities when it is working in conjunction with the Sun. Here in this figure we see the influence of the gigantic mass concentration M34 on the PEG viscosity when it appears in SPGPR. M34 dominates over the Earth's proton ensembles, too, therefore, it removes the gravitation energy from the PEG sample, which causes a shift of the proton–vacuum balance to the dissolving side and decreases the PEG viscosity. There would even be this effect if M34 was on the opposite side of our planet.

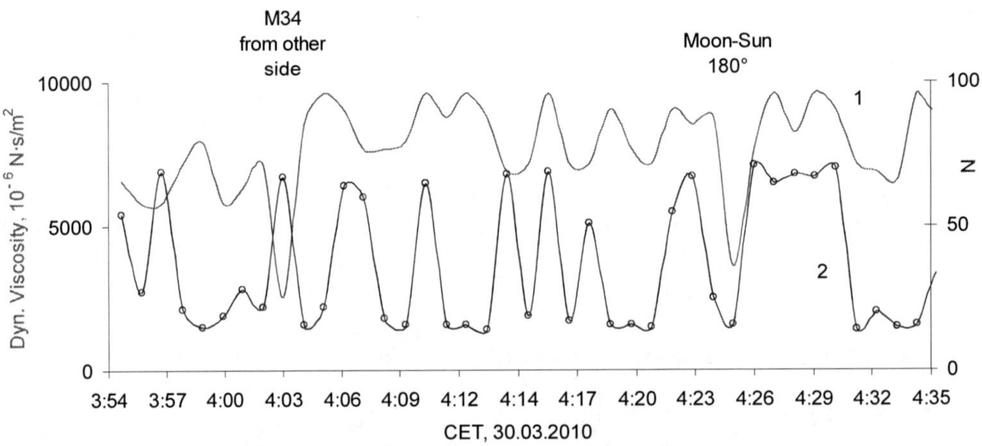

Figure 16.9. Change of the dynamic viscosity (1) and number of cluster kinds (2) in PEG at full moon (Moon–Earth–Sun, 180°), at 289 K. Distance to the Moon: 364,687 km and to the Sun: 149,784,000 km. The normal value, N (without gravitation exciting), is equal to 67. PEG cluster ensemble up to 3 million Daltons.

Additionally, it is visible that the cluster number does not change synchronously with the viscosity. For M34, the LRO destruction in PEG was assumed to cause an increasing of the number of cluster kinds up to the normal conditions as well as a reduction in viscosity. In the case of the Sun–Moon opposition, however, a decreasing of the viscosity is connected with that of the number of cluster kinds. Therefore, at the opposition of the Sun and the Moon, the hydrogen bonds in PEG shall be selectively influenced. At this time, the part of the collapsed clusters was observed to be the highest (100%). The decreased viscosity has therefore been explained by a destruction of the inter-cluster hydrogen bonds (HB), therefore. Since the inter-cluster HB are very tense and mobile, this selective destruction has to be understood from the position of two stimulating events: the proton condensation in HB by the Moon and the proton dissolution in HB caused by the Sun. The Sun has an influence on the inter-cluster HB, while the Moon has an influence on those inside the clusters.

It has to be mentioned that at that time (full moon, 30.03.2010, Figure 16.9), in the closest neighborhood to the measuring place, in Norway, all the computer communication networks were out of order and several hours were necessary for their repair. This event is the result of a reversible LRO destruction of the molecular matter caused by the planets' GR [6, 7, 8] and in the present case, by the full moon phenomenon.

The dependence of the cluster mass on the oscillation frequency (ω, Hz) is described satisfactorily by following the First Zubow equation [6]:

$$m = 10^{11} \cdot \omega^{-2}$$

The first six to nine sequence members of the above investigated clusters and of clusters in natural water [10], in distilled water [11], in polymers [9, 12]; in solutions [22], in alcohols, and hydrocarbons [7, 11]; and in protein and polysaccharide coils (lysozyme, L-arginin, chymotrypsinogen-A, amylopectin [7,9], etc. (see [7])) were well described by the Second Zubow equation [6]:

$$m = (260 \pm 20) \cdot N^{(2.08 \pm 0.05)}$$

Deviations were observed for $N > 9 \ldots 13$ and the dependence of the cluster mass (m) on its number (N) shall be characterized by a polynomial of the sixth order of magnitude.

The existence of such a stable row for quite different substances and for their states cannot be accidental. A convincing explanation could be that it arises in the stationary waves of white noises, which penetrate the Earth [6, 7, 8]. The stability of this row can be destroyed, e.g. under temperature influence that charges white noise with thermal components and changes it to color noise. For example, upon the heating of a water NaCl solution up to 368 K, frequencies and masses of oscillating solvated clusters of ion pairs [23] were shifted to higher masses and upon the speedy quenching of SiC nano particles ($\sim 10^6$ K/s) from plasma of adiabatic compression [24], the harmony of this row was destroyed. The “magical” benzene cations [25, 26] also do not fit in this sequence because they are formed in color noises.

CONCLUSION

Clusters and super cluster formations (cluster associates) were concluded to be the most statistically probable structures in the long-range order of liquid PEG where the smallest cluster is represented by a polymer chain spiral consisting of four to five members.

Clusters of four and seven PEG molecules are the base units in the long-range order, which is built according to the principle of core and shell. Whereas the core consists of terminal hydroxyl groups, the shell is built by polymer chains. Cluster-4 is not stable and it shall be destroyed upon polymer dissolution in water.

Water destroys the long-range order in PEG. At 12±3 wt. % a structure transformation in the long-range order of the solution takes place.

Using the viscosimetry together with the gravitation mass spectroscopy, the nature of the gravitational radiation and the gravitational radiation direction as well the state of the gravitation field in the sample space can be analyzed. The application both of the viscosimetry and the gravitation mass spectroscopy permits us to understand the mechanism of the long-range order destruction in molecular matter and to develop a new communication channel.

REFERENCES TO 16

[1] Cojocariu G.; Natansohn A. Intramolecular Complexation in Aqueous Solutions of an End-Capped Poly(Ethylene Glycol). J. Phys. Chem. B. 2002, vol. 106, pp. 11737-11745.

[2] Wang R.L.C.; Krenzer H.J.; Grunze M. Molecular Conformation and Solvation of Oligo(ethylene glycol)-Terminated Self-Assembled Monolayers and Their Resistance to Protein Adsorption. J. Phys.Chem. B. 1997, vol. 101, no. 47, pp. 9767-9773.

[3] Livadaru L.; Netz R.R.; Kreuzer H.J. Interacting chain model for poly(ethylene glycol) from first principles-stretching of a single molecule using the transfer matrix approach. J. Chem. Phys. 2003, vol. 118, pp.1404-1416.

[4] Atovmyan E.G.; Baturin S.M.; Fedotova T.A. Intermolecular interactions and reactivity of polyalkylene glycols. Vysokomolekulyarnye Soedineniya, Seriya *A*. 1989, vol. 31, no. 8, pp. 685-690.

[5] Alessi, Michael L.; Norman, Alexander I.; Knowlton, Sasha E.; Ho, Derek L.; Greer, Sandra C. Helical and Coil Conformations of Poly(ethylene glycol) in Isobutyric Acid and Water. Macromolecules. 2005, vol. 38, no. 22, pp. 9333-9340.

[6] Zubow K.V., Zubow A.V., Zubow V.A. Ensemble of Clusters – New Form of Molecular Matter, Risks and Chances. Zubow Equations. In Advances in Chemistry Research, vol. 5, 2010, pp. 107-145. Editor J.C. Taylor. Novapublisher. NY.

[7] Zubow K.V.; Zubow A.V.; Zubow V.A. Principles of gravitation stectroscopy. New form of molecular matter. Processes. Fields, Aist Handels- und Consulting gmbH, Berlin, 2010, electronic book (www.zubow.de), p. 854. (in Russian).

[8] Zubow K., Zubow A.V., Zubow V.A. The Phenomenon of Proton Dissolving in Vacuum and of Proton Condensation from Vacuum. Two Forms of Protons, Structure of Nuclei, Electrons and Atoms. J. of modern physics. 2010, vol.1, no.1, pp.175-184.

[9] Zubow K.V., Zubow A.V., Zubow V.A. Using of flicker noise spectroscopy for non destroying analysis of nano structures. Zavodskaja Laboratorija. Diagnostics of materials. 2008, vol. 74, no. 9, pp.40-45 (in Russian).

[10] Zubow A.V, Zubow Kr. V, Zubow V.A. Investigation of the distribution of water clusters in vegetables, fruits and natural waters by flicker noise spectroscopy. Biofizika. 2007, vol. 52, no. 4, pp.585-592 (in Russian).

[11] Zubow K. V, Zubow A.V, Zubow V.A. Cluster structure of liquid alcohols, water and n-hexane. Journal of Appl. Spectr. 2005, vol. 72, no.3, pp. 321-328.

[12] Zubov A.V, Zubov K.V, Zubov V.A. Mechanism of sodium chloride diossolving in water and the process of the solution aging. Russian Journal of Applied Chemistry. 2007, vol. 80, no. 7, pp.1249- 1255.

[13] Baghaw S.A.; Prouzet E.; Pinnavaia T.J. Templating of Mesoporous Molecular Sieves by Nonionic Polyethylene Oxide Surfactants. Science. 1995, vol. 269, no. 5228, pp. 1242-1244.

[14] Deng W.; Bodart P.; Pruski M.; Shanks B.H. Characterization of mesoporous alumina molecular sieves synthesized by nonionic templating. Microporous and Mesoporous Materials. 2002, vol. 52, no. 3, pp. 169-175.

[15] Valange S.; Guth J.-L.; Kolenda F.; Lacombe S.; Gabelica Z. Synthesis strategies leading to surfactant-assisted aluminas with controlled mesoporosity in aqueous media. Microporous and Mesoporous Materials. 2000, vol. 35-36, no. 4, pp. 597- 602.

[16] Tanaev P.T.; Pinnavaia T.J. A Neutral Templating Route to Mesoporous Molecular Sieves. Science. 1995, vol. 267, no. 5199, pp. 865-867.

[17] Ho, Derek L.; Hammouda, Boualem; Kline, Steven R. Clustering of poly(ethylene oxide) in water revisited. Journal of Polymer Science, Part B: Polymer Physics. 2002(2003), vol. 41, № 1, pp. 135- 138.

[18] Hiromi Kitano; Kurao Sudo; Ken Ichikawa; Makoto Ide; Kayuhiko Ishihara. Raman Spectroscopic Study on the Structure of Water in Aqueous Polyelectrolyte Solutions. J. Phys. Chem. B. 2000, vol. 104, no. 47, pp. 11425-11429.

[19] Shinyashiki, Naoki; Asaka, Nobuyuki; Mashimo, Satoru; Yagihara, Shin. Dielectric study on dynamics of water in polymer matrix using a frequency range 106–1010 Hz. Journal of Chemical Physics. 1990, vol., 93, pp. 760-764.

[20] Oesterhelt, F.; Rief, M.; Gaub, H. E. Single molecule force spectroscopy by AFM indicates helical structure of poly(ethylene-glycol) in water. New Journal of Physics [Electronic Publication] (1999), Volume Date 1998-1999, no., 1. Given, Article 6. Publisher: Institute of Physics Publishing. http://www.iop.org/EJ/art/S/Unreg/nj1999001010006/nj9106.pdf.

[21] Kokkotas, Kostas D. Gravitational waves. Acta Physica Colonica. 2007, vol. 38, no.12, pp. 3891- 3923.

[22] Zubow V.A., Zubow K.V., Zubow A.V. The Phenomenon of Water Clusters Distribution in Water and Solvated Clusters of Ion Pairs of NaCl in Solution Near the Wall. The Wall Effect. Journal Chemical Industry (Sankt Petersburg, Ru). 2006, vol. 83, no. 6, pp. 300-306.

[23] Zubow Kr. V., Zubow A.V., Zubow V.A. Oscillation spectra of seed NaCl crystals in aqueous solution. Journal of Appl. Spectr. 2005, vol.72, no. 6, pp. 840-847.

[24] Zubov V.A.; Scholtz R.; Braun M., Zubowa H.-L.; Kolbanovskii Yu. A.; Ovsyannikov A.A.; Shipachev V.S. The High-Temperature Gas-Phase SiC-Based Inorganic Polxmer: Structure and Properties. Russian Journal of Inorganic Chemistry. Russian Journal of Inorganic Chemistry. 1994, vol. 39, no. 11, pp. 1708-1711.

[25] Schriver K.E.; Paguia A.J.; Hahn M.Y.; Honea E.C.; Camarena A.M.; Whetten R.L. Are clusters of nonpolar molecules icosahedral? J. Phys.Chem. 1987, vol. 91, no. 12, pp. 3131-3133.

[26] Mine M.; Mori H.; Ogata M.; Tanaka S.; Tsutsui T.; Miyoshi E. Computational research of the electronic structure of benzene trimer cation by ab initio method. Chem. Phys. Letters. 2007, vol. 438, no. 4-6, pp. 157- 161.

Chapter 17

COMPOSITE MATERIALS

It is interesting to understand the formation of molecular clusters in composite materials. In this chapter, we show how the GMS-method can be applied for the non-destructive analysis of composite materials based on industrial polystyrene (prepared by emulsion polymerization) with Aerosil [1] or with nanodispersed silicon carbide [2]. Composite materials are prepared by mixing a solution of PS (50 wt. %) in *p*-xylene with Aerosil (380 m^2/g) and nano-Si_xC_y. The GMS-spectra of the films were recorded after the solvent was completely removed (Figure 17.1).

As is visible, the introduction of silicon carbide into PS has no significant effect on the long-range order in the polymer; it is almost completely identical. It seems that no interaction has occurred. The interaction of PS with Aerosil, however, is very different. The spectrum of the composite PS-Aerosil is very different from that of the industrial PS but in the area of small clusters ($\log m < 4.5$), the spectrum is similar to that of the pure Aerosil [1] even though the Aerosil concentration is not above 2 wt. %. It is reasonable to assume that the signals in this region are a superposition of the signals caused by PS, Aerosil, and the product of their interaction–composite. The increase in signal intensity of Aerosil in the composite is explained by the fact that the interaction of PS-Aerosil leads to a higher cluster individualization of Aerosil while the PS clusters have lost their individuality. The spectrum was taken after the main restructuring processes were completed (20 days). Up to this time, sweeping changes have been recorded. For instance, up until 10 days, the GMS-spectrum of the composite is very similar to that of PS. The following disappearance of small PS clusters can be caused by the interaction of the aromatic rings (Lewis bases) with strong Lewis acid sites in Aerosil (Scheme I).

A distinctive feature of the composite PS+Aerosil is the increased average cluster mass, the increased part of collapsed clusters, and a slightly reduced number of cluster kinds in comparison with the composite PS+Si_xC_y. Thus, the PS clusters lost their ability to oscillate due to their interaction with Aerosil (reduced *N*), but a part of remaining cluster kinds has partially transformed into the collapsed form (increasing $Д_c$) and the other part raised their individuality as a result of reducing their interaction with the surroundings (increased absolute *f*-value in the formula for M_{GMS} calculation). This behavior of different mass ensembles is unique and shows how each cluster is “fighting” for its potential energy in the new mass ensemble.

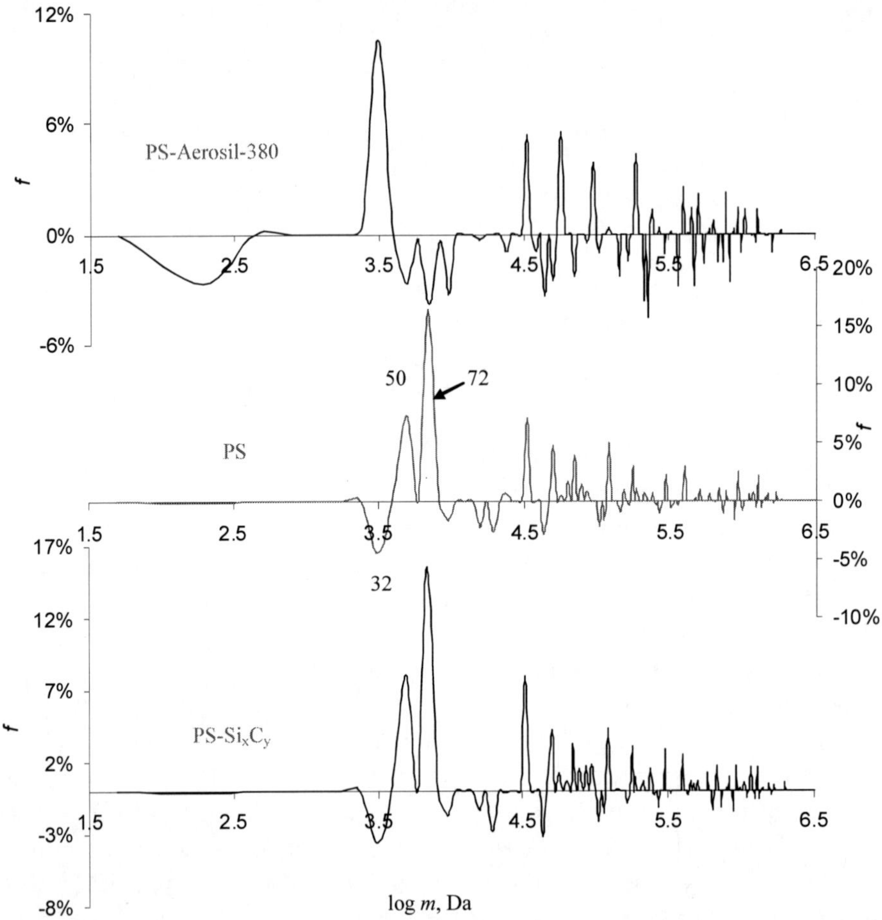

Figure 17.1. GMS-spectra of industrial polystyrene (PS) and its simple compositions with Aerosil and silicon carbide, $p > 1$ N/m^2, 288 K. The PS-Aerosil-380 film (1 ml, 50 wt. % solution of PS in *p*-xylene and 10 mg Aerosil-380) is very fragile and the PS-silicon carbide film (1 ml, 50 wt. % solution of PS in *p*-xylene and 10 mg of silicon carbide) is elastic. The PS film is obtained from a 50 wt. % solution in *p*-xylene. All films are prepared on Teflon. Zubow constant is equal to $6.6 \cdot 10^{-15}$ N/m. The number above the signal corresponds to the number of styrene units in the PS cluster.

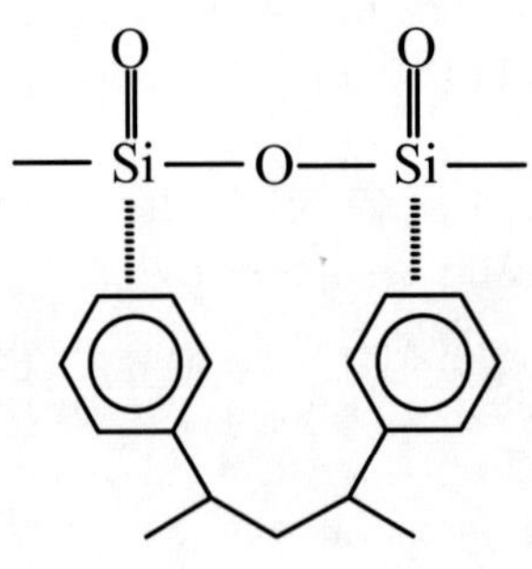

(I)

Table 17.1 shows the main parameters of the GMS-spectra of the investigated composites on the basis of emulsion PS.

Table 17.1. Main parameters of the long-range order in films of emulsion PS and its composites with inorganic fillers.

Sample	M_{GMS}, Da	Д$_c$, %	*N*
PS	248,507	26	60
PS+Si_xC_y	256,312	26	62
PS+Aerosil-380	284,170	54	55

CONCLUSION

Thus, the GMS-spectroscopy is observed to analyze composite materials at low-frequency cluster motions without destroying their structure. Ensembles of clusters in the composite are forcibly formed in white gravitation noises.

REFERENCES TO 17

[1] Zubow K. V, Zubow A.V, Zubow V.A. Use of flicker noise spectroscopy for undestroyed analysis of nano structures. Zavodskaja Laboratoria. Diagnostics of Material. 2008, vol. 74, no. 9, pp. 40-46 (in Russian.

[2] Zubov V.A.; Scholtz R.; Braun M., Zubowa H.-L.; Kolbanovskii Yu. A.; Ovsyannikov A.A.; Shipachev V.S. Structure and Properties of Inorganic Polymer on the Base of Silicon Carbide from High Temperature Gas Phase. Russian Journal of Inorganic Chemistry. 1994, vol. 39, pp. 1708-1711.

Chapter 18

THE STRUCTURAL HETEROGENEITY OF THE PHYSICAL VACUUM

Many theoretical models are known for the formation of the physical vacuum, (PV) however, because of the absence of an experimental base they only give a mathematical description [1, 2, 3]. Here the state of the vacuum has to be understood as isotropical or quasi-isotropical and the hypothetical energy fluctuations in PV should be seen as the reason for heterogeneity and finally for the boson matter. On the other side, until now, there were not any experimental proofs for heterogeneity. When we assume that PV is one of the forms of matter, e.g. of the V-level, then we can conclude some analogies about higher levels of matter (atomic, molecular, space, etc.). All levels are known to be spherical systems, spheres' clusters, and cluster fluctuations even though at this time no experimental result supports this formation of PV. An experiment that proves such a hierarchy of events and the structure of PV must be researched at all structure levels of the matter, especially at levels near to PV, as well as at those that are at the most remote points from each other. The nearest level to PV is the nucleus (N). For the structure analysis of the N-level, the molecule level state (M) must be investigated under passive conditions. In this case, the M-level reflects the state of the N-level in PV. Hence the molecular structure, the molecular clusters, and the ensembles of molecular clusters mirror processes at the N-level.

We have shown [4] that the nature of Newton forces is based on energy minimization of nuclei clusters as a unit ensemble of equal oscillators. For its existence, the N-level gets energy from energy clusters in PV (§3.2).

Nuclei clusters generate special energy histograms that reflect the state of EC in PV. Here the non-excited nuclei and molecular clusters as well as the properties of them shall be described by Zubow equations [5, 6].

A new form of molecular matter—ensemble of clusters—has been discovered by the authors whereas clusters of molecular mass concentrations have to be understood. A cluster ensemble is characterized as an energy carrier for a unit of clusters; ensembles were found to be unique sensors that react on all energy flows known until now from the seismic waves of earthquakes to the gravitational radiation of the planets. For their existence, molecular clusters in an ensemble get their energy from energy clusters in the stationary gravitation waves (EC) of the surroundings. The possibility to form molecular clusters (nuclei clusters) and their ensemble is strongly determined by EC (Figure 3.2.1) [4, 5].

According to Figure 3.2.1, in amorphous substances and liquids, there are both levels of matter organization: molecular and energy clusters and in crystalline matter, energy clusters only. The cluster distribution in an ensemble shall be described by Zubow equations (§3.2.1) even though the existence of clusters in molecular matter is already well known [7, 8]. Clusters that are understood as molecular mass concentrations are experimentally proven in liquids [9, 10].

The formation of a cluster ensemble where each cluster takes a certain energy level, however, has not been completely cleared. To extract a single cluster from this ensemble without destroying the cluster or the ensemble is not possible. For instance, for every polymorph state of liquid water, there is its own ensemble. The clusters were characterized to be thermodynamically stable in a picosecond interval [11]. The energy needed for their stabilization was taken from noises [4, 12], just as that for their destruction was caused by heat fluctuation ($E = \mathrm{k}T$).

The dependence of the cluster number ($K = 1, 2, 3, 4, 5 \ldots$) in the molecular matter on its mass was described (§3.2.1) as follows:

$$m = (250 \pm 20) \cdot K^{(2.2 \pm 0.2)}$$

Therefore, Zubow equations make it possible to get some idea of the nature of clusters and their ensemble as well as their purposeful search in chemical technologies. The clusters in each ensemble are formed at strongly determined masses accordingly to this equation.

The cluster ensembles given in Table 4.3.1 were observed experimentally by the authors with the help of a GMS-device [4]. The clusters in an ensemble were found to compete for energy sources in the ensemble and were visible as a steady rearrangement of the cluster structures [13] where the rearrangement time can last up to some months or even years under normal conditions [14]. Ensembles are able to form mirror-isomers and ensemble-"óboroten" ("wryneck"). They are characterized by the inversion of density (collapsed–expanded) and energy capacity. Furthermore, they compete with each other for the energy potential of white noises ("wall effect") [15]. Large ensembles grow at the expense of small ones [13]. The reason is the potential energy minimization of the new ensemble. It was also concluded [4] that ensembles attract each other at a distance; to the authors, this phenomenon is the background of Newton's Law of Gravitation. In mixtures, cluster ensembles of different substances are not mixable [4] but within one ensemble, a virtual mass transfusion goes between the clusters that reflect the gravitation energy transfusion (fluctuation) on the level of energy clusters in white noises (Figure 3.2.1).

According to Figure 3.2.1 in amorphous substances and liquids, both levels of matter organization—molecular and energy clusters—are present. In contrast, in crystalline substances, only energy clusters are present. A constant convergence and divergence of masses takes place in the M-level (ensemble of molecular clusters), which reflects analogous processes of gravitation energy (fluctuation) at the EC-level in white noises (Figure 18.2). Earlier in his monograph [16, p. 28], Volkov gave the sequence of nuclei orbitals' filling (K_n) as:

$$K_n = 2,\ 8, 20, 28, 50, 82, 126, \text{etc.}$$

This can be ascribed by:

$$N=1.6024\cdot K_n^{2.15}$$

This equation agrees well with the experimental sequences of molecular clusters (second Zubow equation, §3.2.1). If N is assumed to be a value that is proportional to the mass of orbitals in the Volkov model or better yet, to the cluster masses in nuclei (m_N), then the last equation is identical to the Zubow equation. Therefore, the discussion cannot be about orbitals being inside the nuclei but rather about the nucleus as an ensemble of mass clusters on the N-level of the matter.

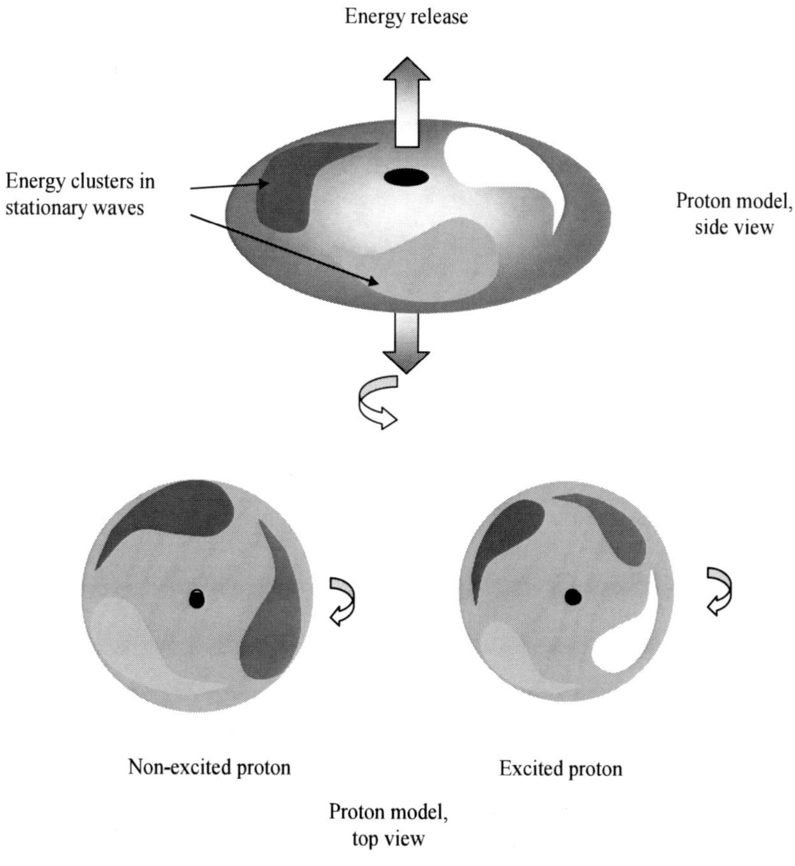

Figure 18.1. Structure models for excited and non-excited protons. Energy clusters as stationary waves (quarks) that rotate around the proton center are shown as colored "drops."

Thus, the following equation:

$$m_N = 1.6024\cdot K^{2.15}$$

reflects the nucleus' structure as an ensemble of mass clusters that are in a constant convergence and divergence (models in Figures 18.1 and 2.3.4). At this stage, the aforementioned mass clusters in the ensemble cannot be extracted, separated, or fractionated by physical methods, as was done for molecular clusters. The model of a rotating proton as an analog to the modern model for the black hole includes the formation of stationary waves consisting of three to four energy clusters or quarks (Figure 18.1). Here the number of quarks depends on the velocity of the rotating energy waves inside the proton. The higher the velocity, the higher the number of virtual particles—quarks. Since the formation of stationary waves in rotating isotropic liquids is a well known phenomenon [17, 18, 19], in the first approximation, the nature of quarks can be understood as virtual particles.

Thus, clusters and their ensembles can reflect the structure of a lower matter form, PV. The present chapter intends to develop an experimental base for the investigation of the PV structure by analyzing structural features of the M-level as well as evaluate the PV energy.

Samples were placed in a box that was maximally protected from outer energy flows (Figure 4.0.1). The measurement method is described in [4, 6, 14, 15]. Celestial body coordinates were taken from the ZET 9 Lite program (www.zaytsev.com). The following materials were chosen as investigative objects: biomatrices of hen's eggs (egg white) and fresh potato tubers, potato starch (powder), distilled water, water solution of NaCl (3.5 wt. %), industrial polyethylene, atactic polystyrenes, collagen (air bladder of a roach), melted quartz (glass), and methanol. The GMS-sensor was placed directly into the biomatrices, weakly pressed into the powders/films (1…2 kPa), or placed into the liquids that were aged for the six months prior [15]. Masses and frequencies of clusters were calculated by Zubow equations using for biomatrices and water the constant $6.4 \cdot 10^{-15}$ N/m, for starch $6.8 \cdot 10^{-15}$ N/m, and for PS, $5.2 \cdot 10^{-15}$ N/m. The weight calibration of the device was carried out with the oscillation of cluster $(H_2O)_{280}$ (5,040 Da) in bi-distilled water at 4,350±100 Hz; for small clusters with the oscillation of lysozyme coils at 2,451 Hz (15,754 Da) for the middle weight area; and with the oscillation of the cluster in polyethylene at 1,002 Hz (94,261 Da) for large oscillators [13]. The energy calibration of the GMS-spectrometer for clusters in liquid styrene in the temperature interval 293…400 K was made by interpolating the formation energy of liquid styrene (15.3 kJ/mol, formation of all clusters) to the evaporation energy at 413 K (38.3 kJ/mol, destruction of all clusters). The sum of all heat effects observed in the measurement cell (0.125 ml) is the base for this calibration in this temperature interval. The measuring error was 10±5%.

Yamauchi [20] suggested a new analogous analysis method of the molecule oscillation with intramolecular and/or intermolecular energy transfer given in time-frequency. With this method based on the short-time Fourier transform the fluctuation of the energy density can also be detected just like with GMS.

A suitable analysis of the M-level was found to be close to isothermal conditions without any energy flows from the surroundings and with the help of passive analysis methods. For that, the substance sample (liquids, powders, gels, polymer films) was placed in the center of an earthed iron box (Figure 4.0.1). GMS earlier described in §4 records the histogram of nuclei clusters (§3) whose lifetime is about some picoseconds. Nuclei clusters were found to be of a gravitational nature that destroys under heat flows. These clusters interact with analogous clusters at a distance that is comparable with a proton's gravitational radius where the interaction phenomenon can be highly increased by the cooperation effects in a group of similar nuclei [21].

According to theoretical physics, gravitational radiation shall be generated only in the case where very heavy masses in space rotate around an axis with a velocity that approaches that of light velocity [22]. This model could be extrapolated to elementary particles even though they would operate under super-light velocity conditions. Then, the reason for the weak gravitational radiation of molecular matter could be understood as a quasi equilibrium of a proton with PV (equilibrium between

proton dissolution in PV 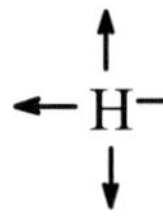 and proton condensation 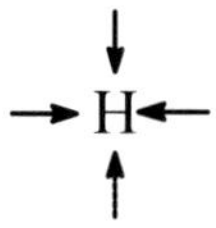 from PV) [4]:

$$P \leftrightarrow E_{\Gamma}$$

The equilibrium, where the proton (P) is the gravitational energy (E_{Γ}) could be shifted to proton dissolving as well as to proton condensing where this equilibrium is highly flexibly. This could be caused by the influence of diverse space masses, seismic shock waves, etc.

WGR was observed for many investigated substances [4] where the energy signals of nuclei and molecule clusters were recorded as a histogram (§3, Scheme I). A histogram reflects the energy state of nuclei clusters inside a molecular cluster. Furthermore, it reflects the equilibrium with analogous ensembles of surrounding pendula (objects, substances, planets).

The proton movement with a super-light velocity has to be understood as a short-life movement. At the level of its dissolving in PV ($P_{р}$), it is like a point, and its condensation ($P_{к}$) from PV is also like a point where there is no straight motion (§21). In this case, the proton jumping model can be given as Scheme I.

$$P_{р} \underset{E_{к}}{\overset{E_{р}}{\rightleftarrows}} P_{к}$$

(I)

For the ideal case, the energy balance $\Delta E = E_{р} - E_{к}$ has to amount to zero:

$$\Delta E = 0$$

However, practically the equation does not work:

$$\Delta E \neq 0$$

Signals, which were recorded in the gravitation spectra, are an indication for an imbalance in Scheme I.

To prove that the energy parameter (E) correlates with the number of protons in the sample, the following experiment has been carried out. Energy clusters in crystalline metals of different weights (iron samples between 1 g and 14 kg) have been investigated under conditions excluding the influences of external energy fields and under constant temperature, atmospheric pressure, and humidity.

In Figure 18.2, it shall be shown how the proton number (log n_p = 0.9933(log m) + 23.452) as well as the energy of the sum of all EC signals (abs (E) = -10,759(log $m)^2$ + 109,450(log m) + 246,285) depended on the mass of the metal samples. The correlation coefficient for both curves (events) amounts to 0.95, indicating that the energy signals were caused by EC. It has to be mentioned that EC is the result of proton jumping in PV that leads to the generation of weak gravitational radiation.

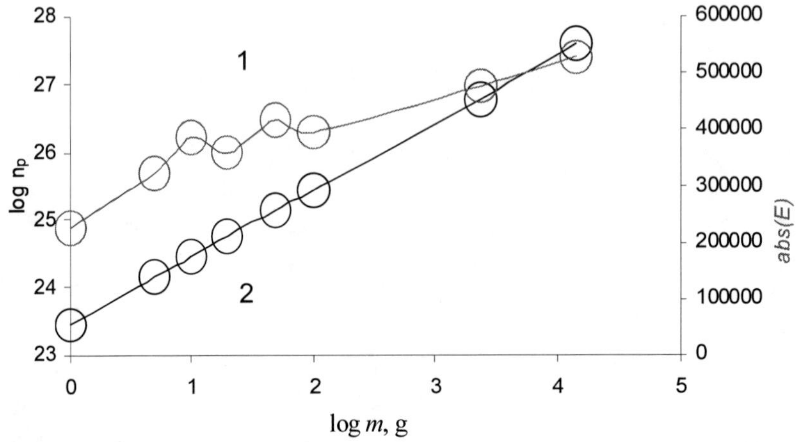

Figure 18.2. Summary energy (abs (E), rel. units) of energy cluster ensembles (1) and the number of protons in iron samples (n_p, 2) versus metal masses. Zubow constant = $5.5 \cdot 10^{-15}$ N/m.

Let us return to the processing energy. The energy imbalance (Scheme I) could be caused by the domination of either the proton dissolving in PV or of the proton's condensation from PV. To understand this dilemma, we have to again discuss the results showing the gravitational radiation influence of the planets on the WGR of molecular matter (Figure 18.3, [4]). After analyzing the obtained results, it was concluded that the WGR energy of the samples (Π_E) had been partially absorbed by celestial bodies.

$$\Delta E = E_p - E_\kappa = -\Pi_E$$

or

$$E_p + \Pi_E = E_\kappa$$

As is visible, a proton dissolving in a PV has been stimulated through the mass cluster ensembles of celestial bodies. The driving force of this dissolving process was found to be the minimization of the potential energy in the newly formed ensemble of masses consisting of celestial bodies and the sample [4].

Here the mass of the planet and the distance to it influence Π_E only. Remarkably, the Π_E value was observed earlier in other experiments on the Earth, for instance, at the interaction of cluster ensembles of different masses that are remote to each other [4] and at the forced change of the H-bond length ((de)compression) [4, 14, 23].

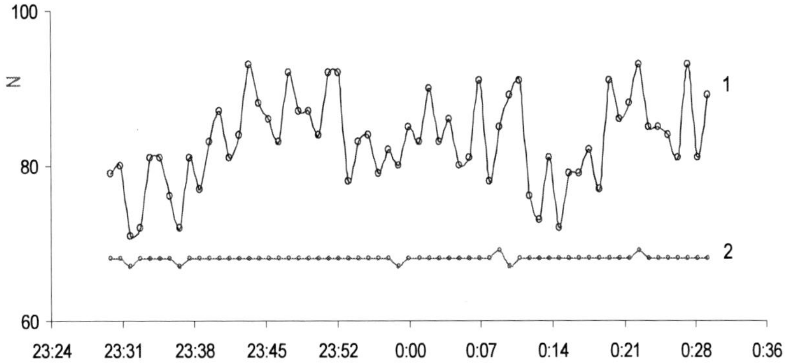

Figure 18.3. Number of energy cluster kinds in copper (1) and of molecular cluster kinds in starch hydrogel (2) versus time. The distance between samples is about 1 cm.

The nucleus model does not go against to the idea of the cluster structure of boson matter. Based on those masses it could be strongly determined by gravitation (Figure 18.1), permanently converging and diverging into each other (Figure 2.3.4). On the other side, the presence of cluster signals in the GMS-spectra indicated that the dissolving/condensing process of the proton is unbalanced and the balance is slightly shifted to the mass dissolving. Here, the dissolving process concerns only very small masses. Therefore, WGR of molecular matter reflects the energy imbalance of this process and can be used for its understanding. Currently, the expanding universe was found to be a result of its density decreasing even though this density decreasing can be understood as a mass decreasing or in our case, as a proton dissolving in a PV, which well agrees with the observed data. Do not confuse the radiation of an activated proton at 21 cm [24] with the WGR of the dissolving/condensing process in a PV.

Let us return to the Zubow equation $m = 10^{11}\cdot\omega^{-2}$. It is similar to the energy spectrum of all particles (N_3) in the radiation belt of the Earth as described with the help of the function $N_3 \sim E^{-1.8}$ [25]. For protons with an energy of ~1 MeV (proton belt with low energy, at most close to the Earth's surface at 50-60°), the last equation can be ascribed as:

$$N_3 = \kappa\cdot(\mathrm{h}\cdot\omega)^{-1.8} \text{ or } N_3 \approx \mathrm{K}\cdot\omega^{-2}$$

That agrees satisfactory with the Zubow equation. While approaching the Earth, the proton flow shall be broken by the atmosphere that leads to its dissipation and absorption. The braking was assumed to cause the gravitational radiation with a corresponding energy distribution. During the day, the front of the radiation approaches the sample place through cups whereas at night it goes through neutral layers of a magnetosphere tail.

In [4] we described how cluster ensembles interact at a distance where the interaction is based on the resonance of their weak gravitational radiation. WGR shall be explained by the proton dissolving in a physical vacuum with its following condensation from the vacuum (influence on the gravitation field, Figure 3.2.1). WGR of samples was supposed to be in resonance with the analogous gravitational radiation of the planets at their culmination. Therefore, it should be of interest to find out how molecular cluster ensembles on Earth are influenced by the gravitational radiation of celestial bodies. If this suggestion is right, one could expect celestial body influence on chemical technologies through the change of cluster ensembles (long-range order) in substances in reactors, which would then lead to changed technological conditions, product qualities, and new unknown risks.

Analyzing molecular clusters, a trivial correlation was observed with the masses of the planets and of the Sun. Additionally, correlations could also be made with their hypothetical gravitational radiation (Table 5.1). In Table 5.1, the base for such resonance (equal frequencies) between water cluster ensembles and celestial body clusters shall be shown. The clusters $(H_2O)_{178}$, $(H_2O)_{280}$, $(H_2O)_{903}$, $(H_2O)_{1351}$, $(H_2O)_{1601}$, and $(H_2O)_{1889}$ were formed in the noises from the galaxy and the Sun or were caused by the interference of noises from single planets, so they do not fit in Table 5.1. It has to be mentioned that the mass of water clusters is proportional to the number of hydrogen bridges in the cluster [4].

In Figure 18.3, for comparison, the time dependence of the number of different cluster kinds in metal (copper) and in starch hydrogel (85 wt. % water) is shown. Here the crystalline metal was used to understand EC in a PV and the hydrogel for that of MC (molecular cluster) in a PV. As is visible, the behavior of curve 2 remarkably differs from that of curve 1; it can be understood as representing the inertia of MC against the quickly moving events in a PV.

In the next section, we discuss how the energy characteristics of EC in crystalline copper and of MC in amorphous hydrogel shall be changed (Figure 18.4). As is already true for molecular cluster kinds, for the energy characteristics of MC, the ensembles in molecular matter have a lower reaction to events in a PV, too. Generally, however, both cluster types were found to weakly react on gravitational radiation from the Milky Way's center. On the other side, the reaction of EC on Pluto at its azimuth of 0° at 00:24...00:27 was very high whereas MC did not react. The GR of Pluto seems to change the PV structure in the sample space but not sufficiently to remarkably influence the LRO of water in hydrogel. The event ascribed with a broad arrow shall be interpreted as the influence of Mercury and Pluto's opposition (planet centers are at one line: Mercury-Earth (sample) - Pluto, 179°29′, see www.zaytsev.com) on the sample that leads to a gravitational shot through the sample.

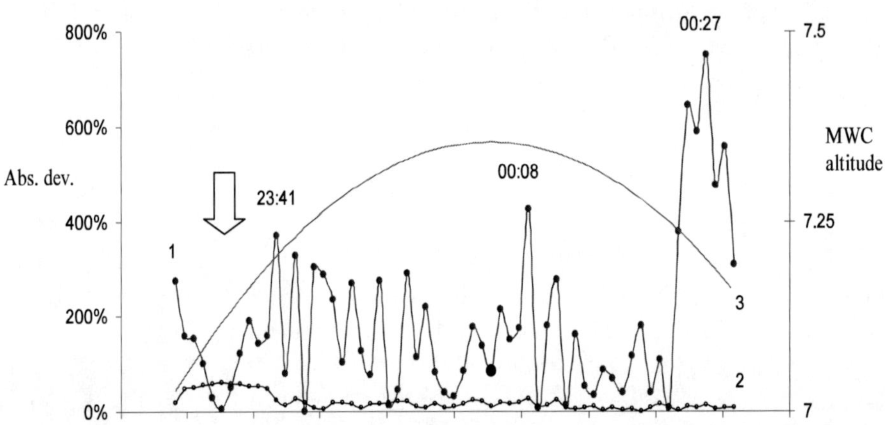

Figure 18.4. Absolute deviation of cluster ensemble energy from the average value at the culmination of the Milky Way's center (black point, MWC). 1 – EC in copper, 2 – MC in starch hydrogel (ensemble up to 2.2 million Daltons), and 3 – altitude of MWC.

Thus, the PV structure was found to be characterized by a broad energy distribution that consists of two components, namely, the base and the variable. Here the last component is highly sensitive to gravitation fields coming from space mass clusters.

Due to the great importance of the energy capacity for a PV, we have to evaluate it. Although there are many theoretical works written about this topic [26, 27, 28, 29], a broad spectrum of theoretical values without any experimental proofs is given. On the other side, the particle nature of a PV was discussed in [30] and the energy distribution in [31], both publications prove our model (Figure 3.2.1).

To get some information on the PV energy, it has been investigated how the gravitational radiation of Jupiter influences the proton jumping energy in H-bonds in starch [4]. The energy loss of H-bonds at Jupiter's culmination, where Jupiter influenced especially small starch clusters (Figure 18.5), was remarkably high. The gravitation energy removed from starch to Jupiter is much higher than that energy moved to the Moon at full moon; this was done by extrapolating from much smaller cluster ensembles (down to 500,000 Dalton). The Jupiter effect gets even clearer. Thus, Jupiter, the planet (gas giant, more than 75% hydrogen) enriched with the highest proton concentration, was observed to have the strongest influence on the smallest cluster ensembles in starch.

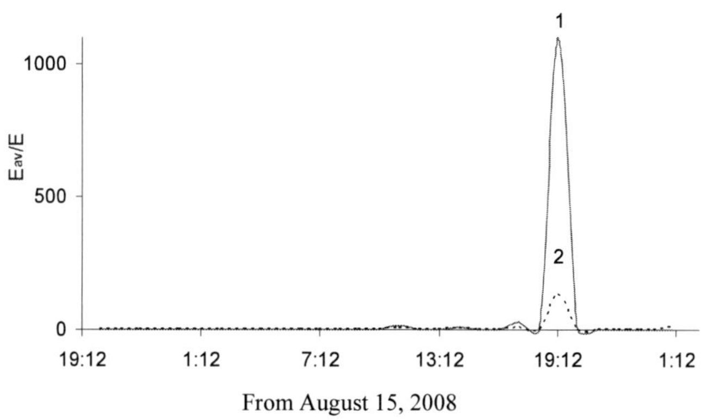

Figure 18.5. Ratio of absolute gravitation energies removed from starch to the average gravitation energy (E_{aver}) that was measured between August 15 and 17, 2008, 1 – cluster ensemble up to 2.2 million Daltons (E_{aver} = 589) and 2 – cluster ensemble up to 223 million Daltons (E_{aver} = 24,258 relative units).

The disintegration of clusters and coils in starch under the influence of Jupiter is nearly inconceivable, not once when this process is reversed from the point of view of the modern physical chemistry of polymers is this possible. To understand the process of reversible disintegration of the long-range order in starch (grape model [32]), GMS were taken before Jupiter's culmination, at its culmination, and after its culmination. In Figure 18.6, the process of reversible disintegration of cluster ensembles in starch is shown. This is a virtual process though. GMS detects WGR energy that is caused by proton jumping [4], but an array of signals (histogram [33]) reflects the matter structure on the level of molecular mass concentrations—cluster and coil (for starch) formations. Let us look at this process formally. The change of signal intensity of a α-D-glycopyranose member (Figure 18.6) is an indication for the forced partial defolding of polymer chains [6]. On the other hand, the almost 6-fold increasing of the *f*-value (for the simplest spiral, Figure 18.6, [6]) indicates an individualization and expansion of this cluster as well as a weakening of its interaction with its surroundings, which proceeds through hydrogen bridged bonds (in §13, Scheme II, only two hydrogen bridged bonds between two meanders in a spiral of a 12 members in a starch cluster are visible).

Chains defolding and the weakening of their interaction between clusters and their surroundings could be an indication for hydrogen bridged bonds' destruction as result of energy removal in the direction of Jupiter. The gravitational energy drain to Jupiter that is connected with the possible coil's disintegration in starch (Figures 18.5 and 18.6) amounts to 2.6 kJ/mol·0.5g/323 = 0.004 kJ, where 323 is the molecular mass of a repeating member in starch. This value is reasonable for the highest limit (100% destruction of all clusters and coils); for the lower one, it is proportionally decreased up to 0.002 kJ, according to a decreased number of cluster kinds from 104 to 59 at Jupiter's rise.

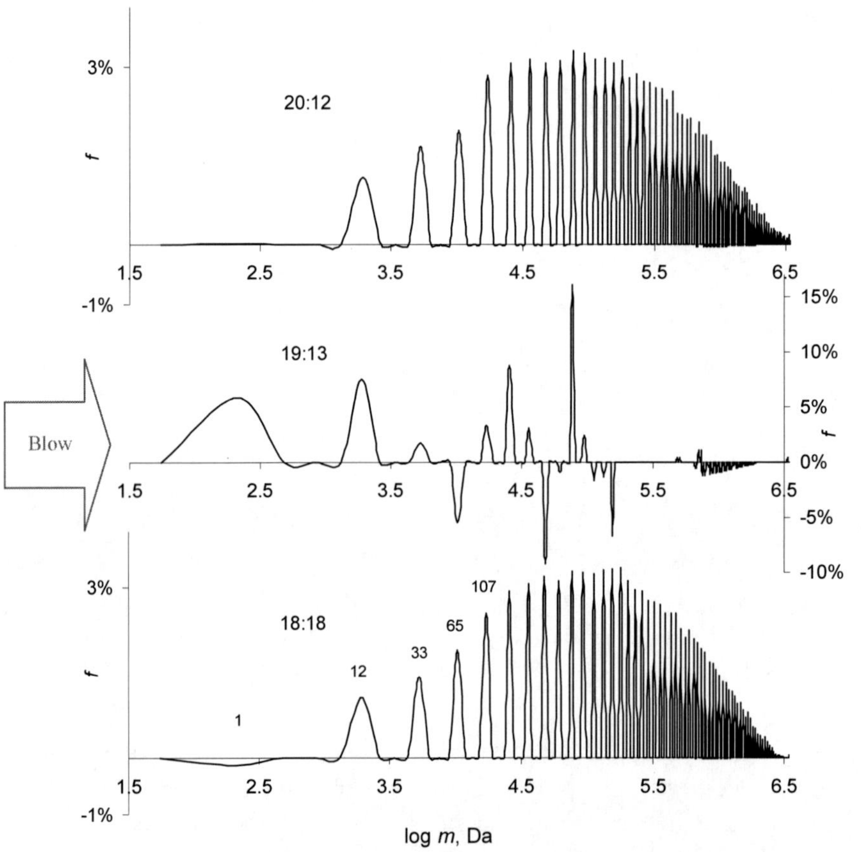

Figure 18.6. GMS of potato starch (powder 1.55 mmol, 0.5 cm^3) at Jupiter's rise. Weak shock wave ($p < 1$ N/m^2), Zubow constant is equal to $6.8 \cdot 10^{-15}$ N/m. The numbers over the signals correspond to the number of α-D-glycopyranose in a cluster. The number of cluster kinds (signals) in the investigated mass ensemble amounts to: 102 at 18:18, 59 at 19:13, and 104 at 20:12 (Central European Time, 16.08.2008). The distance between Jupiter and Earth is $655 \cdot 10^6$ km.

Since Jupiter's gravitational radiation only influences the hydrogen bridged bonds that stabilize the long-range order in polysaccharides, it was of interest to understand how much energy is needed to destroy one hydrogen bridged bond. On average, one repeating member (two members of α-D- glycopyranose) in starch possesses one active hydrogen bridged bond. With help, it interacts with the surroundings. For example, in a spiral of 12 α-D-glycopyranose, the number of hydrogen bridged bonds is equal to six ([6,32], scheme II). They keep the polymer chain in spiral form. In the investigated sample, the number of hydrogen bonds amounts to $N_A \cdot 0.5/323 = 9.32 \cdot 10^{20}$ (N_A – Avogadro constant) but Jupiter's gravitational energy influences the proton jumping through one hydrogen bridged bond and shall be $4.32 \cdot 10^{-21}$ J and $2.16 \cdot 10^{-21}$ J for the higher and lower limits, respectively.

The volume in which a proton jumping from one oxygen atom to another one proceeds through a physical vacuum is calculated to be $l \cdot \pi \cdot D^2/4 = 8 \cdot 10^{-41}$ m^3 where l is length of the proton jumping corridor $\sim 10^{-10}$ m and D is the nucleus' diameter (10^{-15} m). If we consider this volume (dissolving and condensing of proton [4]) as volume of an EC in a physical vacuum (Scheme III), then the density of the gravitation energy that is concentrated in the EC is equal to $5.5 \cdot 10^{16}$ kJ/m^3 and $2.75 \cdot 10^{16}$ kJ/m^3 for the higher and lower limits, respectively (compared with $\sim 10^{-12}$ kJ/m^3 [34]).

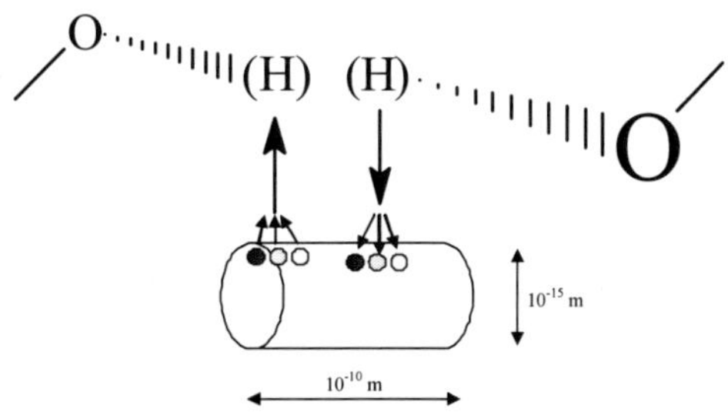

(III)

The diameter of the channel where proton jumping (gravitation string) in hydrogen bridged bonds of starch proceeds is more than 100,000 times smaller than the distance between dissolving and condensing protons. The proton jumping process seems to resemble the reverse high-frequency electrical unloading in gases, though with the difference that the driving force of the process is the energy of a variable gravitation field [4] that supplies the protons with the necessary energy for their dissolution in physical vacuum. The mass transfer, including stages of mass acceleration (dissolution) and mass breaking (condensation) must be paired with velocities that are much higher than light velocity. Only in such a way can WGR from starch be explained so that one can also understand the resonance between the planets' GR and the WGR of molecular matter. The resonance signals have been detected by GMS.

As shown in Figures 18.5 and 18.6, the proton jumping process (dissolution in vacuum) is forcibly broken by GR of Jupiter at its rise. This process has to be understood as increasing the drain of gravitational energy from starch that is connected with the more difficult dissolution of protons from OH-groups in the physical vacuum due to an energy deficit. Another explanation is based on a simple shift of hydroxyl groups under a planet's influence, which leads to an increased distance between hydroxyl groups and the destruction of hydrogen bonds. This is rather improbable because proton jumping is described for much longer distances [35]. Thus, the GR energy of Jupiter is comparable with the energy that is needed for proton dissolution (see Schemes I and IV). Here the hydrogen bridged bonds temporarily disappear (Scheme IV). The energy density of Jupiter's GR on Earth's orbit was found to amount to $(4\pm2)\cdot10^{16}$ kJ/m^3 [4]. This value can reflect the state of vacuum energy. It is reminded that modern people can obtain only $12\cdot10^3$ kJ/m^3 heat energy from the burning of 1 m^3 of hydrogen.

Observing hydrogen bridged bonds' destruction by IR (3300±300 cm^{-1}) was unsuccessful because it is a high-energy method (color noises) with a lower resolution.

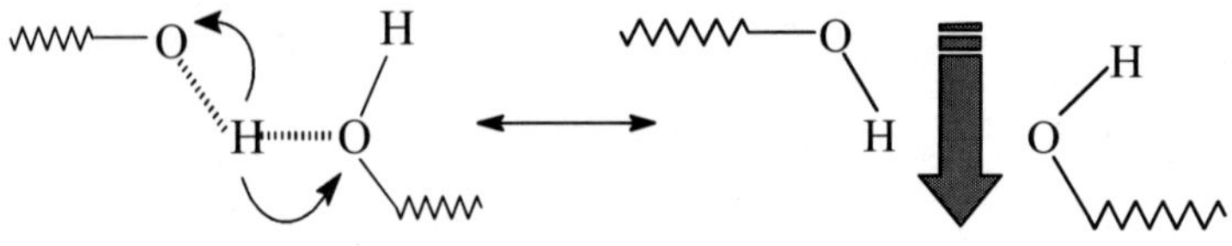

As is visible from Figure 18.6, the energy capacity of this cluster ensemble in starch is highly changed under a planet's influence. All clusters are represented in their expanded form (*+f*) apart from the α-D-glycopyranose members in which both forms are present but the collapsed one is a little more dominant (*-f*). At this moment when the Sun, Earth, and the Moon are on one line (full moon), the gravitation field tension reaches its maximum. Nevertheless, Jupiter's GR destroys the harmony of the cluster ensemble in starch. A possible explanation of this phenomenon could be that Jupiter is a gas giant consisting of more than 90% hydrogen. Furthermore, it has a 318 times more mass than the Earth. The dominating Jupiter as a large mass cluster seems to remove gravitational energy from the smaller mass ensemble from the Earth; it must be reminded that both mass concentrations consist of protons.

Based on the results from Figure 18.6, the risks for living organisms, bio-technologies, microprocessors, and computer networks as well as for organic syntheses and agriculture can be derived not only for one planet but also for the case when energy drains to planets that are on one line are combined. These risks can be systematized and predicted only with the help of astronomers, astrophysicists, chemists, and engineers. This monitoring was concluded to be of high importance for the Earth.

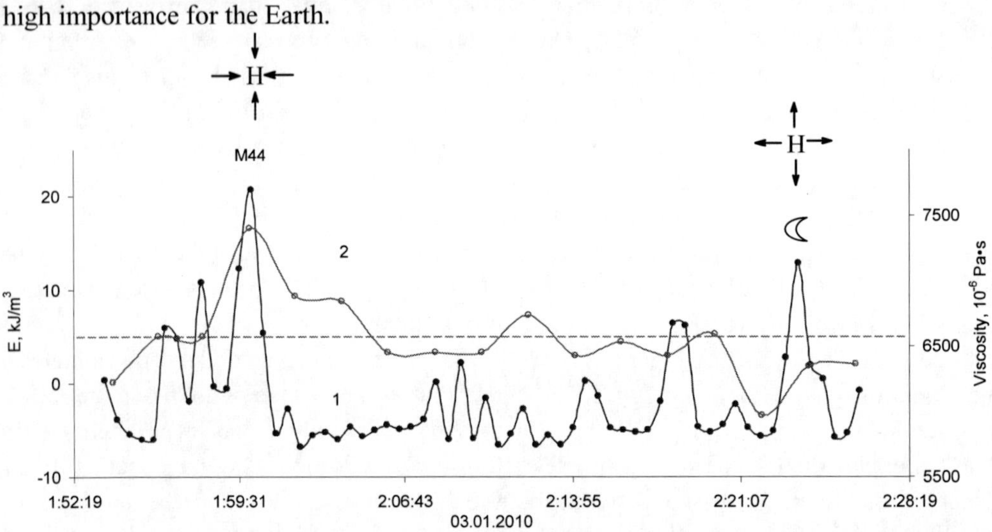

Figure 18.7. Energy change of a cluster ensemble (1) in liquid polyethyleneglycol (PEG-400) and viscosity of PEG-400 (2), 293 K. Dotted line – average viscosity. PEG-400 – pure for gas-chromatography from "Merck" (Germany).

We used celestial bodies to identify the dominating direction in the equation according to Scheme I. If the proton jumping in H-bonds is influenced by celestial bodies, then these effects must be detected by other methods, too, for example, by viscosimetry. The proton condensation from a PV was assumed to increase the H-bonds and connected with that, the viscosity of the sample. In contrast, the proton dissolving in a PV should decrease the H-bonds and consequently, the viscosity.

In Figure 18.7, it has been shown how the nebula M44 and the Moon affect the LRO in a liquid polymer.

As is visible, at 01:59:31 the energy change achieved a maximum right alongside the viscosity. This effect should be explained by the GR of nebula M44, which stimulated the proton condensation from the PV in H-bonds in PEG-400 and therefore increased the liquid viscosity. In contrast, under the Moon's influence, the viscosity decreased. Both events were realized when the sample, celestial bodies, and the Earth's rotating axis were in the same surface—the slide plane of gravitational proton resonance [4].

We observed that the energy drain of the gravitation field depends on the sample nature and on the planets' position [4, 5]. The Sun shows the highest mass between the celestial bodies, which are not far away from the Earth, and at its culmination, it removes gravitational energy (E_{GR}) from the samples, too. The star possesses the highest temperature, which is an indication for exothermic processes; one of them could be the thermo-nuclear synthesis of helium. On the other hand, planets surrounding the Sun are of much lower temperature. As shown in [4], this high temperature in the Sun, necessary for beginning the thermo-nuclear reaction (secondary process), is caused by exothermic proton condensation in the center of the star (primary process). The endothermic proton dissolution in a physical vacuum takes place in the surroundings of the Sun (planets). Here the gravitation energy transfer in physical vacuum into the Sun's center [4, 5] proceeds with a super light velocity ($c_g \sim 15.2 \cdot 10^{35}$ m/s). The Junge equation ($\lambda \geq D \cdot S/R$) (calculation of gravitation wave length according to interference maxima) and the disintegration energy for starch ($E = h \cdot c_g/\lambda = 2.6$ kJ/mol [36]) were applied for this. Owing to the Moon's acceleration, Laplace determined the gravitation velocity that he found to be around $6 \cdot 10^{16}$m/s even though Newton believed that this speed corresponds to only one moment. Unfortunately, the GR velocity does not permit risk monitoring, which makes the forecast more difficult.

Let us return to the PV's energy density, which is equal to $(4 \pm 2) \cdot 10^{16}$ kJ/m^3, and let us analyze it. Based on the gravitational radius of the proton amounting to $2 \cdot \pi \cdot 0.53 \cdot 10^{-10}$ m [21], the gravitational volume can be calculated as $154.69 \cdot 10^{-30}$ m^3. Then the gravitational energy in this space shall be equal to $154.69 \cdot 10^{-30} \cdot (4 \pm 2) \cdot 10^{16} = 850.78 \cdot 10^{-14}$ kJ and the force, which concentrates this energy in the gravitational volume, should amount to 56 N (F = E/L, where L is the diameter of space). The calculated force of 56 N is well agreed upon with the gold Volkov constant (58 N, [16, p. 49]). Thus, in nuclei, the gravitation interactions dominate over electromagnetic ones.

According to the 3D-model for EC (Figure 3.2.1), the PV possesses a heterogeneous structure that consists of a smaller vacuum building energy cluster ensembles. Here, EC cannot be called "graviton." The GR energy drain from the planets to the Sun and proton condensation in the center of the Sun was found to be accompanied by the release of gigantic energies that cause secondary effects, e.g. thermonuclear synthesis and the ejection of boson matter into space. Therefore, the "machine-pump" Sun releases matter as short radiation and particles into space (E_r + m) but this process is not constant (Figure 18.8). Circulation arises from energy and mass that is not constant and depends on the position of the planets and other mass concentrations in the Sun's surroundings.

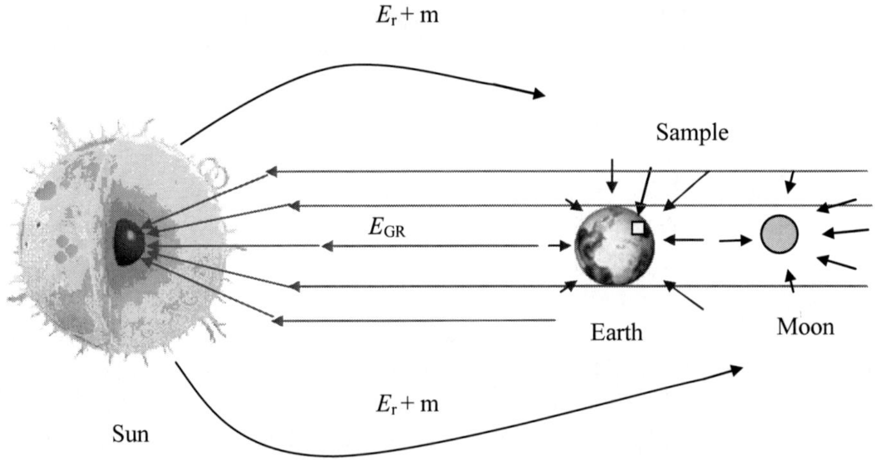

Figure 18.8. Circulation of energy and mass in the system Sun-Earth-Moon. E_{GR} – gravitation energy drain from dissolving protons of molecular planet matter (planet cooling) and counter flow (E_r + m) of energy and mass from exothermic processes in the Sun.

In this highly simplified model, the influence of the other planets is neglected. In [4], it was shown by the authors that the planets interact with each other by the resonance of their gravitational radiation and the fact that they attract each other like gigantic mass concentrations of an ensemble. Some experiments in this work also showed that the energy flow (E_{GR}) does not go evenly as was explained by the fact that this energy flow is influenced by the position of the planets and other mass concentrations in space as well as by their masses and dynamics. An indirect proof for this model could be the poorly reproduced tests with thermo chemical machines to remove gravitational energy [37]. On the other side, the phenomenon of water, the additional energy release from a water drop at its collision with a wall (turbine wheel), is an indicator for the destruction of a molecular cluster ensemble and its harmony with EC (Figure 3.2.1). This evidences the existence of a new molecular matter form—its cluster formation. According to Figure 3.2.1, the destruction of EC should lead to a disharmony of molecular cluster ensembles and to an increased kinetic energy of molecular clusters from which the possible influence of the planets' GR can be derived.

The energy and mass flow instability from the Sun was found to be a risk for chemical technologies especially for nanotechnology and works of nano machines on the Earth and in space (Figure 18.8). This flow instability influence is well known, for instance, in the interruption in electricity supply, micro electronics, computer networks, and chemical and bio-chemical processes as well as in psychological breaking up. Similar risks appear with gravitational energy drain from EC and molecular cluster ensembles (Figure 3.2.1) at the moment of the highest culmination of the Sun, the galaxy's center, and the planets even though the nature and properties of this energy are not yet completely clear. For instance, the T-Mobile (39 million clients) network "black out" on April 21, 2009 at 16:00 (CET) in Germany can probably be explained by the opposition of Saturn-Moon (179º) to the Earth (Saturn-Earth-Moon line).

Table 18.1. Correlation between jets crashes and planet constellation

Date	Place, time	Jet, type	Gravitation shot
20.08.08	Madrid, 14:45	MD-82	*Opposition:* Venus-Uranus *Conjugation:* Venus-Mercury
09.10.2009	Haiti, 19:00	CASA212 UNO	*Opposition*: Uranus-Mercury-Venus-Saturn *Conjugation:* Mercury- Venus-Saturn
27.11.08	Mediterranean, 06:41	A 320	*Opposition:* Saturn-Uranus
25.02.09	Shiphol, 10:40	Boeing 737-800	*Conjugation* Mercury-Jupiter-Mars
20.05.09	Jakarta, 02:59	Hercules	*Conjugation* Earth-Pluto-Center Galaxy
01.06.09	Atlantic, 30°W, 5°S	A330	*Conjugation* Earth-Pluto-Center Galaxy
10.12.2009	Atlantic, 30°W, 5°S	A330*	*Opposition*: Mars-Jupiter-Neptune
30.06.2009	Indian Ocean, 00:54	A 310	*Opposition* Jupiter-Venus *Conjugation:* Venus-Sun
05.12.1945	Bermuda triangle, 16:00	5 USA military jets**	*Opposition* Moon-Lilith
10.04.2010	Smolensk, 10:56	Tu-154	Jupiter in SPGP
12.04.2010	Tripoli, 06:00	A330-200	*Opposition:* Jupiter-Saturn, Uranus-VIRGOHI 21, Uranus-Saturn, Jupiter-VIRGOHI 21
22.05.2010	Mangolore (S.W. India), 03:00 CET	Boeing 737-800	*Opposition:* Jupiter-Saturn, UranusVIRGOHI 21, Uranus-Saturn, Jupiter-VIRGOHI 21, Venus-Pluto, Mars-Lilith

*Repeating case like 01.06.2009 at the same place in the Atlantic (sudden loss of jet steering without crash). [4, 5].

**Strong destruction of LRO in molecular matter especially in hydrosphere when Moon and Lilith (second center of the Moon's orbit) were in opposition leading to full navigational loss [4].

Forced excitement of the PV, EC, and later of MC is of high importance for nanotechnology, e.g., microprocessors, computer networks. In Table 18.1, some data on plane crashes over the two last years that were caused by a sudden loss of modern jet steering and their correlation with the planet's constellation shall be given. The special planet constellation led to conditions that favor gravitation shot between planet centers and crash places. Here the planet centers serve as gravitation lasers that work in opposition or in conjugation with other planets only (gravitation shot).

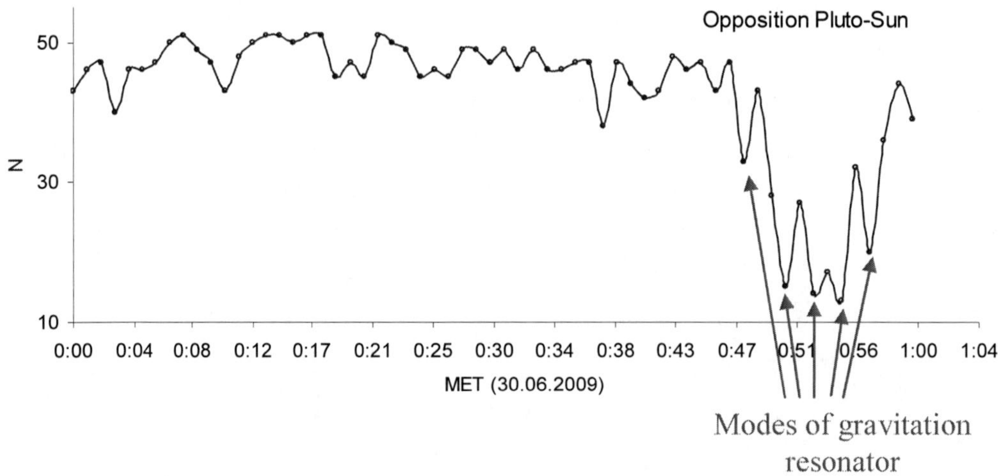

Figure 18.9. Number of cluster kinds in hydrogel at A 310 crash in the Indian Ocean on 30.06.2009 (Table 18.1).

In Figure 18.9, there shall be shown an example of how a gravitational shot influenced the number of cluster kinds in hydrogel. During the highest change of the *N* value Pluto and the Sun were found to be in opposition. At this time in the Indian Ocean, a crash of a modern jet was recorded. Remarkably, the curve behavior at the Pluto-Sun opposition is typical for the usual laser shot. Therefore, the protection of micro-processors and of the LRO in biomatrices of living organisms from strong gravitational excitation in the PV by planets is of high importance.

As the authors believe, these catastrophes appear when Earth regions are hit by the line of a gravitational energy shot (GES, e.g. neutrinos, §24), which connect the centers of large mass concentrations (celestial bodies). The results of GES can be seen in the atmosphere as optical "holes" in clouds (fallstreak, punch hole, see photographs at http://www.google.de/search?
q=fallstreak+hole&hl=de&biw=1400&bih=874&prmd=ivns&tbm=isch&tbo=u&source=univ&sa=X&ei=LNUyTtyYA8GfOsnv6PEL&sqi=2&ved=0CC8QsAQ).

Soft GES will be caused when dissolved protons from the Earth are removed to a large mass ensemble (celestial bodies' conjugation). Here in the ice crystals of clouds, the balance between condensed and dissolved (in a physical vacuum) protons of hydrogen bridges will be shifted to the protons' dissolution, which is accompanied with the lowering of adhesion forces between the molecules of water and, finally, with its evaporation (holes in the atmosphere).

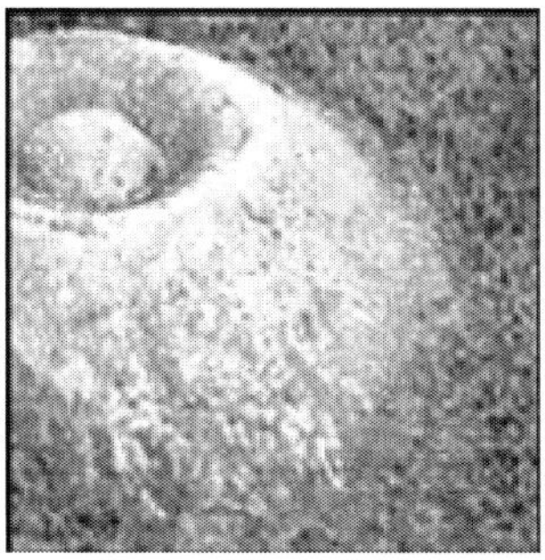

Figure 18.10. Photograph of the Patomskiy crater in East Siberia.

The Patomskiy crater in East Siberia could serve as an interesting example for GES, where the crater is strongly directed to the Earth's center (Figure 18.10) and where regarding the form, it corresponds to an energy shot. Concerning strong GES, they are found to show some similarities with a sudden lowering of the ground, however, the latter cannot be described by a circular perimeter (see http://pandoraopen.ru/2010-06-07/dyra-v-zemle-gvatemala/, Figure 18.11).

Figure 18.11. Crater in Guatemala – the result of GES?

The extraction of protons to larger mass ensembles as a physical model for the formation mechanism of GES is given in [4].

CONCLUSION

The long-range order in molecular matter at the cluster level, the state of energy clusters in crystalline substances, and the interaction of molecular and energy clusters were found to be the experimental base to investigate the heterogeneous structure of the physical vacuum.

By analyzing the planet's interaction with molecular matter on Earth, an idea about the energy of the physical vacuum and its distribution in the space can be developed.

REFERENCES TO 18

[1] Dolan, Brian P.; Szabo, Richard J. Dimensional reduction and vacuum structure of quiver gauge theory, arXiv.org, e-Print Archive, High Energy Physics-Theory, 2009, pp. 1-38, arXiv:0905.4899v1 [hep-th]. Publisher: Cornell University Library, CODEN: AHEPGK http://aps.arxiv.org/PS_cache/arxiv/pdf/0905/0905.4899v1.pdf.

[2] De Felice, Antonio; Suyama, Teruaki. Vacuum structure for scalar cosmological perturbations in modified gravity models, arXiv.org, e-Print Archive, Astrophysics, 2009, pp.1-16, arXiv:0904.2092v1 [astro-ph.CO]. Publisher: Cornell University Library, CODEN: AARSC7 http://aps.arxiv.org/PS_cache/arxiv/pdf/ 0904/0904.2092v1.pdf.

[3] Bopp, Fritz W. Is a rich "vacuum" structure responsible for fermion and weak-boson masses, arXiv.org, e-Print Archive, High Energy Physics—Phenomenology, 2009, pp. 1-18, arXiv:0903.1257v1 [hep-ph]. Publisher: Cornell University Library, CODEN: AHEPFJ http://arxiv.org/PS_cache/arxiv/pdf/0903/0903.1257v1.pdf.

[4] K.V. Zubow, A.V. Zubow, V.A. Zubow. Principles of gravitation stectroscopy.New form of molecular matter. Processes. Fields, Aist Handels- und Consulting gmbH, Berlin, 2010, electronic book (www.zubow.de), p. 854. (in Russian).

[5] Zubow K.V., Zubow A.V., Zubow V.A. Ensemble of Clusters – New Form of Molecular Matter, Risks and Chances. Zubow Equations. In Advances in Chemistry Research, vol. 5, 2010, pp. 107-145. Editor J.C. Taylor. Novapublisher. NY.

[6] Zubow K.V., Zubow A.V., Zubow V.A. Using of flicker noise spectroscopy for non destroyng analysis of nano structures, Zavodskaja Laboratorija. Diagnostics of materials, 2008, vol. 74, no. 9, pp.40-45. (in Russian).

[7] Zakharov V.V.; Brodskaya E.N.; Laaksonen A. Molecular dynamics simulation of methanol Clusters. .J. Chem. Phys., 1998, vol.109, pp.9487-9493.

[8] Daisuke Yokogawa; Hirofumi Sato; Shigeyoshi Sakaki. An integral equation theory for 3D solvatation structure: A new procedure free from 3D Fourier transform. Chemical Physics Letters, 2006, vol. 432, pp. 595-599.

[9] Lian Li; Ogawa T. Clusters and their properties in aqueous solutions of KDP, KCl and sugar. Journal of Crystal Growth. 2000, vol. 211, pp. 286-289.

[10] Zubow K.V., Zubow A.V., Zubow V.A. Cluster structure of liquid alcohols, water and n- Hexane. J. of Appl. Spectr. 2005, vol. 72, no. 3, pp. 321-328.

[11] Guang-Jun Guo; Yi-Gang Zhang; Ya-Juan Zhao; Keith Refson; Gui-Hua Shan. Lifetime of cagelike water clusters immersed in bulk liquid. A molecular dynamics

study on gas hydrate nucleation mechanisms. J. Chem. Phys. 2004, vol. 121, no. 3, pp.1542-1547.

[12] W. Horsthemke; R. Lefever. Noise-Induced Transitions. Theory and Applications in Physics, Chemistry and Biology. Edit. Hermann Haken, vol. 15; Springer-Verlag: Berlin, Heidelberg, NY, Tokio, 1984.

[13] Zubow A.V., Zubow K.V., Zubow V.A. Investigation of the distribution of water clusters in vegetables, fruits and natural waters by flicker noise spectroscopy. Biofizika, 2007, vol. 52, no. 4, pp. 585-592. (in Russian).

[14] Zubow A.V.; Zubow Kr. V.; Zubow V.A. Mechanism of Sodium Chloride Dissolving in Water and the Process of the Solution Aging. J. of chemical Industry (Sant.-Petersburg, Russia), 2007, vol. 84, no. 1, pp. 40-48. (in Russian).

[15] Zubow A.V.; Zubow Kr. V.; Zubow V.A. The Phenomenon of Water Clusters Distribution in Water and Solvated Clusters of Ion Pairs of NaCl in Solution Near the Wall. The Wall Effect. J. of chemical Industry. 2006, vol. 83, no. 6, pp. 300- 306. (in Russian).

[16] J.V. Volkov. Nonlineal Electrodynamics. 3-Ed. Sputnik. M. 2007. (in Russian).

[17] Van Dyke M. An Album of Fluid Motion. The Parabolic Press. Stanford, California, 1982.

[18] Dikarev S.N. Laboratory study of resonant regimes of motions of homogeneous fluid with free surface in tilted rotating container. Izwestija AN USSR Fisika Atmosphery i Okeana. 1990, vol. 26, no. 9, pp. 982-992 (in Russian).

[19] Manin D. Yu., Chernousko Yu. L. An experimental study of the stability of a quasi two- dimensional jet flow produced in rotating fluid by sinks and sources. Izwestija AN USSR/Fisika Atmosphery i Okeana. 1990, vol. 26, no. 5, pp. 483-491. (in Russian).

[20] Yusuke Yamauchi, Hiromi Nakai. Hybrid approach for ab initio molecular dynamics simulation combining energy density analysis and short-time Fourier transform: Energy transfer spectrogram. J. Chem. Phys. 2005, vol. 123, pp. 034101-1- 034101-9.

[21] Oldershaw, Robert L. Amherst College, Amherst, MA, USA. arXiv.org, e-Print Archive, Physics, 2008, pp.1-7, arXiv:0803.1197v1 [physics.gen-ph]. Publisher: Cornell University Library, CODEN: APRHCB http://aps.arxiv.org/ftp/arxiv/papers/0803/0803.1197.pdf.

[22] Braginsky V. B. Gravitational radiation. Fisitsheskij Enziclopeditsheskij Slowar. M.: Sovijet Enyiklopedy. 1984. p.137.

[23] Zubov A.V, Zubov K.V, Zubov V.A. Mechanism of sodium chloride diossolving in water and the process of the solution aging. Russian Journal of Applied Chemistry. 2007, vol. 80, no. 7, pp. 1249-1255.

[24] Loeb, Abraham. Observable signatures of a black hole ejected by gravitational radiation recoil in a galaxy merger. Astrophysics, 2007, pp.1-4, arXiv:astro-ph/0703722. LNASFZ, http://aps.arxiv.org/PS_cache/astro-ph/pdf/0703/0703722.pdf.

[25] Logathev J. I. Radiational layers of Earth. Fisitsheskij Enziclopeditsheskij Slowar. M.: Sovijet Enyiclopedy. 1984, p. 605.

[26] Wilcox, Walter. Thomas-Fermi statistical models of finite quark matter. Nuclear Physics A. 2009, vol. 826, no. 1-2, pp. 49-73.

[27] Urban, Federico R.; Zhitnitsky, Ariel R. The cosmological constant from the ghost. A toy model, arXiv.org, e-Print Archive, High Energy Physics—Theory, 2009, pp. 1-13,

arXiv:0906.2165v1 [hep-th]. Publisher: Cornell University Library, CODEN: AHEPGK http://aps.arxiv.org/PS_cache/arxiv/pdf/0906/0906.2165v1.pdf.

[28] Carneiro, Saulo; Tavakol, Reza. On vacuum density, the initial singularity and dark energy, arXiv.org, e-Print Archive, Astrophysics, 2009, pp.1-8, arXiv:0905.3131v1 [astro-ph.CO]. Publisher: Cornell University Library, CODEN: AARSC7 http://aps.arxiv.org/PS_cache/arxiv/pdf/0905/0905.3131v1.pdf

[29] Klinkhamer, F. R.; Volovik, G. E. Gluonic vacuum, q-theory, and the cosmological constant. Physical Review D: Particles, Fields, Gravitation, and Cosmology. 2009, vol. 79, no. 6, pp. 063527/1-063527/8.

[30] Zhang, Wenzhuo. Particle-like property of vacuum states, *arXiv.org, e-Print Archive, Physics.* 2009, pp.1-12, arXiv:0904.4021v1 [physics.gen-ph]. http://aps.arxiv.org/PS_cache/arxiv/pdf/0904/0904.4021v1.pdf.

[31] Takahashi, Fuminobu. A possible solution to the strong CP problem. Progress of Theoretical Physics. 2009, vol.121, no. 4, pp.711-725.

[32] Zubow Kr. V.; Zubow A.V.; Zubow V.A. Low Frequency Movement of Cluster-12 in Potato Amylopectin During Growth. Influence of white Noises. Journal Chemistry of raw plant material. Altay State University (Himija rastitel'nogo syr'ja). 2009, no. 2, pp. 81-88. (in Russian, http://chem.wood.ru or http://www.chem.asu.ru/chemwood).

[33] Vasilief S.A. On the some field of the Earth in view of its internal movements. In Degassing of the Earth Geodynamics: Geofluids, Oil, Gas and their Paragenesis. Proceedings of the All- Russian Conference, April 22-25, 2008, Moscow. Moscow: GEOS, 2008, pp. 576-579 (in Russian).

[34] Annett J.F. Superconductivity, superfluids, and condensates. In. Particle Astrophysics; Editor, Perkins D.H. ; Second Edition. Oxford University Press, 2009, pp. 120-123.

[35] Isaev A. N. Co-operation interactions of hydrogen bonds in the processes with proton jumping and water molecules participation. Modelling of biochemical systems. Russian Chem. Journ. (J. of russian chem. soc. Mendeleev D.I.), vol. LI, no. 5, 2007, pp. 34 -48. (in Russian).

[36] Kozlov S.S., Krivandin A.V., Shatalova Olga V., Noda T., Bertoft E., Fornal J., Yuryev V.P. Structure of Straches Extracted from near-isogenic Whaet Lines. Part II. Molecular organization of amylopectin clusters. Journal of Thermal Analysis and Calorimetry. 2007, vol.87, no. 2, pp. 575-584.

[37] Williams R.S. “How We Found the Missing Memristor", IEEE Spectrum, December, 2008, pp. 25-31.

18.1. Turbulences in a Physical Vacuum Generated by the Moon’s Motion

A method recording the heterogeneous structure and energy of a PV is described in one of our earlier works [1]. Furthermore, in [2], it is suggested how distances to other gravitational excitations in a PV, which are caused by celestial bodies, can be determined. Moving objects are described by the laws of hydrodynamics and they are available from the van Dyke atlas [3]. For celestial bodies, it is reasonable to imagine these invisible gravitational excitations in a PV, too. Their movement, e.g. of the Moon (~1 km/s), must cause special vortices in a PV but for its multiple rotations around the Earth, it can generate even a corridor of gravitational traces that are formed by the superposition of vortices. These vortices' superposition as stationary waves follows the Moon. Let us call the media in which the Moon moves as physical vacuum and its gravitational excitations as the vortices of dark matter (DM). The foregoing can be represented by the following model shown in Figure 18.1.1.

Figure 18.1.1. Model for the moving Moon in a PV in which traces of DM vortices are formed. Hypothetical first rotation around the Earth.

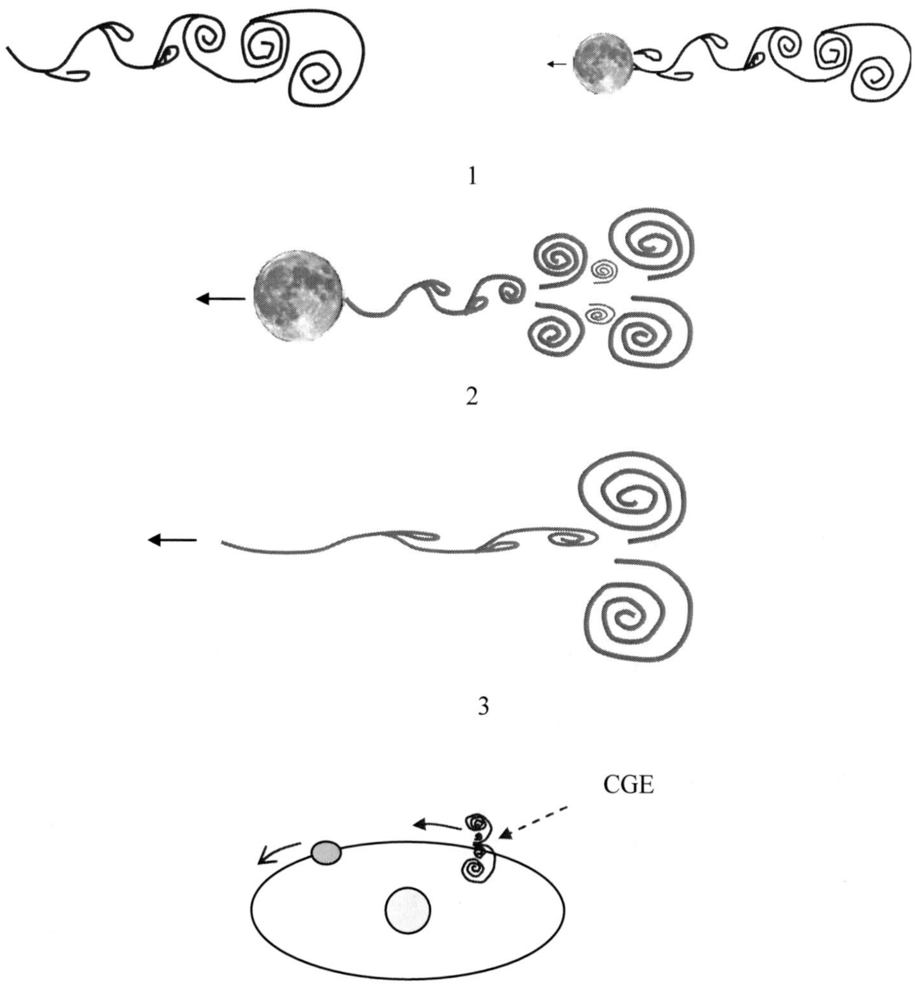

Figure 18.1.2. Model of CGE formation in a PV on the Moon's orbit. 1 – hypothetical Moon's impact on the turbulences, which were previously created by it, 2 – formation of new stable cumulative turbulences (gravitational anomaly, GA), 3 - cleavage of the Moon from GA, and 4 – top view.

Model

When the Moon impacts its first vortices for the second time impacts, then the secondary vortices penetrate the primary ones (1, Figure 18.1.2) leading to new whirls with cumulated energies from both events [4]. According to Maxworthy [5] in these new turbulences, stronger coherent structures should be expected. After multiple Moon rotations around the Earth, the new accumulated turbulences move as a unit behind the Moon, though they move slower than it (2, 3). This unit, as concentration of gravitational excitation (CGE) in a PV (4), resonates with CGE in the sensor space but only if the unit and sensor meet the slide plane of gravitational proton resonance (SPGPR, plane going through the Earth's rotating axis and the sensor place on the Earth's surface) [2].

The medium, in which the turbulences are formed, is a PV. The PV has to be seen as an impulse space of energy [6] characterized by a high density of energy ($\sim 5 \cdot 10^{19} J/m^3$, [1, 2]) while that of baryonic matter is very low ($\sim 10^{-29}$ g/m^3). The GA excitation by the Moon should increase each time it passes through the GA and will decrease until its next impact. According to this model, in the average traffic corridor of the Moon around the Earth, GA should move in the same direction but much slower. GA is expected to release gravitational radiation, which arises under rotational forces inside the vortices and thereby influences the dissolved form of protons in a PV [14]. This radiation can be recorded with the gravitational mass spectrometer developed by the authors.

Next, this model should be proved experimentally. For this, a location in space must be chosen that is on the Moon's orbit and follows the Moon (4, Figure 18.1.2). This location is found empirically and is called "Lilith" (http://en.wikipedia.org/wiki/Lilith).

The impact of the present work is to investigate GA using the gravitational mass spectrometer [2].

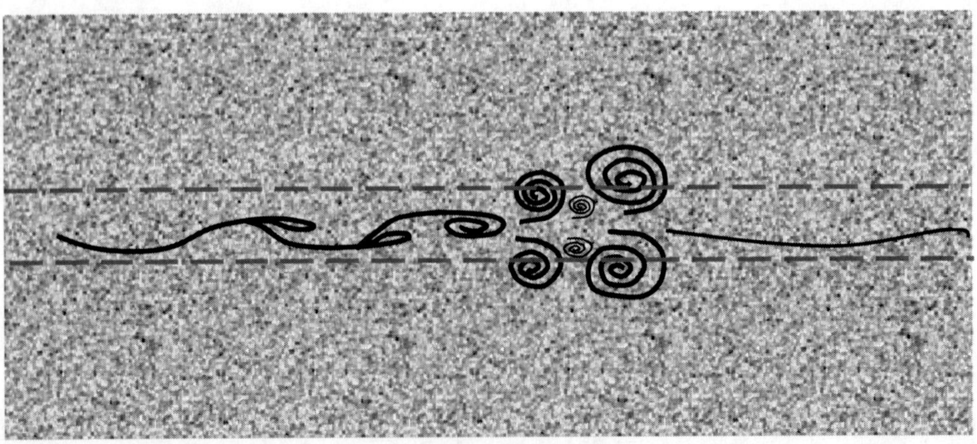

Figure 18.1.3. Model for GA in the impulse space of a PV. The hypothetical traffic corridor of the Moon is marked with dotted lines.

Agarose hydrogel (97 wt. % water) modeling the biomatrix is chosen as the investigation object (GMS-sensor). Here the GMS-sensor is placed into the agarose hydrogel and the measuring procedure of GMS is given in [7, 8, 9, 10]. The masses of molecular clusters and their frequencies are calculated according to the Zubow equations [11] using the Zubow constant, $6.4 \cdot 10^{-15}$ N/m. The long-range order in water at the molecular cluster level is analyzed during the period when the GA appeared in SPGPR in North Germany (53°34′54″N and 12°47′02″E). The GMS-sensor is placed into an earthed iron box that is protected from noises, heat, and mechanical fields at most [11]. The box itself is in a building far from industrial centers and anthropogeneous noise sources. To understand the method, some curves that reflect the GR energy flow (sum of cleaned signals) are given in [2, 11]. The molecular mass clusters are found to be formed in energy clusters of stationary gravitational waves of space and they reflect the state of the gravitational field in the sample space. To find the correct celestial bodies' position the program ZET 9 (www.astrozet.net) is applied; for the direction of the GMS-sensor to the GA, SPGPR is used [2, 12].

In Figure 18.1.4, the reaction of a water cluster ensemble in agarose hydrogel is given when the sensor and the GA hit SPGPR. The curve is characterized as a typical wave; the energy rise in the moment of GA appearance in SPGPR is followed by an adequate energy drain.

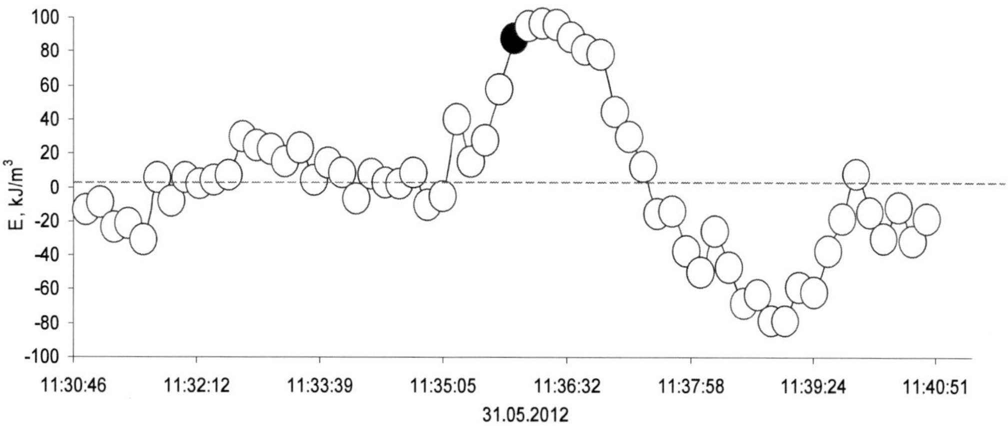

Figure 18.1.4. Energy of water cluster ensembles in hydrogel scanned when a gravitational anomaly from the Moon's turbulences in a PV meets SPGPR. The moment when the Black Moon (Lilith, www.astrozet.net) hits SPGPR is marked with a black point. The red dotted line denotes the evaporation energy of water (complete destruction of the water cluster in the ensemble $2.45 \cdot 106$, kJ/m^3).

The GA draws energy from the collapsed, high-energy water clusters [13] where the balance between collapsed and expanded clusters shall be shifted to the expanded forms (Figure 5). This process, however, is typical for the beginning, only then is the balance moved to the domination of collapsed clusters ($Д_c$ > 50 %).

Consequently, the GA is characterized by gravitational radiation, which promotes the dissolution of protons in hydrogen bonds in a PV [14] leading to cluster expansion in the water of the hydrogel sensor. As is visible from Figure 18.1.4 the rise of the part of collapsed clusters (arrow) is equivalent to their loss in the previous period (broad arrow). The process is comparable to the high and low tides of gravitational energy that are not caused by the Moon itself, but rather by its gravitational excitation in a PV. The foregoing can be thought of as balance between baryonic matter (BM) and matter in a PV [14]:

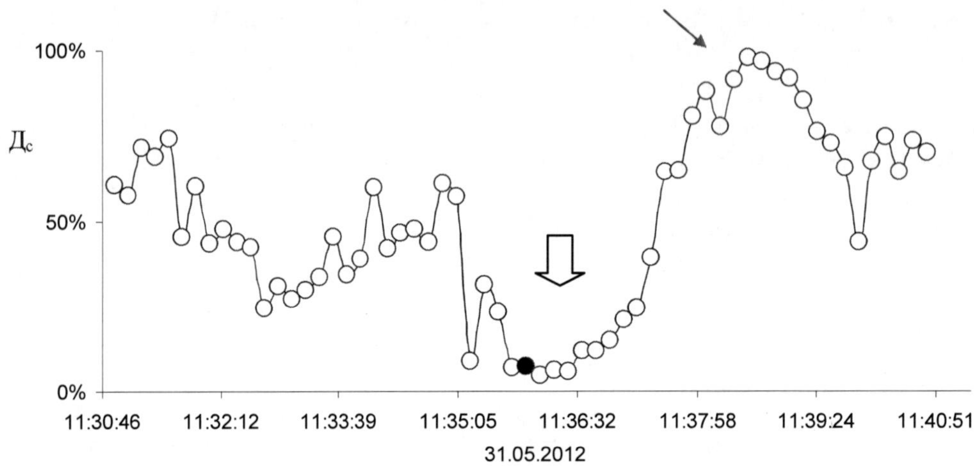

Figure 18.1.5. Part of collapsed clusters in GMS-sensor under GA influence. Black point denotes the moment when the GA (Lilith, www.astrozet.net) hits SPGPR.

$$PV \leftrightarrow BM$$

In the visible world the balance is shifted to BM:

$$PV \rightleftharpoons BM$$

In a PV, however, it is dissolved into BM:

$$PV \leftrightharpoons BM,$$

The dissolution of BM in the sensor PV (space between molecules) is accompanied with the transfer of the dissolved proton into a GA, where the reverse process takes place—proton condensation from a PV (in space of vortices). A similar phenomenon was previously reported by us in the investigation of the GR of the Sun and the gas giants [11, 12]. Consequently, the forced destruction of a PV structure by celestial bodies leads to the condensation of protons from a PV and thus to a release of the excess energy in the form of baryonic matter in PV vortices. This process, however, is reversible and the recovery of the PV structure leads to the slow dissolution of the baryonic matter in the disappearing GA vortices (Figure 18.1.9). Vortices can be considered as a local fast rotation of matter in a PV that is under the influence of torsion forces; baryonic matter is able to condense into them [15].

At the moment of the Moon's entry into a CGE (May 19, 2012), the curves presented in Figures 18.1.4 and 18.1.5 are not characterized by extremes, but for some signals of individual clusters, in GR, the "windows" of formation dynamics of CGE vortices in a GA can be followed. In Figure 18.1.6, the reaction of a fragmented water cluster on a GA is further shown with the resonance hierarchy of events proceeding in the GA. These events confirm the model of the vortices' development in a PV as stationary waves following the Moon (Figures 18.1.1 and 18.1.2). The reaction of the sensor water clusters on the initial vortices is weak (left of the arrow). On the following ones, however, it will be stronger (right of the arrow).

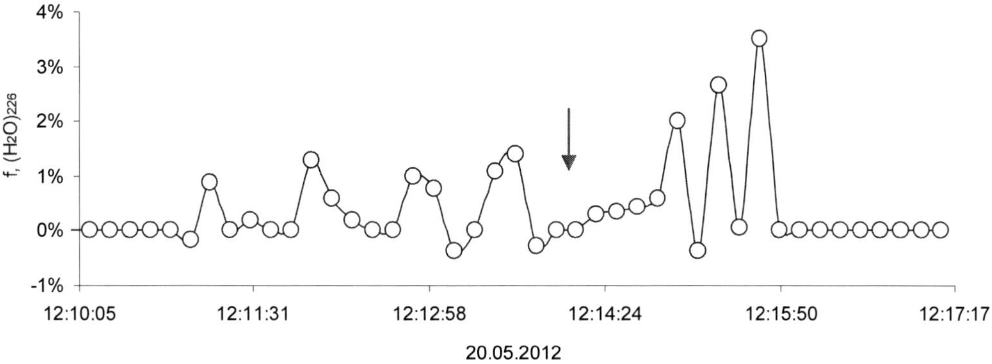

Figure 18.1.6. Reaction of $(H_2O)_{226}$ on a GA (arrow) immediately after the Moon passed it. Comments in the text.

The influence of the GA on the water cluster density is different, so $(H_2O)_{226}$ is loosened (positive *f* values) while $(H_2O)_{25}$ is strongly compacted. That means there are at least two mechanisms regarding how the protons of hydrogen bridges between water molecules in the sensor react to GR from a GA. Scanning the GA three days later showed no signals of $(H_2O)_{226}$, however, new signals, such as $(H_2O)_{4241}$ (1,131 Hz, Figure 18.1.7) did appear. According to the shape of the curve in Figure 18.1.6, the signals are divided in two groups that serve as a proof for the model given in Figures 18.1.2 and 18.1.3. Based on Figures 18.1.2 and 18.1.3 as well as the missing signals of cluster $(H_2O)_{226}$ (Figure 18.1.6), the gravitational perturbations of a vortex and its aging will be modeled. Using the integral characteristics of the water cluster ensembles in the hydrogel sensor, however, no clear answers resulted because of the low signal intensity of the GA.

The GA scanning over a long period shows that the gravitational energy excitation spectra of water clusters in an ensemble are not constant. This instability depends on the fast movement of large rain clouds and GR flows between other celestial bodies. Figure 18.1.8 shows an average resonance energy spectrum of the GA obtained from the results of a daily scanning between May 20 and June 20, 2012.

As shown in the Figures 18.1.8...18.1.10, the GA is found to age. This aging process is characterized by a plateau from the 5th to 20th days after the energy of the water cluster ensemble strongly decreases, obtaining a negative energy value. This indicates that the condensation process from the PV (+E) proceeds reversibly while being connected with dissolution of the baryonic matter in the PV (-E). It has to be mentioned that these processes are allowed by the PV theory [15]. Furthermore, this rehabilitates the German astronomer Georg Waltemath, who discovered in the presence of a group of independent postal officials in 1898, the “Black Moon” against the background of the Sun (http://en.wikipedia.org/wiki/Other_moons_of_Earth#Waltemath.27s_moons). The PV condensate in GA vortices seems to consist of protons as well as of hydrogen atoms that are partially absorbing the sunlight. The astronomer, however, mistakenly viewed it as a new Earth satellite.

After the Moon's “first” collision with the GA (May 20, 2012), the energies remarkably changed and approached the values of June 13 (the Moon's “second” collision with the GA). These results are proof of the GA model given in Figures 18.1.2 and 18.1.3.

The splitter cluster composed of 226 water molecules reacts on the Moon's second collision (June) with the GA, as it did in May (Figure 18.1.10). In this cluster, the expanded structures begin to dominate under the GR influence.

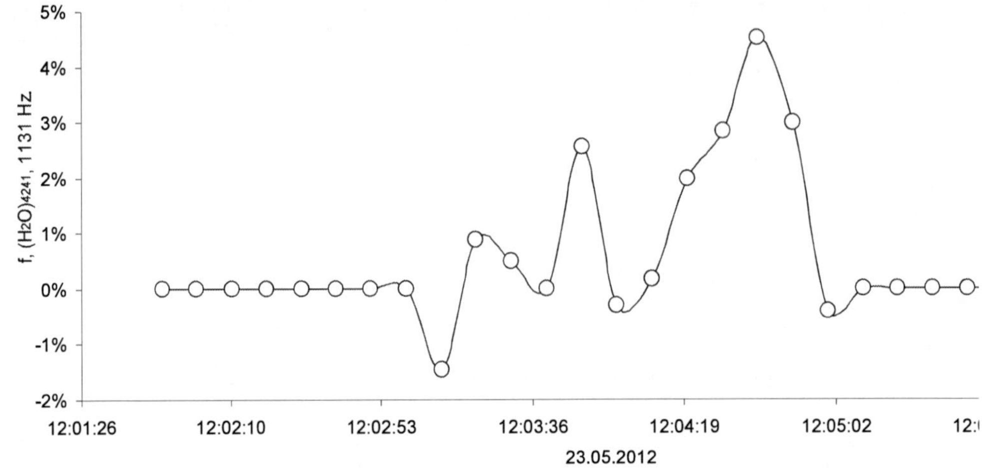

Figure 18.1.7. The reaction of cluster $(H_2O)_{4,241}$ when the GA meets SPGPR.

A broad range of vortices in the turbulent region of the GA leads to a large number of water clusters in the sensor reacting to it, where in some clusters expanded structures dominate and in other ones, the collapsed ones do. The main water clusters in the ensemble, however, are not explicitly defined. This fact is also confirmation for the GA model in Figure 18.1.2.

On May 18, 2012, the Sun was in opposition to the center of our galaxy (GC, 179.4°). According to [16], this mass constellation forms a neutrino flow directed to the GC. Due to the fact that this neutrino flow is strong, GR from the GA is overlapped (Figures 18.1.4 and 18.1.8) where some clusters show the GR flow from the Sun when the GA hits SPGPR (Figure 18.1.11). This "window" of gravitational radiation is clearly visible for a large water cluster of micelle origin.

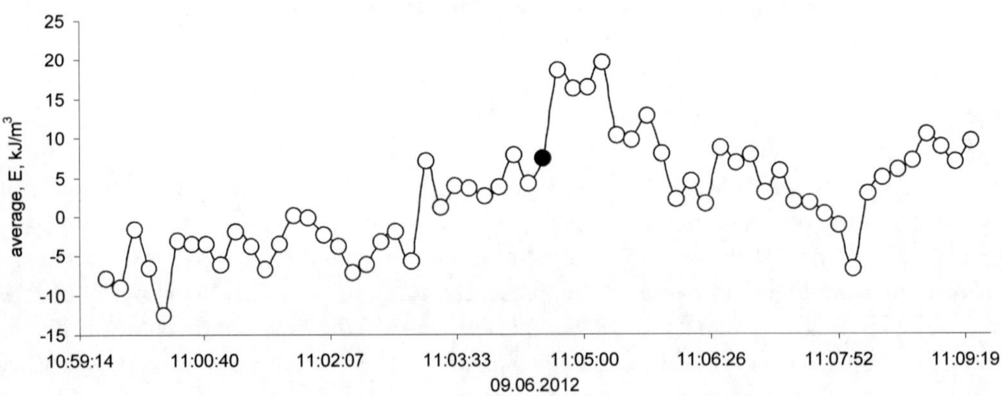

Figure 18.1.8. Average energy of water cluster ensemble in hydrogel at GA scanning from May 20 to June 20, 2012 (June 9 was taken as a base point). Scanning time – 10 minutes every day. Black point denotes the moment when the GA (Lilith, www.astrozet.net) hits SPGPR.

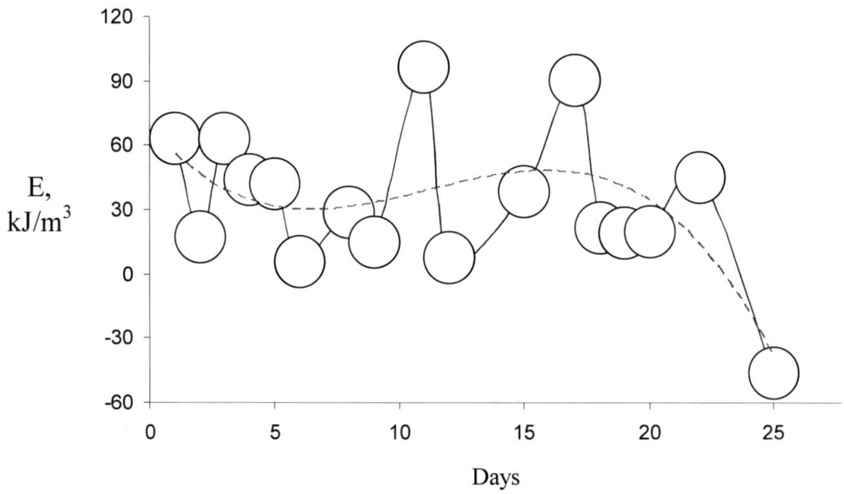

Figure 18.1.9. Reaction of water cluster ensembles in the sensor on GR weakening from Lilith GA (red dashed line – smoothing function: $E = -0.045t^3 + 1.5t^2 - 14.8t + 68$). Measurements started on May 20, 2012 (first point on the left), only the maximum values in the range of ±2 minutes from the hypothetical position of Lilith in the sky were taken into account, t – time.

Upon the appearance of the Earth in the GR stream from the Sun to the GC, the signals of the base water cluster ($(H_2O)_{12}$ [17, 18]) are characterized by a diffuse minimum of the curve (13:10:10), whose course is described by the smoothing function $f = 2316.9t^2 - 2545.8t + 699.31$, where t is time. This cluster ensemble behavior could indicate that GR flows from other celestial bodies strongly influence the structural heterogeneity of the GA in a PV.

Recently a cloud of dark matter in the constellation of Scorpius (http://www.eso.org/public/news/eso1303/) was also discovered.

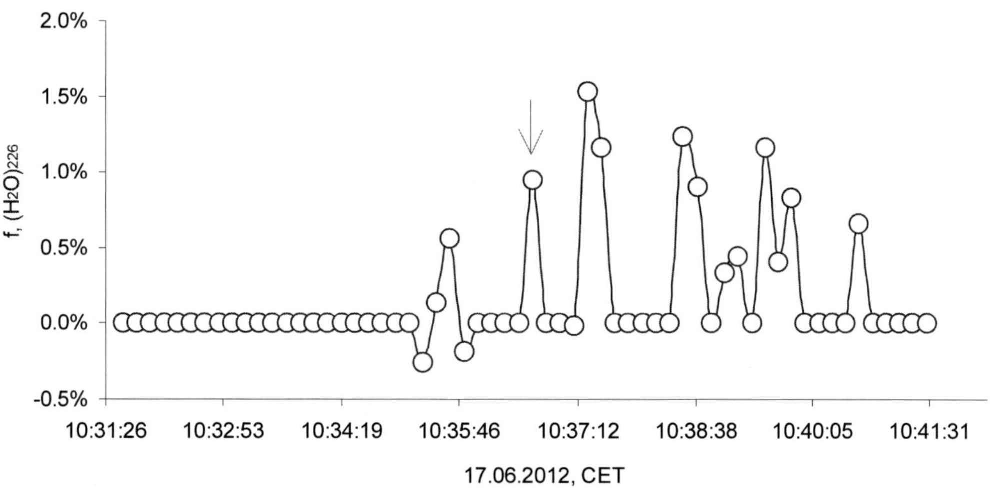

Figure 18.1.10. The reaction of cluster $(H_2O)_{226}$ on GR from a GA after the passage of the Moon (arrow).

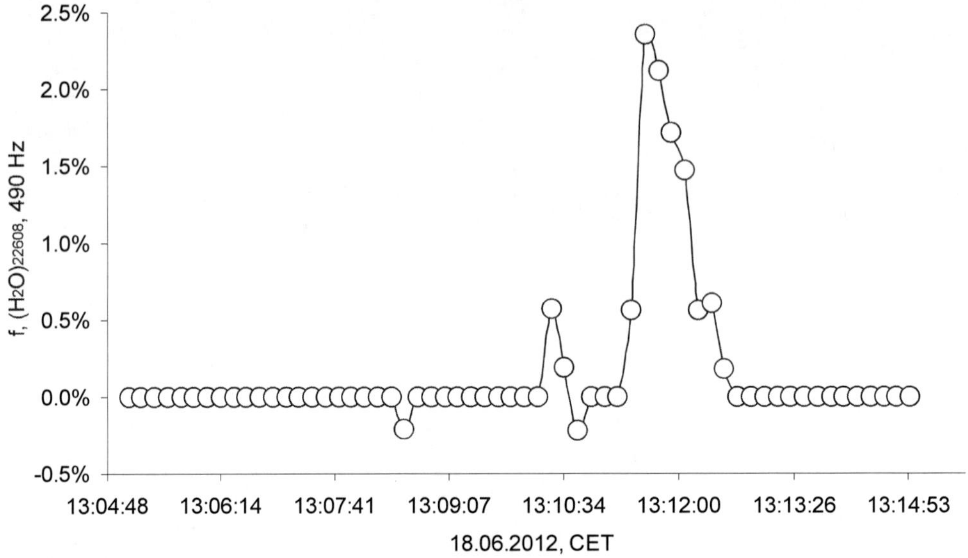

Figure 18.1.11. Reaction of the large water cluster (490 Hz) in agarose hydrogel on GR flow from the Sun, which is directed to our galaxy's center (opposition of 179.4°).

Conclusion

The Moon's orbiting of the Earth is characterized by the turbulences of dark matter in a physical vacuum.

Vortices in turbulences of the physical vacuum influence the balance between the dissolved form of the proton in the physical vacuum and its baryonic form. This shift of balance is accompanied by gravitational radiation that changes the long-range order of water in the Earth' hydrosphere.

Vortices in turbulences of the physical vacuum from the Moon's orbiting form a stable gravitational anomaly; its celestial coordinates match those of Lilith.

Thus, Lilith (Black Moon) is concluded to influence the hydrosphere and biosphere of the Earth, leading to a changed behavior within living organisms.

References to 18.1

[1] Zubow K., Zubow A. V., Zubow V. A. Experimental Platform for the Investigation of the Structural Heterogeneity of the Physical Vacuum. Vacuum Energy Risks and Chances. In Horizons of Physics. Editor A. Reimer. NY. 2012, vol. 277, pp. 87-110. https://www.novapublishers. com/catalog/product_info.php?products_id=29310.

[2] Zubow A., Zubow K., Zubow V. A. New Experimental Method for Determination of Distances to Celestial Bodies and their Masses. In Horizons of Physics. Editor A. Reimer. Nova Science Publishers, Inc. NY. Vol. 279. 2012, in print https://www.nova publishers.com/cata-log/product_info.php?products_id=34674.

[3] Milton van Dyke. An album of fluid motion. The parabolic press, Stanford, California, 1982.

[4] Yamada H, Matsui Tatsuya. Preliminary study of mutual slip-through of a pair of vortices. Phys. Fluids. 1978, vol. 21, pp. 292-294.

[5] Maxworthy T. Comments on Preliminary study of mutual slip-through of a pair of vortices. Phys. Fluids. 1979, vol. 22, no. 1, p. 200.

[6] Volkov, Y. Nonlinear electrodynamics. 3rd ed. Moscov. Sputnik. 2007. p. 198 (in Russian).

[7] Zubow K. V., Zubow A. V., Zubow V. A. Using of flicker noise spectroscopy for non destroying analysis of nano structures. Zavodskaja Laboratorija. Diagnostics of materials. 2008, vol. 74, no. 9, pp.40-45. (in Russian).

[8] Zubow K. V., Zubow A. V., Zubow V. A. The Dynamics of Low Frequency Movements of Molecular Clusters in the Hardening Process of Epoxide Resins. Chem Promislennost Segodnja. 2008, no. 9, pp. 12-21. (in Russian).

[9] Zubow K., Zubow A. V., Zubow V. A. Water Clusters in Plants. Fast Channel Plant Communications. Planet Influence. Journal of Biophysics Chemistry. 2010, vol.1, no.1, pp.1-11.

[10] Zubow K. V., Zubow A. V., Zubow V. A. Cluster structure of liquid alcohols, water and n-Hexane. J. of Appl. Spectr. 2005, vol. 72, no. 3, pp. 321-328.

[11] Zubow K. V., Zubow A. V., Zubow V. A. Ensemble of Clusters – New Form of Molecular Matter, Risks and Chances. Zubow Equations. In Advances in Chemistry Research, vol. 5, 2010, pp. 107-145. Editor J. C. Taylor. Nova Science Publishers, Inc. NY. https://www.novapublishers. com/catalog/product_info.php?products_id=27769.

[12] Zubow K., Zubow A. V., Zubow V. A. Scanning of the Sun and other celestial bodies with help of gravitation spectroscopy. Optic and Photonics Journal, 2011, № 1, pp.15-23.

[13] Zubow A., Zubow K., Zubow V. A. Molecular Water Clusters in River and Lake of Northern Germany and their Dynamics. Celestial Bodies Influence. Physical Chem., 2012, vol. 5, no. 5, in print. USA.

[14] Zubow K., Zubow A. V., Zubow V. A. The Phenomenon of Proton Dissolving in Vacuum and of Proton Condensation from Vacuum. Two Forms of Protons, Structure of Nuclei, Electrons and Atoms. J. of modern physics. 2010, vol.1, no.1, pp.175-184.

[15] Shipov G. I. Theory of physical vacuum. Theory, Experiments, and Technologies. Moscow, Nauka, 1997. (http://publ.lib.ru/ARCHIVES/ SH/SHIPOV_Gennadiy_Ivanovich/_Shipov_G.I.html).

[16] Zubow K., Zubow A. V., Zubow V. A. New gravitational channel of communication. Signals of ETI. In Horizons of Physics. Editor A. Reimer. NY. 2012, vol. 280, in print, Chapter ID: _8738_Horizons in World Physics. USA.

[17] Bogdanov E. V., Manturova G. M. Equiclaster model of water. Biomedicinskaja Radioelektronika. 2000, no. 7, pp.19-28. (in Russian).

[18] Annika Lenz, Lars Ojamäe. On the stability of dence versus cage-shaped water clusters: Quantum-chemical investigations of zero-point energies, free energies, basis-set effects and IR spectra of $(H_2O)_{12}$ and $(H_2O)_{20}$. Chemical Physics Letters. 2006, vol. 418, pp. 361-367.

Chapter 19

SCANNING OF THE SUN AND OTHER CELESTIAL BODIES WITH THE HELP OF GRAVITATIONAL SPECTROSCOPY

Even in the early nineties of the last century, a group of Russian scientists under the leadership of Professor Lavrentijev found out that a photo sensor and micro-organisms react to celestial bodies before they visibly appear. The reasons for these phenomena were not explained [1, 2, 3, 4] and they were additionally ignored by the scientific society. Using GMS, it is possible to analyze the gravitational radiation of celestial bodies in far space and to record the GR velocity [5, 6].

To better understand the structure and properties of the Sun's GR (directed from celestial bodies to the Sun [6, 7]), GR was scanned with the help of GMS; explaining these results is the aim of the present chapter.

Agarose hydrogel (97 wt. % water) modeling of the biomatrix was chosen as the investigation object (GMS-sensors). Here the GMS-sensor was placed into the liquids; the measuring procedure of GMS was given in publications [8, 9, 10, 11] and in §4.3. The masses of the molecular clusters and their frequencies were calculated according to the Zubow equation [11] using the Zubow constant, $6.4 \cdot 10^{-15}$ N/m. The long-range order in the GMS-sensors at the molecular cluster level was analyzed during the period when the celestial bodies appeared in SPGPR [5] in North Germany (53°34′54″N and 12°47′02″E). The GMS-sensors were placed into an earthed iron box that was protected from noises, heat, and mechanical fields (Figure 4.0.1). The box itself was in a building far from industrial centers and anthropogeneous noise sources. To understand the method, some curves that reflect the GR energy flow (sum of cleaned signals) are given in [6]. Molecular mass clusters (nuclei concentrations) in the GMS-sensors were found to be formed in energy clusters of the stationary gravitational waves from space and they reflect the state of the gravitation field in the sample space. The gravitation energy flow of celestial bodies resonates with the weak gravitational radiation (WGR) of protons in the sensor, which shall be recorded. The algorithm for the signal extraction is given in [10] and in Figure 4.2.2. To find the correct celestial bodies' position, the program ZET 9 (www.astrozet.net) was applied.

In Figure 19.1, the GMS-scanning results of the Sun and Mercury at the moment when they appeared in SPGPR are shown.

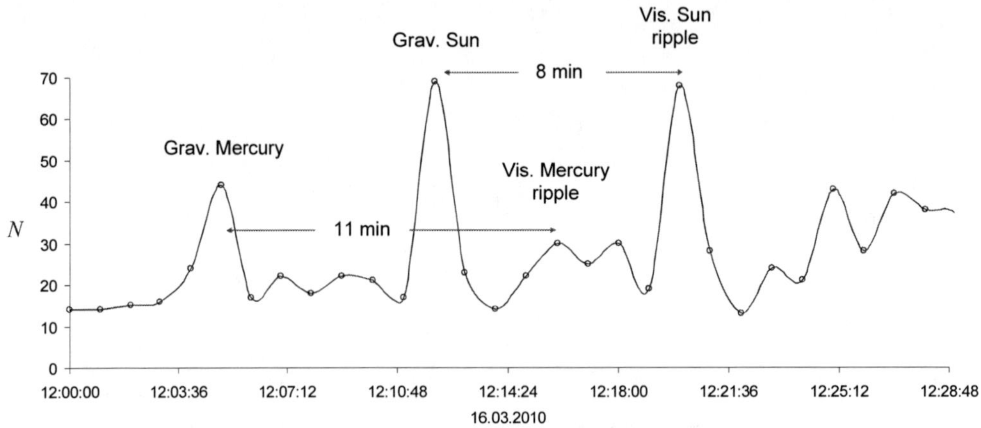

Figure 19.1. Reaction of the number of water cluster kinds (N) in agarose hydrogel on GR of the Sun and Mercury at their appearance in SPGPR. Distance to Sun = $1.49 \cdot 10^8$ km and to Mercury = $2.02 \cdot 10^8$ km. For a cluster mass ensemble up to 3 million Daltons.

As is visible, N achieved extreme values at moments when the visible and gravitational signals of the Sun and Mercury appeared in SPGPR. That agreed with the two-component GR structure: main gravitational wave and a ripple caused by the main wave. Here the ripple spread with light velocity [12].

To discuss the signal character in the area of the Sun's gravitation in detail, a higher GMS resolution has been applied (Figure 19.2). According to Figure 19.2, the Sun's GR was found to be heterogeneous and to be represented by a group of signals. Regarding the geometric parameters, the visual and gravitational Sun's discs were assumed to be the same. On the left and right side from the main signal there was observed a group of weaker signals. To understand the nature of these signals, we analyzed the average molecular mass (M_{GMS}), which is a more sensitive characteristic for a cluster ensemble up to 3 million Daltons (Figure 19.3).

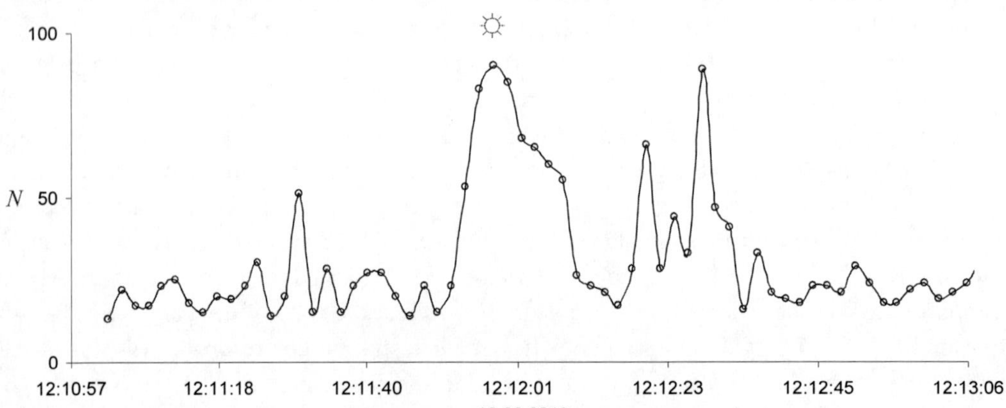

Figure 19.2. Signal structure in the area of the Sun's gravitation (Figure 19.1). Scanning step of 2 s. The center of the Sun's gravitation in SPGPR is marked with a ☼.

The M_{GMS} value reflects the effective interaction of clusters with their surroundings [6, 7], which is realized by hydrogen bonds (HB, regarding Pauling "naked" protons [13]). This interaction is shown in Figure 19.4 at the moment when the Sun's gravitation appeared in SPGPR (see Figure 19.3, too). At 50 times higher, M_{GMS} is an indication for the clusters' individualization, decreasing their interaction with their surroundings by intercluster HB, which practically disappeared completely at moment 1 (diapason II). The protons of HB were dissolved in the physical vacuum and they left the sensor room in the direction of the Sun [6, 7, 14]. Moment 1, however, cannot be ascribed to the center of the Sun's gravitation. At a higher resolution of 0.5 s as well as at moment 3, it became an accompanying function of moment 2, which for its part became the central one. The position of moments 1 and 3 to moment 2 is nearly symmetrical, possibly serving as evidence for a layered structure of the Sun's nucleus (onion structure).

A better understanding of the three diapasons is made easier when models are used (Figure 19.4).

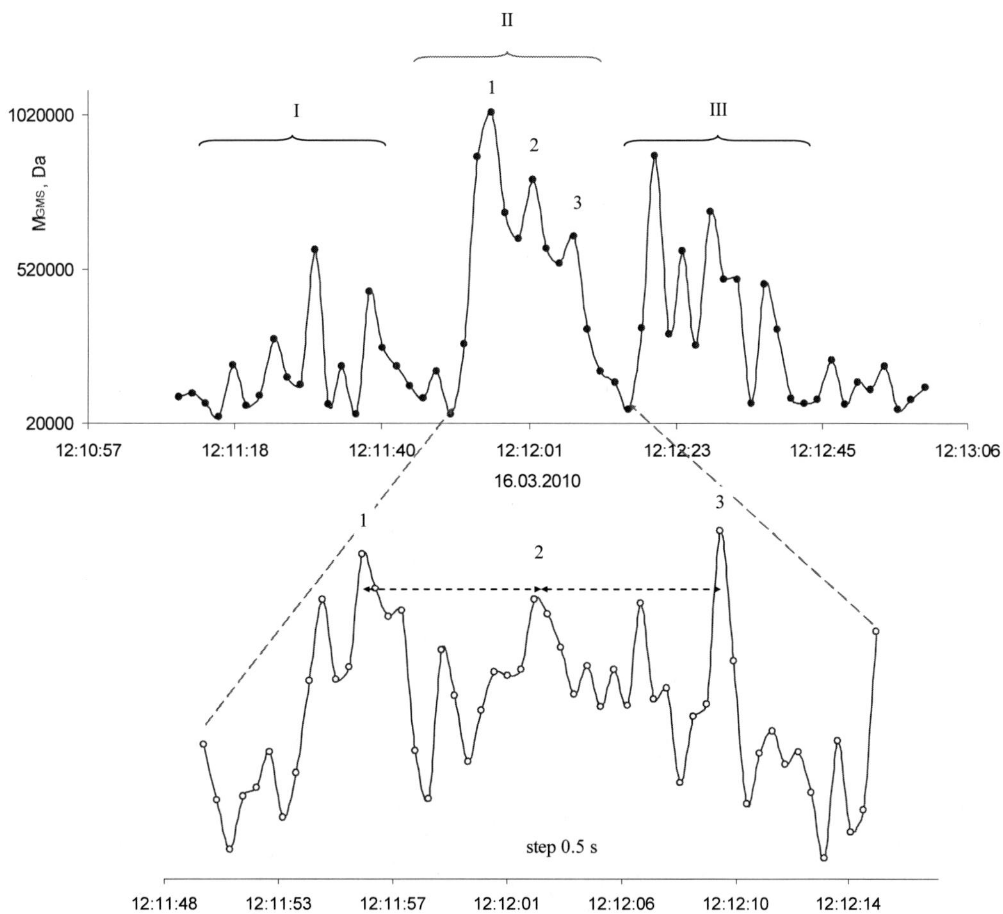

Figure 19.3. The change of the average molecular mass of water clusters in hydrogel at the time when signals of the Sun's gravitation appeared (Figure 19.1). The spectrum (Figure 19.2) was analyzed in 2 second and 0.5 second steps.

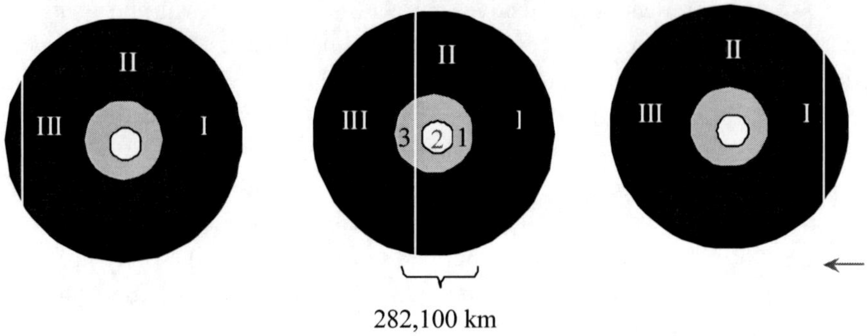

Figure 19.4. Model of the Sun's gravitational disk that appeared in SPGPR (white vertical line) and scanned the events in the time diapasons I, II, and III (Figure 19.3). The horizontal arrow means theSPGPR movement. The black, grey, and white discs were ascribed to the Sun's gravitation, the Sun's nucleus, and the Sun's sub-nucleus (2), accordingly.

The intercluster HB protons were dissolved most strongly in diapason II [6, 7] (Figure 19.3) where three signals were observed. These signals could be an indication of at least one sub-center (2). The size of the center (1+2+3) and sub-center (2) can be calculated roughly by applying the scanning times of the Sun's gravitation. They were found to amount to 26 and 4 seconds or 282,100 and 43,400 km, accordingly. Now we shall return to Figure 19.3. The signals of diapasons I and III seem to be of the same nature and they could be caused by energy fluctuations inside the nucleus. Here, however, the nucleus' surface has to be understood as a diffuse and unstable one. Processes that take place in the nucleus (diapason II) proceed in diapasons I and III, too. Analogous processes that take place inside the nucleus (diapason II) were supported to proceed in the Sun's sub-center (2) as well. To be explained alongside that, there exists a resonance interaction between the HB protons of the sensor and the sub-center matter, which could be an indication that both are of the same nature. Then, according to [15] and to this conclusion, one can imagine the matter of the Sun's sub-center as condensed quarks-gluon matter. The process of the quarks' condensation to new protons in the Sun's sub-center is characterized by an exothermic effect that is sufficient for the thermonuclear synthesis of helium [14]. The fluctuation processes in the nucleus has to be understood as the result of fluctuation dynamics of gravitational flows caused by a fast changing constellation of celestial bodies as well as by events far in space. Thus, the nucleus of the gravitational Sun was concluded to consist of neutrons (1 and 3, Figure 19.4). This sufficiently proves the position of quark stars as an intermediate stage from neutron stars to black holes [16].

As it was shown earlier, gravitational velocity was influenced by the constellation of celestial bodies from far in space, e.g. galaxies and their centers with regard to the Sun [5]. If the light velocity from the Sun to the Earth would be taken for the velocities' calibrations then it should be possible to calculate the velocity of the main gravitational wave in near space by applying, for instance, the Earth, Sun, and gaseous giants. For this, the time characteristics of the gravitational Sun and the influence of the gaseous giants on it shall be used. In Figure 19.5 it shall be shown how the energy of water cluster ensembles in agarose hydrogel behaves when the Sun and gaseous giants appear in SPGPR. Here the gaseous giants and the Sun almost lie on a line in opposition to each other: Sun-Earth-Jupiter-Uranus. As is visible, water cluster ensembles in the hydrogel sensor lose energy at the appearance of celestial bodies in

SPGPR which has to be understood as the gravitational energy drain from the sensor space to giant mass concentrations in the near space [6, 7]. These effects were characterized as endothermic ones and they therefore support the suggested model of HB proton dissolution in a physical vacuum [14].

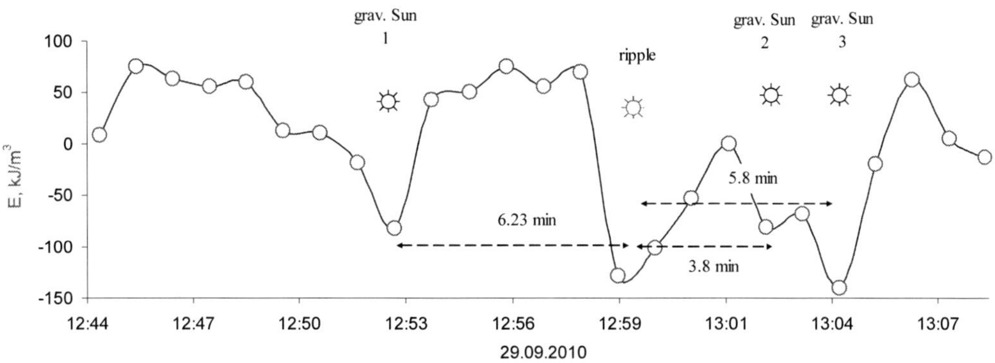

Figure 19.5. The influence of celestial bodies on the energy of a water cluster ensemble in agarose hydrogel (97 wt. % water, cluster ensemble mass up to 3 million Daltons). Distance to the Sun = 1.00173 au, to Jupiter = 3.96499 au, and to Uranus = 19.09932 au. Gravitational velocity: main gravitational wave = $1.15 \cdot 10^9$ m/s (gravitational Sun 1), gravitational ripple = $3 \cdot 10^8$ m/s (visible Sun, light velocity), gravitational Sun 2 (GR inhibited by Jupiter) = $2.06 \cdot 10^8$ m/s, and gravitational Sun 3 (GR inhibited by Uranus and partially by Jupiter) = $1.77 \cdot 10^8$ m/s. The GMS device was calibrated according to the evaporation energy of water at 298 K ($2.45 \cdot 10^6$ kJ/m^3) at which point the water cluster structure is completely destroyed.

As is shown in Figure 19.5, the gaseous giants influence the GR velocity. Depending on the position of the observer and of other surrounding mass concentrations, the GR velocity seems to change permanently in space. Large mass concentrations in near space were observed to destroy the main gravitation flow leading to the appearance of new GR with different velocities. Super light velocities of GR were calculated with a dependence on the opposition angle of the gaseous giants Jupiter and Uranus (Figure 19.6). As is visible in Figure 19.6, the curse of the curve can be described as a parabola with a minimum at an angle that does not correspond to 180° (opposition of the Sun and planets). The minimum is shifted a little to smaller angles (175...176°), which is explained by the fact that the gaseous planets, the Earth, and the Sun do not lie on a line. We observed that at destroyed planets' oppositions, the GR velocity was highly increased, e.g. at conjugation.

The generation of GR by celestial bodies has to be understood as a process in which gravitation lasers release the energy that was accumulated by those celestial bodies. After modeling this process the Earth was like a large gyroscope (Figure 19.7) and the atomic nuclei of the Earth like elemental gyroscopes. Here in the non-exited state, the elemental gyroscopes rotate around their own axes that are perpendicular to the rotation axis of the Earth. On the other side, the model of the rotating proton being analogous to the modern black hole model describes the formation of three to four stationary waves (quarks, Figures 18.2 and 19.7) where the quarks' number depends on the rotation velocity of the stationary wave inside the proton [6]. The formation of stationary waves in rotating isotropic liquids is already known well [17, 18, 19], which permits us to understand quarks as virtual particles or perhaps as energy clusters.

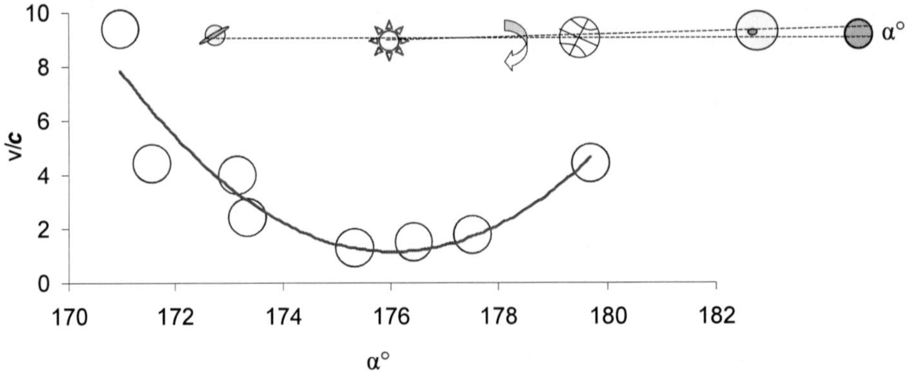

Figure 19.6. The influence of the opposition angle (α°) of Jupiter, Uranus, and the Sun on the ratio of gravitational velocity (v) to light velocity (*c*) (Figure 19.5). The results were obtained at oppositions of the Sun to the gaseous giants (Neptune, Jupiter, Uranus) as well as at the conjugation of Sun–Saturn between August 6, 2009 and September 29, 2010.

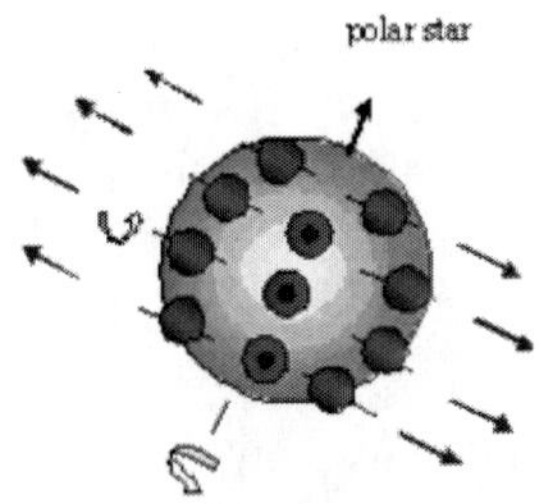

Figure 19.7. Structure model for approximate orientation of proton axes to the Earth's axis (right side). The energy clusters as stationary waves rotating around the proton center (axis) are shown as colored drops. The right model discusses the Earth as a large gyroscope and the protons as elemental ones. The gravitation energy flows are marked with blue arrows and blue points.

In this model, the protons were regarded as analogous to a "black hole" [20]. Here protons absorbed the gravitational energy from the surrounding physical vacuum. Then the energy shall be partially concentrated and released as an energy jet [6, 7]. The process strongly depended on the surroundings of the Earth and the process balance could be shifted to stronger oscillator ensembles, e.g. to celestial bodies that are larger than the Earth [5].

Thus, all Earth protons were activated by the rotation of the Earth where one part of the gravitation energy was released in the direction of the centrifugal force. This effect was observed at the rotation of simple bodies, too, §10.10. The second part of the gravitation energy was directed to the Earth's axis and concentrated there (right model in Figure 19.7). With the help of this model the existence of a gravitation laser was concluded to be possible in principle. Here, for the gravitational energy supply of the laser, the energy concentrated near the Earth's axis as well as that which arose by the resonance interaction of the sensor WGR with GR of celestial bodies in SPGPR were used. If this model is right, then gravitational laser shots have to be expected. That means we have to find experimental signals (modes) of a gravitational resonator that are typical for all laser radiations [21]. In one of our earlier works, we recorded and described [22] such a gravitational shot.

It would be reasonable that the GR release shall be highly increased when SPGPR hits a celestial body that also appears in the ecliptic, e.g. Jupiter, Uranus, or the Sun (Figure 19.5). Here WGR of the sensor is either greatly weakened or reinforced depending on the resonance between the sensor nuclei and larger cluster dimensions of the Earth and the celestial bodies in SPGPR.

The signals of the gravitational Sun (Figure 19.3) can be considered as gravitational resonator modes, too, even though the gravitation shot is not only straight but it can also be diverted by celestial bodies, e.g. by the Moon. In Figure 19.8, the modes of the Moon's gravitational shot through the Earth (gravitation mirror) directed to the Sun on September 18, 2010 are given. As is shown, the signal of the gravitational Sun appears eight minutes earlier in SPGPR than that of the visible Sun. By comparing the data of Figures 19.5 (gravitational energy drain) and 19.8 with the exothermic processes of the gravitational energy pumping in the sensor space, the double planet Earth–Moon was found to act as a single gravitational energy donor for the Sun. Here the gravitational energy flow from the Moon is first directed to the Earth (gravitational mirror [23]) and then to the Sun. In this case, the Moon was near the ecliptic plain (2°17') in which the most important gravitational events inside the Sun's system proceed. The energy of the modes coming from the gravitational resonator Moon (I...VI, Figure 19.8) was too weak to destroy the long-range order in the water of the sensor. This is explained by the fact that the gravitational shot only partially hits the sample.

According to Ignatiev [23], the mass of the gravitational mirror amounts to $3.8 \cdot 10^{-3}$ of the Earth's mass or $2.27 \cdot 10^{22}$ kg (for an average density of 5520 kg/m^3) and its volume comes out to $4.1 \cdot 10^{18}$ m^3. Using a simple calculation we have shown that a paraboloid is a concave mirror formed by the Earth's crust; it is of an average thickness of 33 km and the Earth's surface $A = \pi \cdot D^2$, where D = 12742 km. The calculated mirror mass (hemisphere) was found to be equal to $A \cdot 33 \cdot d \cdot 10^6/2 = 4.6 \cdot 10^{22}$ kg and the mass of the paraboloid amounts to $2.3 \cdot 10^{22}$ kg [24]. In accordance with our results, those found by Ignatiev are an indication for the correctness of the proposed models (Figure 19.7).

As we have described in [22], gravitational shots are able to destroy computer networks and even to cause aircraft crashes. A picture of the Patomskiy crater in Siberia could be an example for a highly concentrated gravitational shot (http://www.panoramio.com/user/559346/ tags/patomski%20krater).

First, the structure of a gravitational shot at the time when the Moon appears in the ecliptic shall be discussed. The number of water cluster kinds in agarose hydrogel at the appearance of the Moon in SPGPR and in the ecliptic of the Sun's system is given in Figure 19.9. At the moment when the satellite enters the ecliptic plane, the sample protons resonate with those of the Moon and Sun and are attended with a highly decreased number of water cluster kinds in agarose hydrogel. It has to be mentioned that at this time the largest volcano on the island of Java and two volcanoes on Kamchatka were very active. Furthermore, strong earthquakes and a tsunami in the Indonesian region (Java area) took place. Under the influence of the Moon, the liquid magma seems to rise until it reaches the Earth's surface and leads to an overheating of water in underground reservoirs [6]; these are connected with the catastrophes. The character of signals on the curve with 2 seconds' resolution is typical for a laser irradiation that can be understood as the resonance modes of a gravitational resonator.

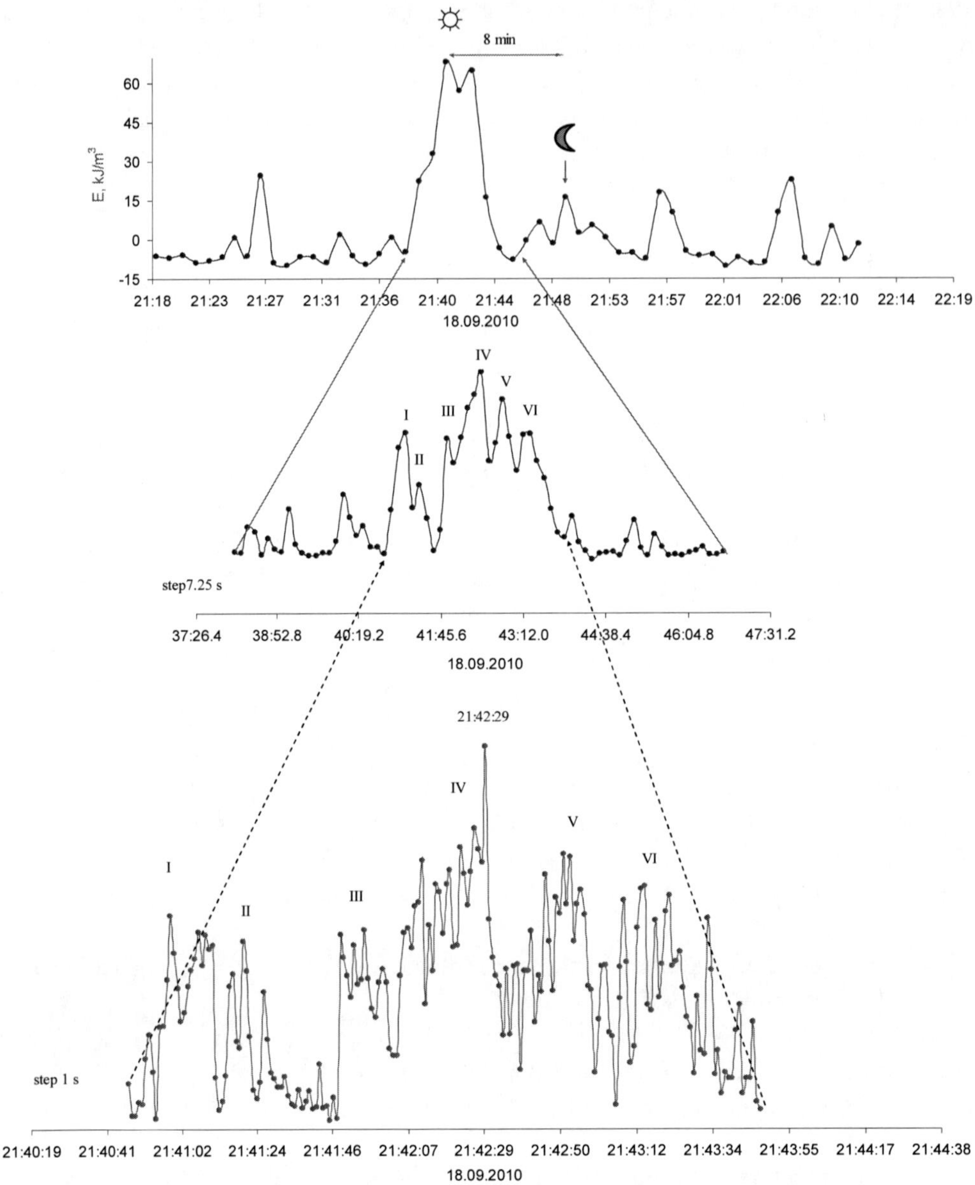

Figure 19.8. Gravitational shot from the Moon (geocentric latitude 2°17’) directed to the sensor (agarose hydrogel). Water cluster ensemble up to 3 million Daltons. I…VI modes of the gravitation resonator.

Conclusion

The Sun was found to absorb the gravitational radiation of the planets surrounding it. The structure of the gravitational radiation was represented by two flows: (i) the main gravitational flow directed to the Sun and spreading out with a super-light velocity and the (ii) traces of the main gravitational wave as ripples with light velocity propagation.

The Sun shall be recorded by two signals, first by the signal from the gravitational Sunand second by that of the visible Sun. The gravitational Sun is characterized by a multiple layer structure: corona, nucleus, and sub-nucleus. The matter in the sub-nucleus is in resonance with that of protons of hydrogen bonds of water in agarose hydrogel indicating that they are of a similar nature (quarks-gluon matter).

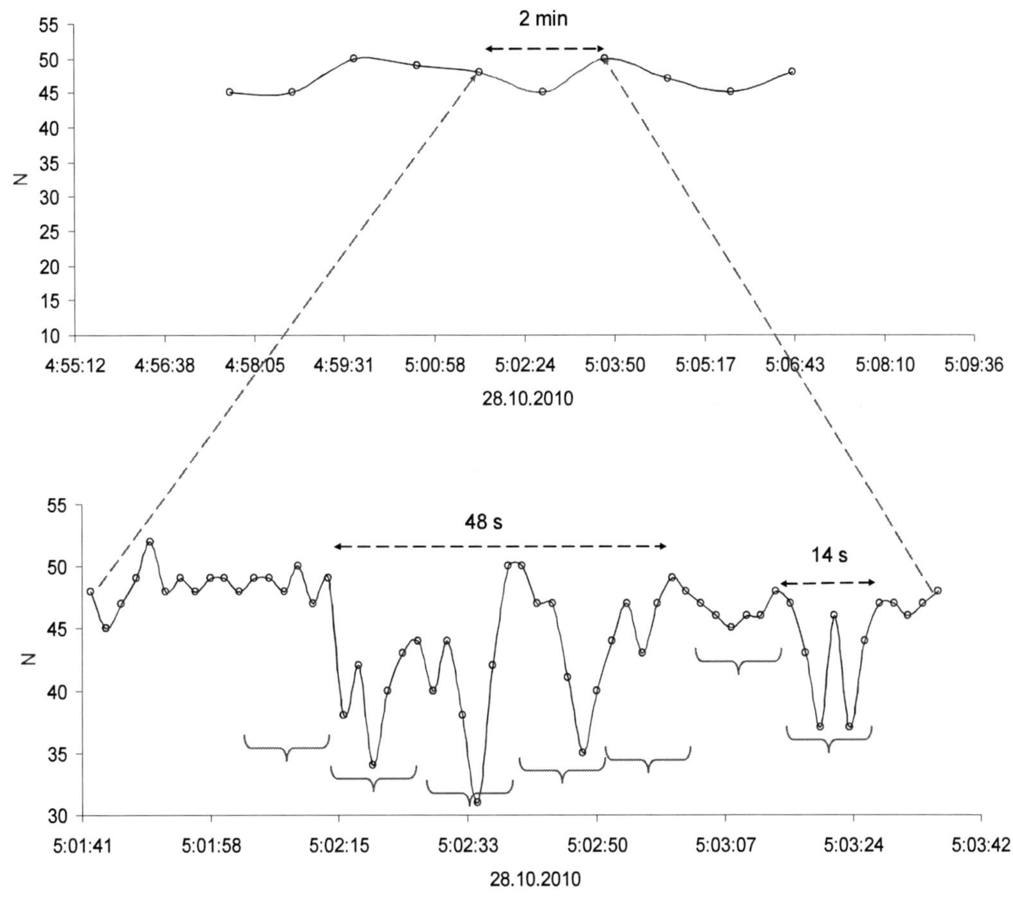

Figure 19.9. Change of the number of water cluster kinds in agarose hydrogel at the appearance of the Moon in SPGPR and in the ecliptic (geocentric latitude 0°0'). The modes of the gravitational resonator are marked with horizontal brackets.

The thermonuclear synthesis in the Sun is a secondary reaction, obtaining energy from proton condensation in the sub-nucleus.

Gravitational radiation velocity in space permanently changes depending on the observation position and on the constellation of mass concentrations with regard to the observer.

The absorption of gravitation energy by the Sun can be reinforced by celestial bodies and achieve a level of gravitational shots with catastrophic consequences.

REFERENCES TO 19

[1] N. A. Kozyrev, V.V. Nasonova. In. Displaying of space factories on the Earth and stars. M., L., 1980, pp. 76-84. (in Russian).

[2] Lavrentijev M.M, Eganova I.A., Luzet M.K. About planet influence on a resistor at distance. Doklady academii nauk. Fisika. 1990, vol. 314, pp. 352-355. (in Russian).

[3] Lavrentijev M.M, Gusev V.A., Eganova I.A., Luzet M.K, Fominyh S.F. About registration of real sun constellation. Doklady academii nauk. Fisika. 1990, vol. 315, pp. 368-370. (in Russian).

[4] Lavrentijev M.M., Eganova I.A., Medvedev V.G., Oleynik V.K., Fominyh S.F. About scanning of sky by Kozyrev sensor. Doklady academii nauk. Astronomy.1992, vol. 323, pp.649-652. (in Russian).

[5] Zubow A., Zubow K., Zubow V.A. New Experimental Method for Determination of Distances to Celestial Bodies and their Masses. In Horizons in World Physics. Editor: Albert Reimer. Nova Science Publishers, Inc. NY. Vol. 279. 2012, in print. https://www.novapublishers.com/catalog/product_info.php?products_id=34674

[6] K.V. Zubow, A.V. Zubow, V.A. Zubow. Principles of gravitation spectroscopy. New form of molecular matter. Processes. Fields. Aist Handels- und consulting GmbH, dep. R&D, Berlin, 2010, electronic book (www.zubow.de), p. 854. (in Russian).

[7] Zubow K.V., Zubow A.V., Zubow V.A. Ensemble of Clusters – New Form of Molecular Matter, Risks and Chances. Zubow Equations. In Advances in Chemistry Research, vol. 5, 2010, pp. 107-145. Editor J.C. Taylor. Novapublisher. NY.

[8] Zubow K.V., Zubow A.V., Zubow V.A. Using of flicker noise spectroscopy for non destroying analysis of nano structures. Zavodskaja Laboratorija. Diagnostics of materials. 2008, vol. 74, no. 9, pp.40-45. (in Russian).

[9] Zubow K.V., Zubow A.V., Zubow V.A. The Dynamics of Low Frequency Movements of Molecular Clusters in the Hardening Process of Epoxide Resins. Chem Promislennost Segodnja. 2008, no. 9, pp. 12-21. (in Russian).

[10] Zubow K., Zubow A.V., Zubow V.A. Water Clusters in Plants. Fast Channel Plant Communications. Planet Influence. Journal of Biophysics Chemistry. 2010, vol.1, no.1, pp.1- 11.

[11] Zubow K.V., Zubow A.V., Zubow V.A. Cluster structure of liquid alcohols, water and n- Hexane. J. of Appl. Spectr. 2005, vol. 72, no. 3, pp. 321-328.

[12] Kokkotas, Kostas D. Gravitational waves. Acta Physica Polonica. 2007, vol. 38, no.12, pp. 3891-3923.

[13] L.Pauling, P. Pauling. Chemistry. W.H. Freeman and company San Francisco. 1975.

[14] Zubow K., Zubow A.V., Zubow V.A. The Phenomenon of Proton Dissolving in Vacuum and of Proton Condensation from Vacuum. Two Forms of Protons, Structure of Nuclei, Electrons and Atoms. J. of modern physics. 2010, vol.1, no.1, pp.175-184.

[15] Schertler, Klaus; Leupold, Stefan; Schaffner-Bielich, Jurgen. Neutron stars and quark phases in the Nambu-Jona-Lasinio model. Physical Review C: Nuclear Physics. 1999, vol. 60, no. 2, pp.025801/1-025801/11.

[16] Nakazato, Ken'ichiro; Sumiyoshi, Kohsuke; Yamada, Shoichi. Astrophysical implications of equation of state for hadron-quark mixed phase: Compact stars and

stellar collapses. Physical Review D: Particles, Fields, Gravitation, and Cosmology. 2008, vol. 77, no. 10, pp. 103006/1- 103006/12.

[17] An Album of Fluid Motion. Assembled by Milton van Dyke. The Parabolic Press, Stanford California. 1982, p. 83.

[18] Dikarev S.N. Laboratory study of resonant regimes of motions of homogeneous fluid with free surface in a tilted rotating container. Izwestia AN. Fizika atmospheri i okeana. 1990, vol. 26, no. 9, pp. 982-992 (in Russian).

[19] Manin D.Yu., Chernousko Yu. L. An experimental study of the stability of a quasi two-dimensional jet flow produced in rotating fluid by sinks and sources. Izwestia AN. Fizika atmospheri i okeana. 1990, vol. 26, no. 5, pp. 483-492 (in Russian).

[20] Leonow V.S. Discoveryof gravitation waves by professor Weinik. M.. Agroprogress, 2001.

[21] Zabotinsky M.E. In Fisitsheskij Enziclopeditsheskij Slowar. M.: Sovijet Enciclopedia. 1984. p. 339.

[22] Zubow K., Zubow A.V., Zubow V.A. Experimental Platform for the Investigation of the Structural Heterogeneity of the Physical Vacuum. Vacuum Energy Risks and Chances. In Horizons in World Physics. Editor A. Reimer. NY. vol. 277. 2012, pp. 87-110. https://www.novapublishers.com/catalog/product_info.php?products_id=29310

[23] Ignatiev, A. Yu.; Volkas, R. R. Geophysical constraints on mirror matter within the Earth, Physical Review D: Particles and Fields. 2000, vol. 62, no. 2, pp. 023508/1-023508/7.

[24] Rogdestwensky V.N. In Fisitsheskij Enziclopeditsheskij Slowar. M.: Sovijet Enciclopedia. 1984. p. 200.

19.1. Axions from the Sun, Achernar, and the Gas Giants

To register axions, diverse experiments are carried out (http://en.wikipedia.org/wiki/Axion), [1]. It is believed that the high penetration of these neutral particles is an obstacle to their detection by conventional sensor systems. Water and water solutions serve as a sensor in which, according to Volkov [2], absorption centers composed of protons (3H) can be realized. In these protons, the conditions for the formation of fermion loops—circular electrical current of Fermi-particles, shall be given where the fermion loops act as absorption centers for axions. According to these views, fermion loops can be formed into molecular clusters, e.g. into solvated clusters of ion pairs (SCIP). We have previously detected such currents in SCIP [3], but their properties were not investigated. On the other side, the hypothetical axion as a neutral particle can interact with water in another way, namely, through the "naked protons" [4], which usually serve as hydrogen bridges between water molecules. The interaction axion-proton can be realized at the level of gravitational forces.

The intention of the present work is to apply the gravitational mass spectroscopy developed by the authors [5] in detecting axions based on their interaction with molecular water cluster ensembles in hydrogel (biomatrix model), in water solutions, and in a 3.5 wt. % NaCl solution (sea water model).

The GMS-sensor (agarose hydrogel containing 97 wt. % water), NaCl 3.5 wt. % water solution, citric acid 0.6 wt. % solution in water, 37 wt. % HCl water solution, atactic polystyrene) is used as a GMS-sensor. The GMS-sensor is located inside liquids in an iron box that is isolated from electromagnetical, mechanical, and seismic energy flows and it is kept in a laboratory room [5] (53°34'54"N, 12°47'02"E). The analysis of the long-range order in water at the cluster level proceeds according to the method described in [6, 7]. The visible constellation of the Sun is determined using the program, ZET 9 (www.astrozet.net). The Sun axions' registration is carried out in 10 second steps. The pressure on the GMS-sensor is calculated using the experimental data for all noise kinds (dB) in the sensor space (log P = dB – 5.7).

Figure 19.1.1 shows how the pressure hypothetical particles axions act on the GMS-sensor before and after sunrise. There is a broad band at 16,020±10 Hz, which appears in all investigated liquids at daytime (Figure 19.1.2). During the whole observation time in the summer of 2012, the band intensity remained constant (- 46±3). This is not true, however, for the shoulders of the band, which changed even during a day. Long-term observations of the band intensity at 16,020 Hz from July to September 2012 show its irregular reduction. This decrease can be described by the smoothing function: log P = - 0.25*t* - 45.25, where *t* is the time in days beginning with July 18, 2012 (Figure 19.1.10) up to September 1, 2012; this reduction is correlated with a decrease in the altitude of the Sun as well (correlation coefficient of 85%).

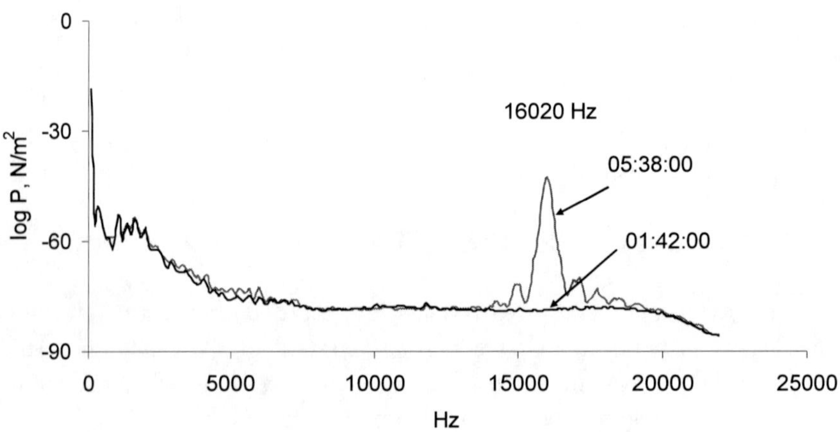

Figure 19.1.1. Axions' pressure on GMS-sensor in a citric acid water solution before and after sunrise (rise of the Sun's center at 05:09:58, CET, July 18, 2012).

A more detailed analysis of the signal intensity at the time when the gravitational and visible Sun's discs hit SPGPR (plane going through the Earth's rotating axis and the sensor location on the Earth surface) [8] make no difference, which means that only the visible Sun emits this radiation. Visible light, however, cannot penetrate the GMS-sensor box; the radiation could therefore only be invisible and neutral in sun axions. The energy and wave length of the main signal, calculated according to $E = h\nu$ (h – Planck constant, ν – frequency), amounts to $1.06 \cdot 10^{-29}$ J and 18.7 km, correspondingly. All of the aforementioned can evidence in favor of the axions as particles with high penetrability that are responsible for gravitational and electromagnetic interactions (http://en.wikipedia.org/wiki/Axion). According to Ljubičić and Rusov [9, 10], the mass of an axion is found to be around 100...30

eV. To verify it, additional experiments at another time and with other molecular clusters are carried out.

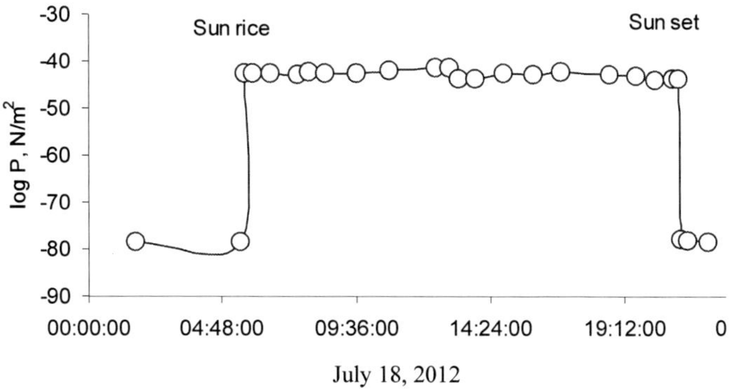

Figure 19.1.2. Axions' pressure on GMS-sensor in a citric acid water solution (16,020 Hz, $6.6 \cdot 10^{-11}$ eV) at daytime.

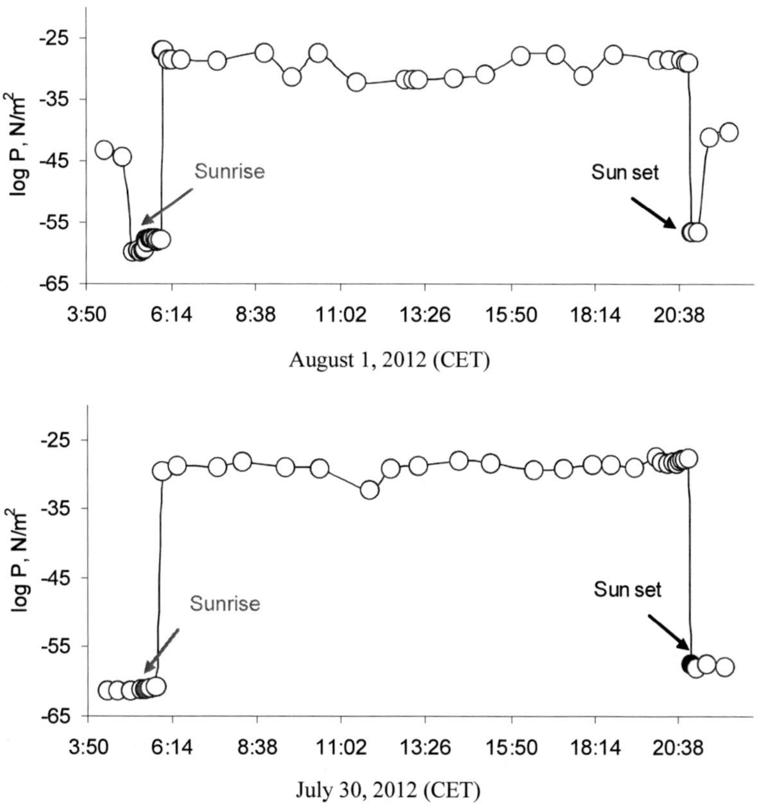

Figure 19.1.3. Axions' flow pressure (16,020 Hz) on GMS-sensor in NaCl solution at daytime.

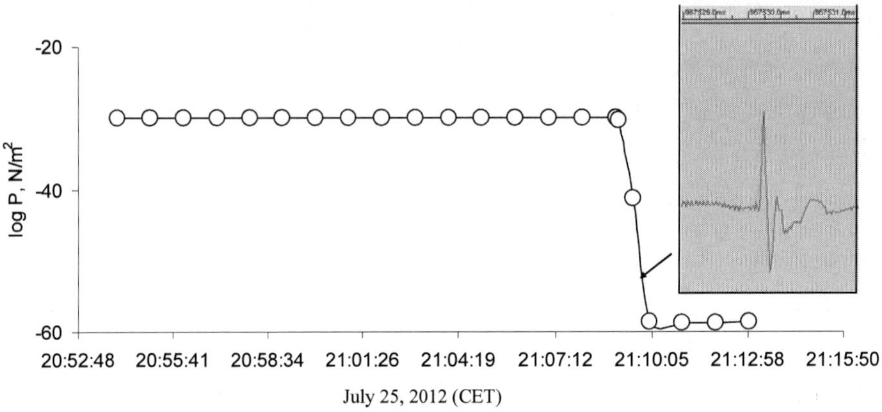

Figure 19.1.4. Axions' flow pressure (16,020 Hz) on GMS-sensor in NaCl solution at sunset (arrow). Right shows the oscillogram of the noise damping process. The average peak intensity before sunset is about -29 N/m^2.

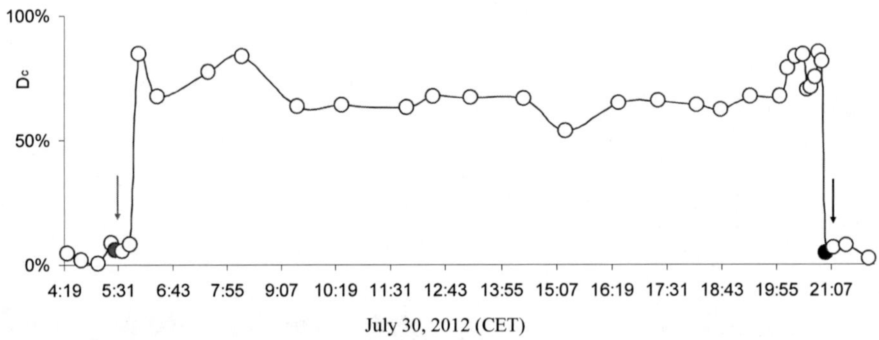

Figure 19.1.5. Part of collapsed clusters (D_c) in a 3.5 wt. % NaCl solution at daytime. Arrows indicate the time of sunrise and sunset.

Figure 19.1.3 shows how the pressure of the sun's axion flow act on the GMS-sensor in a NaCl water solution during two different days (for understanding and comparing the signal intensities with those in the citric acid water solution see Figure 19.1.2). As is visible, the signal intensity of the sun's axions increases sometime after the Sun rises (~30 min.). In contrast, it disappears very quickly when the visual center of the Sun sets. At sunrise and sunset the influence of the axions is assumed to be different. Before sunrise, the energy from the axions will probably be pumped into the molecular clusters in the Earth's hydrosphere. This pumping process needs time because of an energy hierarchy between the molecular clusters in an ensemble and the permanent necessity to supply the axions' absorber with energy (see Figure 19.1.6 for the base water cluster). At the center of sunset, the last condition is not met and the signal at 16,020 Hz rapidly disappears; here much less time is required than at sunrise (Figure 19.1.4). Axions are therefore concluded to be emitted from the star's center.

The oscillogram curve of the noise pressure relief indicates a shock mechanism, which has to be understood as breaking up the absorption axions' centers and then as releasing energy (rise of noises from the transforming water cluster ensemble); then the process changes to a strong pressure decrease that is connected with the typical damping of this oscillation circuit. Reasonably, this process has to increase the number of expanded clusters in the solution as evidenced by the data shown in Figures 19.1.5, 19.1.8, and 19.1.9. The sunset is accompanied by loosening molecular clusters.

Another property of axions is their ability to be reflected from clouds. The model of this phenomenon is shown in Figure 19.1.6. To prove this model, the following experiment has been conducted (Figure 19.1.7).

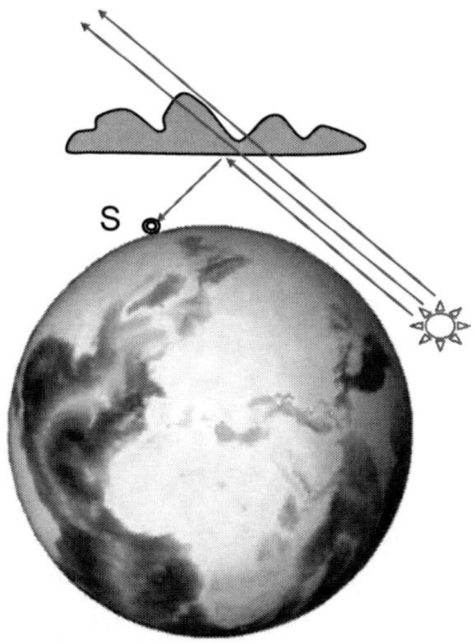

Figure 19.1.6. Model for the reflection of axions from the bottom surface of clouds. S – GMS-sensor.

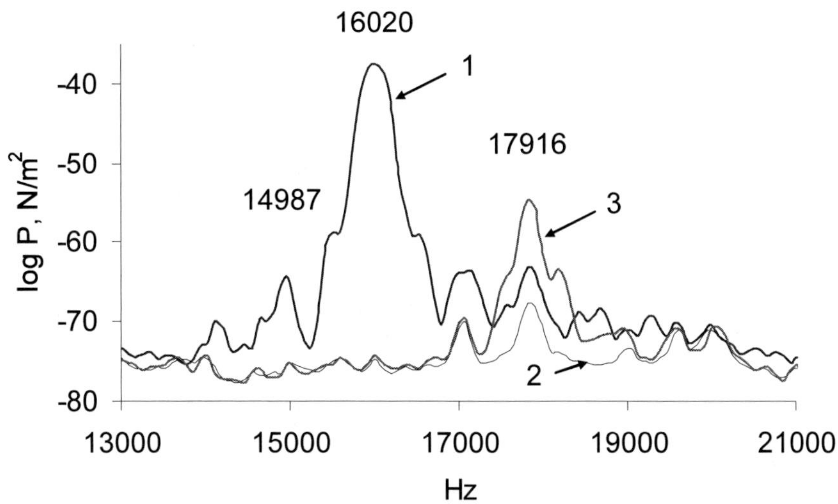

Figure 19.1.7. Axions' flow pressure on GMS-sensor during sunset with stratocumulus clouds (http://en.wikipedia.org/wiki/Stratocumulus_cloud). Polystyrene GMS-sensor in an isolated box inside a laboratory building. August 31, 2012. Sunset center at 19:53:00 (CET). 1 – impact of axions on PS clusters (average value of 10 measurements) until the center of the Sun has set, including the signals 10 minutes later; 2 – signal level at 20:02:00; and 3 – signal level at 20:03:00.

As follows from Figure 19.1.7, stratocumulus clouds are able to reflect axions. The disappearance of the main signal at 16,020 Hz at 20:02:00 (curve 2) is accompanied with an increase of the signal intensity at 17,916 Hz (20:03:00), which disappears after a few seconds. The effect should be explained by the mechanism of shock noise damping (Figure 19.1.4) remodeling the energy hierarchy of clusters in the polymer. The same cannot be said about the signal at 14,987 Hz, which simply disappears.

It is of interest to understand how axions interact with molecular clusters. According to the first Zubow equation [5], the bands at 16,020 and 17,916 Hz belong to the molecular cluster masses of 3,896 and 3,116 Daltons (Da), correspondingly. For aqueous solutions, these are the splitter clusters $(H_2O)_{216}$ and $(H_2O)_{178}$; for the polystyrenes it is $(St)_{37}$ and $(St)_{30}$, respectively (St is the repeating styrene unit) [5].

The densification, for example, of the base water cluster is to be understood by the fact that the interaction of surface cluster protons with the surroundings is weakened. The weakening is caused by a balance shift between baryonic protons (*p*) and their dissolved form in the physical vacuum [11] (Figure 19.1.8).

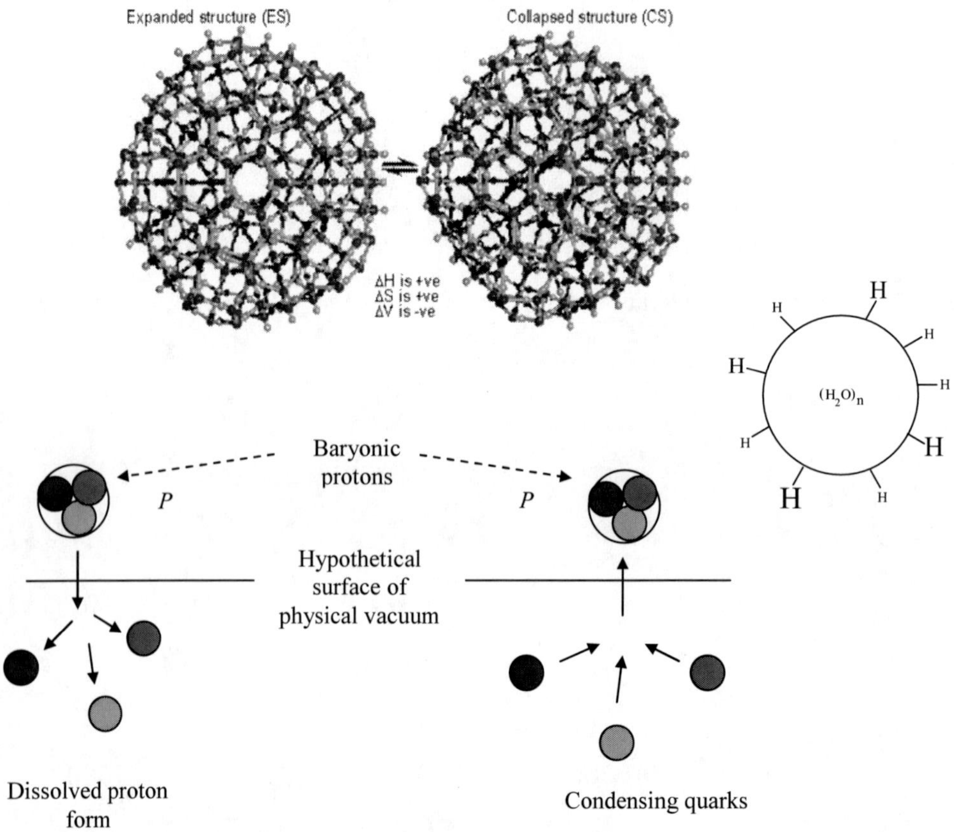

Figure 19.1.8. Scheme illustrating the formation of collapsed and expanded water clusters as result of the balance between baryonic protons and their dissolved form in PV. Right – model of a water cluster with surface hydrogen atoms; capital letter – mainly baryonic protons, and lowercase letter – dominance of protons' dissolution in a PV. For a better understanding, the balance between the expanded and collapsed forms of the water cluster composed of 280 molecules is given above (with the kind permission of Professor Chaplin, http://www.lsbu.ac.uk/water/index.html).

Interacting with baryonic protons (P) and axions (A) [11] forms the transition complex ((A-P)$^{\#}$) that will be later destroyed to the dissolved proton form (P_s):

$$A + P \leftrightarrow (A\text{-}P)^{\#} \leftrightarrow P_s + A.$$

A destroyed balance influences the information state of every cluster [12] (interaction with surroundings) and changes the density [5] and energy situation of each water cluster in the ensemble. This is proved by the base water cluster (Figure 19.1.9) and the whole ensemble as a unit (Figure 19.1.5). At the action of Sun axions, the nature of the water cluster interaction with the surroundings strongly changes (-f) [5]. In the ensemble of water clusters the collapsed structures are beginning to dominate. The destruction of some hydrogen bonds [13] between water molecules on the cluster surface can be interpreted as the partial transition of "naked protons" into their dissolved state in a PV (Figure 19.1.6). This logic implies another generation for Sun axions. According to [5], proton matter circulates in the Sun system and its dissolved form condenses in the center of our star with the release of energy, which is sufficient to initiate nuclei fusion while the endothermic proton dissolution in a PV is realized in the planets around the Sun. Then the role of the axions in this circulation can be seen as a proton "solvent." The dissolution process of protons in a PV, which is stimulated by axions, is satisfactorily confirmed with our experiment.

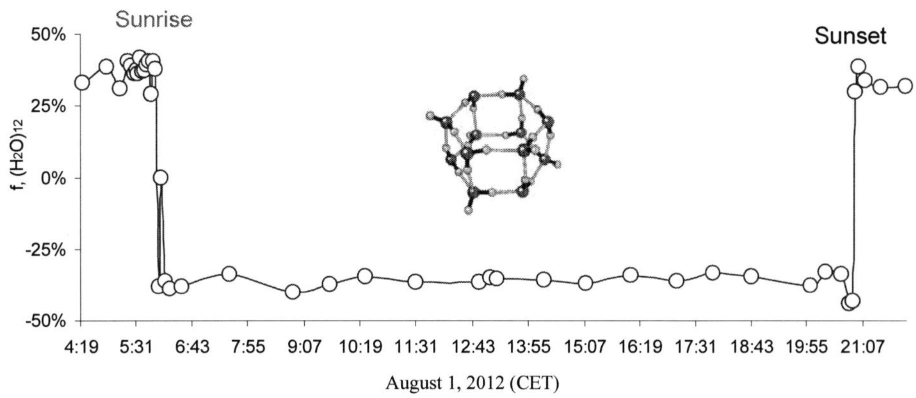

Figure 19.1.9. The influence of sunrise and sunset on the density of the base water cluster (model provided by Professor A. Lenz) in 3.5 wt. % NaCl water solution. Positive f values indicate the dominance of expanded water clusters (see comment to Figure 19.1.8), the negative, that of collapsed ones.

The influence of celestial bodies on the signal intensity at 16,020 Hz is given in the table.

As follows from the table, the mass concentrations in the space affect the flow of the Sun's axions. When our galaxy's center and the Sun are in opposition to us then the flow of axions increases, while at conjugation, the flow weakens. Conjugation of the Moon and the Sun leads to the weakening of the axion flow, apparently as a result of the interception of the axions by the mass of our satellite. When Saturn and the Sun are conjugated to the Earth, a weakening of the axion flow is also observed. What will the impact of Jupiter will be? As in the former three cases, a weakening of the axion flow is expected, however, the opposite happens; it unexpectedly increased (+2). The phenomenon indirectly indicates that this gas giant is an extinct star in whose center similar processes as that of the Sun's center are taking place. The observed data confirms the nature of the axions, which are responsible for gravitational interactions. It should be noted that in the period from August to October 2009, the axion signal at 16,020 Hz has not been found.

Table 19.1.1. The influence of the constellation of celestial bodies on the signal intensity of the Sun's axions at 16,020 Hz. Without the influence of celestial bodies, the average value of the signal is equal to log P = -17.0 N/m^2. Sensor - agarose hydrogel in SPGPR.

Celestial body	log P, N/m^2 at 16,020 Hz	Deviation from average	Constellation with Sun	Data
Milky Way center	-13.3	3.7	opposition	18.06.2012
Milky Way center	-19.8	-2.8	conjugation	20.12.2011
Jupiter	-15.0	2.0	conjugation	13.05.2012
Saturn	-19.8	-2.8	conjugation	12.10.2011
Moon	-19.5	-2.5	conjugation	01.07.2011
–	-17.0	0	–	–

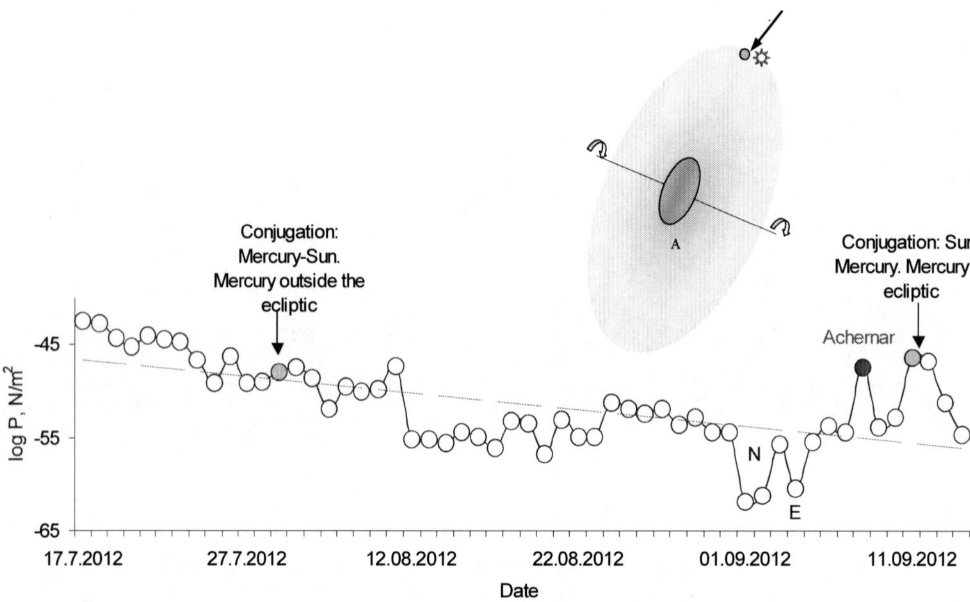

Figure 19.1.10. Axion flow (16,020 Hz) pressure on GMS-sensor (citric acid water solution) when Achernar's axions hit the Earth and the model of this event (September 8, 2012). Influence: A – star Achernar, N – Neptune, E – Egg Nebula, RAFGL 2688. Dotted line – average function log P = -0.25t - 45.25 (only by September 1, 2012 is it valid). The slash arrow shows the location of the GMS-sensor on the Earth.

On September 8, 2012 the GMS-sensor had been hit by an axion flow from the fast rotating star giant, Achernar (http://en.wikipedia.org/wiki/Achernar, 139 ly), resulting in highly increased pressure on the sensor (~7 orders of magnitude, Figure 19.1.10). Therefore, just like the Sun, Achernar is able to generate axions, which are emitted in the centrifugal direction and whose mechanism is the same as in the center of the Sun. In this period, however, Achernar was on the other side of the Earth (southern hemisphere) and its direct axion flow should not reach the sensor. The signal is, therefore, interpreted as the sum of the direct Sun's axions and Achernar's axions that are reflected from the Sun. A similar effect is observed at full moon (Figure 19.1.13).

It is of interest to analyze the axion signals from Mercury. At its close conjugation with the Sun and its deviation from the ecliptic of > 4.5°, the signal intensities of the Sun's axions do not change but when Mercury hits the ecliptic plane (remote conjugation), the signal intensities sharply rises. This may be evidence of the hypothesis that this planet is the core of a former gas giant (http://lasp.colorado.edu/education/outerplanets/solsys_planets.php) [14] in which occurred similar processes like those found in the Sun's center. The axion flow from the Sun is reinforced by that of Mercury to Earth.

Figure 19.1.11 shows the results of the search for axion signals from Jupiter from 22:00:00 (August 20, 2012, gas giant rise at 23:48:23, CET) to 5:00:00 (August 21, 2012). Besides other signals, a weak signal at 16,020 Hz was seen, which is much weaker in intensity compared to the Sun's (Figure 19.1.2). Consequently, similar processes could be proceeding in the center of the planet as those in the center of the Sun, which means that Jupiter is a protostar (http://www.newton.dep.anl.gov/askasci/ast99/ast99616.htm).

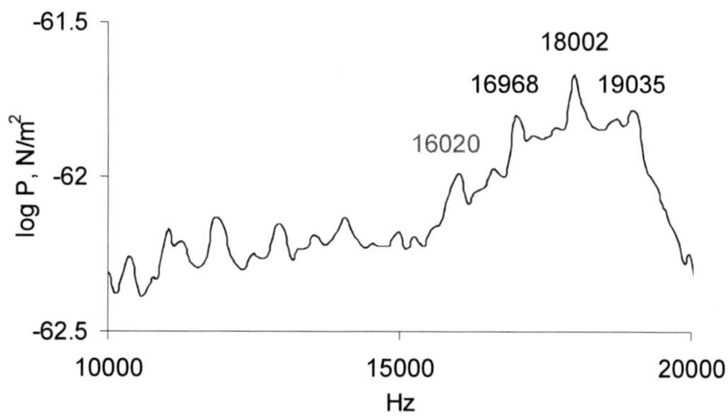

Figure 19.1.11. Flow pressure of Jupiter axions on GMS-sensor in a citric acid water solution. Average value of 20 GMS-spectra recorded from 22:00:00 (August 20, 2012, CET) to 5:00:00 (August 21, 2012) with an interval of 20 minutes. The distance to this planet 5.18658 au.

The GMS-spectrum of the gravitational Jupiter disk entering SPGPR (01:33:05, October 13, 2011) is composed of signals at 16,451; 18,088; 19,552; 20,844; and 20,930 Hz, while in the spectrum of the visible Jupiter (in SPGPR at 02:06:00, October 13, 2011), the signal at 18,088 Hz is missing. It is therefore concluded to be responsible for the gravitational interaction of Jupiter's axions with the water cluster ensembles in agarose hydrogel.

Scanning the other gas giants, Neptune also reveals a number of axion signals (Figure 19.1.12). The weak signal at 16,020 Hz may indicate that the nearby signals in this region are caused by axions, too, however, by another of their type. The strong signal at 19,552 Hz is also present in the spectrum of Jupiter. The ratio of signal intensities of axions from gas giants is not constant and depends on the constellation of the planets. Thus, processes proceeding in the center of Neptune are believed to be similar to those in the centers of the Sun and Jupiter.

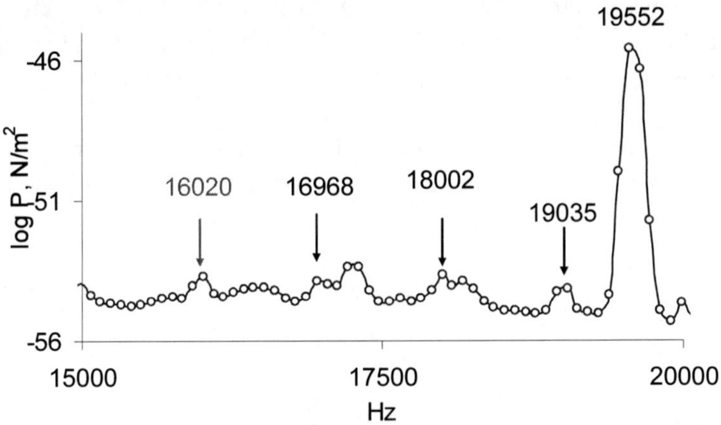

Figure 19.1.12. Flow pressure of axions from the visible disk of Neptune on GMS-sensor in agarose hydrogel. Average values of 30 spectra taken from 1:09:09 to 1:19:09 (CET), August 24, 2012, measurements taken every 20 s. The distance to the planet 28.98393 au.

Axion emissions from gas giants can become a platform for understanding the existence of other life forms, which use heat coming from the planet's center [15].

In the frequency range from 15,000 to 22,000 Hz, the signals of the Sun's axions reflected from the full moon appear while the signal intensities of axions during the other moon phases are strongly decreased (Figure 19.1.13). These "Moon's" signals correlate with existing or missing signals from the Sun's axions and depend on the altitude and azimuth of the Moon, i.e., on the reflexion geometry. The mechanism of this phenomenon requires a special investigation, which is not the subject of the present work.

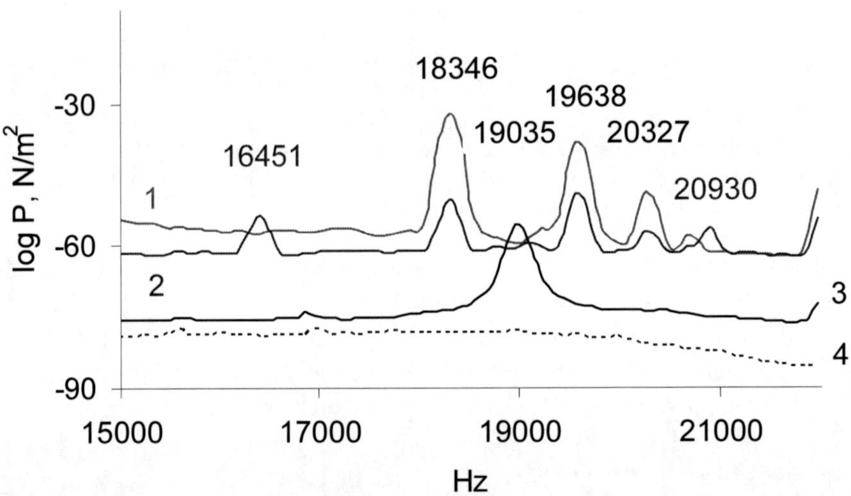

Figure 19.1.13. Spectra of sun axions reflected from the visible disk of the Moon. 1 – full moon on February 8, 2012 (altitude 47°, azimuth 357°, scanning at 00:11:14, CET) agarose hydrogel; 2 – December 14, 2011 (altitude 52°, azimuth 0.87°, 02:55:51, CET), agarose hydrogel; 3 – January 19, 2008 (altitude 63°, azimuth 347°, 21:16:00, CET) agarose hydrogel; and 4 – August 28, 2012 (altitude 21°, azimuth 0.18°, 22:52:00, CET) citric acid water solution.

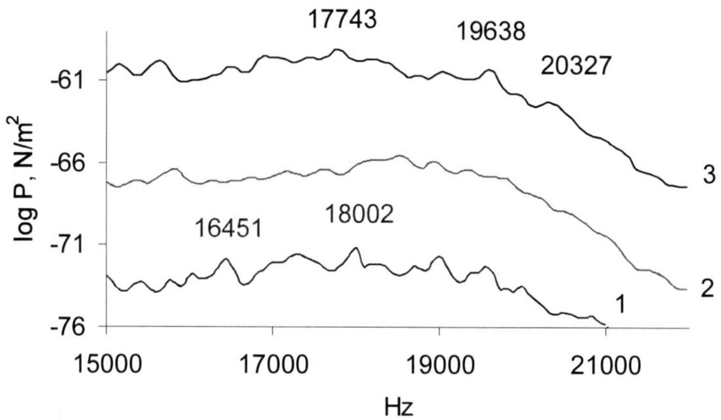

Figure 19.1.14. Influence of a hypothetical axions' flow from a twisted space on a 37 wt. % HCl solution in water. Frequency range for axions. 1 – pressure on GMS-sensor without graser influence, 2 – pressure on GMS-sensor when the twisted space is of the same direction as the axis of the rotating body, and 3 – pressure on GMS-sensor when the twisted space is perpendicular to the axis of the rotating body (the sample is in the ecliptic to the body rotation). Broad band gravitational graser, distance to the sample of 5 cm. August 30, 2012, 22:00:00, CET.

The axion emission from the Sun's center can be of a torsion nature. If we assume that in the Sun's center exists a giant vortex (http://en.wikipedia.org/wiki/Sun - Core; http://www.disclose.tv/action/viewvideo/64019/Sun_Vortex_2/) [16] that is emitting axions by twisting the space (http://en.wikipedia.org/wiki/Torsion_field) [10], then the assumption should be proved experimentally [17, 18]. For this, we used a high-speed body graser [19] on whose centrifugal vector (model ecliptic of the Sun) the sample of the investigated liquid is placed. Figure 19.1.14 shows how a flow of hypothetical axions from a twisted space influences the hydrochloric acid solution.

As is visible in Figure 19.1.14, the level of the noise pressures depends on how the sample is positioned in relation to the hypothetical axion flow. The highest values are observed when the twisted space is perpendicular to the axis of the rotating body. There are peaks that are typical of axions, but due to the low graser selectivity, the signals are not distinctive. Differing from the standard generation of axions by an electron jumping in ^{57}Fe [10], this experiment is an indication for another axion generation mechanism, namely, the highly rotating center of the Sun.

To test the hypothesis that molecular clusters as molecular traps absorb axions, an experiment using a crystal sensor without a molecular cluster structure is carried out. Using sensors based on crystalline metals (Au, Pt, Cu), a very weak signal at 16,020 Hz is recorded. Therefore, axions are concluded to be absorbed by molecular matter protons even though the process depends on the substance's nature.

Thus, the sensor signal for polystyrene (C-H acid, $pK_a \sim 58$, [20]) is significantly weaker than that of citric acid (O-H acid, $pK_a \sim 4$), which is much smaller than that of concentrated hydrochloric acid (Cl-H acid, $pKa \sim -1.7$). A broad spectrum of "naked" protons in water and aqueous solutions seems to favor the absorption of axions. On the other hand, the absence of "naked" protons in crystalline metals as well as the reaction of some metals on axions indicate that axions can be partially absorbed by nucleus protons. For example, nucleus protons of polystyrene are observed to absorb 83% of the Sun's axions while the axion absorption in gold is much lower (17%).

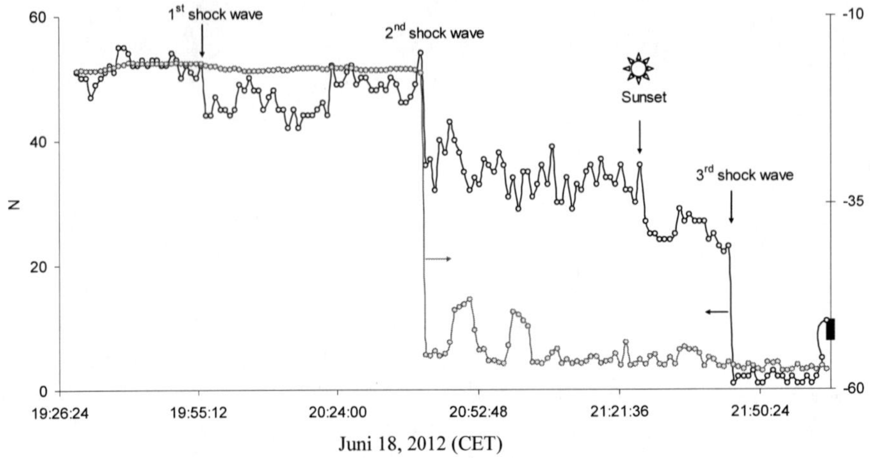

Figure 19.1.15. Number of water cluster kinds (N) in agarose hydrogel and signal intensity of the Sun's axions (16,020 Hz) during the arrival of the Sun's wind front in Northern Germany.

June 8, 2011 at 20:00 in Europe, a shock wave from the Sun's winds was expected (http://www.youtube.com/watch?v=hpgzNKufn7g); it was therefore decided to study its impact on the Sun's axion flow. Figure 19.1.15 shows how the number of water cluster kinds in agarose hydrogel and the signal intensity of the Sun's axions at 16,020 Hz change during the arrival of the Sun's wind front.

Outbreak in the Sun on July 7th dramatically affected the water cluster ensemble, which is recorded on July 8th at the time of the shock front's arrival on Earth. The figure shows that the ensemble of water clusters in the hydrogel is exposed to at least three shock waves and the impact of each destroys it. Every time the number of cluster kinds is strongly decreased. The signal intensity at 16,020 Hz especially reduces when the water cluster ensemble is destroyed by the second shock wave. The third shock wave is after sunset but its influence could not be examined. Thus, the interaction of the axions with the protons of hydrogen bridges in water is confirmed to be of a gravitational character, where the signal at 16,020 Hz is to be understood as that of unstable, exited molecule clusters.

On July 8th in the axion spectra of water clusters, there are additional signals: 5,081 Hz; 7,321 Hz; 15,503 Hz; and 18,346 Hz. While the first three signals actively react to the arrival of the Sun's wind shock waves, the last one is inactive. It has to be mentioned that small clusters selectively react to the shock wave of Sun's wind. Therefore the fragmentation cluster $(H_2O)_{25}$ clearly responds to the arrival of the second and third Sun's wind shock wave whereas the base cluster $(H_2O)_{12}$ [21, 22] only responds to the third one. Therefore, clusters are characterized by a different absorption capacity for axions and they can be used to understand the nature of these particles. Such cluster selectivity negates the model of capturing electrically neutral axions through ring currents in hypothetical structures like 3H [2].

From September 1-2, 2012, the intensity of the Sun's axion flow is highly decreased from log P = -54.3 to log P = -61.9 N/m^2 (Figure 10). Then it returns to the baseline on August 31,

2012 (-54.3 N/m^2). At the same time, a cloud phenomenon in the atmosphere has been observed (Figure 19.1.16).

This phenomenon is very similar to the interference of coherent radiation well known from physics. Let us discuss this phenomenon in detail. The interference of the axion flows from the Sun (source S_1, azimuth 53°, altitude 33°) and Neptune (source S_2, azimuth 244°, altitude -32°) is suggested to have caused it in the skies of Northern Germany at 16:00 (CET) on September 2, 2012 (Figure 19.1.16).

To calculate the distance (Δx) between the interference bands (Figure 19.1.16), the following equation is used:

$$\Delta x = L\lambda/d$$

where L is the distance from us to the Sun (dominant flow) in km, λ = 18.7 km, and d is the base distance between S_1 and S_2 ($4.5 \cdot 10^9$ km is the distance between two axion sources). The calculated value (0.61 km) satisfactorily agrees with the one found experimentally. Thus, the interference pattern shown in Figure 15 really reflects the interaction of the Sun and Neptune's axion flows. Here solid water crystals seem to sublimate into the vapor state (transparent bands in the cloud). This is caused by axions, which interact with the protons in hydrogen bonds between the water molecules. As a consequence, the balance of protons with a PV is shifted to the dissolving side leading to weakened hydrogen bonds between the water molecules (Figure 8, the scheme clusters $(H_2O)_n$).

Figure 19.1.16. Photo of clouds over the GMS-sensor location at 16:00 (CET) on September 2, 2012. The distance between the bands is about 0.7±0.1 km. North of Germany. Comments are in the text.

CONCLUSION

- A new method to record axion flows from celestial bodies has been suggested.
- Axions are found to be absorbed selectively, mainly by protons of hydrogen bridges in water and water solutions. The process is of a gravitational nature.
- Molecular clusters can be used for selective axion absorption.
- The axion energy is really low and its registration is possible only in an indirect manner—by measuring the noises of cluster ensembles in liquids and the reactions of individual clusters on these elemental particles.
- Axions are generated in vortices of the star centers where the dissolved proton form is condensed.
- Axion mass is much lower than $6.6 \cdot 10^{-11}$ eV.
- The space is permeated by an axion "highway" that is useful for information transport.
- Axions are able to be reflected from celestial bodies and Earth's clouds.
- Axion flows are characterized to form an interference pattern in the atmosphere.
- Axion emission from gas giants could be a platform for the existence of life based off of thermal energy coming from the cores.

REFERENCES TO 19.1

[1] http://www.phys.washington.edu/groups/admx/ADMX_Files/Rybka_patras2011.pdf
[2] Volkov J. V. Technic of axion radiation registration with help of water sensor. In "System "Planet Earth": 300 years Lomonosov M.V.1711-2011. In *System "Planet Earth": 300 years Lomonosov M.V.1711*, Monograph. Moscov.: LENAD, pp.216/221, (2011) (in Russian).
[3] Zubow K., Zubow A.V., Zubow V.A. Molecular Cluster Distribution in Sea Water Near the Wall. Antifouling aspect. Journal of Clusters, 2011, v.22, no. 4, pp. 603-619.
[4] Pauling L., Pauling P. Chemistry. W.H. Freeman and company San Francisco.
[5] Zubow K.V., Zubow A.V., Zubow V.A. Ensemble of Clusters – New Form of Molecular Matter, Risks and Chances. Zubow Equations. In Advances in Chemistry Research, vol. 5, 2010, pp. 107-145. Editor J.C. Taylor. Nova Science Publishers, Inc. NY. https://www.novapublishers.com/catalog/product_info.php?products_id=27769
[6] Zubow Kr. V., Zubow A.V., Zubow V.A. Cluster structure of liquid alcohols, water and n- Hexane. Journal of Appl. Spectr. 2005, vol. 72, №3, pp. 321-328.
[7] Zubow Kr. V., Zubow A.V., Zubow V.A. Oscillation spectra of seed NaCl crystals in aqueous solution. Journal of Appl. Spectr. 2005, vol. 72, №6, pp. 840-847.
[8] Zubow K., Zubow A.V., Zubow V.A. Scanning of the Sun and other celestial bodies with help of gravitation spectroscopy. Optic and Photonics Journal. 2011, № 1, pp.15-23.
[9] Ljubičić, A.; Kekez, D.; Krečak, Z. VII Latin American Symposium on Nuclear Physics and Applications. AIP Conference Proceedings, 2007, Vol. 947, pp. 407-410.
[10] DOI: 10.1063/1.2813842.

[11] Rusov V.D., Linnik E.P., Kudela K, Mavrodiev S.Cht., Sharph I.V., Zelentsova T.N, Beglaryan R., Smolyar V.P., Merkotan K.K. Axion mechanism of the Sun luminosity and solar dynamo - geodynamo connection. arXiv.org > astro-ph > arXiv:1009.3340, 2010.

[12] Zubow K., Zubow A.V., Zubow V.A. The Phenomenon of Proton Dissolving in Vacuum and of Proton Condensation from Vacuum. Two Forms of Protons, Structure of Nuclei, Electrons and Atoms. J. of modern physics. 2010, vol.1, no.1, pp.175-184.

[13] Zubow K., Zubow A.V., Zubow V.A. Water Clusters in Plants. Fast Channel Plant Communications. Planet Influence. Journal of Biophysics Chemistry, 2010, vol.1, no.1, pp.1- 11.

[14] Yoshiteru Yonetani. A severe artifact in simulation of liquid water using a long cut-off length: Appearance of a strange layer structureChem. Phys. Letters. 2005, vol. 406, pp. 49 - 53.

[15] http://www.quora.com/Could-Mercury-be-the-dead-core-of-a-gas-giant-planet

[16] Zubow K., Zubow A.V., Zubow V.A. Signals from Extraterrestrial Intelligences. Horizons in World Physics, editor. Nova Science Publishers, Inc. NY. 2012, vol. 280. Chapter ID: _8738_Horizons in World Physics, in print.

[17] Maercker M., Mohamed S., Vlemmings W.H.T. et al. Unexpectedly large mass loss during the thermal pulse cycle of the red giant star R Sculptoris. Nature, 2012, vol. 490, pp. 232-234, doi: 10.1038/nature11511. http://www.eso.org/public/images/eso1239a/

[18] http://video.yandex.ru/#search?text=%D0%A8%D0%B8%D0%BF%D0%BE%D0%B2%2C%20%D1%82%D0%BE%D1%80%D1%81%D0%B8%D0%BE%D0%BD%D0%BD%D1%8B%D0%B5%20%D0%BF%D0%BE%D0%BB%D1%8F&where=all&filmId=I74kGL%2Fs8gE%3D.

[19] Shipov G. I. Theory of physical vacuum. Theory, Experiments, and Technologies. Moscov, Nauka,1997. (http://publ.lib.ru/ARCHIVES/SH/SHIPOV_Gennadiy_Ivanovich/_Shipov_G.I..html http://www.trinitas.ru/rus/002/a0231004.htm; http://www.trinitas.ru/rus/doc/0231/007a/02311016.htm; http://www.trinitas.ru/rus/doc/0231/004a/02311041.htm.

[20] Haldna J.L. Encyclopedia of Chemical. Moscov, Soviet Encyclopedia, 1990, vol. 2, pp. 393- 395.

[21] Bogdanov E.V., Manturova G.M. Equiclaster model of water. Radioelektronika, Biomedicinskaja Radioelektronika, 2000, vol. 7, pp.19-28, (in Russian).

[22] Lenz A., Ojamäe L. On the stability of dence versus cage-shaped water clusters: Quantum- chemical investigations of zero-point energies, free energies, basis-set effects and IR spectra of $(H_2O)_{12}$ and $(H_2O)_{20}$. Elsevier, Chemical Physics Letters. 2006, vol. 418, pp. 361-367.

Chapter 20

EARTH'S GRAVITATIONAL DIPOLE: GRAVITATIONAL COMPASS

The presence of an iron core in the crystalline state in the center of the Earth is generally accepted at the beginning of the 21st century [1, 2]. It can be spherical or represented as a giant crystal with a shape closer to spherical. The cause of the magnetic field is now understood but the question of where the gravitational field comes from remains unanswered. Analyzing the dynamics of the Earth's geomagnetic field variations, Olson [3] assumed the presence of a rapidly changing planet gravitational dipole, which is very sensitive to the surroundings.

Earlier, for remote monitoring of earthquakes' gravitational radiation (§9.2), we used the new Zubow gravitational spectrometer (GMS). GMS was later applied to record the gravitational radiation from celestial bodies in the slide plane of gravitational proton resonance (SPGPR, plane going through the Earth's rotating axis and the sensor place on the Earth surface, §8.13.1).

The purpose of this chapter is to investigate the source of gravitational radiation from the Earth, its structure, and its properties. To do this, a gravitational sensor based on agarose hydrogel (§5.6.7) is placed in a light and sound protected box, which in turn is installed in a grounded iron box in a thermostated room. The long-range order (LRO) of water is recorded for a cluster ensemble with a mass up to 3.3 million Daltons, while the water temperature and atmospheric pressure during the measurements remain constant (293 K). As investigation objects, the following systems were chosen: Earth-Sun, Earth-Moon, Earth-Mars, Earth-Venus, Earth-Mercury, Earth-Pluto, and Earth-center of our galaxy. The analysis of LRO is carried out when the celestial bodies appear in SPGPR. The energy calibration of the GMS proceeds according to the evaporation energy of water at 293 K (44.1 kJ/mol or $2.45 \cdot 10^6$ kJ/m^3). That means at this temperature, for the complete decomposition of all cluster ensembles needs 44.1 kJ/mol. To monitor the constellation of celestial bodies the Zaitsev program ZET 9 Lite (www.zaytsev.com) is applied. During the observation, the Sun's activity is characterized as weak (http://www.swpc.noaa.gov/weekly/). For the analysis of the polyethylene glycol viscosity the cylindrical viscosimeter "Reotest-2" (Germany) was upgraded with modern electronics and software for the shear stress is applied. PEG has a molecular weight of 400 Daltons (pure for gas chromatography, Merck, Germany).

The model. Suppose that in the Earth, in addition to the giant magnet, is a gravitational dipole, which surrounds the magnet axis like a broad belt. The whole structure, including both the magnet and gravitational dipole, is nearly spherical (Figure 20.1).

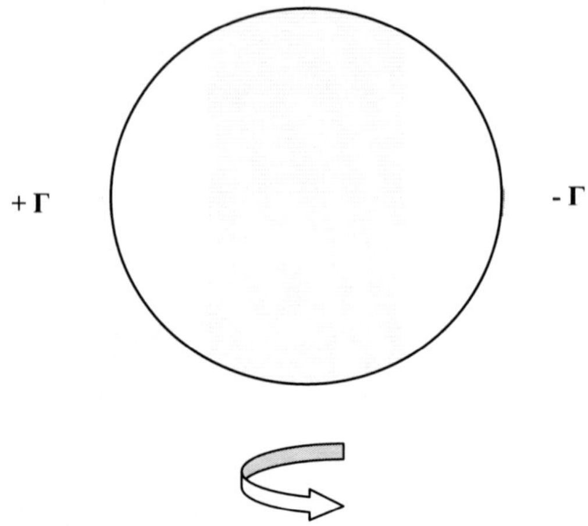

Figure 20.1. Model of the Earth. -Γ…+Γ gravitational dipole – “the broad belt,” the main section, which runs through the core center, is located perpendicular to the observer and to SPGPR. SPGPR in this figure divides the dipole into the two gravitational monopoles +Γ and -Γ and the axis of the Earth's rotation (not shown).

In this model, we are interested in the gravitational dipole, only. If it exists, then the celestial bodies will act differently with it depending on which side of SPGPR they are (Figure 20.2). In this case, the gravitational sensor is able to register differences of gravitational fields on both sides of SPGPR. It has to be mentioned that the GMS-spectrometer records molecular clusters, which get their formation energy from energy clusters (EC) of the physical vacuum. The molecular clusters are using the gravitational energy of EC for their self-organization [4]. EC, in turn, are formed in stationary gravitational waves of protons and their concentrations in space, and the gravitation waves themselves are the result of an imbalance in the equilibrium proton-physical vacuum (dissolving in a vacuum and condensing from a physical vacuum with super light speed, [5]). A dominated dissolution process leads to **-Γ** field, while the dominance of condensation favors a +**Γ** field.

The GMS-sensor and one gravitational pole are considered a unit, which interacts with a celestial body. If the interaction on both sides of SPGPR is equal, then there is no dipole, but if it is different, a dipole is assumed to exist (Figure 20.2). This model contradicts the theoretical models of gravitational dipoles as a particle-antiparticle pair in vacuum [6], as dark matter and dark energy [7], as two rapidly rotating huge masses [8], and as radiation of gravitational dipole by a pulsar [9]. It agrees well, however, with the views of Huang [10], who theoretically concluded that gravity is more stable when it is created by a dipole and not by a point. Furthermore, the presence of gravitational dipoles is discussed for neutron stars and super massive black holes [11], but to conceive this for the Earth is difficult. On the other hand, Dolgov discovered in 1990 that gravitational dipoles are possible in scalar fields with unusual properties and configurations [12]. Thus, our proposed model is new; therefore, it needs to be proven experimentally.

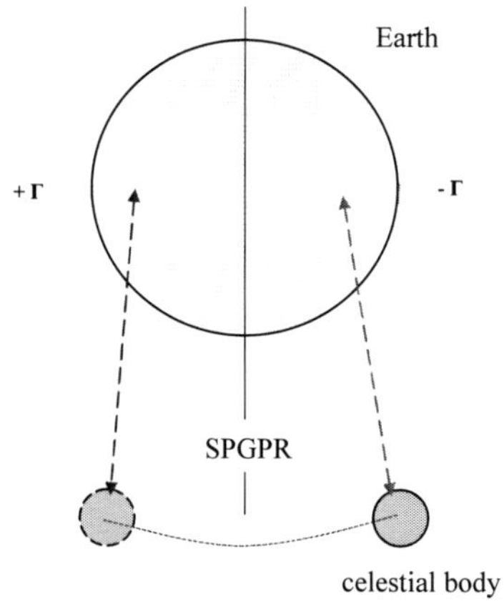

Figure 20.2. Model for a possible interaction of a celestial body with the gravitational earth dipole. Two situations of "mirror" locations of celestial bodies relative to SPGPR, left and right of SPGPR and against the gravitational dipole (+**Γ**...-**Γ**, Figure 20.1).

Agarose hydrogel containing 97 wt. % water is used as sensor for the gravitation mass spectrometer. The sensor is located inside in an isolated box in a room [2], Figure 4.0.1, with the coordinates 53°34'54"N; 12°47'02"E. The analysis of the long-range order in water of hydrogel at cluster level proceeds according to the method described in [13]. The visible constellation of celestial bodies is determined using the program ZET 9 (www.astrozet.net). The scanning of celestial bodies by SPGPR [14] is carried out in 3...20-second steps when the anthropogenic influence and gravitational radiation from other celestial bodies were minimal as well as the weather was clear. The energy spectrum of a water cluster ensemble is calibrated using the evaporation constant of water at the measuring temperature (44.1 kJ / mol).

Figure 20.3 shows how the energy of water cluster ensembles changes at the culmination of the Sun, the center of the Milky Way, and Jupiter and when they appear in SPGPR.

As shown in Figure 20.3, the curves show a completely different picture after SPGPR; for example, they change the sign. This confirms the correctness of the model shown in the Figures 20.1 and 20.2. A celestial body interacts differently with the Earth's induced dipole. Here the Earth's dipole seems to be subordinated to the dipole of the Sun. Before the culmination of the Sun and its SPGPR crossing, the part of collapsed clusters is around 80% but it is reduced to 15% after the crossing.

This phenomenon does not always appear. Apparently, the influence of other celestial bodies' dipoles is the condition for its existence. The sign change of the energy is suggested to be an indication of the instability of the Earth's observed dipole, e.g. when scanning the center of the Milky Way, which shows because of its mass heterogeneity, no abrupt change is seen in water cluster properties. When scanning Jupiter, a rarely occurring recovery of the dipole sign in a short time has been observed, therefore confirming the high instability of the Earth's gravitational dipole.

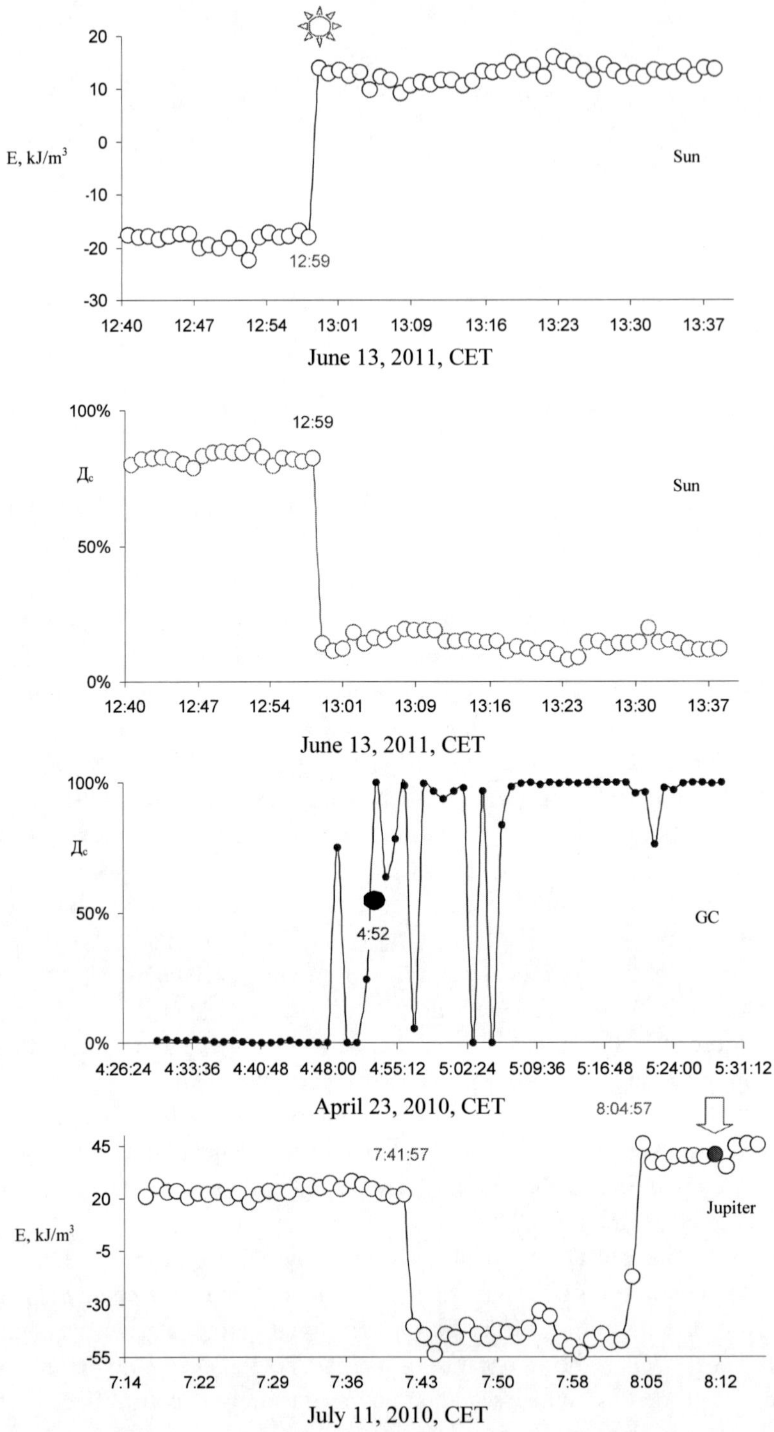

Figure 20.3. Examples for abrupt changes of the main characteristics in a water cluster ensemble in the sensor upon polarity reversal of the Earth's dipole. The developments in SPGPR (points) are given where the Sun, the galaxy's center (GC, black point), and Jupiter (arrow) are scanned.

Since the sensor is much closer to the gravitational dipole of the Earth than a celestial body, it is reasonable to assume that changes in the proton balance within the physical vacuum in the sensor reflect similar processes in the gravitational dipole of the Earth. Compared with a celestial body, the sensor is much closer to the gravitational dipole, so they are considered a unit. The nucleus ensemble of the gravitational dipole dominates that of the sensor.

If the Earth's gravitational dipole consists of crystalline iron, the equilibrium reaction between the iron protons with the physical vacuum is more energy-intensive (compact) than that of the protons in the hydrogen bonds in the water sensor. According to this logic, we can assume with some restrictions that the energy events in the GMS-sensor and the Earth's gravitational dipole are comparable. Then the replacement of the signs of the gravitational dipole (Figure 20.3) can be understood in terms of the forced influence of the celestial bodies on the balance of the proton with a vacuum.

If the celestial body emits a positive gravity ($+\Gamma_{cb}$, dominance of proton condensation on its dissolution) and if its $+\Gamma_{cb}$ potential is higher than the potential of a monopole in the Earth, then the sign of the Earth's monopole will be changed to $-\Gamma$ and in the Earth, the process of dissolution of protons in a vacuum will be the dominating one. If in the "game" three or more celestial bodies (opposition, conjugation) are involved, the gravitational charge on the Earth's monopole will adapt to the energy conditions created by participants. The opposite can also be expected if, for example, the gravitational field of the Earth influences the dipole of another planet.

Figure 20.4 shows the number of cluster kinds of water molecules in hydrogel when the Moon is crossing SPGPR. One can see the difference in the number of cluster kinds before the SPGPR crossing (~65) and after it (~18). This phenomenon can be explained by the structuring influence of the Earth's monopole on LRO of water. The reduced number of cluster kinds is caused by the destruction of hydrogen bridges (HB). The Earth's dipole attracts gravitational energy from the "naked" protons, which are accompanied with a shifting of the proton-physical vacuum balance to increased proton dissolution.

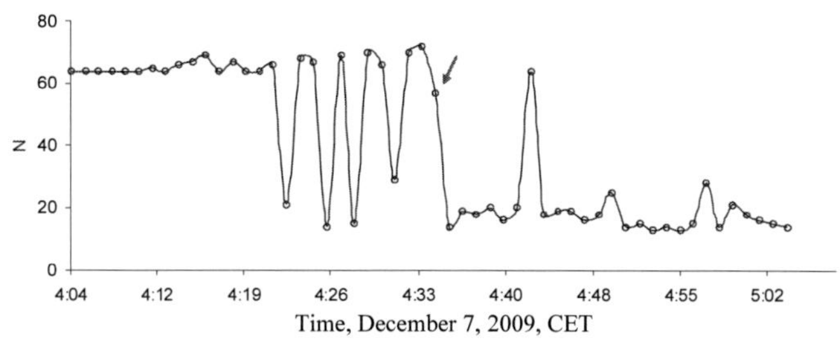

Figure 20.4. Number of water cluster kinds in the sensor at the Moon crossing SPGPR (indicated by arrow).

After the moon has passed through SPGPR, it comes into the area where the proton dissolution is dominant; it is the $-\Gamma$ pole of the Earth's gravitational dipole. Before crossing SPGPR, the Moon was in the area of $+\Gamma$ pole influence.

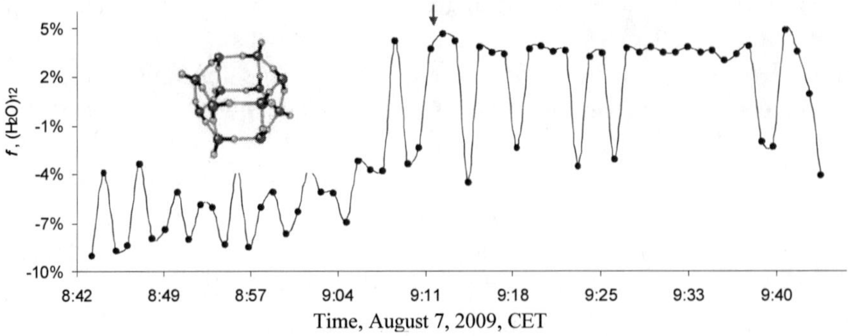

Figure 20.5. The energy part of the base water cluster $(H_2O)_{12}$ in the cluster ensemble in hydrogel at Mars' crossing of SPGPR (arrow). The model for the base water cluster in the collapsed form has been kindly provided by Professor A. Lenz (Sweden).

Figure 20.5 shows how the part of the basic water cluster changes at Mars' crossing of SPGPR. Density differences of the cluster before and after crossing SPGPR are visible.

From 08:42 to 09:15, the sensor records the cluster predominantly in a collapsed form. Here the hydrogen bonds are mainly concentrated within the clusters ($f < 0$ %). In this case, the point when Mars crosses SPGPR does not coincide with the azimuth and the moment of the Sun crossing the world's meridian (13:14). Any further celestial body seems to influence the orientation of the dipole. This case can serve as proof of the disk/bar nature of the gravitational dipole (Figure 20.1) and of its directional flexibility to external influences. After crossing SPGPR the expanded structures begin to dominate ($f > 0$%).

It becomes even clearer when the number of all the parts of the collapsed water clusters is analyzed. Figure 20.6 shows the state of the clusters in the sensor at the Sun crossing SPGPR.

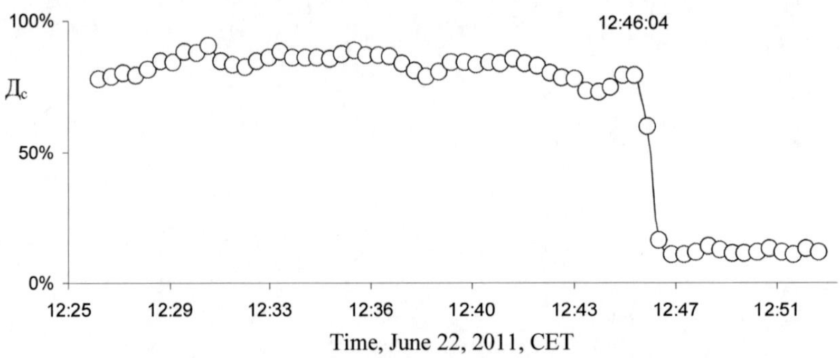

Figure 20.6. Part of collapsed clusters ($Д_c$) at the Sun's crossing of SPGPR.

In this figure, the dual character of the dipole is seen. To the left of the point at 12:46 (SPGPR), all water clusters in the hydrogel are in the collapsed energy-rich form because of strong hydrogen bridges. In contrast, the right side of the curve shows the almost complete destruction of the collapsed clusters due to the weakening of HB between the water molecules. Hence, the Earth's gravitational dipole has a significant influence on the structure of water and thus on the processes in the hydrosphere and even the climate. A similar process was of course already observed nine days earlier (Figure 20.3), but it is characterized by an even sharper decrease on the part of the collapsed water clusters in the sensor.

Figure 20.7 shows the dependence of the dynamic viscosity of polyethylene glycol (PEG-400) on the time of day. It is visible that until SPGPR crossing the signal form is very different from that after its crossing. Signals become narrower and stronger. A number of unexplainable signals of viscosity reduction could be caused under the influence of a specific constellation of celestial bodies as well as of an impulsive change of the total gravitational field. The behavior of the dynamic viscosity of the polymer can be understood from the position of the structural change of HB that affects the adhesion of water molecules. Furthermore, the broad peaks of the viscosity reduction up to SPGPR and the narrow ones after it are an indication of the heterogeneity of the gravitational fields of both sides of the Earth's dipole on December 6, 2009. We can assume that in the first case, the broad peaks are due to the slowly developing processes of HB destruction because of the insufficient gravitational energy. In the second case, on the contrary, the field of the monopole is stronger and the influence of the celestial bodies on the HB protons is impulsive. This phenomenon can be understood from the standpoint of different volumes of poles and as a consequence of the different masses of protons involved in the process of dissolution/condensation. Hence the gravitational dipole at the center of the Earth is labile.

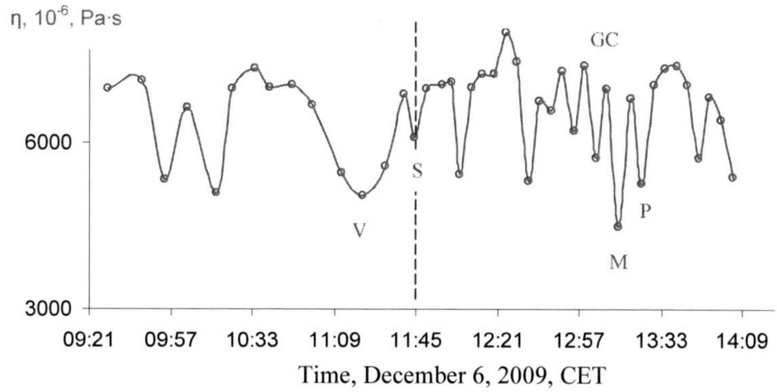

Figure 20.7. Dynamic viscosity (η) of PEG-400 versus the day time. Appearance in the SPGPR: V – Venus, S – Sun, M – Mercury, and P – Pluto. Dotted line – Sun in SPGPR.

In Figure 20.8 is given the diagram illustrating the results of the experiment. During the rotation of the Earth, SPGPR scans the space on both sides of our planet, and the sensor can be at different distances from the gravitational monopoles. Those nearest to the sensor monopole will have a dominant influence on the water cluster structure. At the moment of the sign change of the gravitational influence on the sensor, the LRO structure (water) changes abruptly as well.

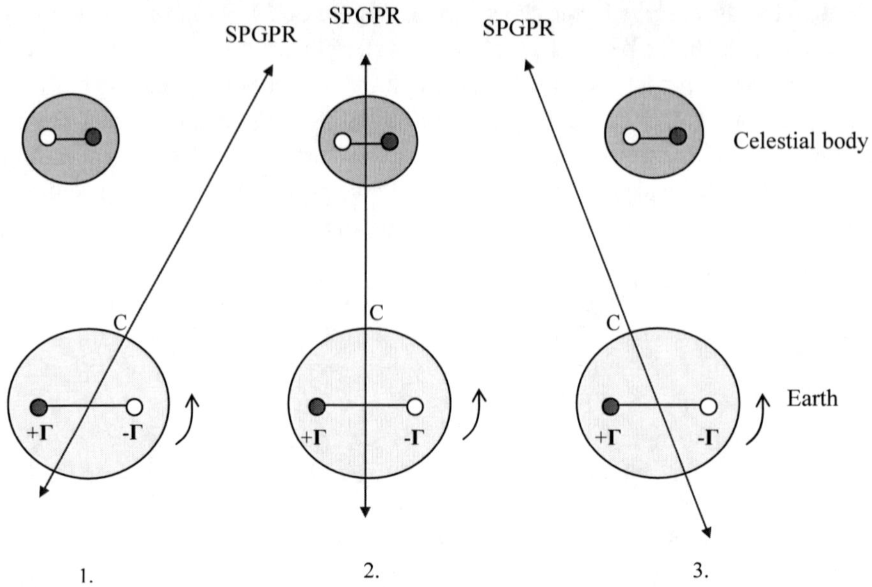

Figure 20.8. Scheme monitoring the center of the Earth's dipole. C – sensor. 1 – the state where SPGPR does not match the line connecting the centers of the Earth and a celestial body, for example, the Moon (the **-Γ** monopole dominates over the sensor); 2 – the state where SPGPR coincides with the line connecting the centers of the Earth and a celestial body (equal influence from both poles on the sensor); and 3 – the state in which the +**Γ** monopole dominates over the sensor.

CONCLUSION

Thus, in the Earth there is a gravitational dipole.

It is the main source of gravitational radiation from the planet.

The structure of the dipole could be that of a virtual disk or bar, whose rotation axis agrees with that of the Earth.

The gravitational dipole is a vector in that it reflects the middle direction of the influencing space masses.

The gravitational dipole of the Earth can be used as a gravitational compass.

The gravitational potential and the signs of the dipole depend on the equilibrium of protons with the physical vacuum and on single celestial bodies and their constellation.

The gravitational dipole of the Earth has the strongest influence on the long-range order in molecular matter, especially in water, through resonance with the protons of hydrogen bridges.

The forced change of hydrogen bridges in the water of hydrogel caused by the Earth's gravitational dipole will affect the physical chemical properties of the molecular matter (viscosity, elasticity, rate of chemical and biological reactions, etc.), in particular of water in the hydrosphere. In addition, it will indirectly have an impact on climate, health, and the behavior of humans and animals.

REFERENCES TO 20

[1] Hemley, Russell J.; Cohen, Ronald E. Structure and bonding in the deep mantle and core. Philosophical Transactions of the Royal Society of London, Series A: Mathematical, Physical and Engineering Sciences. 1996, vol. 354, no.1711, pp. 1461-1479.

[2] Cote, Alexander S.; Vocadlo, Lidunka; Brodholt, John P. The effect of silicon impurities on the phase diagram of iron and possible implications for the Earth's core structure. Journal of Physics and Chemistry of Solids. 2008, vol. 69, no. 9, pp. 2177-2181.

[3] Olson, P. Gravitational dynamos and the low-frequency geomagnetic secular variationProceedings of the National Academy of Sciences of the United States of America. 2007, vol. 104, no. 51, pp. 20159- 20166.

[4] Zubow K.V., Zubow A.V., Zubow V.A. Ensemble of Clusters – New Form of Molecular Matter, Risks and Chances. Zubow Equations. In *Advances in Chemistry Research*, vol. 5, 2010, pp. 107- 145. Editor J.C. Taylor. Novapublisher. NY. USA

[5] Zubow K., Zubow A.V., Zubow V.A. The Phenomenon of Proton Dissolving in Vacuum and of Proton Condensation from Vacuum. Two Forms of Protons, Structure of Nuclei, Electrons and Atoms, J. of modern physics, 2010, vol.1, no.1, pp.175-184.

[6] Hajdukovic, Dragan Slavkov. Dark matter, dark energy and gravitational proprieties of antimatter. arXiv.org, e-Print Archive, Physics (2008), 1-8, arXiv:0810.3435v1 [physics.gen-ph]. Publisher: Cornell University Library, http://aps.arxiv.org/ftp/arxiv/papers/0810/0810.3435.pdf.

[7] Blanchet, Luc; Le Tiec, Alexandre. Dipolar dark matter and dark energy. arXiv.org, e-Print Archive, Astrophysics (2009), 1-4, arXiv:0901.3114v1 [astro-ph.CO]. Publisher: Cornell University Library, CODEN: AARSC7 http://aps.arxiv.org/PS_cache/arxiv/pdf/0901/0901.3114v1.pdf.

[8] Jeong, Eue Jin. Non-Newtonian force experienced by gravitational dipole moment at the center of the two mass pole model universe. Physica Scripta. 1999, vol. 59, no. 5, pp. 339-343.

[9] Sivaram, C. Constraints on gravitational dipole radiation from pulsars. arXiv.org, e-Print Archive, Astrophysics (2008), 1-5, arXiv:0803.1896v1 [astro-ph]. Publisher: Cornell University Library, CODEN: AARSC7 http://aps.arxiv.org/ftp/arxiv/papers/0803/0803.1896.pdf.

[10] Huang, Wung-Hong. Thermal giant graviton with non-commutative dipole field. Journal of High Energy Physics (online) (2007), (Nov.), No pp. given. Publisher: Institute of Physics Publishing, CODEN: JHEPFG ISSN: 1029-8479. http://www.iop.org/EJ/article/1126-6708/2007/11/015/jhep112007015.pdf

[11] Scharre, Paul D.; Will, Clifford M. Testing scalar-tensor gravity using space gravitational-wave interferometers. Physical Review D: Particles and Fields. 2002, vol. 65, no. 4, pp. 042002/1- 042002/8.

[12] Dolgov A.D. Gravitational dipole. Letters to journal of experimental and teoretical physics. 1990, vol. 51, no. 7, pp. 346-349, in Russian.

[13] Zubow Kr. V., Zubow A.V., Zubow V.A. Cluster structure of liquid alcohols, water and n-Hexane Journal of Appl. Spectr., 2005, vol.72, no.3, pp. 321-328.

[14] Zubow K., Zubow A.V., Zubow V.A. Experimental Methods for the Determination of the Super Light Velocities of the Gravitation. Nature, Structure and Properties of Gravitation Waves. In Horizons of Physics. Editor A. Reimer. NY. vol. 277. 2012, pp. 109-118.

Chapter 21

PROTON DISSOLUTION IN A VACUUM AND PROTON CONDENSATION FROM A VACUUM

Proton jumping via hydrogen bridges (H-bonds) or via even a group of atoms in a molecule has been illustrated as follows in Scheme I:

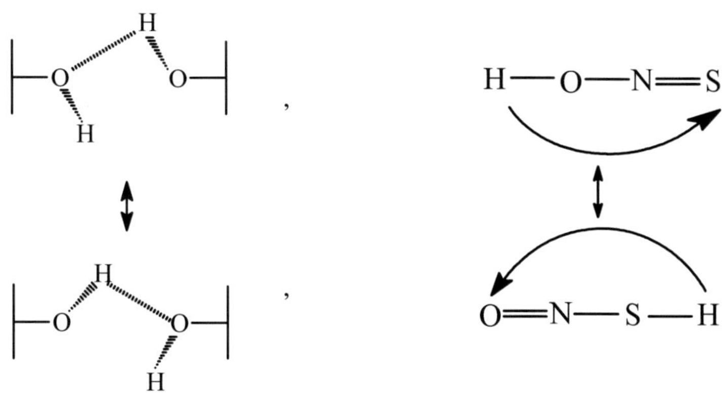

I

The hydrogen atom that is linked with an oxygen atom simultaneously builds an H-bond (~0.2 nm) with other molecule fragments or other molecules; it is able to jump to the second oxygen atom. Instead of oxygen, other atoms like nitrogen, sulfur, etc., can be applied to this scheme. Inner molecular jumps in low molecular substances can proceed at distances longer than 0.2 nm but in high molecular ones, the scale is even on the umpteenth nm.

This effect has been observed for many organic and inorganic substances and it is well described [1, 2, 3] even though a proton jumping through electron orbitals of atoms and molecules seems to be more or less improbable. Furthermore, because of the unavailable mechanism models and analysis methods, it is difficult to understand this effect. On the other side, hydrogen bridges were found to be important for the stabilization of the conformation of synthetic macromolecules and biomolecule coils (proteins, starch). Destruction of hydrogen bridges leads to a change of the conformation of protein macromolecules, to changed directions of biochemical reactions, and finally to other behavior from organisms. In our earlier publications we informed on fields of unknown nature that are present in molecular systems with hydrogen bridges. For instance, in [4] we described the "wall effect"; near the wall there is a special gravitation excitation (water, salt solutions) that seems to be connected with hydrogen bonds and proton jumping. We developed the idea about the proton jumping via the vacuum to the nearest energy vacancy place in space. According to this hypothesis, proton jumping is connected with the acceleration and breaking of masses in the proton leading to weak gravitational radiation. To identify a gravitational wave in a real time it is necessary to "shift" the interference of the gravitation pair (mutual compensation) at the time or to expand the events in the space. For that, we have to look for the resonance of gravitation waves between micro and macro objects, for example, in gravitation pairs between clusters in the molecular matter and masses of planets. To prove this hypothesis we developed a gravitational mass spectrometer by which we registered molecular mass clusters as well as their interaction with the surroundings [5, 7].

It was the aim of the present chapter to understand the mechanism of proton jumping in molecular matter as well as to find out the reasons that cause unknown fields within it.

As investigation objects, the following substances were chosen: water, agarose hydrogel (97 wt. % water), salt solutions (NaCl, 3.5 wt. %), starch, polyethylene, atactic polystyrene (aPS, average molecular mass of 15 kDa), and biomatrices of eggs and potato tubers. The GMS-sensor was placed directly in the biomatrices, pressed (1-2 kPa) into powders or films, or installed in liquids that were aged for six month. The GMS method is described in [5, 6, 7, 8]. Masses and oscillation frequencies of clusters in the samples were calculated by the Zubow equations using the Zubow constant $6.4 \cdot 10^{-15}$ N/m for biomatrices and water, $6.8 \cdot 10^{-15}$ N/m for starch, and $5.2 \cdot 10^{-15}$ N/m for aPS. Samples were placed in an iron (grounded) box that was isolated from light, anthropogeneous noises, and heat/mechanical fields. The box itself was in a building far from industrial centers. Some curves, reflecting the energy flow (sum of cleaned signals) that reached the GMS-sensor in the samples during August 15-16, 2008 are shown in Figure 10.6.1.4. The base line (dotted line) was obtained as follows. After the noises were registered by the GMS-sensor in a vacuum (10^{-4} N/m^2), the real signals (first) were extracted. An analogous procedure was done for the sample without the vacuum (second). The real sample signals were calculated by subtracting the second signals from the first ones [6], and 265,784 represents the sum of all signals in the mass interval up to 4.3 billion Da.

The existence of cluster ensembles in molecular matter showing both wave and corpuscle properties is undoubted [9]. Clusters oscillate and are affected by noise. These oscillations, e.g. in liquids, are similar to a "structure transfusion" and are connected with a constant virtual mass jumping in the dynamic balance between noise energy stabilizing clusters [10] and heat (kT), which destroys the clusters.

In Figure 10.6.1.4 (§10.6), the planet's influence on the energy of cluster ensembles in molecular matter was shown. As is visible, the molecular cluster ensemble energy is sensitive to the planet's constellation and this energy, for instance, of a gravitational nature, was found to be detectable by the GMS method.

Let us discuss two pairs of peaks at 09:00 and 15:00, and at 11:00 and 17:00. Both peaks are on a line that is perpendicular to the line connecting the centers of the Sun and the Earth, and of the Milky Way and the Earth, accordingly. The peaks can be interpreted as interference of the gravitational waves and with the help of the Junge equation the wave length can be calculated and compared with literature references. The calculation was made using:

$$\Lambda \geq D \cdot S/R$$

where λ is the average gravitation wave length; R is the distance from the gravitation source to sample (from the Sun 150 million km, from the Milky Way's center 10,000 pc); D is the distance between the peaks, amounting to D = 2· r· sin(90°-53°)·sin(120°/2); r = 6,370 km (radius of the Earth); and S is the diameter of the source (the Sun: 1,392,000 km, the galaxy's center: 500 pc). Here the role of two slits is played by the Earth's atmosphere and the role of display is the area that connects the peaks with each other and stands vertically to the gravitation source. Thus, λ ≥ 62 km (gravitation waves of Sun) and λ ≥ 330 km (gravitation waves of the galaxy's center) are found. The gravitation wave length calculated by Hickey [11] was found to be ~100 km. Furthermore, Smirnov described a device for the registration of gravitation excitation in the frequency interval 10 Hz...40 kHz [12], which includes all investigated cluster oscillations in molecular matter. Therefore, we can note with a great probability that the GMS-device registers directed gravitational radiation (GR).

Now we want to analyze how GR influences the long-range order in water, whose structure was already proved experimentally [8]. Inside the clusters, the water molecules are linked with each other by two hydrogen bridges but on the cluster surface, the water molecules possess only one hydrogen bridge (HB) for their interaction with their surroundings. If the proton jumping in the HB proceeds with the emission of weak GR then we have to expect a resonance between the weak GR of molecular matter and the GR of the planets. The proof of this assumption is given in Figure 21.1.

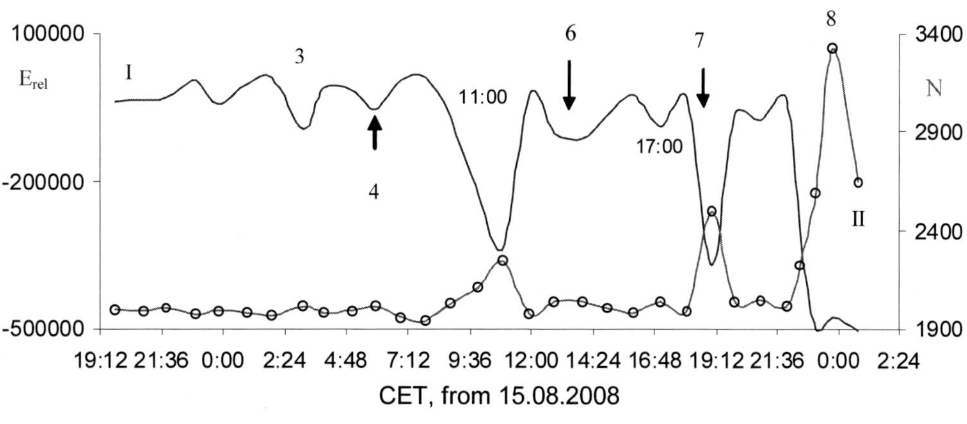

Figure 21.1. Behavior of the GR energy curve registered by GMS-sensor in water (I) and of the number of cluster kinds (II) in dependence on the events in the near space, cluster ensembles up to 4.3 billion Da. For comments, see Figure 10.6.1.4.

As is visible from Figure 21.1, the number of clusters in water reacts to the planets' GR. This effect is shown most strongly at full moon (23:16, point M) and at the eclipse of the Moon (the Moon is covered up to 80% by the Earth). The GR of Jupiter at its rise has a strong influence on the cluster number in water, too. To understand the structure of GR, we investigated the GR of Jupiter (Figure 21.2). The GR of Jupiter consists of two parts that were ascribed to the signals of "gravitational Jupiter" and of "visible Jupiter." The sample was fully protected from visible light; therefore, the signal of "visible Jupiter" can be interpreted as ripples of main Jupiter's GR [7, 11]. That means we have the right to use the correlation between the visible and gravitational constellation of the planets.

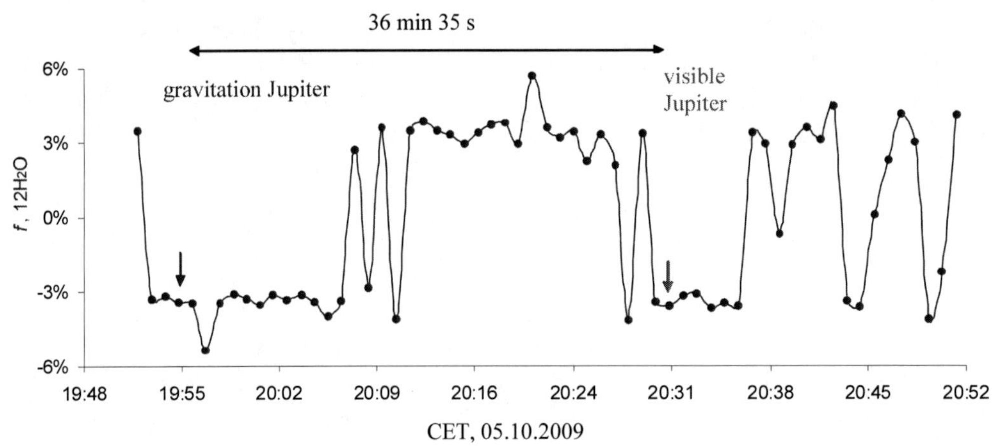

Figure 21.2. Signal intensities of the base water cluster in agarose hydrogel when the flows of both Jupiter's gravitational energy kinds are in the slide plane of gravitational proton resonance (SPGPR). Distance to Jupiter = 658,629,000 km; altitude 19.7°. SPGPR – plane going through the Earth's rotating axis and the sample place on the Earth's surface.

In Figure 21.1 (§2.1), computer models for the simplest water clusters, consisting of the base water cluster $(H_2O)_{11\pm1}$ [13, 14], and experimentally observed methanol clusters are given [8]. The base water cluster belongs to an ensemble of dominating clusters that were forcibly formed in the white noises: $(H_2O)_{11\pm1}$, $(H_2O)_{100}$, $(H_2O)_{178}$, $(H_2O)_{280}$, $(H_2O)_{545}$, $(H_2O)_{903}$, $(H_2O)_{1,351}$, $(H_2O)_{1,601}$, $(H_2O)_{1,889}$, etc.

Jupiter at its rise (Figure 10.6.1.4) and culmination (Figure 21.2) strongly influences the long-range order in water through its gravitational radiation and destroys the cluster ensemble temporarily. On the other side, the destruction of the water cluster ensemble has been observed during the gravitational interference (signal at 11:00) and even at the GR of Uranus and sunrise (Figure 21.1).

Cluster ensembles in methanol react analogously to the GR flow of the planets where the highest destruction of hydrogen bridges takes place at Jupiter's rise. The behavior of polyethylene, however, which does not have any hydrogen bridges, is something differing from the strongest cluster ensemble's destruction at the Moon's eclipse and at the GR interference of the galaxy's center and the WGR of polyethylene (11:00 and 17:00).

In the Table 5.1 (§5.1), the GR frequencies of celestial bodies and molecular water clusters that were calculated by the Zubow equation are given. Let us now discuss the hydrogen bridges (HB) in the amorphous part of starch (Figure 21.3). As shown in Figure 21.3, HB, reflecting conformation changes in the starch coils, react on the GR of celestial bodies. In Figure 18.7 (§18), the review of the GMS-spectra of starch powder at Jupiter's rise will be given (see Figure 10.6.1.4, too) and the destruction of the hydrogen bridges that form the globular structure of this polymer can be seen. This destruction process proceeds reversibly and is an indication for the special resonance interaction between gravitational radiations. On the other side, the dominating influence of the planets' GR on a cluster ensemble in molecular matter often leads to a simultaneous energy inversion of the density of all clusters. In starch, the conformation of macromolecule coils are known to be stabilized by

HB and its destruction, which is caused by the resonance with the planets' GR, should be modeled by a forced compression. With pressure, the length of HB shall be changed and the oxygen influence on the "naked" proton becomes stronger. We found that by pressing the starch powder mechanically (1…5 kPa), the hydrogen bonds were forcibly and reversibly destroyed, too. Consequently, mechanical pressure is comparable to the planets' GR.

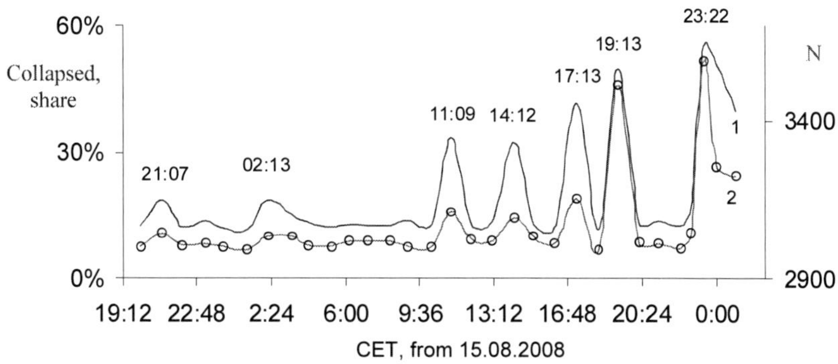

Figure 21.3. Share of collapsed clusters (1) and of cluster kinds' numbers (2) in the amorphous part of potato starch observed with the GR influence of planets (Figure 10.6.1.4).

Thus, the existence of weak gravitational radiation in molecular matter that is caused by super light mass transfer in hydrogen bridges has been concluded. It is hard to imagine this process as a simple jump on a distance of 0.2 nm or longer, a process in which the proton must penetrate a "net" of electron clouds of atoms and molecules. There is another way for this movement, however, namely, the transition of the proton into the vacuum (dissolution) connected with the arising of its equivalent simultaneously in the next free energy place. This process of super fast dissolution and condensation of mass could cause the WGR from molecular matter [14]. Under the examined substances, the ones that stand out are those that do not have any HB (aPS, polyethylene). The proton jump in HB is therefore a special case of the general phenomenon for the proton jump from the atomic nucleus into a vacuum and vice versa.

To evaluate the velocity of the proton jumping into a vacuum the results of the GR wave length analysis were applied (Figure 10.6.1.4, §10.6) using in the first approximation the Planck equation, $E = h \cdot c_\Gamma \cdot / \lambda$, where h is Planck's constant; λ is the wave length, m; and c_Γ is the velocity, m/s. The energy value (E) can be considered as representative of the disintegration of energy in the starch coils [15]. If we assume that this value amounts to ~2 kJ/mole, then:

$$c_\Gamma = E \cdot \lambda / h > 15 \cdot 10^{35} \text{ m/s}$$

This velocity is $5 \cdot 10^{27}$ times higher than the light velocity and the GR frequency is equal to $c_\Gamma \cdot 1/\lambda = 2.32 \cdot 10^{31}$ s^{-1}, which does not agree with the gravitation frequencies published in [16, 17, 18, 19] as well as with the conclusions on the GR influence on the long-range order in molecular matter. This situation, however, can be cleared with the results of Kokkotas [20], according to whom at the moving of masses their gravitational waves initiate ripples of smaller ones (gravitational noise) that spread with light velocity. The gravitational noise we observed to be the reason for the forced oscillation of clusters in molecular matter and matter in space but the clusters themselves are the result of stationary gravitational waves. Then, the energy dependence of the main gravitational waves on their wave length can be given as:

$$E = z \cdot \lambda$$

where $z \sim (2...3) \cdot 10^{-2}$ J/m, $\lambda \sim 65{,}000...100{,}000$ m

Of course, the proton moving with super light velocity has to be understood as a short living time at the level of the proton dissolving in a vacuum (P_p) and proton condensing from a vacuum (P_K). Then, the model for the proton jump can be shown in Figure 19.1.8. In this figure, the distance between dissolving and condensing protons can be forcibly changed, for example, by a simple sample compression [7] or by resonance with analogous processes from planets. This nucleus model does not contradict the idea concerning the cluster formation in baryonic matter, according to which the mass transformation proceeds in line with strictly deterministic gravitational laws [7]. Physicists have to explain the stability of the great and thick accumulations of positively charged particles in a very small volume of the nucleus. One tried to find a force that holds these particles together but it was not found. Now as we understand it, the proton is an oscillating substance (dissolving and condensing) so the nucleus formation of chemical elements and their isotopes should be clearer. In nuclei, the proton's dissolving is in an energy balance with the condensing of the neighboring proton (see scheme). In this case, the probability that two positive charges are besides each other gets lower but the gravitational interaction in the nucleus becomes more dominant.

The instability of radioactive isotopes has been explained by the destruction of the oscillation harmony in the nucleus. Radioactive isotopes are characterized by a high potential energy of their nuclei (as a pendulum system) and their minimization is connected with nucleus disintegration.

The molecule stability of hydrogen (H-H) can be better explained by this model, according to which the protons alternately dissolve and condense and the electrons permanently catch up with the positive charges (condensed proton). Here the electrons permanently rotate around the condensed proton and its "Fata Morgana" (dissolved proton), leading to multiple canonical structures, for instance, H-H.

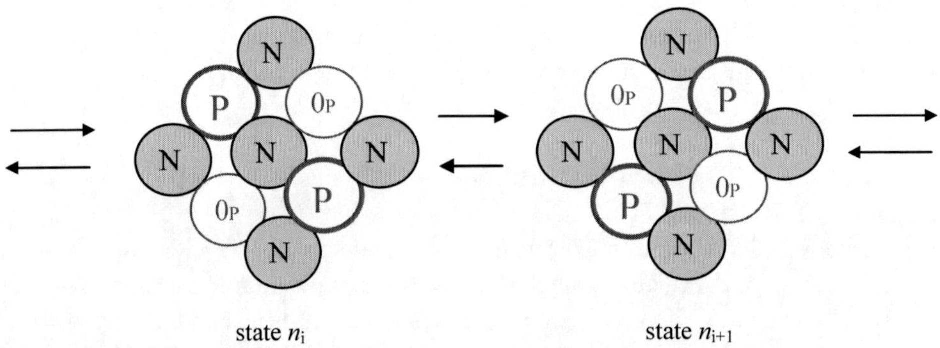

Figure 21.4. Statistical model for the nucleus formation of beryllium (four protons (P) and five neutrons (N)). P – proton condensed from vacuum and 0p – place of the dissolved proton in vacuum. The most probable states are shown.

For the beryllium nucleus, two canonical structures shall be shown. For example (Figure 21.4), this scheme is analogous to the one that describes the formation of clusters in molecular matter [9].

The electron clouds around the nuclei can be understood as stationary shock waves that arose from condensing protons from a vacuum. The shock wave front is in balance with the energy state of the vacuum $1.4 \cdot 10^{16}$ kJ/m^3 [7, 9]. This front (electron orbital) is permanently generated by high frequency shock waves from the condensing proton and at proton dissolution; it does not manage to dissipate in the vacuum and therefore saves its spherical form (model for hydrogen atom, Figure 21.5) probably because of a higher inertia of the dissipation process. A proof for this model could be that the electron orbital in the hydrogen atom oscillates (known as electron transition from one energy state to another one). In accordance with this logic, the energy released at proton condensation should correspond with the vacuum energy, namely, with ~$1.4 \cdot 10^{16}$ kJ/m^3. For multiple proton nuclei, the spherical shock wave construction shall be changed to differently formed ones, known as "orbital hybridization" in chemistry. Here the hybridization reflects an energy minimization process. On the other side, the state of free electrons can be understood as an energy cluster of excited matter from a vacuum that exists in movement form only. According to this model of the nucleus, the molecule is less caused by the unification of nuclei through electron orbitals than by a forced cluster formation of protons in stationary gravitational waves from space [7]. The concentration of atomic nuclei in a cluster represents a molecule but that of any molecule is a molecule cluster. Starting out from this model, the electron orbital can be seen as a shock wave front in the compressed vacuum, which is another form of matter. Under special conditions, the shock wave front could be concentrated in a cluster of negative energy—electron—particle. We therefore have to work with two forms of matter that are not mixable with each other: vacuum matter and matter that shall be released or absorbed upon the condensation/dissolution of a proton.

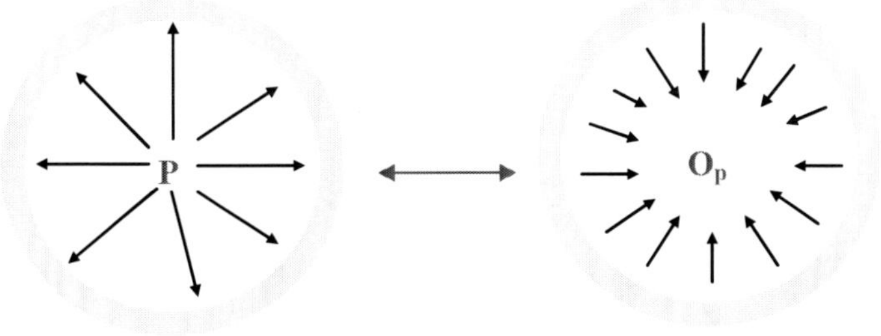

Shock wave (electon cloud – orbital) around condensing proton from vacuum

Shock wave evolution (electon cloud – orbital) around dissolving proton in vacuum

Figure 21.5. Model for the formation of a spherical electron orbital in the hydrogen atom. This orbital is caused by a shock wave front that arises from the pulsation of the condensing proton from the vacuum (left) and from the forward movement at a proton dissolving in a vacuum (right). The structure and energy of the vacuum are described in [7, 9].

The GMS-spectra observations of water at the period when the Sun is in SPGPR could be additional proof for the dissolution of the proton and for its existence in two forms (dissolved and condensed). The results of these observations shall be given in Figure 21.6.

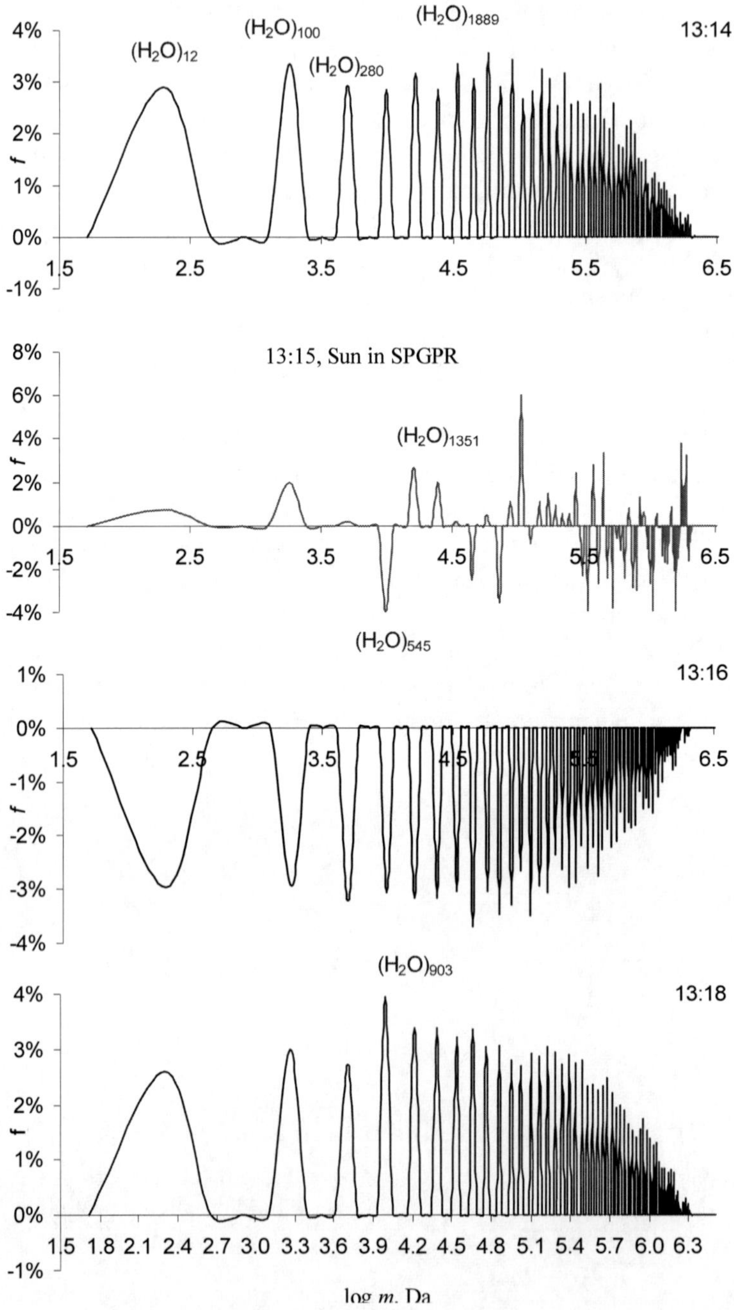

Figure 21.6. Dynamics of the proton state in hydrogen bonds between water clusters in agarose hydrogel at the period when the Sun is in SPGPR (August 7, 2009).

As is visible from Figure 21.6, all base skeletal water clusters are present in the water of the agarose hydrogel [21]. The GMS-spectra are identical up to the entrance of the Sun in

SPGPR (13:14) and after its exit from SPGPR (13:18). The molecular clusters are given in their expanded forms, indicating a strong inter-cluster interaction through the hydrogen bonds. In these bonds, the so-called "naked" protons [22] are in a condensed state. At the entrance of the Sun in SPGPR (13:15), the protons disappear. This is explained by their forced dissolution into a physical vacuum and drainage into the direction of the Sun's center [7].

Thus, the structure of hydrogen bonds between molecular clusters has been destroyed and that means the inter-cluster hydrogen bonds possess the weakest protons that shall be removed to the Sun's center first. In the following two minutes in the long–range order of water, another form of hydrogen bonds was observed where water clusters became collapsed (13:16...13:17). A weakened proton is present here in the inter-cluster hydrogen bonds that return into the normal condition at 13:18, again, the same state before the entrance of the Sun in SPGPR. Therefore, two forms of "naked" protons and a state without any protons were concluded to be present in the hydrogen bonds (Figure 21.7).

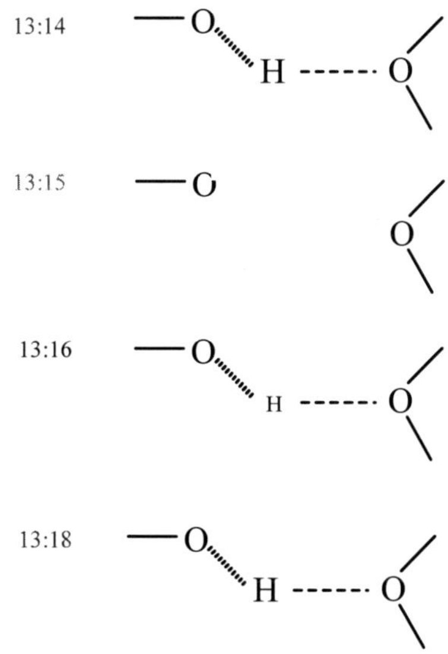

Figure 21.7. Scheme illustrating the inter-molecular cluster states of hydrogen bonds when the Sun hits SPGPR.

If the dissolution of protons in the physical vacuum takes place under the influence of the neutrino flow (ν, in the scheme of §24) from the celestial bodies, then the neutrinos as a solvent of baryonic protons can be seen as a part of them. The basis of this assumption is the universal rule: similar dissolves into similar.

At this time, to differentiate between the dissolution and condensation state of protons is impossible because they proceed with a super light velocity, however, the focus can be on the canonical structures in their harmonic resonance. It does not have to be excluded that neutrons take place in this family of pendula, too. The gravitational strength that holds the nucleus as a unit together is the result of the minimization of the potential energy of this mass ensemble [7].

As is shown in Figure 21.7, the proton's dissolution is accompanied by the disappearance of its electron (stationary front of shock wave). Here, in the matter, only one free radical of the so-called "free electron," as it was understood by the natural scientists of the 20th century, remains. If only the proton should dissolve, then the matter should have a gigantic negative charge but this is not the case. This simple logic supports the suggested electron model as a phenomenon of a stationary shock wave front (Figure 21.5). Our universe consists consequently of two, not mixable, highly dispersed and competitive forms of matter: matter of a vacuum and matter of a proton.

CONCLUSION

Thus, the mass transfer of the protons of hydrogen bonds or atomic nuclei in a vacuum (dissolution) proceeding with a super light velocity as well as the almost simultaneous appearance of protons from a vacuum (condensation) at the nearest vacant place in space are concluded to be the reason for the weak gravitational radiation of molecular matter.

The structure of the hydrogen atom reflects the surface state between the two forms of matter: matter in a vacuum and matter consisting of dissolved protons. These two forms could exist independently of each other.

The electron orbitals in atoms and molecules are stationary shock waves in a vacuum coming from two fast running processes (dissolving/condensing of protons) in atomic nuclei.

REFERENCES TO 21

[1] Isaev A. N. Co-operation interactions of hydrogen bonds in the processes with proton jumping and water molecules participation. Modelling of biochemical systems. Russian Chem. Journ. (J. of russian chem. soc. Mendeleev D.I.). 2007, vol. LI, no. 5, pp. 34 -48. (in Russian).

[2] Baisa D. F., Chesnokov E.D., Ovcharenko A.I., Vertegel I.G. NMR study of proton dynamics in ferroelectric $KIO_3 2HIO_3$ crystal. Functional Materials. 2008, vol. 15, no.1, pp. 19-22.

[3] Pablo Jaque, Alejando Toro-Labbe, Peter Politzer, Paul Geerlings. Reaction force constant and projected force constants of vibrational modes along the path of an intramolecular proton transfer reaction. Chem. Phys. Letters. 2008, vol. 456, pp.135 – 140.

[4] Zubow A.V., Zubow K.V., Zubow V.A. The Phenomenon of Water Clusters Distribution in Water and Solvated Clusters of Ion Pairs of NaCl in Solution Near the Wall. The Wall Effect. J. of chem. Industry (Sankt Petersburg). 2006, vol. 83, no. 6, pp 300- 306. (in Russian).

[5] Zubow K. V., Zubow A.V., Zubow V.A. Using of flicker noise spectroscopy for non destroying analysis of nano structures. Zavodskaja Laboratorija. Diagnostics of materials. 2008, vol. 74, no. 9, pp.40-45. (in Russian).

[6] Zubow K. V., Zubow A.V., Zubow V.A. The Dynamics of Low Frequency Movements of Molecular Clusters in the Hardening Process of Epoxide Resins. Chem. Promislennost segodnja. 2008, no. 9, pp. 12-21. (in Russian).

[7] Zubow K. V., Zubow A.V., Zubow V.A. Principles of gravitation stectroscopy.New form of molecular matter. Processes. Fields. Aist H&C gmbH, Berlin, 2010, electronic book (www.zubow.de), p. 854. (in Russian).

[8] Zubow K. V., Zubow A.V., Zubow V.A. Cluster structure of liquid alcohols, water and n-hexane. J. of Appl. Spectr. 2005, vol. 72, no. 3, pp. 321-328.

[9] Zubow K.V., Zubow A.V., Zubow V.A. Ensemble of Clusters – New Form of Molecular Matter, Risks and Chances. Zubow Equations. In Advances in Chemistry Research, vol. 5, 2010, pp. 107- 145. Editor J.C. Taylor. Novapublisher. NY.

[10] W. Horsthemke, R. Lefever. Noise-Induced Transitions. Theory and Applications in Physics, Chemistry and Biology. Edit. Hermann Haken. Springer-Verlag, Berlin, Heidelberg, NY, Tokio, vol. 15, 1984.

[11] Hickey, Michael P.; Yu, Yonghui. A full- wave investigation of the use of a "cancellation factor" in gravity wave -OH airglow interaction studies. Journal of Geophysical Research, [Space Physics]. 2005, vol. 110, no. 1, pp. A01301/1-A01301/15.

[12] Smirnov V.N., Egorov N.V., Shchedrin I.S. A New Detector for Pertrubations in Gravitational Field. Progress in Physics. 2008, vol.2, no. 4, pp. 129-133.

[13] Bogdanov E.V., Manturova G.M. Equiclaster model of water. Biomedicinskaja Radioelektronika. 2000, no. 7, pp.19-28. (in Russian).

[14] Braginsky V. B. Gravitational radiation. Fisitcheskij encikloped. Slowar. Moscov: Sovietskaja Enciclopedija. 1984, p.137

[15] Kozlov S.S., Krivandin A.V., Shatalova O.V., Noda T., Bertoft E., Fornal J., Yuryev V.P. Structure of starches extracted from near-isogenic wheat lines. Journal of Thermal Analysis and Calorimetry. 2007, vol. 87, no. 2, pp.575-584.

[16] Di Virgilio, A.; Bigotta, S.; Barsotti, L. *et al.* Experimental upper limit on the estimated thermal noise at low frequencies in a gravitational wave detector. Physical Review D: Particles, Fields, Gravitation, and Cosmology. 2007, vol. 76, no.12, pp. 122004/1-122004/10.

[17] Katsavounidis E., Ballmer S. For how long will gravitational waves remain hidden? Physics Letters A. 2005, vol. 347, no.1-3, pp. 33-37.

[18] Stebbins R.T., Bender P.L. A lunar gravitational wave antenna using a laser interferometer. In AIP Conference Proceedings. 1990, vol. 202, pp.188-201.

[19] Gondek-Rosinska, Dorota; Limousin, Francois. The final phase of inspiral of strange quark star binaries General Relativity and Quantum Cosmology. Preprint Archive, (Los Alamos National Laboratory, 2008, arXiv:0801.4829v1) pp.1-13. http://aps.arxiv.org/PS_cache/arxiv/pdf/0801/0801.4829v1.pdf

[20] Kostas Kokkotas D. Gravitational waves. Acta Physica Polonica. 2007, vol. 38, no.12, pp. 3891- 3923.

[21] Zubow A. V., Zubow K.V., Zubow V.A. Investigation of the distribution of water clusters in vegetables, fruits and natural waters by flicker noise spectroscopy. Biofizika, 2007, vol. 52, no. 4, pp. 585-592. (in Russian).

[22] L. Pauling, P. Pauling. "Chemistry". W.H. Freeman and Company San Francisco, 1975.

Chapter 22

THE NATURE OF THE ELECTRON AND GRAVITATIONAL RADIATION

After the discovery of three types of matter dominating the universe, dark matter (24…25%), dark energy (70…72%), totally ~95…96% [1, 2 ,3], and baryonic matter (~4…5%) [2], the last one was concluded to be the product of the first two types. Modern physics therefore describes the electron as a simple space polarization, as a point or even as a shock wave [4] whose size is difficult to determine (from 10^{-15} to 10^{-25} m) and whose lifetime changes in a broad interval from fs up to ns. The earlier discovery of the proton balance with a PV [4] indicates that the electron is nothing else but a shock wave (§21). The quick proton oscillation between the baryonic matter and a PV creates a stationary spherical shock wave (SSW); thus its characterization as an electron orbital (EO) according to Bohr's atom model is misleading. SSW was observed to be in an energy balance with the energy density of the PV ($5 \cdot 10^{19}$ J/m^3 [4]).

The aim of the present chapter was to develop an idea of the nature of the electron using mathematical calculations.

Let us now discuss the electron-free model of the hydrogen atom (Figure 22.1). The proton oscillation between PV and baryonic matter proceeds with a frequency with which the "electron orbital" (SSW, earlier called "outer electron orbitals") oscillates between 10^{21} and $13.7 \cdot 10^{22}$ Hz [5, p. 387]. Using these frequencies and the energy density of the PV, the atomic radius of hydrogen can be calculated. The radius will be determined by the balance between the SSW energy, which is generated by the proton condensing from the PV and the PV energy that prevents the spreading of SSW. It is easy to substitute the energy by its density, which permits us to calculate the atomic radius. The energy density of PV was taken from the experimental results [4], $E_W = 5 \cdot 10^{19}$ J/m^3, hence the energy generated by the oscillation of the proton can be described by the following equation:

$$E_p = h\nu = 6.63 \cdot 10^{-34}\ (\text{J} \cdot \text{s}) \cdot 4.5 \cdot 10^{22}\ (\text{Hz}) = 2.98 \cdot 10^{-11}\ \text{J}$$

where h is the Planck constant and ν is the proton oscillation frequency that is close to the SSW oscillation (experimentally obtained). The SSW energy density (E_{dssw}) was obtained from the following equation:

$$E_{dssw} = E_p/V = 2.98 \cdot 10^{-11}/(1.33\pi r^3)$$

where V is the atomic volume with the radius, r.

According to the equation $E_{dssw} = E_w$, the SSW radius was calculated to be $5.3 \cdot 10^{-11}$ m, which agrees well with the Bohr atomic radius.

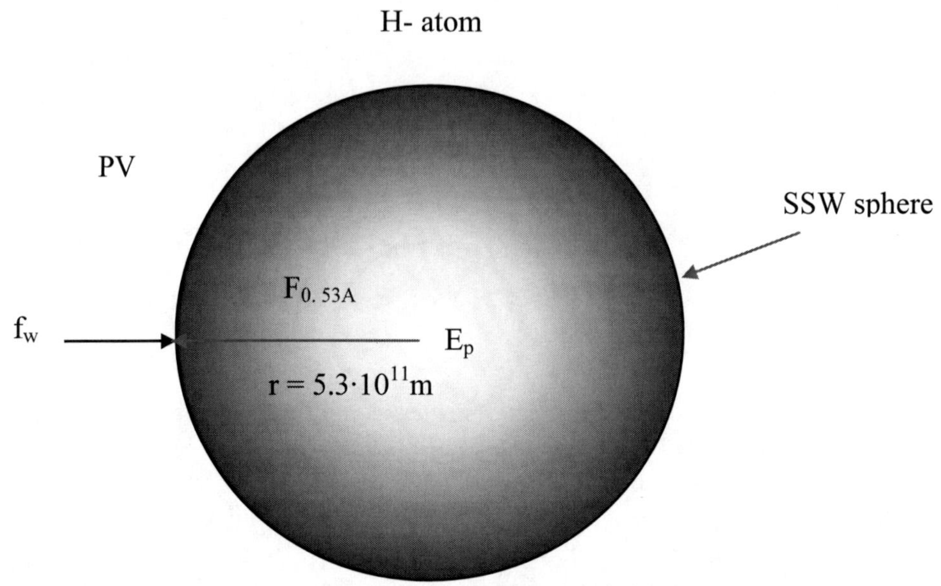

Figure 22.1 The Zubow model for the electron-free hydrogen atom. The SSW, oscillating between the PV and baryonic matter, is represented as a sphere. On the surface of the sphere, the pressure generated by the PV (external pressure f_w) is equal to the expanding pressure ($F_{0.53A}$) resulting from the shock waves of the proton.

Furthermore, the experimental data for the proton jumping frequency between a PV and baryonic matter ($4.5 \cdot 10^{22}$ Hz [5]) as well as for the energy density of the PV ($5 \cdot 10^{19}$ J/m^3 [4]) are consistent with the Bohr atomic radius. Thus, a SSW sphere with a radius of $5.3 \cdot 10^{-11}$ m can be formed only under these conditions.

Therefore, the electron orbital in the Bohr atomic model is suggested to be a stationary shock wave. The multiple energy levels of the SSW depends both on the proton oscillation state and the energy density of the surrounding PV. Different kinds of stationary shock waves in atoms and molecules explained by energy minimization and different forms of EO, e.g. *s*, *p*, *d*, *sp*^{3}, *f*, etc., are the result of stationary shock wave super positions.

The velocity with which the proton dissolves in physical vacuum (within the limits of atom size) can be calculated using the SSW size and proton jumping frequency. It amounts to $V = 5.3 \cdot 10^{-11} \cdot 4.5 \cdot 10^{22} = 23.85 \cdot 10^{11}$, m/s and is faster than the speed of light. Thus, the proton dissolving in a PV and its simultaneous condensation from a PV with super light velocity are concluded to be the reason for weak gravitational radiation from atomic matter. According to the theory of gravitation, gravitational radiation emerges when very large bodies rotate around each other with a velocity near that of light [6]. Thus, for the hydrogen atom, where the mass of the nucleus is very small, the proton must move with super light velocity.

Protons have to be understood as high frequency impulses of dark energy in a physical vacuum (dark matter) while electrons are shock waves from these impulses. Then, neutrons are low frequency impulses of dark energy in a physical vacuum that generate remarkably weak shock waves. All these processes are the base of baryonic matter.

CONCLUSION

There is no electron in the hydrogen atom but a spherical stationary shock wave generated by the proton.

REFERENCES TO 22

[1] Blennow, M. *et al.* Aidnogenesis via Leptogenesis and Dark Sphalerons. MPP-2010-125, IFIC/10-32, FTUV-10-0909. *arXiv*:1009.3159v1 [hep-ph]. 2010. 16Sep.

[2] Cueva, F., Nucamendi, U. Reconstructing the interaction term between dark matter and dark energy. *arXiv*:1007.2459v1 [gr-qc]. 2010. 15 Jul.

[3] Frampton, P. H. Black Holes as Dark Matter. AIP Conference Proceedings (Quantum Theory). 2010, vol. 232, pp. 53-57.

[4] Zubow, K. *et al.* The Phenomenon of Proton Dissolving in Vacuum and of Proton Condensation from Vacuum. Two Forms of Protons, Structure of Nuclei, Electrons and Atoms. J. of modern physics. 2010, vol. 1, pp.175-184.

[5] Dubovik, V. M. *et al.* Demonstration of magnetic electrical radiation under labor, geophysical and cosmological conditionsand its quantum nature. In System "Planet Earth": 300 years Lomonosov M.V.1711-2011 (ed. Fedorov, A.E.) pp. 379-393 (Moscov, LENAD, 2010), in Russian.

[6] Braginsky, V.B. Gravitation radiation. In Fizsitsheskij Enziclopeditsheskij Slowar (ed. Prochorov, A.M.) p.137, (Moscov, Sovijet Enciclopedia, 1984), in Russian.

Chapter 23

Method for Determining Distances to Celestial Bodies and their Masses

The search for new alternative experimental methods to determine distances to celestial bodies is a top problem of modern physics.

The phenomenon of protons dissolving in a physical vacuum was discovered in 2009 by the authors [1] and could become a platform for the new experimental method to determine distances to celestial objects (§21). Thus, the aim of this chapter is to find a new method for the determination of distances to celestial objects with the help of the gravitation mass spectroscopy developed by the authors [2].

Agarose hydrogel containing 97 wt. % water is used as a GMS-sensor for the gravitation mass spectrometer. The sensor is located inside in an isolated box in a room ([2], Figure 4.0.1) with the coordinates 53°34'54"N, 12°47'02"E. The analysis of the long-range order in the water of hydrogel at cluster level proceeds according to the method described in [3]. The visible constellation of celestial bodies is determined using the program "ZET 9" (www.astrozet.net). The scanning of the celestial bodies by SPGPR ([2, 4], Figures 19.4 and 24.11) is carried out in 0.01...20-second steps when the anthropogenic influence and gravitational radiation from other celestial bodies were minimal as well as the weather was clear. The energy spectrum of a water cluster ensemble is calibrated using the evaporation constant of water at the measurement temperature (44.1 kJ/mol). For the calculations, the average energy of a water cluster ensemble in the period between the visible and the gravitation signal of a celestial body [4, 5] is used. To calculate the distance to a celestial body, the shifting of the balance between the protons in the baryonic state with their dissolved form in the physical vacuum [1] when the sensor and celestial body are simultaneously in the SPGPR is applied. The potential energy of an expanded feather (model of undamped proton oscillation between dissolved and baryonic form in the direction to a celestial body) has been set equal with that of Newton's second law. At maximum deviation of the oscillator from its equilibrium state (x), the expansion force of a hypothetical feather (F_f) must be equal to the gravitational force between a proton and a celestial body (F_g) with the subtraction of opposite forces, e.g. proton attracting towards the Earth and other celestial bodies (Scheme III).

$$F_s = F_g \qquad \text{I}$$

For the simplest case, the equality of potential energies is as follows:

$$x \cdot (\gamma m_p M_{CB}/R^2) = kx^2/2 \qquad \text{II}$$

where k is the feather constant, R is the distance between the proton (p) and the celestial body (CB), m_P is the proton mass, M_{CB} is the mass of celestial body, and γ is the gravitational constant.

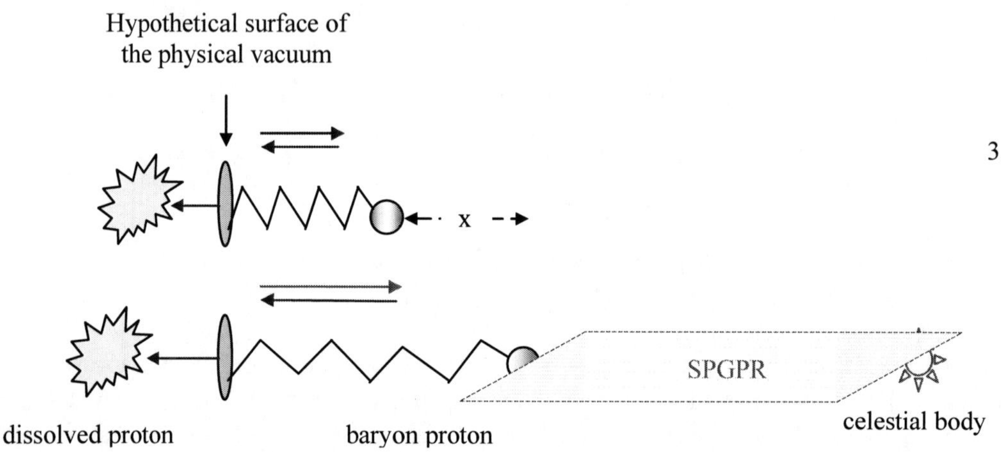

3.

III

Based on the 3rd Zubow equation (2;4) and the selection of real values for x and k, the distances between the pendulum and celestial bodies and the masses of the bodies are found and compared with known data.

Weak gravitational radiation from molecular matter is found to be the base for the GMS-sensor. After numerous experiments we have found that there is a resonance between the gravitational radiation of molecular matter, e.g. water and that of the celestial bodies at their appearance in SPGPR [4].

Figure 23.1 shows examples for Mars, the Sun, and a concentration of stars, which were scanned with a hydrogel sensor in the time interval when the celestial object appeared gravitationally and visibly in SPGPR. The average value of energy ($E_{av} \cdot C_{[H2O]act} = 1.1 \cdot 10^{-39}$ (kJ/proton) $\cdot 5.3 \cdot 10^{-13} \cdot 55.6$) is applied to determine the distance to the planet:

$$R = (2\gamma m_p M_M/kx)^{0.5} \qquad \text{IV}$$

where $k = 2000 \cdot E_{av} \cdot C_{[H2O]act}/x^2$ and $C_{[H2O]act} = 0.0278 \cdot kx^2/(E_i + E_{i+1} + \ldots E_N)$, N is the number of water cluster kinds in an ensemble, and E_i is the formation energy of cluster i.

The experimental determination of x, R, and the part of water molecules involved in the interaction with the gravitational radiation of Mars for different times and distances of the planets are shown in Table 23.1 and Figure 23.2.

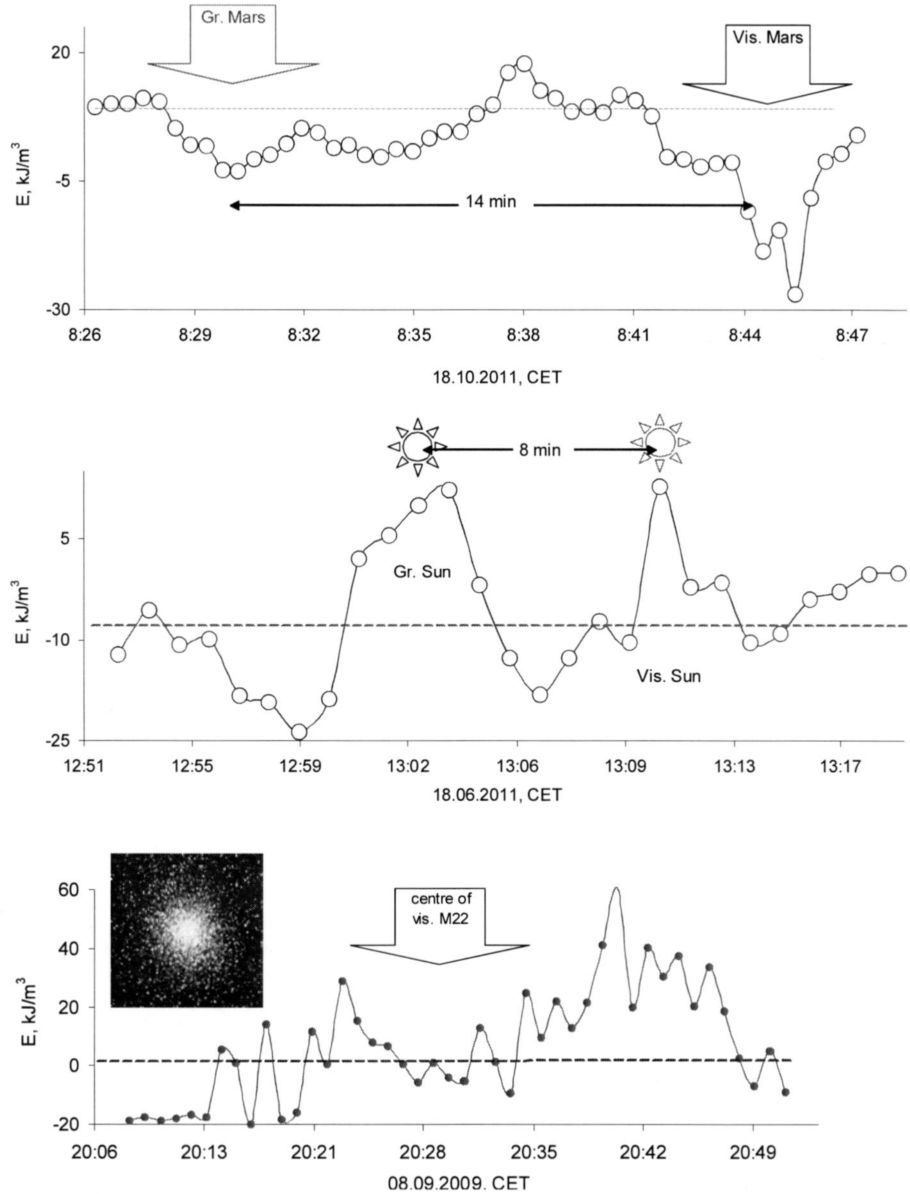

Figure 23.1. Some energy spectra of a water cluster ensemble in hydrogel at the gravitational and visible appearance of a celestial object in SPGPR: Mars (step 25 s.), $R = 2.53544 \cdot 10^{11}$m, $k = 8.79 \cdot 10^{-15}$ N/m, $x = 6.41 \cdot 10^{-11}$m; Sun (step 60 s.), $R = 1.523985 \cdot 10^{11}$m, $k = 1.752 \cdot 10^{-16}$ N/m, $x = 1.87 \cdot 10^{-10}$m; and M22 (for comments, see Figure 23.6), dotted line – base line.

As is shown, it is principally possible to develop a new method to determine the distances to celestial objects using GMS. Note that for the other celestial bodies, such as the Sun, Moon, Jupiter, Saturn, and the galaxies, k, x, and $[H_2O]_{act}$ are different and are dependent on R, the celestial bodies' mass (M), and the mass' heterogeneity in the celestial body (Figure 23.3). The fact that protons of the molecular matter are extracted by the celestial bodies could help us to understand the phenomenon of gravitational shots [5]. This extraction process can be enhanced or weakened by a certain constellation of planets.

Table 23.1. The results of determining the distance to the planet Mars by scanning with agarose hydrogel

Observation time	R_k, km known	R_f, km found	Deviation $(R_k-R_f)/R_k$
23.10.2011	247,801,500	247,805,194	-0.001%
21.10.2011	250,119,000	250,131,824	-0.005%
19.10.2011	252,408,000	252,424,117	-0.006%
18.10.2011	253,543,500	253,526,749	0.007%
16.10.2011	255,789,000	255,801,427	-0.005%
15.10.2011	256,902,000	256,893,213	0.003%
14.10.2011	258,007,500	258,012,113	-0.002%

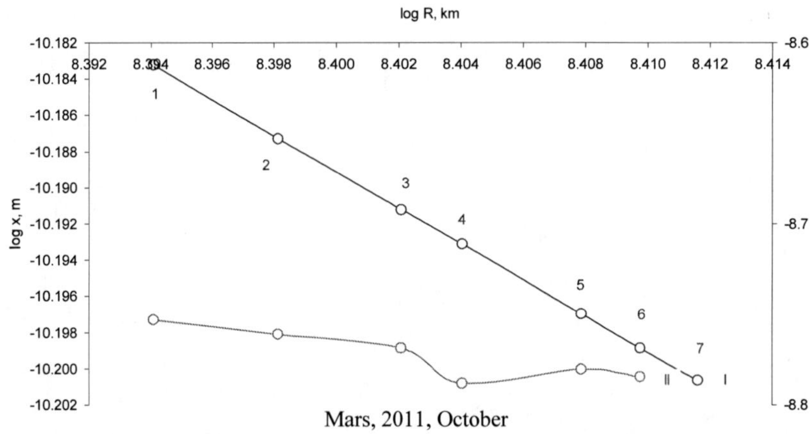

Figure 23.2. Recorded distance to Mars (R) versus amplitude x and part of the water molecules involved in the gravitational interaction with Mars. October 2011: 1 – 23rd, 2 – 21st, 3 – 19th, 4 – 18th, 5 – 16th, 6 – 15th, and 7 – 14th. Sensor – agarose hydrogel (97 wt. % water), 293 K, log ($C[H_2O]_{act}$) – concentration of water molecules involved in the gravitation interaction with Mars (curve II). Curve I is described by the linear equation log x = -0.9961·(log R) – 1.8223. R deviates no more than 0.001% from the known value (see Table 23.1).

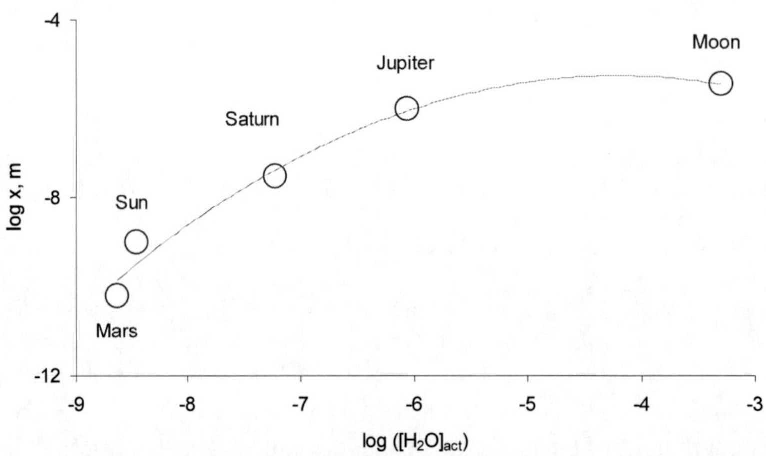

Figure 23.3. x – shifting to celestial bodies that are dependent on the concentration of water molecules involved in the gravitational interaction with them.

For the Sun's system, the dependence of proton shifting on the concentration of active water molecules will be described by the next equation:

$$\log x = -0.2223(\log([H_2O]_{act}))^2 - 1.9564\log([H_2O]_{act}) - 9.3897$$

Larger mass concentrations cannot be described with this equation because of their high mass heterogeneity. The Sun, galaxies, gas giants, and the center of the Milky Way are on a line, which is described by the function: $\log k = 0.722 \cdot \log(M/R^2) - 22.564$ (Figure 23.4). According to this line, the feather constants of the proton oscillators attain minimal values for the most distant mass concentrations.

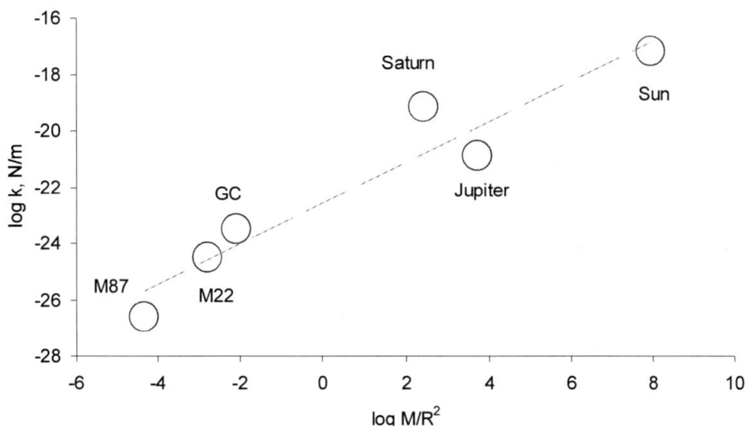

Figure 23.4. The influence of the distance to celestial bodies on the feather constant of a proton oscillator. GC – center our galaxy.

For large and very distant mass concentrations it is more convenient to use the part of the activated water molecules instead of the feather constant (Figure 23.5).

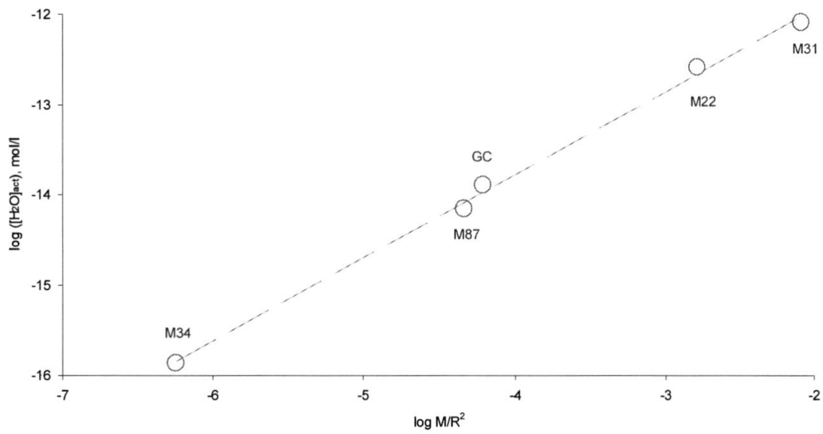

Figure 23.5. Part of the activated water molecules in the agarose hydrogel sensor that are dependent on the nature of huge cluster masses in the SPGPR. M34 (04.09.2009, 100 stars with a whole mass of $9.95 \cdot 10^{31}$ kg, average distance $1.3245 \cdot 10^{19}$ m), M22 (08.09.2009, $1.39 \cdot 10^{37}$ kg), M31 ($2.39 \cdot 10^{42}$ kg), GC – center of our galaxy (24.06.2011, $5.97 \cdot 10^{36}$ kg), and M87 (04.11.2011, $1.19 \cdot 10^{43}$ kg). This dependence is described by: $\log([H_2O]_{act}) = 0.9209 \cdot \log(M/R^2) - 10.096$.

All data is summarized in Table 23.2.

Table 23.2. Experimentally determined distances to clusters of large mass ensembles in remote space

Mass ensemble	R, km (known)	R, km (found)	Time
Milky Way center	3.12213E+20	3.12484E+20	24.06.2011
Andromeda (M31)	1.707E+22	1.70876E+22	06.09.2009
M34	1.32454E+19	1.32459E+19	04.09.2009
M22	9.25800E+19	9.25894E+19	08.09.2009
M87	5.1089E+23	5.10887E+23	04.11.2011

This table satisfactory shows agreement between the found and known distances to remote huge celestial objects.

Finally, an energy spectrum of a water cluster ensemble in the gravitation sensor upon scanning M87 (Figure 23.6) shall be discussed. For the visible part of M87 (24.5 s), there are three pronounced signals where the middle signal belongs to the galaxy's center and the two side symmetrical ones of nearly the same intensity belong to the invisible ring around the GC. All these signals show large clusters of masses that are not detected in the classical electromagnetic spectra. The mass of the GC can be calculated by Equation 4 using the value found for R (Table 23.2), the proportion of activated water molecules (Figure 23.5), and the water cluster formation energy of 34 kJ/m^3 (Figure 23.6). For the M87 galaxy center (diameter of 4 seconds), the following parameters are obtained: $M = 2.09 \cdot 10^{35}$ kg, $k = 2.726 \cdot 10^{-42}$ N/m, and $x = 6.56 \cdot 10^{-8}$ m. Hence only ~$18 \cdot 10^{-7}$ % of the whole mass of the galaxy is concentrated in the center.

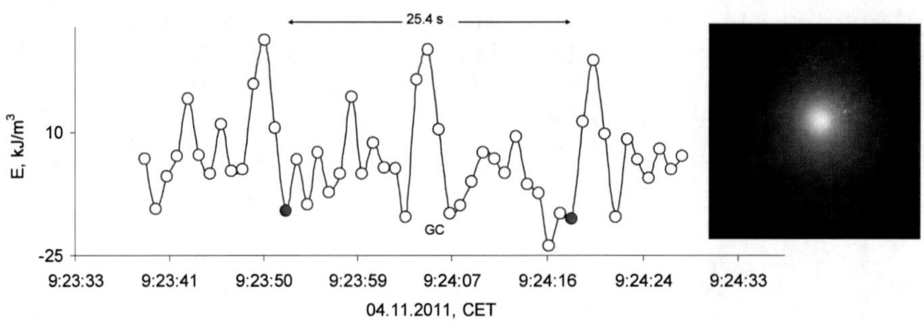

Figure 23.6. Energy spectrum of a water cluster ensemble at the appearance of the visible M87 in SPGRP. Shaded dots indicate the visible area in the electromagnetic spectrum of M87 (http://en.wikipedia.org/wiki/Messier_87).

Conclusion

Thus, the method of gravitational resonance of the protons is concluded to be useful for determining distances to celestial objects as well as for determining their masses.

The experimental data proves the reality of electron-free atoms, where the "electron" is to be understood as a shock wave rather than a particle. Furthermore, the method allows us to get an idea of the mass distribution in the object observed and analyzed for its heterogeneity.

REFERENCES

[1] Zubow K., Zubow A.V., Zubow V.A. The Phenomenon of Proton Dissolving in Vacuum and of Proton Condensation from Vacuum. Two Forms of Protons, Structure of Nuclei, Electrons and Atoms. J. of modern physics. 2010, vol.1, no.1, pp.175-184.

[2] Zubow K.V., Zubow A.V., Zubow V.A. Ensemble of Clusters – New Form of Molecular Matter, Risks and Chances. Zubow Equations. In Advances in Chemistry Research, vol. 5, 2010, pp. 107-145. Editor J.C. Taylor. Novapublisher. NY.

[3] Zubow Kr. V., Zubow A.V., Zubow V.A. Cluster structure of liquid alcohols, water and n- Hexane. Journal of Appl. Spectr, 2005, vol. 72, no. 3, pp. 321-328.

[4] Zubow K., Zubow A.V., Zubow V.A. Experimental Methods for the Determination of the Super Light Velocities of the Gravitation. Nature, Structure and Properties of Gravitation Waves. In Horizons of Physics. Editor A. Reimer. NY. 2012, vol. 277, pp.131-152.

[5] Zubow K., Zubow A.V., Zubow V.A. Experimental Platform for the Investigation of the Structural Heterogeneity of the Physical Vacuum. Vacuum Energy Risks and Chances. In Horizons of Physics. Editor A. Reimer. NY, 2012, vol. 277, pp. 87-110.

Chapter 24

Extraterrestrial Intelligence (ETI): A New Channel of Gravitational Communication

If the existence of ETI with a higher level of development is possible, as is believed by the authors [1], then in our galaxy (200...400 billion stars), a life form could be on 30% of the planets. It is naïve, however, to believe that their communication with us would be based on electromagnetic rays, which propagate with light velocity. On the other side, the recently discovered super light velocity and the properties of gravitational radiation (GR) let us assume that ETI would use such communication platforms [2, 3, 4]. It is reasonable to believe that these platforms will be utilized by ETI very efficiently but they will need large energy sources This problem can be solved by applying the energy of the physical vacuum [5] and the principle possibility "to put" the information on a main gravitational wave [3]. To direct their information flow to the Earth, ETI can probably use the SPGPR of their planet [3, 4, 5]. This logic implies our limited ability to "catch" these signals with the Earth's SPGPR, perhaps, even for many ETI. Additionally, the problem becomes more complex because contact with ETI is not constant in time and space. There is a narrow signal vector where ETI would meet the SPGPR of the Earth, where all the factors involved are in constant motion; hence, the communication window is very short. Nevertheless the traffic must be loaded with the highest information volume. There remains no choice but to observe GR from the surrounding events in the passive mode.

Another problem is the signal decoding from ETI, but here we might expect , that certain rules are respected in the creation of messages, such as the generation of the transmitter call sign code, the receiver code, protocol messages, etc. For established contacts between ETIs, entering into their real information exchange without understanding their previous contacts would be extremely difficult.

The aim of this chapter is to search for ETI signals in the gravitational waves' interval.

Agarose hydrogel containing 97 wt. % water is used as sensor for the gravitation mass spectrometer. The sensor is located inside an isolated box in a room ([4], Figure 4.0.1) with the coordinates 53°34'54"N, 12°47'02"E. The analysis of the long-range order in the water of hydrogel at cluster level proceeds according to the method described in [6]. The visible constellation of celestial bodies is determined using the program "ZET 9" (www.astrozet.net). The ETI signals in SPGPR are recorded in 10- and 60-minute scanning intervals when the anthropogenic influence and gravitational radiation from other celestial bodies were minimal and the weather was clear. To fully exclude the appearance of pulsars or unknown celestial bodies in these "windows," however, is impossible.

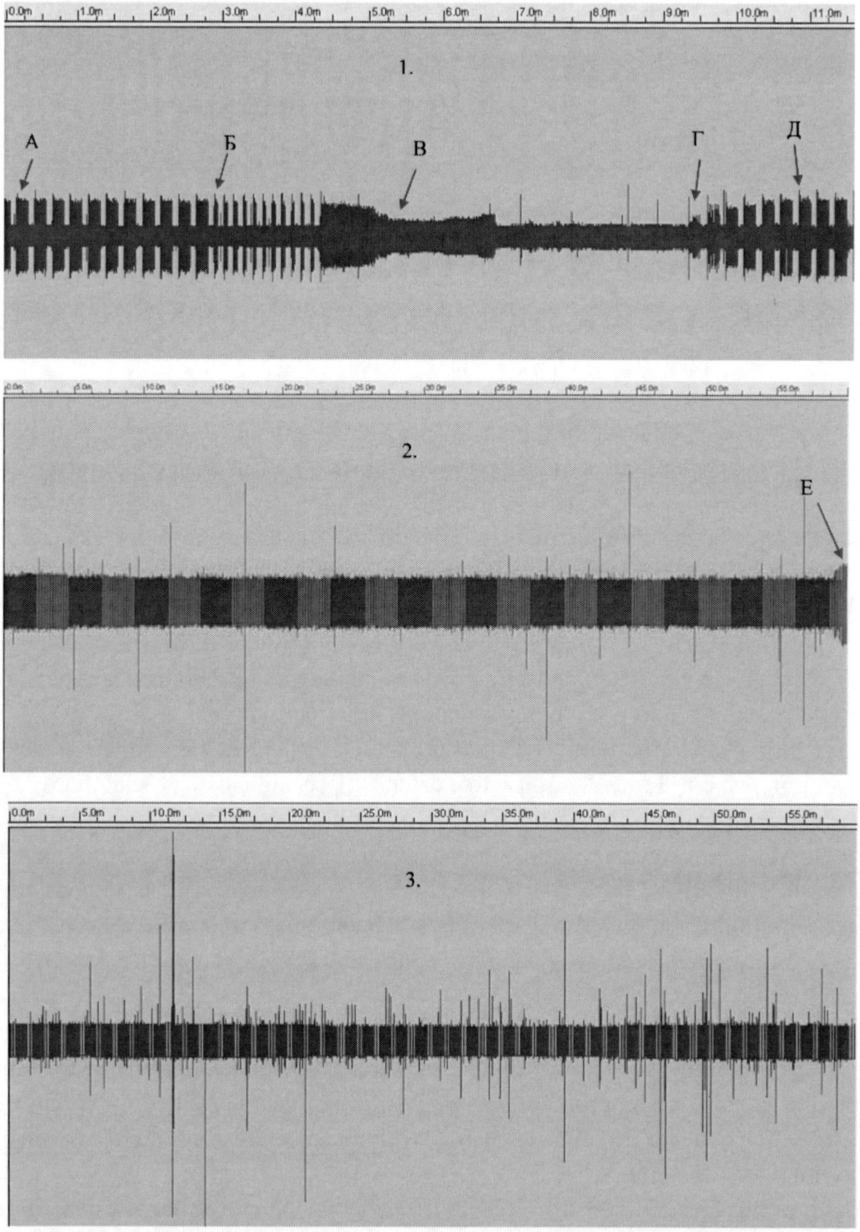

Figure 24.1. Review blocks of ETI-1 (Zubow-1) signals scanned on February 19, 2012 from 00:47:00 to 01:58:00 (1) and on March 1, 2012 from 09:30:00 to 10:30:00 MET (2) and the background spectrum of gravitational white noises on March 4, 2012 from 00:20:00 to 01:20:00 (3). Coordinates of the sensor: 53°34'54"N, 12°47'02"E.

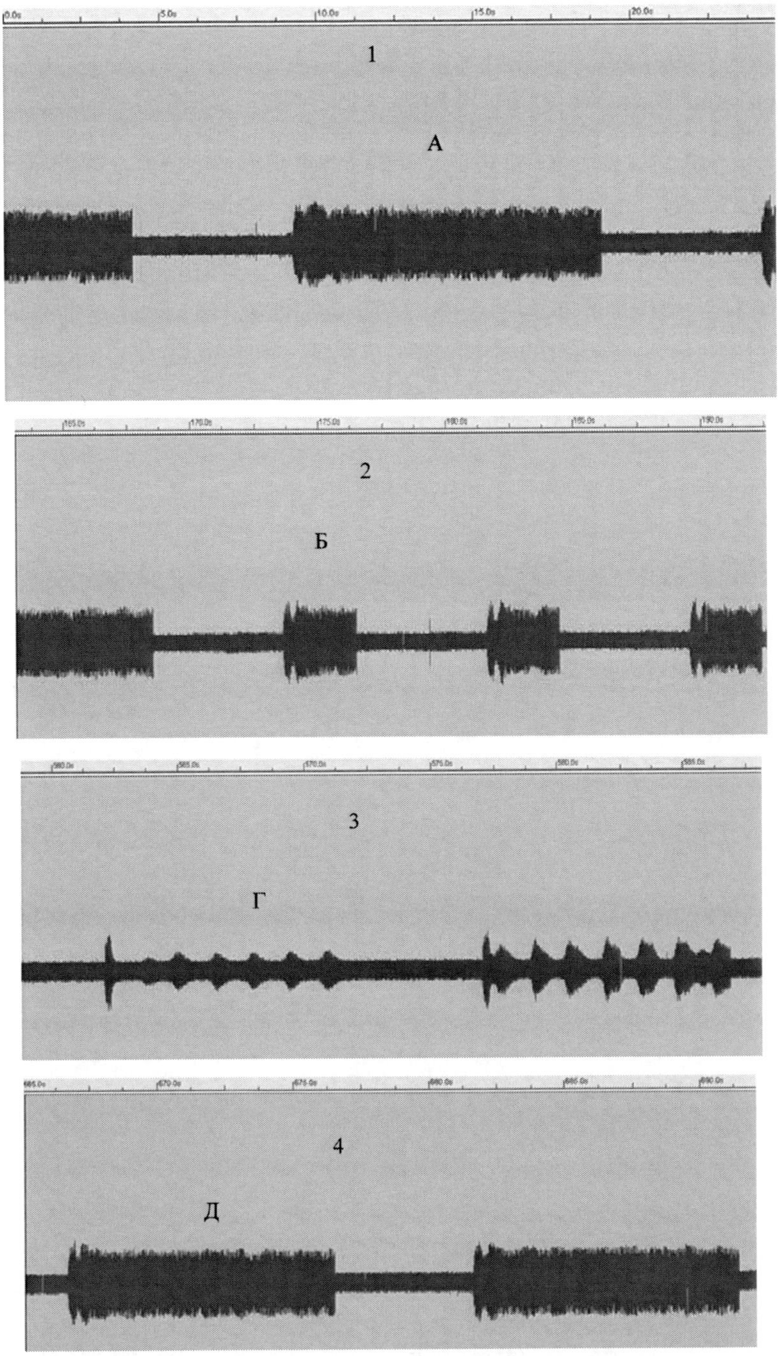

Figure 24.2. Spectra of some individual ETI-1 signals (Figure 24.1) in a more expanded form. 1 – signals of class **A**, 2 – signals of class **Б**, 3 – signals of class **Г**, and 4 – signals of class **Д**.

Figure 24.1 provides an overview block of periodical signals caught in SPGPR of the Zubow gravitational spectrometer within 11 minutes of scanning. The signals are presented in a concise form for us to better understand their structural periodicity and to analyze logically how to separate them from natural events on Earth and in space. As is visible from this figure, the block of signals (1) differs distinctly from the gravitational background of white noises (2 and 3). The signals **А**...**Д** (1) are obviously not of natural origin, but take the character of intelligence. The signal **E** in spectrum 2 is described as the "head" (Figures 24.2 and 24.3) of a new block of signals, beginning 10 h later than the first one; this signal for technical reasons could not be continued. The background (3) is considered as a spectrum of random signals in the spectrum (1), however, we see a certain regularity of an unnatural origin, namely, frequency, size of periodicity, structural formation of the whole information block, and order in each individual signal (Figure 24.2).

Eleven signals of class **A** are characterized by a constant length of 10 s and the distance between them is 5 s. The signals of class **Б** (11) have a length of 3 s and a distance of 5 s. The signals of **B** seem to be a superposition of several information blocks. For signals **A** and **Д**, there are certain similarities in the form while the signals of class **Г** (only two) are completely different in that they are characterized by a wavy shape (Figure 24.3).

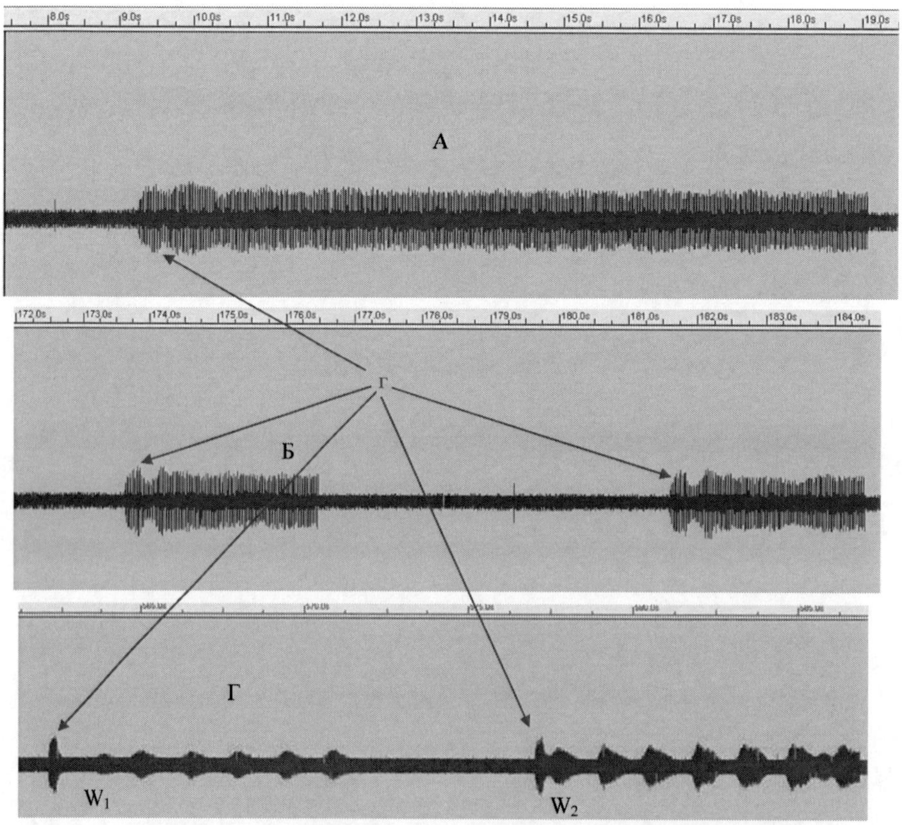

Figure 24.3. Signals **A**, **Б**, and **Г** at a higher spectrum resolution, **г** - "head" of information blocks.

As seen in Figure 24.3, all information blocks are described by a typical "head" for recognition even though for the ends of the blocks a recognition marker has not yet been found. For example, the ending of the first block of **Г** is diffuse in comparison with that one **A** and **Б** and even with the end of the second block of **Г**. Consequently, they are not important.

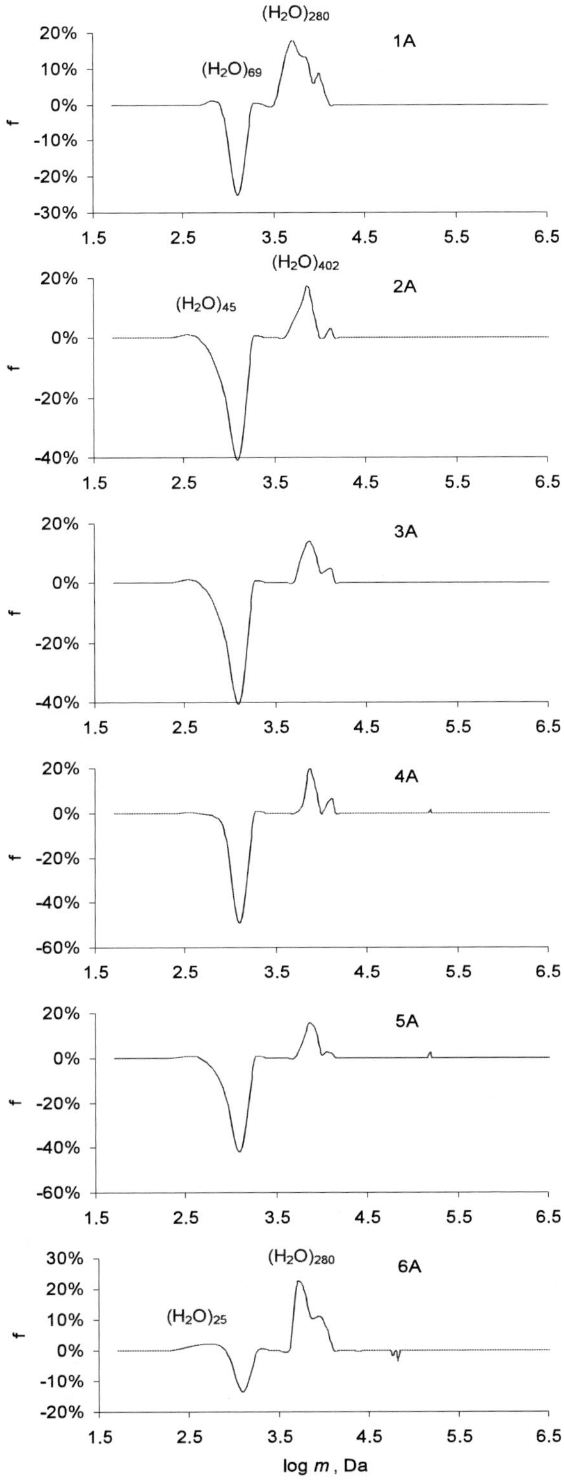

Figure 24.4. GMS-spectra of water in hydrogel (sensor) during the influence of gravitational radiation from ETI-1. Information blocks of class **A** (Figure 24.1). Mass ensemble up to 3 million Daltons, Zubow constant is $6.4 \cdot 10^{-15}$ N/m.

The form and regularity of the signals may describe the structure of the ETI-1 hydrosphere, which seems to be located inside a giant planet, e.g. gas giant. The large number of signals of class **A** could be an indication that ETI-1 is isolated by numerous dense layers on the outside, which is typical for gaseous giants. Then follow the shorter signals of class **Б**, which could provide information about the somewhat looser gas atmosphere than that in **A**. The signals of class **Г** should be ascribed to liquid water (freshwater and oceans), of which there are only few. Whereas the signals of class **Д** should belong to underground water.

Striking is here the special shape and size of the signal of class **B** that, in our opinion, reflects the state of water in the biosphere. This state is heterogeneous and belongs to various water forms of the biosphere. The nature of the signal blocks' sequences indicates that the living ETI-1 form is located between the vapor (**Б**) and liquid phase (**Г**) of water, where it is nearer to the bottom of the layer **Б**. Water of the layers **Г** and **Д** is close to the center of a gas giant and is unsuitable for the biosphere (large distance between **B** and **Г**).

Since the gravitational waves only allow the formation of specific clusters in water of the sensor [12], ETI-1 signals could include important information about the state of the gravitational field of the transmitter. To find out which water clusters are allowed to be formed in the ETI-1 gravitational field, the gravitational mass spectrometer has been applied. In Figure 24.4 the GMS-spectra for the first six signals of class **A** are given. With the exception of the Chaplin cluster of 280 molecules (http://www.lsbu.ac.uk/water/index.html), all other clusters are non-typical for the Earth; they are formed in energy-rich colored noises. The smallest cluster allowed in ETI-1 is composed of 25 molecules while the basic cluster of 12 water molecules [7, 8] disappears.

This may be an indication for other cluster structures for ETI-1, which must be logically caused by other gravitational fields. Since we know that the cluster structure (long-range order of water) is the base platform for the biosphere, we consequently conclude the construction of other biomolecule domains [9]. The small water clusters of 25, 45, and 69 molecules are rarely on the Earth. They are formed under the influence of strong energy fields [10, 11]. In the information block of the transmitter, these clusters are present but they mainly appear as collapsed structures ($f < 0$), while an "island" of medium clusters (226 to 627 water molecules, 2,941 to 4,902 Hz) is in the expanded form.

The situation changes when going to the short blocks of the information, class **Б** (Figure 24.5). **Б** consists of 12 blocks indicating the presence of 12 components in the atmosphere. Here the small clusters dominate in the expanded form, with the exception of the cluster $(H_2O)_{69}$ and the "island" of medium clusters remains in the expanded form. Also in this case, there is no base cluster. The short block $\mathbf{Б}_p$ (2 s) is to be specially discussed because it separates (0.3 s) class **Б** from class **B**. ETI is probably protected by this layer from star radiation. The percentage of this layer in the overall atmosphere amounts to 2.1 (ratio of the layer thickness to the complete atmosphere length in s). In the Earth's atmosphere, consisting of only the four main components: N_2, O_2, H_2O, and Ar, the ozone layer is observed to be about 0.5% (5 km of 1,000 km).

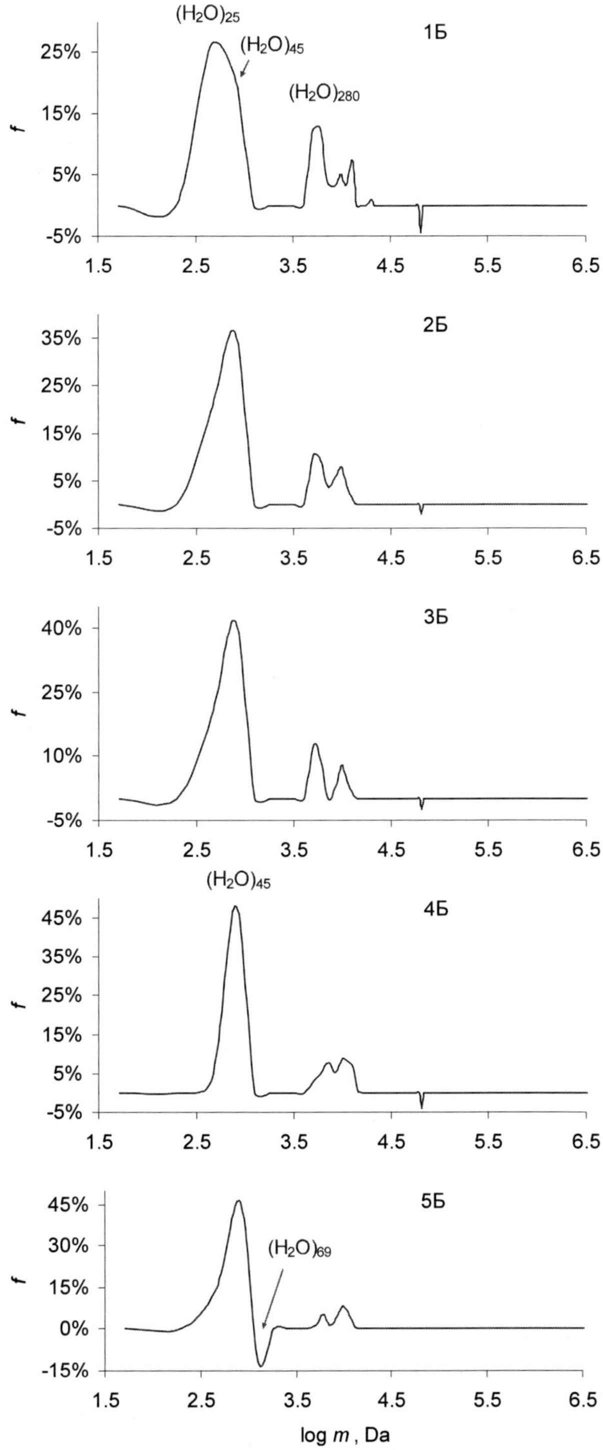

Figure 24.5. GMS-spectra of water in hydrogel during exposure to gravitational radiation from ETI-1. Information blocks of class **B** (Figure 24.1). The ensemble of water clusters is about 3 million Daltons.

If, in fact, ETI-1 uses water as the biosphere, then such a water cluster hierarchy can give an idea of the cluster formation in biomatrices and even on biochemical processes at the domain and coil levels of biomolecules [9].

Let us consider the signals of the biosphere (**B**) shown in Figure 24.6. At least five signal fragments in **B** can be chosen: $\mathbf{B}_1$, $\mathbf{B}_2$, $\mathbf{B}_3$, $\mathbf{B}_4$, and $\mathbf{B}_5$. At a higher resolution, however, these signal fragments are heterogeneous and they carry specific information that should be analyzed separately.

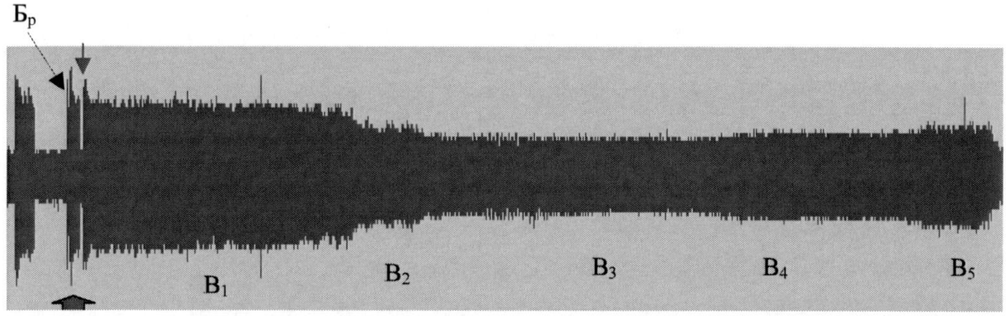

Figure 24.6. Signal fragments of classes **Б** and **B** (Figure 24.1). On the left side are the last two signals of class **Б**. The narrow arrow describes the distance between the signals of classes **Б** and **B** (0.3 s). Regarding the broad arrow and $\mathbf{Б}_p$, the comment is given in the text.

The signals in $\mathbf{B}_1$ seem to be closer to the surface of the gas giant than those of $\mathbf{B}_5$. If that is true, then $\mathbf{B}_1$ should reflect the lighter forms and $\mathbf{B}_5$ the heavier ones of the biosphere elements. The short distance between the last signal of class **Б** and the beginning of class **B** is an indication that the atmosphere and biosphere are very close and that the biosphere needs the atmosphere, for example, for breathing. The ratio of signals in **B** can help us to understand the hierarchy of the distribution of water in the biosphere and thus also of the distribution of the main elements such as flora and fauna and their densities. It can be assumed with high probability that $\mathbf{B}_1$ represents the main part of the biosphere "floating" between the gaseous and liquid phases of water (the signals of class **Г**). Furthermore, it appears that the biosphere is protected by layer $\mathbf{Б}_p$ from strong cosmic radiation. $\mathbf{B}_1$ could be water in plants, supplying the other elements of the biosphere with energy. Signal $\mathbf{B}_3$ is ascribed to water in a low water biosphere (e.g. ice) between flora and "dense" fauna ($\mathbf{B}_5$), being at a considerable distance from each other. The analysis of the water's long-range order (LRO) in **B** is shown in Figure 24.7. As seen, it is represented in the ETI-1 biosphere by the cluster of 45 water molecules that are mainly in the expanded form. On Earth, these clusters correspond to Mercury [12] and their appearance in the ETI-1 biosphere can be understood as the influence of the gas giant's core.

It is believed that Mercury was previously also a gas giant [13]. A wide variety of water clusters is found only in $\mathbf{B}_4$. In the $\mathbf{B}_1$ layer, which is the closest to the planet's surface and exposed to the radiation of the stars, the highest content of collapsed clusters (Table 24.1) as well as cluster splitters of $(H_2O)_{25}$ [14] are found. Based on earlier investigations [10], giant shadows in this layer are assumed to be present. In the underlying layers the star's ray shall be focused as in a lens, resulting in the destruction of the LRO. Generally, the biosphere is relatively stable even though compared to us it is in the "colored noise" classification.

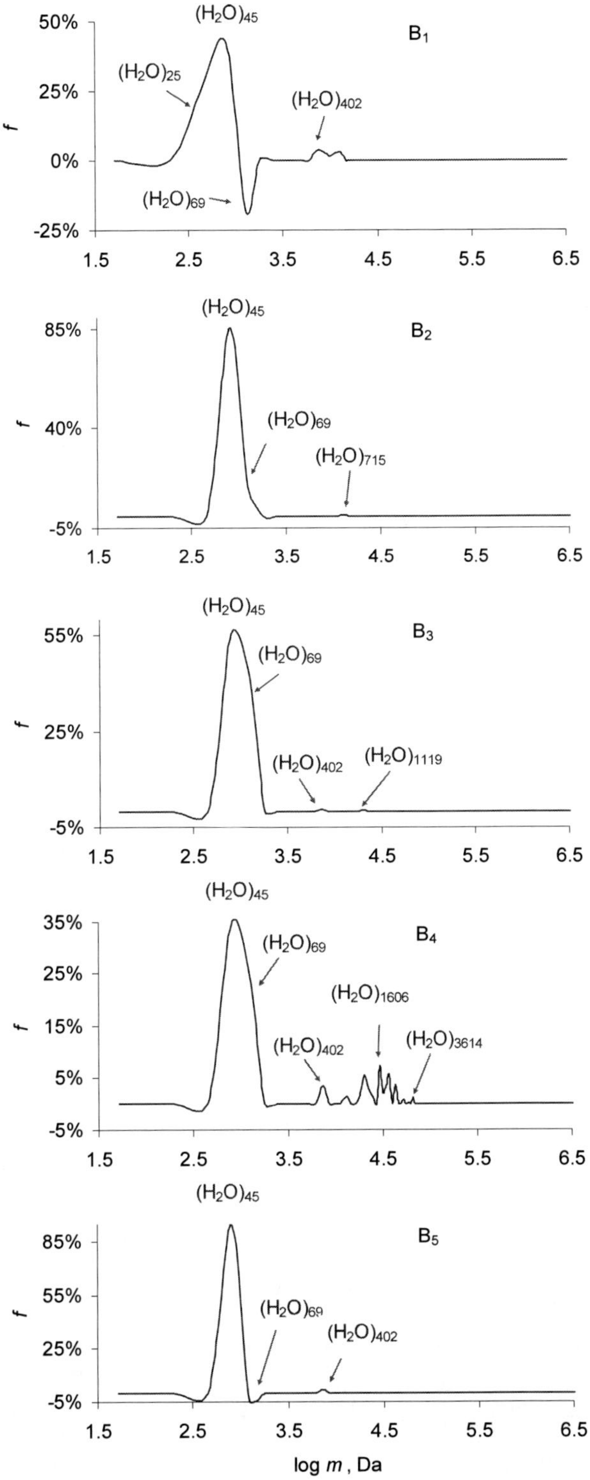

Figure 24.7. GMS-spectra of water in class **B,** mass ensemble up to 3 million Daltons.

Table 24.1. Main parameters of water's LRO in the sensor hydrogel reflecting the LRO in the ETI-1 biosphere. M_{GMS} – average cluster mass in Daltons, *N* – number of cluster kinds in a cluster ensemble in the mass range up to 3 million Da. $Д_c$ – part of collapsed (energy-rich) water clusters. E_{rel} – relative energy of a cluster ensemble

Part	M_{GMS}, Da	*N*	$Д_c$, %	E_{rel}	Min. cluster	Max. cluster
B_1	1,993	8	18%	26	$(H_2O)_{25}$	$(H_2O)_{715}$
B_2	968	3	0%	28	$(H_2O)_{45}$	$(H_2O)_{715}$
B_3	1,108	4	0%	18	$(H_2O)_{45}$	$(H_2O)_{1,119}$
B_4	12,445	16	0%	30	$(H_2O)_{45}$	$(H_2O)_{3,614}$
B_5	934	3	4%	20	$(H_2O)_{45}$	$(H_2O)_{1,119}$

Two signal blocks of class **Г** (Figure 24.3) indicate two types of water. For both types of water, the GMS-spectra of the first signals behind the head of the protocol are shown in Figure 24.8.

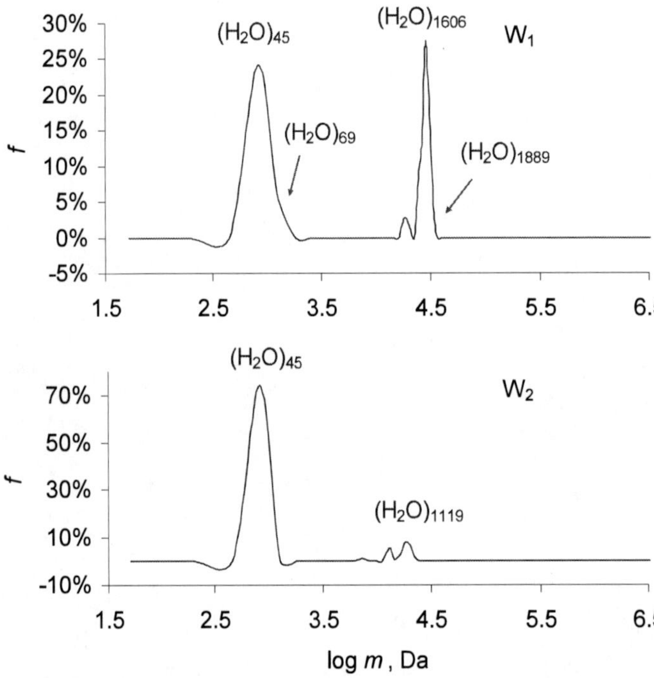

Figure 24.8. GMS-spectra of the first signals of class **Г**: $\mathbf{W}_1$ and $\mathbf{W}_2$ (Figure 24.3).

Analyzing the GMS-spectrum of $\mathbf{W}_1$, the first layer from the surface is shown to consist of liquid water with the base water cluster of 45 molecules. Additionally, it has the typical Chaplin water cluster of 1889 molecules (http://www.lsbu.ac.uk/water/index.html), which decomposes and therefore cannot be found in the second spectrum ($\mathbf{W}_2$). The second spectrum ($\mathbf{W}_2$) could be an indicator for salt water (sea water) in which the base cluster $(H_2O)_{45}$ is usually pushed out of the solution; this is caused by solvated clusters of ion pairs of salt. This leads to a rapid increase of the base water cluster signal intensity [15]. Compared with the first signal block (surface fresh water, six water types), the second one for the aqueous solutions is characterized to be longer and more heterogeneous. Eight signals after the head of the protocol (**r**) are recorded, which shall be ascribed to eight types of solutions (**Γ**, Figure 24.3). This means that the seas are larger and deeper than surface waters. Both types of waters are likely but only in the color noise that is not on the Earth.

Unfortunately, at present, the world's science is unable to resolve issues regarding color noises, but their understanding will help to develop a strategy for the investigation of biomatrices, develop the state of water in color noises, and find analogs on the Earth or in the near space.

Figure 24.9 shows a scheme when the Earth enters into communication traffic with ETI-1.

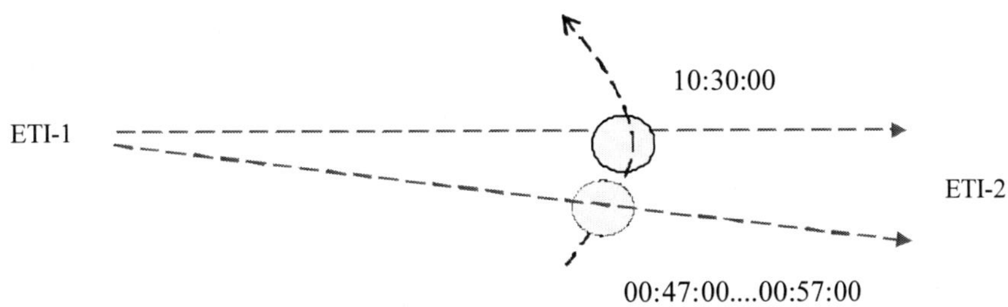

Figure 24.9. Model of the Earth's entrance into communication traffic with ETIs on February 19, 2012.

The observation of the second communication session 10 h later provides insight into the mechanism of the signals' sending. In this case, the transmitter must rotate and probably use its own SPGPR. Unfortunately, the signal 10 h later is not complete but it is clear that it belongs to the beginning of the communication, perhaps, to the recurrence of the information block at 00:47:00...00:57:00 (MET). Many of our attempts to detect the signals of this block again (Figure 24.1) have proven unsuccessfully. That means ETI-1 sends signals to ETI-2 under defined conditions only. Our Earth could have been accidentally between ETI-1 and ETI-2 or maybe not. If it was not an accident, then the message is directed deliberately to us. While at this time the Earth's SPGPR crossed a number of stars in the Northern hemisphere and the nebula Foramen (NGC 3372) in the Southern one. To send its information with super light velocity, the transmitter uses a specific constellation of celestial bodies. It has succeeded in setting a neutrino information block on the main gravitational wave and it has applied the gravitational acceleration in the direction of the giant mass cluster NGC 3372 [2, 3, 12].

To determine the coordinates of the transmitter more accurately a repeated scanning of this hemisphere region one year later by applying our special equipment is needed in the hopes of once again getting signals from ETI [16]. We are now, however, already able to make a preliminary evaluation. The line that NGC 3372, Earth, and ETI-1 are on most likely indicates the position of the transmitter (ETI-1). The hemisphere with the longitude of 20.7° and latitude of 59.5° in the constellation Cepheus could be in such a position, Figure 24.10. At this time the stars near α Umi Polaris (431 ly) are in SPGPR, too, but they do not lie on this line, which means the ETI signals cannot get from them. The exit of the Earth out of this line leads to the conclusion that no further signals are received.

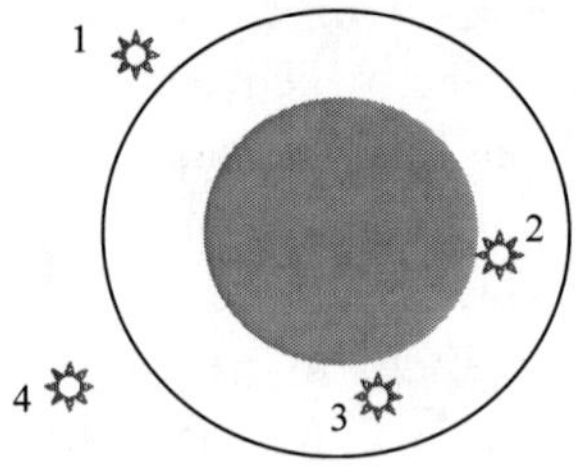

Figure 24.10. The place of ETI-1 in Cepheus constellation. Surroundings stars: 1 – HD212495 (Right Ascension (RA) 22^h23'20.91", Declination 62°28'36"), 2 – HD 216595 (RA 22^h53'28.71", Decl. 60°9'36"), 3 – HD215907 (RA 22^h47'47.92", Decl. 58°32'30"), and 4 – δ cep (RA 22^h29'10.27", Decl. 58°24'54.7"). Green circular area 70% and white one 90% hit probability.

Let us briefly consider a possible mechanism of how the information block shall be installed on the main gravitational wave, which is realized when giant masses, such as the Earth and galaxies are in conjugation to ETI-1. The role of the massive galaxy could be taken by the nebula NGC 3372 (Figure 24.11).

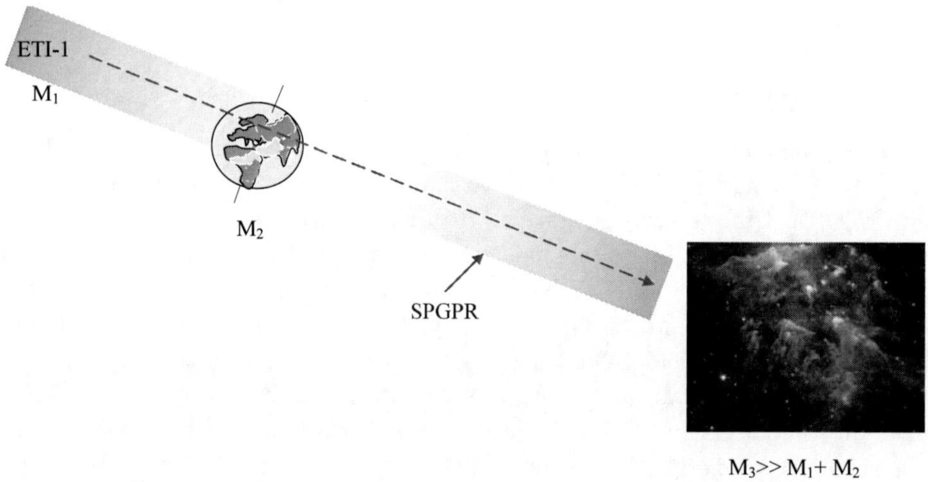

Figure 24.11. Constellation of ETI-1, Earth, and NGC 3372 in SPGPR when the information is registered by the gravitational sensor.

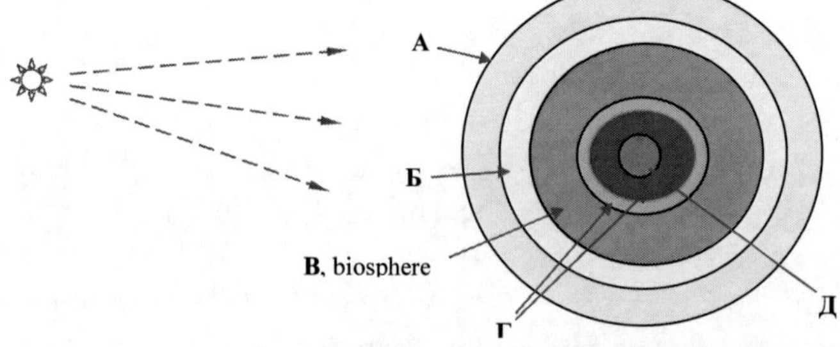

Figure 24.12. Model for the existence of ETI-1 in a multilayered planet. The classes of signals are described in the Figures 1, 2, 3, and 6. **А** – multilayer flows of different light gases; **Б** – atmosphere; **В** – biosphere; **Г** – hydrosphere (only condensed state), fresh and salt solutions reservoirs; and **Д** – underground waters in surface layers of the planet's core.

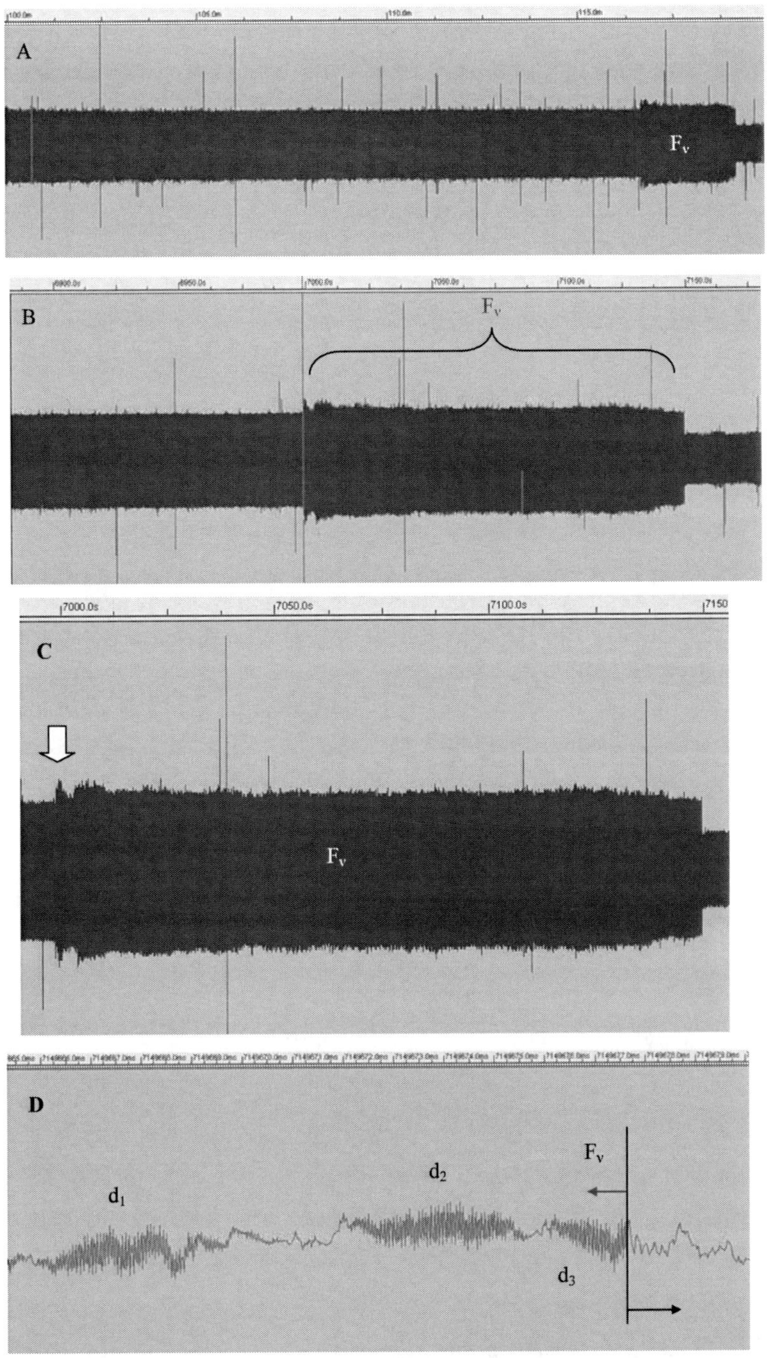

Figure 24.13. Spectra of an information block against the background of white noises when GR_{MGW} flows hit SPGPR. Sensor – agarose hydrogel. March 27, 2012. The vertical line corresponds to 19:29:55 CET. A – signal against the white noise background; C – signal with the selected head (arrow), the call protocol; and D – communication end and the beginning of white noises. F_v – GR_{MGW} flow.

The acceleration of the main gravitational wave (MGW) from ETI-1 to NGC 3372 depends on the ratio of all masses involved in the conjugation M_1 - ($M_2 + M_3$) [2, 3, 4]. ETI-1 is assumed to have a strong, controlled GR emission generator to create and install an information block inside its own SPGPR when the SPGPR meets the MGW. First, the acceleration of GR_{MGW} (e.g. neutrino, **ν**) occurs, which then selectively interacts [17,18] with "naked" protons [19] in water hydrogen bridges (***p***) of the gravitational sensor. Hence, the ETI-1 information is transmitted to us in the form of water clusters:

$$\boldsymbol{p} + GR_{MGW}(\boldsymbol{\nu}) \leftrightarrow [\boldsymbol{p}\text{- } GR_{MGW}(\boldsymbol{\nu})]^{\neq}$$

Thus, ETI-1 (**B**) is concluded to be inside a gas giant (Figure 24.12) where it finds livable conditions. The gas giant could also be a cooled star.

ETI-2 (Zubow-2) signals, from a constant, already existent communication process, are different from those of the first contact. For example, we found the signal (F_v = 150.5 s) with the typical "head" (Figure 24.13). From this figure (D), it is visible how the information is put on white noises (d_1, d_2, and d_3).

Figure 24.14 shows the GMS-spectrum of the LRO in water of the hydrogel sensor when it is exposed to a GR_{MGW} flow (Figure 24.13, F_v). There are signals that are not typical for water cluster ensembles on the Earth. The GR_{MGW} flow could be caused by the Gemini constellation with the coordinates 98.6±0.1° (longitude) and 0.7±0.1° (latitude) that moves in the direction to M22 through the Earth at 19:23:35 CET. Another case is possible, too; namely, the GR_{MGW} flow is coming from a space with the coordinates 277.9±0.1° (longitude) and 0.4±0.1° (latitude) and is moving in the direction of nebula NGC 2217 even though it is already at 19:08:55. Both events will take place only if the main gravitational wave is curved and the coordinates of M22 and NGC 2217 are correct.

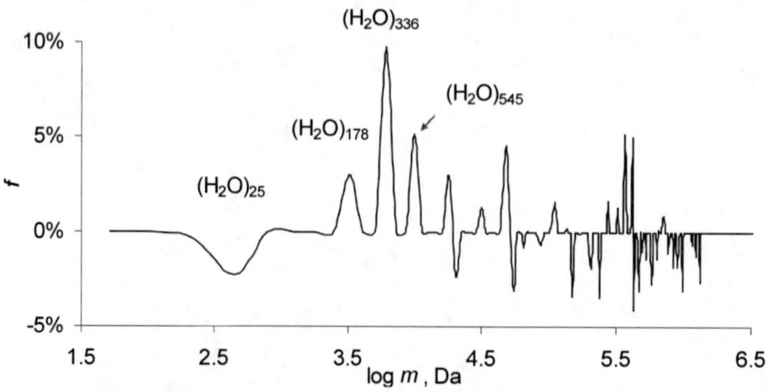

Figure 24.14. GMS-spectrum of water in the hydrogel sensor under the influence of a GR_{MGW} flow, F_v (Figure 24.13).

Another example that was accidentally discovered for neutrino signals, which are loaded on the gravitational waves, is shown in Figure 24.15. They are similar to those of a natural pulsar of gravitational signals but because of a head protocol (r) at the beginning of each block, they should have been caused artificially.

The reaction of water clusters in the sensor on the gravitational radiation in signal 2 (Figure 24.15) is shown in Figure 24.16.

All clusters in this figure except the base $(H_2O)_{12}$ are fragmentations that are untypical for the white noises of our planet. They arise, however, when water is exposed to electromagnetic and mechanical fields. These signals are assumed to be originated by an artificial pulsar gravity coming from Uranus (ETI-3) that hits SPGPR at this time. The ETI-1 model (Figure 24.12) should be applied to the gas giant Uranus, too. In Table 24.2, the main characteristics of ETI are given.

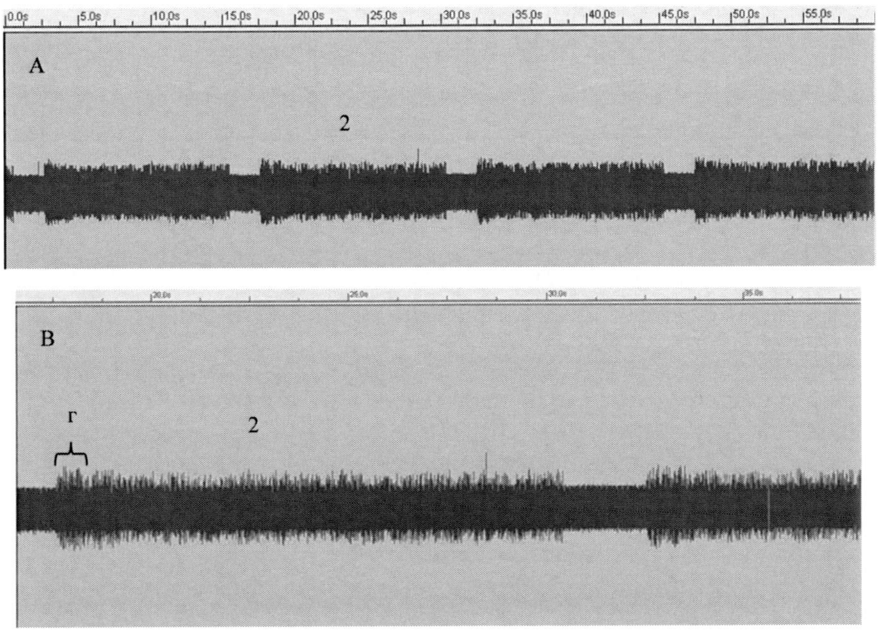

Figure 24.15. Periodic signals recorded by an agarose sensor on April 11, 2011 at 9:23:34 (CET). B – signal 2 in a more expanded form (r – head protocol, see Figure 24.3). At the moment when Uranus and the galaxy clusters of the Virgo constellation are in an opposite position (179°).

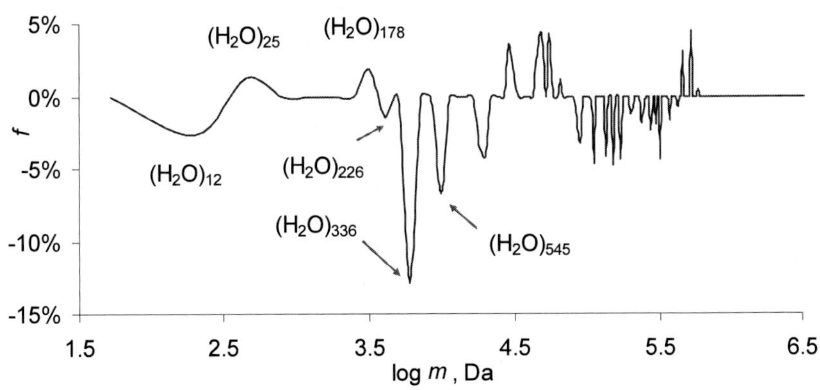

Figure 24.16. GMS-spectrum of water in the hydrogel sensor under the influence of GR 2 (Figure 24.15).

Table 24.2. Main characteristics of ETI

ETI, No.	Name	Position, Longitude/Latitude	Type	Level of development	Communication type	Activity	Discovery time
1	«Zubow-1»	21±2° / 60±3° (Cepheus const.)	Water	Higher	Gravitation	High	19th February, 2012, Germany
2	«Zubow-2»	98.6±0.1°/0.7±0.1° (Gemini const.)	Water	Higher	Gravitation	High	27th March 2012, Germany
3	Uranus	Uranus	Water	Higher	Gravitation	High	09.2012
4	«Zubow-3» (§24.2)	M87	Water	Higher	Gravitation	High	October November 2012,
5	«Zubow-4»	M87	Water	Higher	Gravitation	High	Germany (§ 24.2)

We are going to create an international account for all observed ETI; please inform us immediately if you get ETI signals, too.

CONCLUSION

- The signals are assigned, with a high probability to the first extraterrestrial intelligence «Zubow-1», which is more highly developed than us.
- «Zubow-1» knows about us.
- The form of the signals could be an indication that it is not its first attempt to contact us.
- It is in principle possible to contact «Zubow-1» or at least log in a communication between ETIs.
- Regarding the level of the development, it is concluded that the advantage of «Zubow-1» is ca. 1,000 years compared with us.
- ETI communication is mainly based on gravitation. In space, there seems to be a large number of ETI that are in various development and communication stages.

REFERENCES

[1] Bonfils X., Delfosse X., Udry S., et al. The HARPS search for southern extra-solar planets? XXXI. The M-dwarf sample. Astronomy & Astrophysics manuscript no. main c ESO 2011 November 24, 2011, arXiv: 1111.5019v2 [astro/ph.EP] 23 Nov 2011.

[2] Zubow K., Zubow A.V., Zubow V.A. Scanning of the Sun and other celestial bodies with help of gravitation spectroscopy. Optic and Photonics Journal, 2011, № 1, pp.15-23. USA.

[3] Zubow K., Zubow A.V., Zubow V.A. Experimental Methods for the Determination of the Super Light Velocities of the Gravitation. Nature, Structure and Properties of Gravitation Waves. In Horizons of Physics. Editor A. Reimer. NY. vol. 277. 2012, in print. https://www.novapublishers.com/catalog/product_info.php?products_id=29310

[4] Zubow A., Zubow K., Zubow V.A. New Experimental Method for Determination of Distances to Celestial Bodies and their Masses. In Horizons of Physics. Editor Richard Schortemeyer III. Nova Science Publishers, Inc. NY. Vol. 279. 2012, in print. https://www.novapublishers.com/catalog/product_info.php?products_id=34674

[5] Zubow K., Zubow A.V., Zubow V.A. Experimental Platform for the Investigation of the Structural Heterogeneity of the Physical Vacuum. Vacuum Energy Risks and Chances. In Horizons of Physics. Editor A. Reimer. NY. vol. 277. 2012, in print. https://www.novapublishers.com/catalog/product_info.php?products_id=29310

[6] Zubow Kr. V., Zubow A.V., Zubow V.A. Cluster structure of liquid alcohols, water and n- Hexane. Journal of Appl. Spectr. 2005, vol.72, no. 3, pp. 321-328.

[7] Bogdanov E.V., Manturova G.M. Equiclaster model of water", Radioelektronika, Biomedicinskaja Radioelektronika. 2000, vol. 7, pp.19-28. (in Russian).

[8] Lenz A., Ojamäe L. On the stability of dence versus cage-shaped water clusters: Quantum-chemical investigations of zero-point energies, free energies, basis-set effects and IR spectra of $(H_2O)_{12}$ and $(H_2O)_{20}$. Chemical Physics Letters. 2006, vol. 418, pp. 361-367.

[9] Zubow K., Zubow A.V., Zubow V.A. New Non-Destructive Method for the Analysis of Domains' Distribution in Proteins and Biomatrices. Nature of Domains. SAP. American Journal of Biochemistry, 2012, vol.2, no. 2, pp. 14-26.

[10] Zubow A., Zubow K., Zubow V.A. Molecular Water Clusters in River and Lake of Northern Germany and their Dynamics. Celestial Bodies Influence. Physical Chem. USA, 2012, in print.

[11] Zubow K., Zubow A.V., Zubow V.A. Molecular Cluster Distribution in Sea Water Near the Wall. Antifouling aspect. Journal of Clusters, 2011, vol.22, no. 4, pp. 603-619.

[12] Zubow K.V., Zubow A.V., Zubow V.A. Ensemble of Clusters – New Form of Molecular Matter, Risks and Chances. Zubow Equations. In Advances in Chemistry Research, vol. 5, 2010, pp. 107-145. Editor J.C. Taylor. Novapublisher. NY.

[13] http://ufo-z-51.narod.ru/merk.html , in Russian

[14] K.V. Zubow, A.V. Zubow, V.A. Zubow. Principles of gravitation spectroscopy. New form of molecular matter. Processes. Fields. "Aist Handels- und consulting gmbH", dep. R&D, Berlin, 2010, electronic book (www.zubow.de), p. 854. (in Russian).

[15] Zubow Kr. V., Zubow A.V., Zubow V.A. Oscillation spectra of seed NaCl crystals in aqueous Solution. Journal of Appl. Spectr. 2005, vol.72, no. 6, pp. 840-847.

[16] Zubow K., Zubow A.V., Zubow V.A. The Phenomenon of Planets' Influence on the Long- Range Order in Polyethylene glycol and Its Dynamic Viscosity. Research and Reviews in Polymer. An Indian Journal. 2011, vol 2 , no.2, in print.

[17] Zazepin G. T., Kopysov J.S., Smirnov A. J. Зацепин Г.Т., Neutrino. In Fisithesky Enciclopedithesky Slowar. Ed. Prochorov A.M. Moscow, Soviet Enciclopedy.1984, pp. 448- 451.

[18] http://en.wikipedia.org/wiki/Neutrino
[19] Pauling L., Pauling P. Chemistry.W.H. Freeman and Company San Francisco. 1975.

24.1. FIRST CONTACT WITH THE INHABITANTS OF URANUS

In November 2011 the authors received an information signal block of gravitational radiation, which was sent from Uranus [1]. At that time, not all signals could be completely understood; it was therefore decided to take up contact with Uranus.

The aim of the present work is to develop a more simple equipment to generate GR in the direction of Uranus to receive signals from living beings on Uranus.

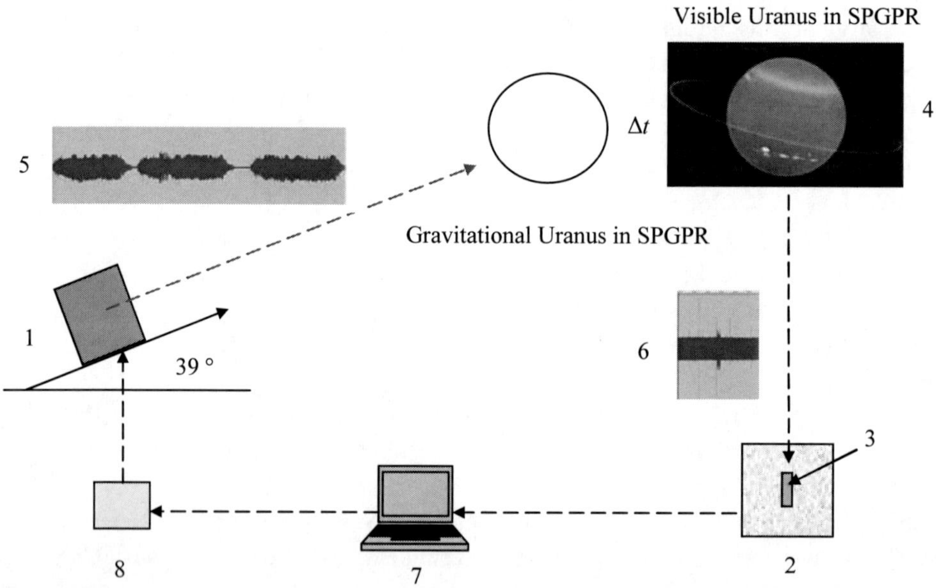

Figure 24.1.1. Principle equipment and work scheme. 1 – graser (see Figure 24.1.9), 2 – receiver, 3 – GMS-sensor (see Figures 4.0.1 and 24.1.9), 4 – visible Uranus (http://en.wikipedia.org/wiki/Uranus), 5 – graser signals (audio format), 6 – signals from Uranus (audio format), 7 – GMS-spectrometer, 8 – control panel. Measurement time from 25 to 28 June 2012. The time interval (Δt) between gravitational and visible Uranus in the SPGPR amounts to 2 h 47 min 40 s (June 25, 2012).

Agarose hydrogel (97 wt. % water) is chosen as a receiver (GMS-sensor). The measuring procedure of gravitation mass spectra is given in [2, 3, 4, 5]. The long-range order in water at the molecular cluster level is analyzed during the period when Uranus appeared in the SPGPR in North Germany (53°38′N and 12°35′E). The GMS-sensor is placed into an earthed iron box that is protected from noises, heat, and mechanical fields at most [6]. The box itself is in a building far from industrial centers and anthropogeneous noise sources. To find the correct position for Uranus, the program ZET 9 (www.astrozet.net) is applied; for the direction of the GMS-sensor to Uranus, the SPGPR is used [1, 7]. A simple graser, constructed for the laboratory (excitation of physical vacuum energy in graser space in the direction of Uranus, up to $32 \cdot 10^{11}$ kJ), serves as a transmitter. Information about the principles of grasers and their energy source is available in [1, 8, 9, 10]. The graser excites the structure of the physical vacuum (PV), and this excitation is, according to [11], independent of the distance. GR from graser is directed to the gravitational disk of Uranus [1, 12] but the arriving signals are received from the visible planet in SPGPR. The principle scheme of the equipment is shown in Figure 24.1.1.

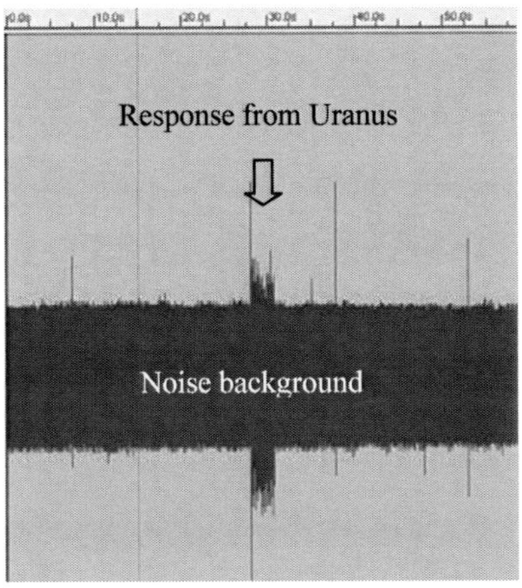

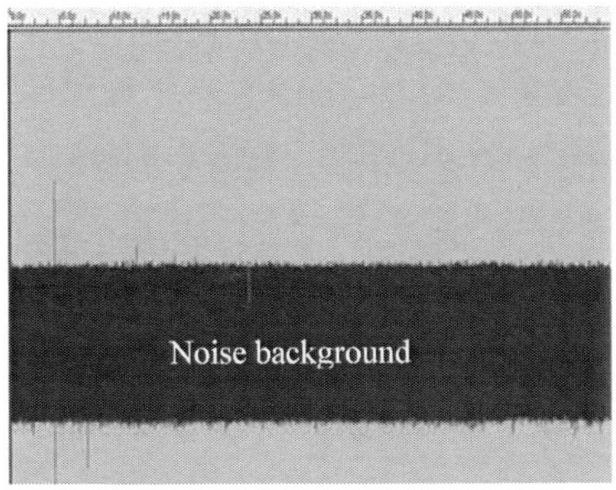

June 26, 2012 from 7:20:20, CET

Figure 24.1.2. Uranus' response to the graser signal sequences of different lengths (systemless). The time of the signal (ping, 3 s) is comparable with the apparent diameter of Uranus from Earth (http://www.webcitation.org/60qT3owNB). Right side: noise background without pings from Uranus if no signals had been transmitted by us, one day later (audio-format).

On June 25, 2012 a signal sequence (1 min from 04:36:33 to 04:37:33 CET) is sent by the graser in the direction of gravitational Uranus (39° altitude, 04:37:03), which at that time is in the SPGPR. The time between turning on and off is around 2…5 s. As expected, after 2 h 47 min 40 s (appearance of visible Uranus in SPGPR at 7:24:43 CET), the response from it (7:24:43 CET) is received (Figure 24.1.2, left). Below is shown how the gravitational sensor reacts on the response (3 s).

Applying this short signal the inhabitants of Uranus inform us that they received our signals; we have to see it as their fingerprint.

Conclusion 1

- The graser signals are received by the inhabitants of Uranus immediately, however, their response is transmitted to us with light velocity because of the absence of a large mass concentration behind the Earth. On November 4, 2011 a galaxy cluster in the Virgo constellation played the role of this mass concentration.
- The ping is found to be the fingerprint for Uranus.
- The GR jet from graser is spreading to Uranus with super light velocity, practically immediately.
- The signal transmission from Uranus to Earth proceeds but with light velocity (2 h 47 min 40 s). Where do the inhabitants of Uranus get the time for the signal processing (receiving, analysis, decision, and transmission)? A possible answer can be that the

ping comes from a space station on one of Uranus' satellites (27) or from a space station in the L1 Lagrangian point (http://en.wikipedia.org/ wiki/Lagrangian_point).

On *June 27, 2012* a signal sequence with total length of 80 s and small intervals of 2…3 s was directed to gravitational Uranus in the SPGPR, and then a weak broad signal with the same length from visible Uranus was received (Figure 24.1.3). This response is understood as conformation that our signals were received.

Figure 24.1.4. Graser signals sent to gravitational Uranus on June 28, 2012. For better understanding only three signals are given; furthermore, they are in audio-format. The length of the middle signal is 10 s. Comments are in the text.

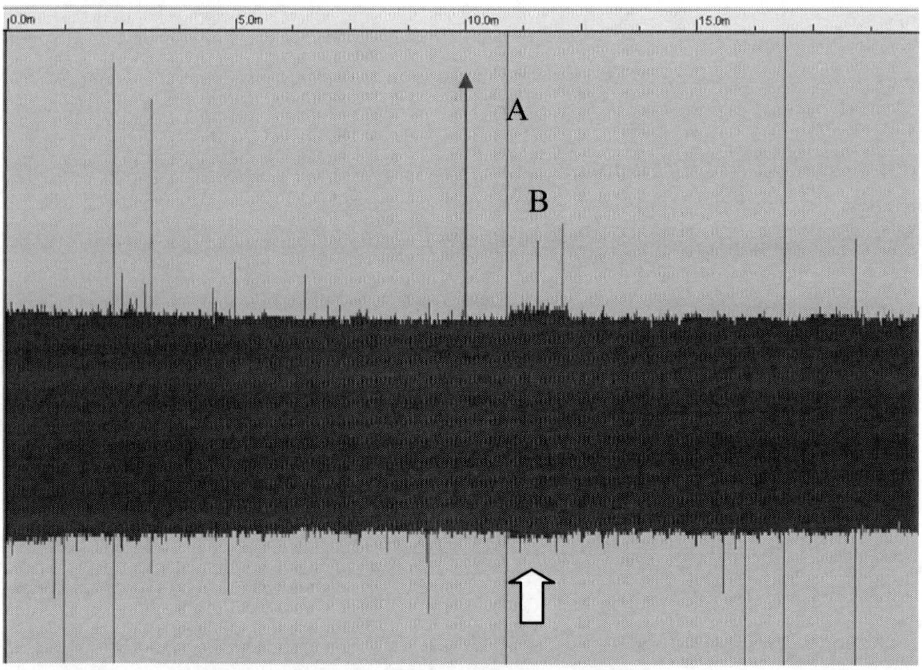

Figure 24.1.5. Response from Uranus on June 28, 2012 (marked with broad arrow). The broad band of weak signals from 7:03:04 is the answer on the graser shot sequence at 4:25:49. The time difference of 2 h 47 min 15 s equals the time that is needed for the light from Uranus to travel to the Earth (07:13:04, Figure 24.1.1). The distance to the planet is 20.06906 au (audio-format). Comments are in the text.

On *June 28, 2012* at 4:25:49 graser shots in the direction of gravitational Uranus's disk (altitude 39°, SPGPR) proceeded: (i) PV activation by graser 20 s before start, (ii) break of 1 s and first shot for 5 s (4:25:49), (iii) second break (3 s), shot (10 s), and (iv) third break of 10 s, etc., for a total of 78 s (Figure 24.1.4). The response from visible Uranus (in SPGPR) is similar to our signal (Figure 24.1.5).

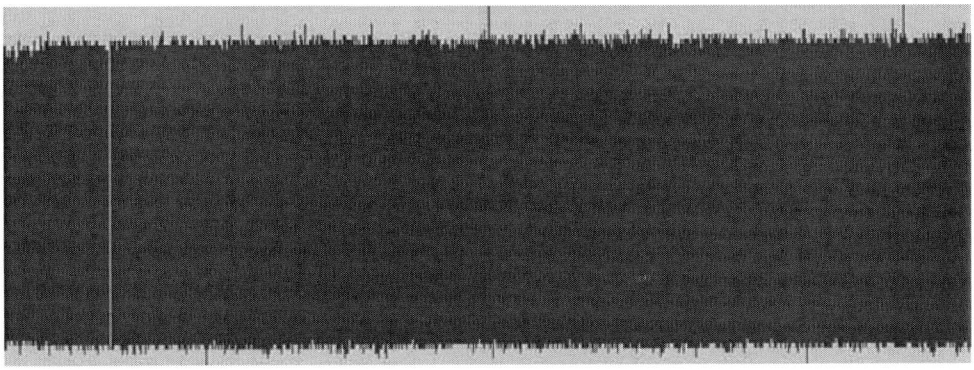

Figure 24.1.6. Signal from Uranus (Figure 24.1.5) at high resolution.

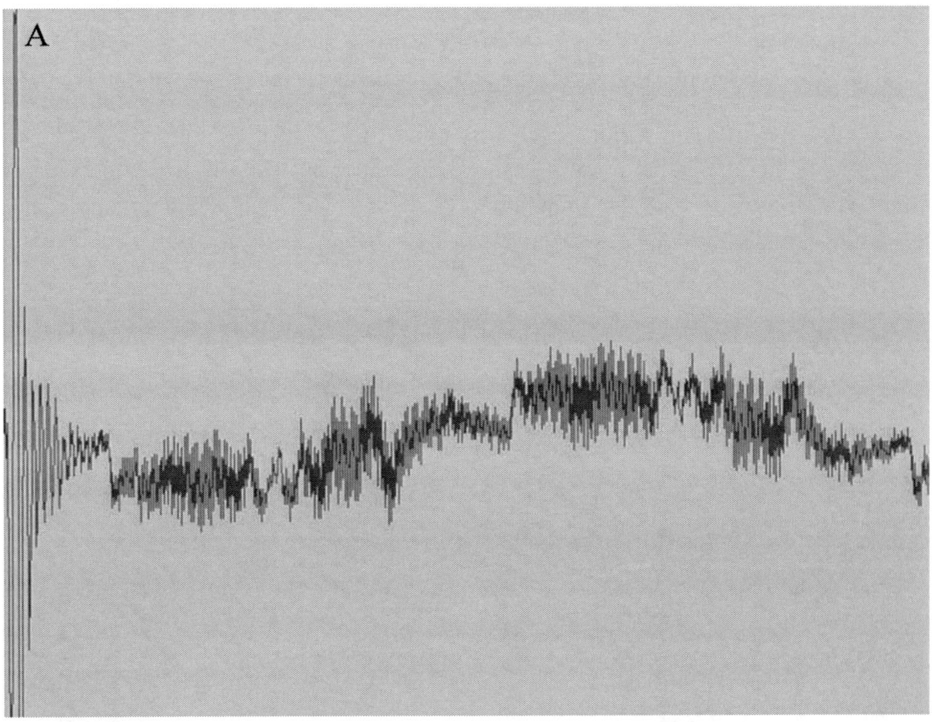

Figure 24.1.7. Noise background. The signal resolution amounts to 1 ns.

Concerning the data in Figure 24.1.5, the answer from Uranus seems to be sent directly. The expected signal must be at 10 min (red arrow) but the signal's arrival time is 1 minute later (A). This could mean that the inhabitants of Uranus need 1 min for signal processing or for an answering machine; the signal is transmitted from a space station on a Uranus satellite, which is further away from the Earth. In another case (June 27th), the signal came 20 s earlier (arrow, Figure 24.1.3), which has to be understood as signal transmission from a space station that is nearer to us than Uranus itself.

Despite the higher resolution of the signals (Figure 24.1.6), we are not yet able to interpret them correctly.

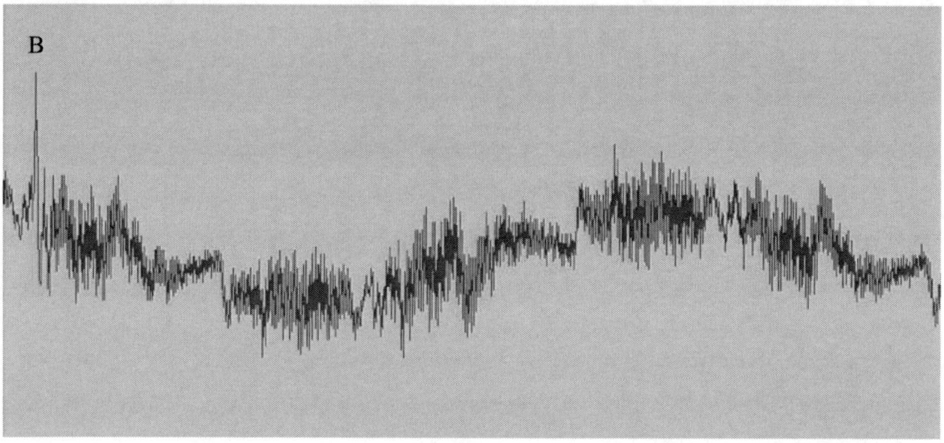

Figure 24.1.8. Noise background and signals from the inhabitants of Uranus at point B in Figure 24.1.5.

Part of the spectrum in Figure 24.1.5 at point A is given in Figure 24.1.7, and for comparison, the other one at the point B is shown in Figure 24.1.8.

Analyzing Figure 24.1.8, the inhabitants of Uranus are concluded to apply a signal resolution that is higher than that one we use.

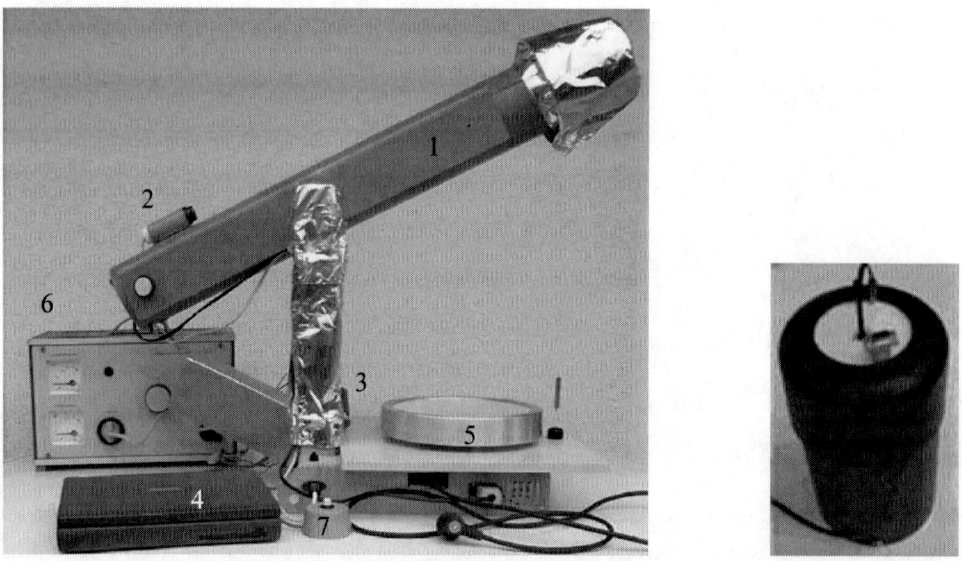

Figure 24.1.9. The world's first graser (left) and GMS-sensor for gravitational connection with UI, see Figure 24.1.1. 1 – main graser, 2 – supporting graser, 3 – time relays, 4 – computer, 5 – compass, 6 – power supply, and 7 – manual switch.

CONCLUSION 2

- At Uranus there is a highly developed civilization, which in contrast to us, is far more advanced in the fields of high technology, astrophysics, astronomy, and gravitation.
- The main method of communication with the inhabitants of the planet of Uranus is similar to the communication of bats.

- Regarding the level of the development, it is concluded that the advantage of the inhabitants of Uranus is ca. 150…250 years compared with us.

  The photo in Figure 1 shows that the inhabitants of Uranus could have survived a few disasters caused by themselves, e.g. wars. Unlike us (oxidative medium), they live in a reductive environment (alkanes, hydrogen, helium,) under giant atmospheric pressure and are really far from the Sun. Further they do not use the Sun's energy and thus are no threat to us. Our oxygen supplies could be interesting for the inhabitants of Uranus, e.g. for classical combustion processes.
- We are able to inform the inhabitants of Uranus that we have begun to understand GR, to operate it, and are ready to contact them.
- In the case of a war on Uranus, we have to decide with which site contact has to be established.
- First contact with the inhabitants of Uranus could be on a neutral territory, e.g. Mars.
- The worldwide first graser, generating GR signals with super light velocity was developed in our laboratory.

REFERENCES TO 24.1.

[1] Zubow K., Zubow A.V., Zubow V.A. Signals from Extraterrestrial Intelligences. Horizons in World Physics, editor Albert Reimer. Nova Science Publishers, Inc. NY. 2012, vol. 280, in print.

[2] Zubow K.V., Zubow A.V., Zubow V.A. Using of flicker noise spectroscopy for non destroying analysis of nano structures. Zavodskaja Laboratorija. Diagnostics of materials. 2008, vol. 74, no. 9, pp.40-45. (in Russian).

[3] Zubow K.V., Zubow A.V., Zubow V.A. The Dynamics of Low Frequency Movements of Molecular Clusters in the Hardening Process of Epoxide Resins. Chem Promislennost Segodnja. 2008, no. 9, pp. 12-21. (in Russian).

[4] Zubow K., Zubow A.V., Zubow V.A. Water Clusters in Plants. Fast Channel Plant Communications. Planet Influence. J. of Biophysics Chemistry. 2010, vol.1, no.1, pp.1-11.
[5] Zubow K.V., Zubow A.V., Zubow V.A. Cluster structure of liquid alcohols, water and n- Hexane. J. of Appl. Spectr. 2005, vol. 72, no. 3, pp. 321-328.
[6] Zubow K.V., Zubow A.V., Zubow V.A. Ensemble of Clusters – New Form of Molecular Matter, Risks and Chances. Zubow Equations. In Advances in Chemistry Research, vol. 5, 2010, pp. 107-145. Editor J.C. Taylor. Nova Science Publishers, Inc. NY. https://www.novapublishers.com/catalog/product_info.php?products_id=27769
[7] Zubow K., Zubow A.V., Zubow V.A. Scanning of the Sun and other celestial bodies with help of gravitation spectroscopy. Optic and Photonics Journal, 2011, № 1, pp.15-23.
[8] http://www.second-physics.ru/node/28
[9] http://www.trinitas.ru/rus/doc/0231/007a/02311016.htm
[10] http://www.trinitas.ru/rus/doc/0231/004a/02311041.htm
[11] Shipov G.I. Theory of physical vacuum. Theory, experiments and technologies. Moscov, Nauka, 1997 in Russian.
[12] Zubow A., Zubow K., Zubow V.A. New Experimental Method for Determination of Distances to Celestial Bodies and their Masses. In Horizons of Physics. Editor A. Reimer. Nova Science Publishers, Inc. NY. Vol. 279. 2012, in print. https://www.novapublishers.com/catalog/product_info.php?products_id=34674

24.2. Listening to the Communication between ETIs

In §24 and in [1], we discovered a new channel of gravitational communication between ETIs, and in §24.1, the first data of our communication with the UI (Uranus-inhabitants) was represented. According to UI, our development is characterized as level 1, while they give themselves level 8 in a 10-level system, where 10 means the highest development. This data was obtained in September 2012, when the main gravitational wave from Uranus to the Sun passed through the Earth (opposition of Uranus and Sun), hence providing conditions for simultaneous communication with us. In October 2012 the situation changed and Uranus' opposition to the Sun was broken. The Earth was on a new line connecting Uranus with a galaxy cluster in the Virgo constellation (most likely M87) that now MGW arose between Uranus and M87 (Figure 24.2.1). MGW between Earth and Uranus was always present because of the greater mass of this gas giant, which was slowly removing the mass from the Earth [2]; the same situation is present between the giant M87 and the dwarf Uranus. As UI informed us [2], they know six highly developed ETIs. Therefore, it is interesting to listen to the communication between UI and M87. If UI is looking for ETIs, then a galaxy cluster in the Virgo constellation would be an interesting challenge for them.

The aim of this chapter is to scan the gravitational noises of Uranus and M87, since we are in the slide plane of gravitational proton resonance (plane going through the Earth's rotating axis and the sensor place on the Earth's surface).

Agarose hydrogel (a model of biometrics) containing 97 wt. % water is used as sensor for the gravitation mass spectrometer (see §24.1). The spectrometer is at a place [3] with the coordinates 53°34'54"N, 12°47'02"E. The analysis of the long-range order in the water of hydrogel at the cluster level proceeds according to the method described in [4]. The GMS-sensor is placed into an earthed iron box that is protected from electromagnetical noises, heat, and mechanical fields [3]. The box itself is in a building far from industrial centers and anthropogeneous noise sources. To understand the method, some curves that reflect the GR energy flow (sum of cleaned signals) are given in [3, 5]. The shifting of the energy balance between the proton in the baryonic state and its dissolved form in the physical vacuum [2,5,6] is the platform for the GR resonance registered by GMS. The visible constellation of celestial bodies is determined using the program "ZET 9" (www.astrozet.net). The ETI signals are recorded in 20- or 60-minute scanning intervals when the celestial bodies meet the SPGPR [7].

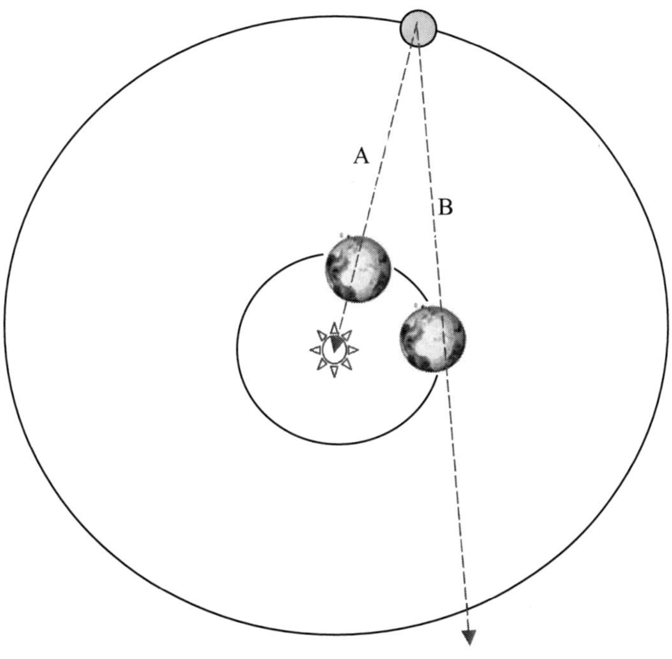

Figure 24.2.1. Scheme illustrating the mass distribution in space when MGW arises between them. A – Sun-Uranus opposition and B – Uranus-M87 opposition (Autumn 2012, > 177°, http://de.wikipedia.org/wiki/Messier_87).

The negotiations of two partners, previously unknown to each other, began with a greeting followed by a communication procedure, which depends greatly on the developing level of both parties, their openness, understanding the communication language, and the ethics of communication (for example, no heckling, listening to the end, etc.). Here, each partner pursues its own aim. A third party cannot participate in this communication; a third party can, however, carefully analyze the peculiarity of this process (language, speed of information transmission, density, clearness, etc.). It is reasonable to believe that the mentioned factors should be seen by the third party (observer) analyzing the negotiations between ETIs. Let us try to briefly analyze a large information volume, which we have received between October-November 2012 when the GMS-sensor was on the line connecting Uranus and M87 (Figure 24.2.1). Note that we tried to scan gravitational noises from M87 and Uranus twice a day, so that one of them was close by the Earth mass and the signals of one ETI should dominate over those of the other. This did not always work due to the fact that the ETIs use gravitational radiation to which our planet is transparent. ETIs on the giant M87 are called, according to [1], Zubow-3 and Zubow-4, but for ease of understanding, we describe them as M87, although in theory this galaxy can have many advanced civilizations [8].

The website http://www.zubow.de/pdf/News.pdf provides the complete packet for listening to the communication between Zubow-3 and Zubow-4 in the Virgo constellation (M87) on the one hand and UI (U-1 and U-2) on the other hand. Zubow-3 was the first who sent signals to our galaxy, which were received in October 2012 by UI. UI used first the U-1 transmitter for the information exchange (§24.1); later the work was connected to the more powerful and developed transmitter U-2 on Uranus' orbit.

Figure 24.2.2 shows the signals from Zubow-3 as a spectrum of gravitational noises (GN) when visible M87 appears in the SPGPR (protocol of October 2, 2012, http://www.zubow.de/pdf/News.pdf).

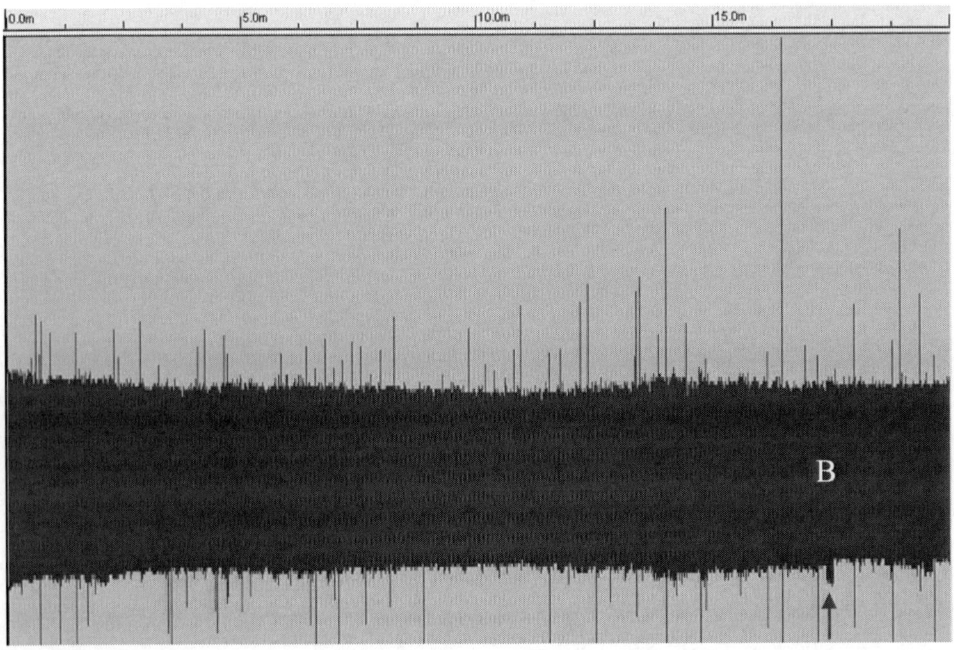

Figure 24.2.2. GN spectrum from M87 in the SPGPR. October 2, 2012, daylight, opposition Uranus-M87 > 175.7°. Signal marked by an arrow belongs to a GN reduction type (direction to bottom). Comments are in the text.

Figure 24.2.2 shows a typical GN spectrum that is overloaded with information; it is divided by many punctuation marks. This information splitting is produced by pulsed grasers and the gravitational signals are directed towards our solar system. Further, they are not typical for UI (§24.1) and belong more likely to Zubow-3 in M87. Considering the expanded spectrum we understand that it is very dense and generated with at least nanosecond grasers. The resolution of our devices in microsecond steps is insufficient to extract real signals so we have to analyze forms of GN (Figure 24.2.3). On the other hand, their interpretation will require a new generation of computers, new mathematics, and new algorithms.

In Figure 24.2.3, we see the dense information block (B) in which the signals are mainly directed downwards, which means damping GN. Block B is not a typical UI signal and should be ascribed to Zubow-3. Additionally, detailed signals of a pulsed 3-modal and a multimodal graser applied for text separation are given.

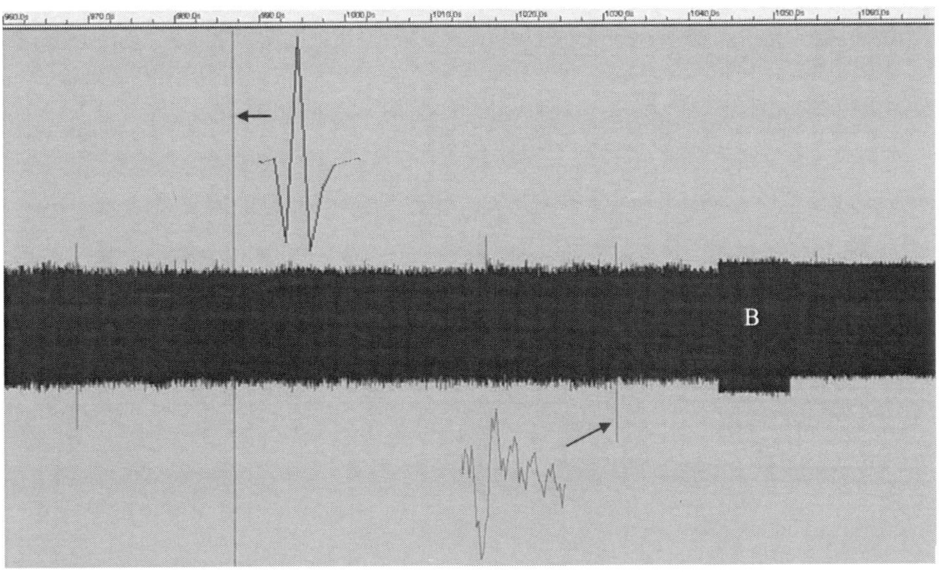

Figure 24.2.3. A fragment of the spectrum in Figure 2, from 1,033 to 1,060 s.

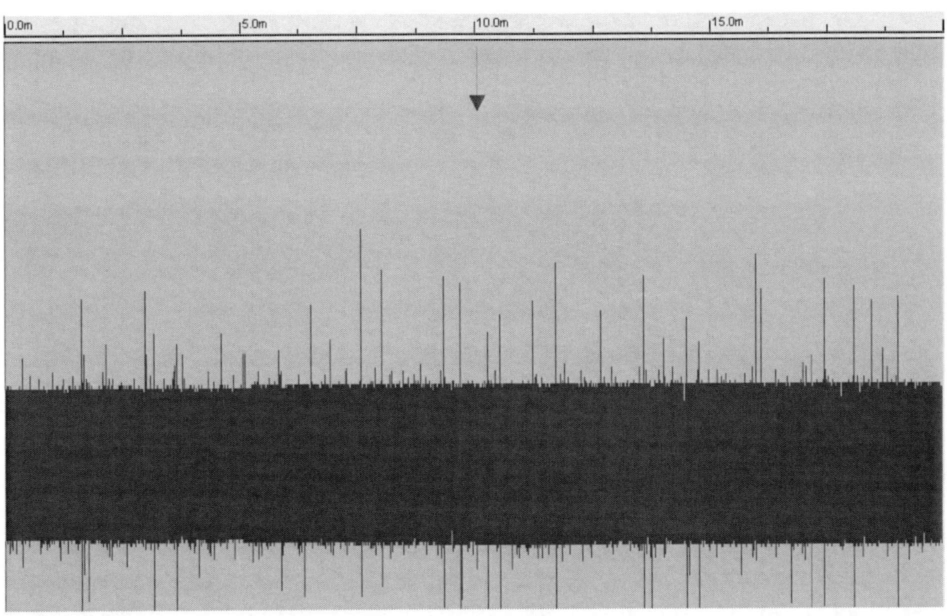

Figure 24.2.4. GN spectrum of gravitational Uranus disk in the SPGPR (arrow). October 21, 2012 (fog). Gravitational disk in the SPGPR at 20:47:16, altitude 28.3°. Opposition to M87 > 177°. Scanning start at 20:37:16.

In the night from 2 to 3 October, the gravitational and visible disks of Uranus were scanned at 21:55:19 and 00:34:11, respectively. The signals typical for U-1 and U-2, however, were not detected; they were silent. Like M87 (Figures 24.2.2 and 24.2.3) the GN spectra are characterized by periodic signals with a multiple text division using very strong impulses. Response signals from UI to Zubow-3 were only obtained on October 21 (Figure 24.2.4).

As is visible in Figure 24.2.4, there are many hidden periodic structures of artificial origin but analyzing the more resolved spectrum, the typical structure of information blocks from U-1 to Zubow-3 (Figure 24.2.1) are found. Let us consider them in more detail.

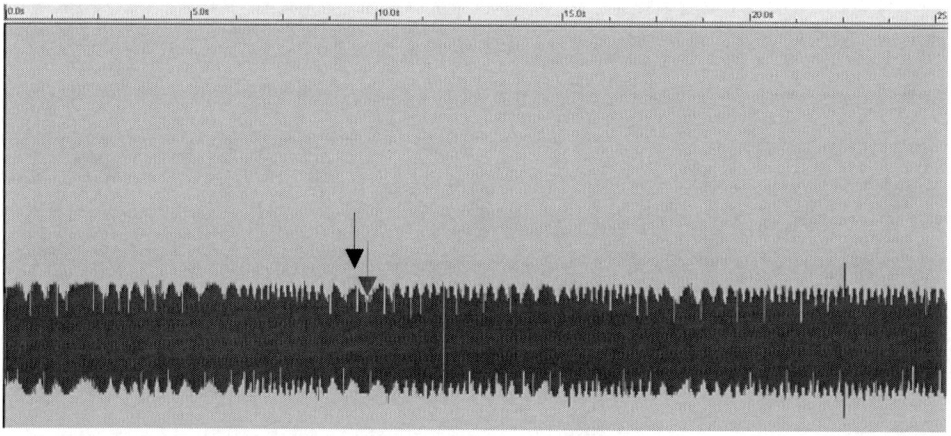

Figure 24.2.5. The first 25 seconds of the spectrum in Figure 4.

As already said (§24.1), the signals in Figure 24.2.5 belong to U-1, see §24.1. A continuous GN background (compare with Figure 24.2.4) can be seen. It is superimposed by information of high density in the form of signals (arrows) of different duration and generation, known for U-1.

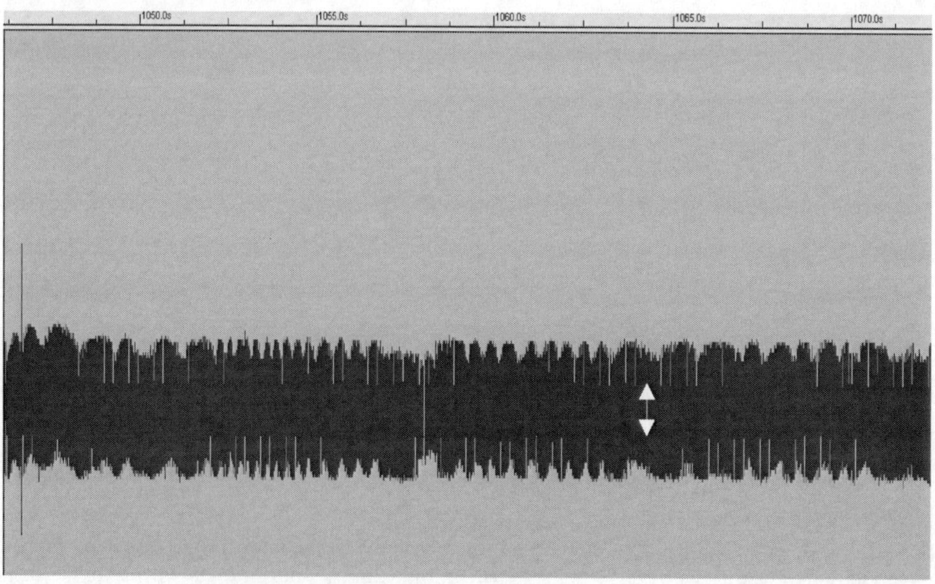

Figure 24.2.6. Expanded GN spectrum from Figure 4 between 1,047 and 1,072 s.

The GN background level is the same as 9,572 s later (arrow, visible Uranus), see below. Hence the spectrum in Figure 24.2.4 reflects high-density information, which was previously already sent to us by U-1 in a second time interval that was simple and understandable (§24.1). Only a computer system that is 200 years in advance of us can handle this large amount of information; it is therefore not provided to us. Besides us, Zubow-3 was on the communication traffic; that means it is the receiver of the information packet.

After 9,572 s, GN is scanned repeatedly at the moment when the visible Uranus disk hits the SPGPR (Figure 24.2.7). Here the UI signals do not appear but gravitational noises are strongly superposed with Zubow-3 information.

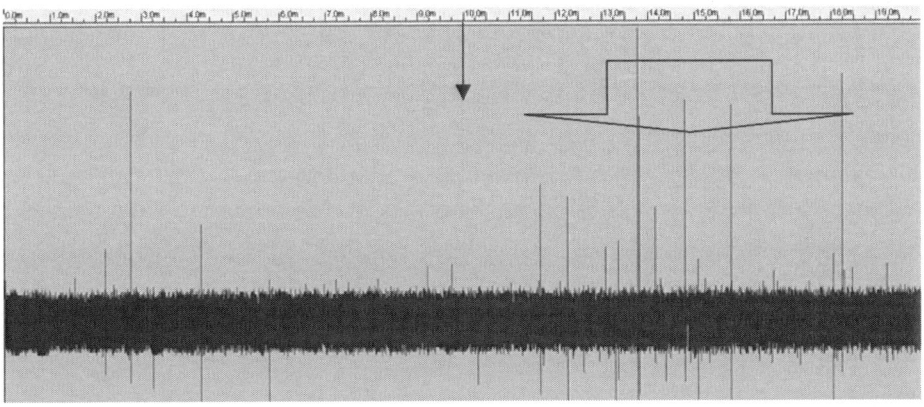

Figure 24.2.7. GN spectrum of Zubow-3 (in the SPGPR on the other Earth side) when the visible Uranus disk hits the SPGPR. October 21, 2012.

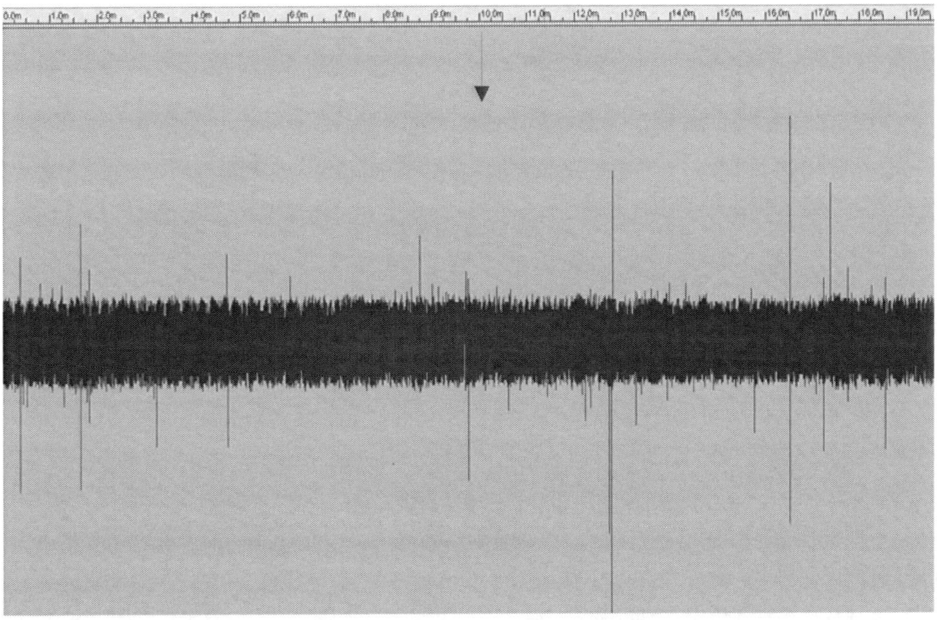

Figure 24.2.8. GN spectrum of the visible Uranus disk in the SPGPR (arrow). October 25, 2012, clear weather. Uranus altitude 38°, 19.17327 ly. Uranus visible disk in SPGPR at 23:10:346, CET. Opposition to the M87 > 177 °. Scanning from 23:00:34.

The spectrum in Figure 24.2.7 is not similar to a typical UI spectrum (Figure 24.2.4); it is more enriched with information, but has less in signal intensity. Due to that, the receiver is close by the Earth's mass and the transmitter is possibly far away. Furthermore, it is saturated by a large number of division pulses (broad arrow) (http://www.zubow.de/pdf/News.pdf).

The same characteristics belong to the GN spectrum given in Figure 24.2.8; the specific signals from our neighbor are not found. Hence the UI response is considered to be received by Zubow-3 who changed the transmission character a little (Figure 24.2.8).

In the resolved spectrum (Figure 24.2.9), we can see signal blocks with artificial origin; for their identification new recording equipment, mathematics, and methods of interpretation are needed. These signals are also part of Zubow-3, which is on the MGW [3] between M87 and Uranus. Super light velocity information transfer can be understood in relation to the neutrinos [9, 10] but in the communication between the two galaxies, the neutrino movement may not be essential. It seems most likely that the phases move with super light velocity [11]. For example, the gravitational waves could be forming energy clusters [3] in the physical vacuum. These clusters serve as energy sources for the formation of molecular clusters [3] that will be recorded by the GMS-sensor. Phases of waves propagate almost immediately [11]; on the other hand, the gravitational velocity is assumed to be $> 2 \cdot 10^{10}\, \boldsymbol{c}$ [12, 13].

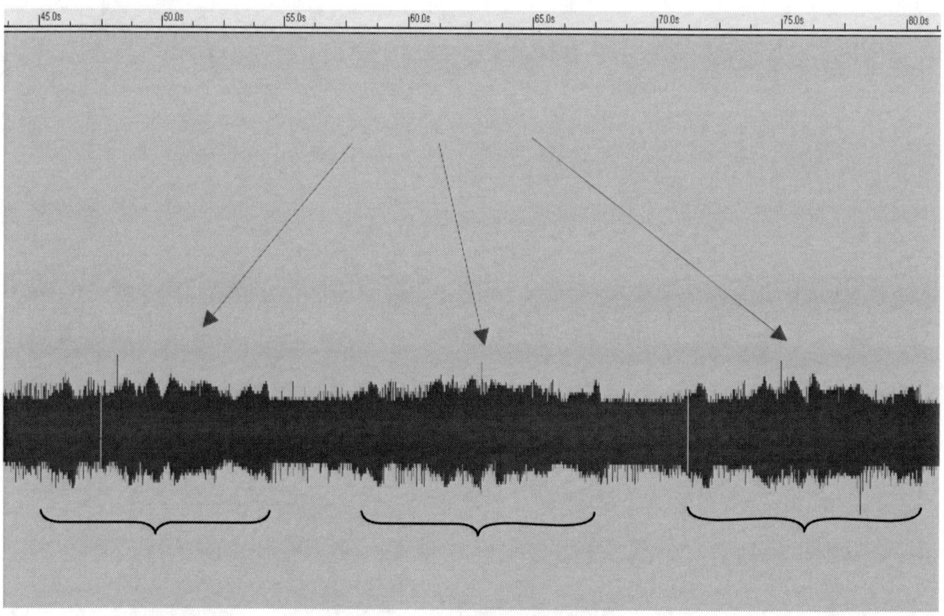

Figure 24.2.9. Expanded GN spectrum (Figure 24.2.8, from 42 to 81 s). Arrows and brackets show repeating information blocks, which are untypical for UI.

Scanning of GN on October 27, 2012 at the moment when the gravitational and visible Uranus disk appears in the SPGPR shows the silence of both parties even though they give each other signals generated by their gravitational lighthouses (see http://www.zubow.de/pdf/News.pdf). Here a situation of hanging arises. Who should start talking first to get "clean" information for the GN analysis?

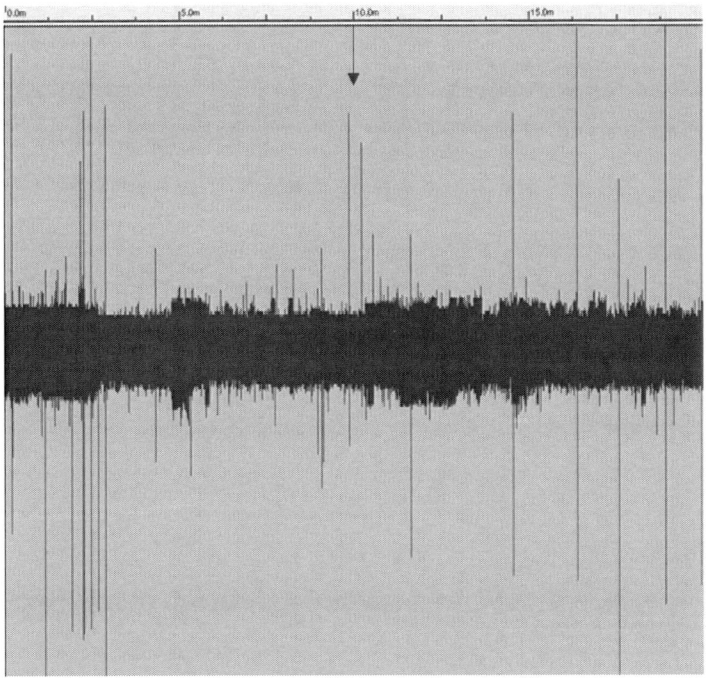

Figure 24.2.10. GN spectrum of gravitational Uranus disk in the SPGPR (arrow). November 3, 2012. The spectrum is a superposition of at least two GN spectra of U-1 and M87. Gravitational disk in the SPGPR at 8:43:39, altitude 28°. Opposition to M87 > 177°, opposition to Venus 177.6°. Scanning start at 18:43:49. Agarose hydrogel.

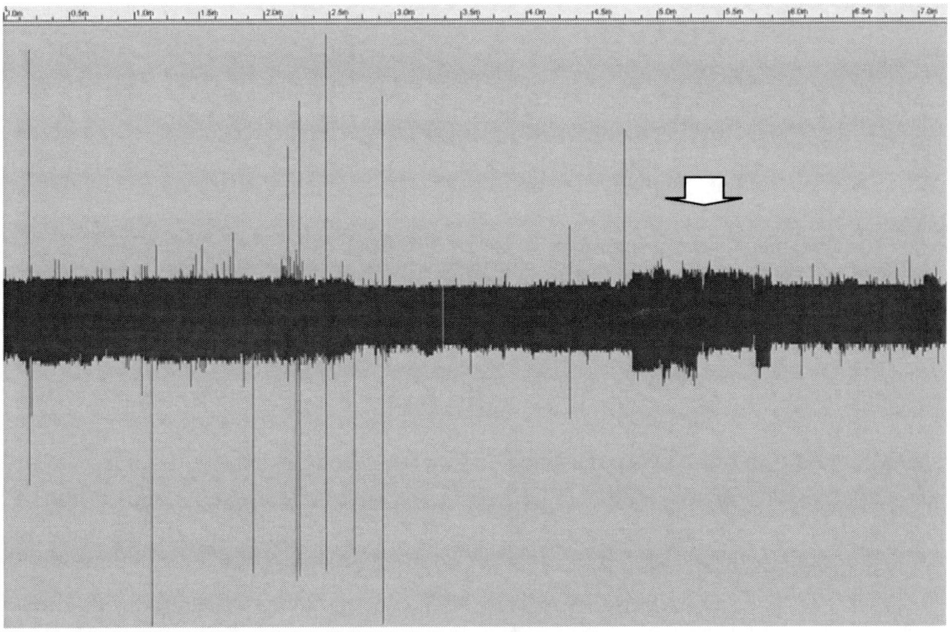

Figure 24.2.11. Fragment of the GN spectrum in Figure 24.2.10 from 0 to 7 minutes. A number of dense information signals, untypical for U-1 and Zubow-3, are visible (bazaar).

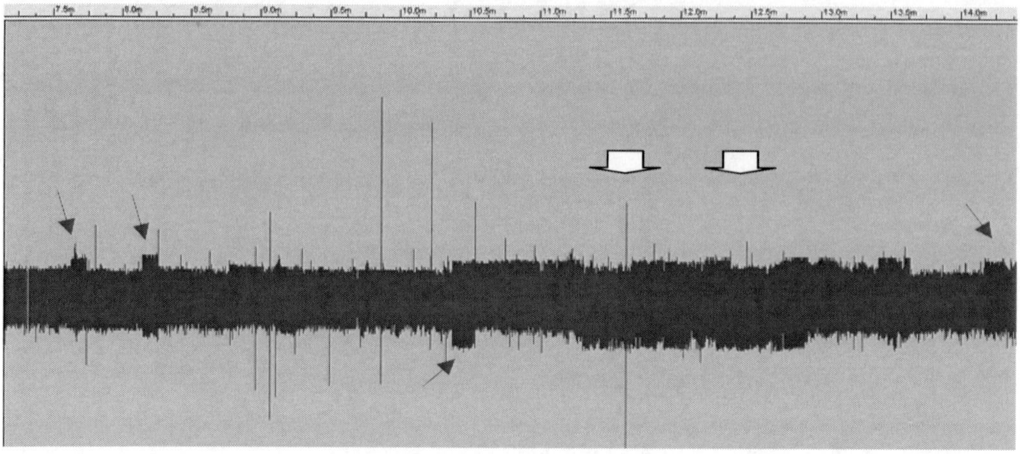

Figure 24.2.12. Fragment of the GN spectrum in Figure 24.2.10 from 7 to 14 minutes. November 3, 2012 (bazaar). A number of dense information signals of artificial origin (arrows) and superposition signals (broad arrows) are visible.

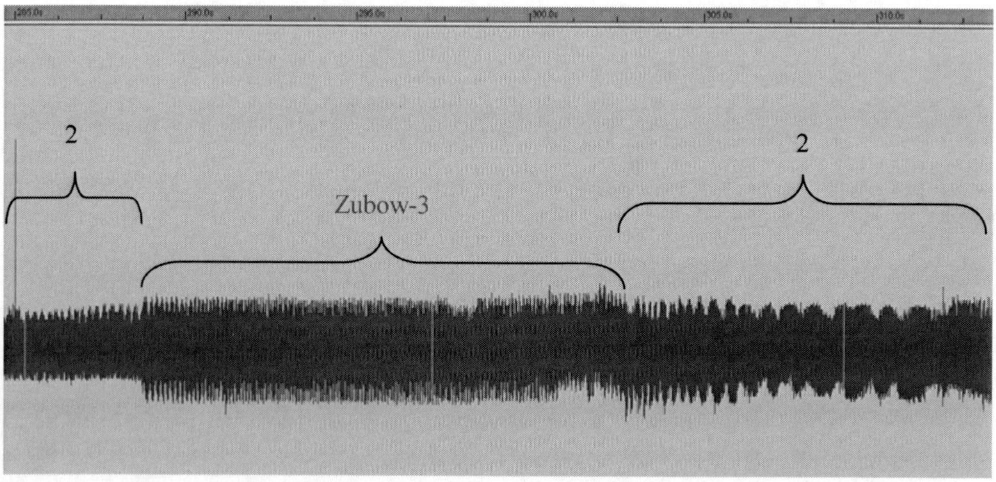

Figure 24.2.13. Fragment of the spectrum in Figure 24.2.10 from 285 to 315 s. A number of dense information signals with clear interruption from the side of Zubow-3 are visible. The information block from U-1 (block 2) is broken by Zubow-3 (block 1). Hence the bazaar is not finished yet and the parties are looking for its solution.

Typical GN spectra of U-1 are provided by the protocols of October 30th and November 1, 2012, for Zubow-3 the spectra come from October 31st, November 2nd and 3, 2012 (http://www.zubow.de/pdf/News.pdf). They enable us to understand the writing style of each party. A superposition of spectra that is caused by the simultaneous signal generation of both parties, "bazaar," is observed on November 2, 2012. It takes time and effort to develop the ethics of negotiation. The main events of the conflict are decided on November 3, 2012 though when the conflict is partially solved; it is yet to occur in later discussions. It is interesting to follow the procedure of how the two sides solve this ancient conflict.

After 24 h, the sides settle down and take the position of waiting. Then Zubow-3 generates the first GN signals.

Figure 24.2.14. Fragment of the GN spectrum in Figure 24.2.10 from 332 to 360 s. Here the signals of the Zubow-3 are mainly directed downwards—damping of GN from Zubow-3 transmitter (arrow). The communication interruption of U-1 (asterisk) by Zubow-3 is continued.

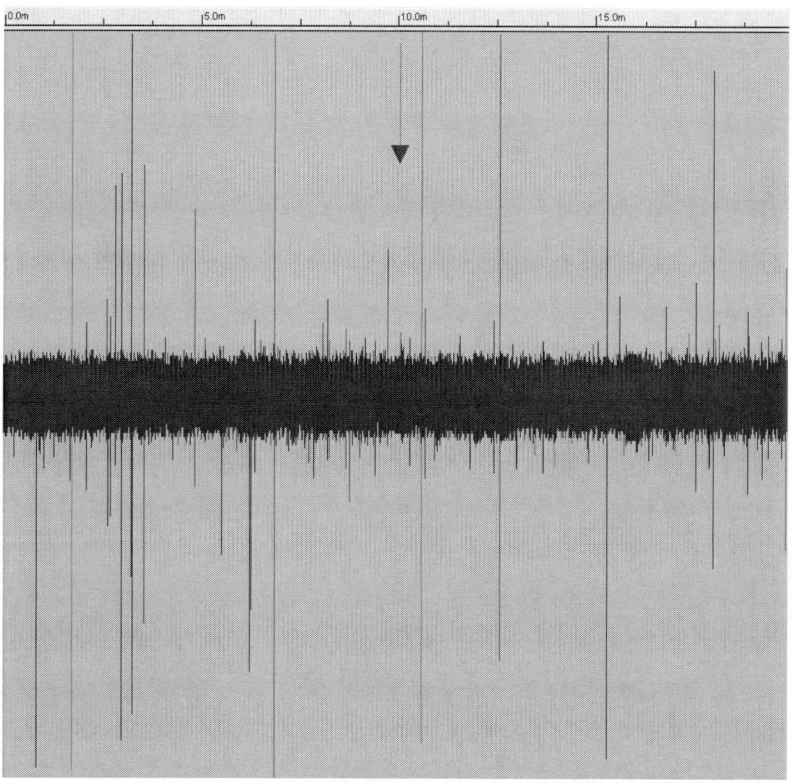

Figure 24.2.15. GN spectrum when M87 meets the SPGPR (arrow). Agarose hydrogel. Opposition M87-Uranus > 177°. Scanning start 6:41:35, November 4, 2012.

The spectrum in Figure 24.2.15 is highly loaded with many punctuation marks from grasers; let us consider them in detail.

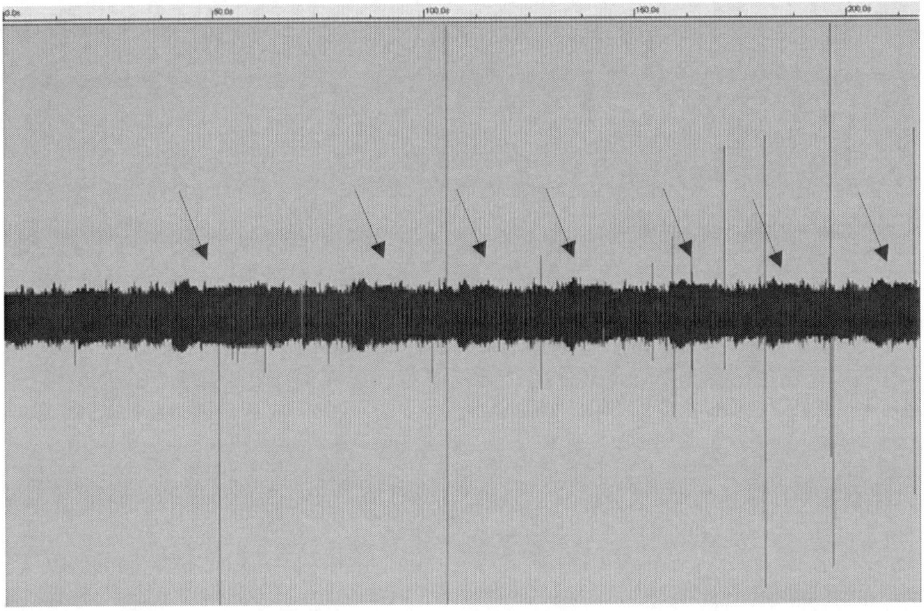

Figure 24.2.16. Fragment of the GN spectrum in Figure 24.2.15 from 0 to 200 s. A number of very dense periodic signals are visible. The signals are not typical of U-1 and seem to belong to Zubow-3.

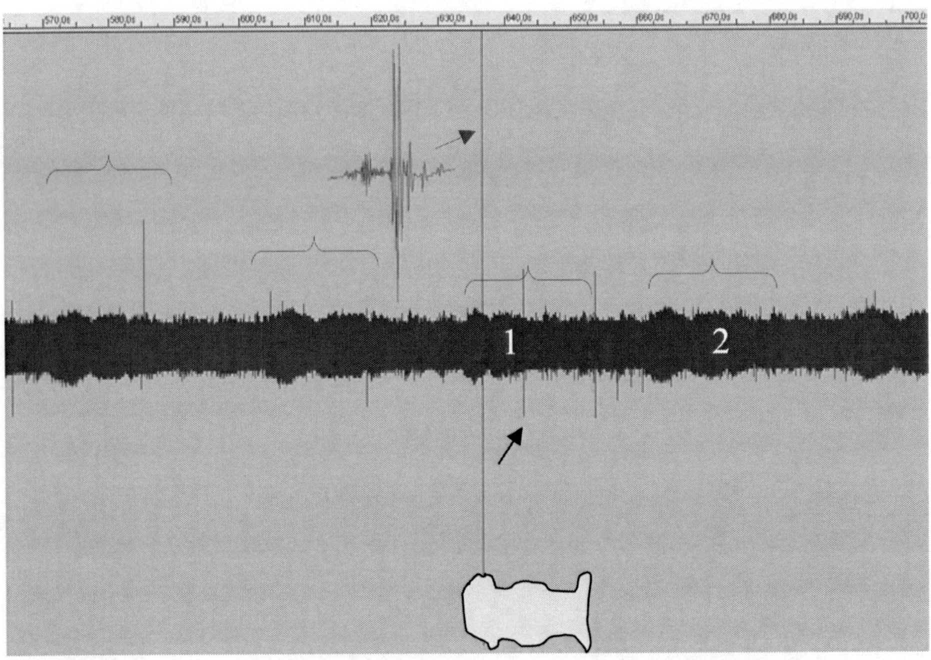

Figure 24.2.17. Fragment of the GN spectrum in Figure 24.2.15 from 560 to 700 s.

A number of dense artificial signals, atypical of U-1, are visible. The signals are assigned to Zubow-3. To understand the very strong graser pulse, it is shown in a more resolved form. The energy of this pulse is comparable with lightning energy. Information block 1 shows some similarities with the side view of a "man" while the pulse structure resembles a noise like a bell ringing that probably plays the role of a sound generator in the head (mouth). The second pulse (arrow 2) should belong to a body place, where the genetic information shall be transferred. The next "man" (2) could be a representation from top etc. (Figure 24.2.17).

Then, after a pause, both sides begin again to "bazaar,"; only on November 10, 2012 do they seem to approach a dialogue. The signals then will be more apparent. UI achieves it first; using the more powerful transmitter U-2 (see §24.1). They sent the following information in the direction of Zubow-3, Figure 24.2.18.

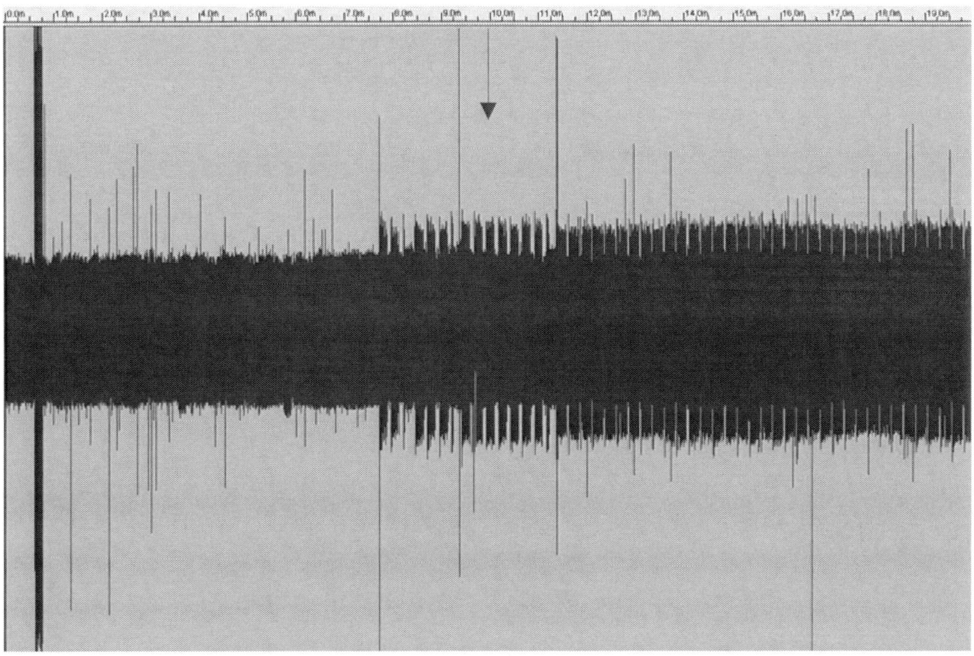

Figure 24.2.18. GN spectrum from M87. Opposition Uranus to M87 > 177.2°. November 10, 2012. Scanning start at 8:57:54. Arrow – M87 hits SPGPR.

The nature and sequence of the signals from U-2 indicates an understanding regarding the negotiation procedure between Zubow-3 and UI. Since the Zubow-3 signals are weak, dense, and complex, U-2 gives its signals in a simpler form like that which was already given to us previously (§24.1), which means they go down the "ladder of development" by a couple hundred years. The inhabitants of Uranus are therefore characterized by a higher level of development than Zubow-3 in M87. Logically, the information exchange should begin as follows, "Who are you? Where are you from?" neighbors/surroundings, and so on, indeed, the analysis of the spectrum in Figure 24.2.18 corresponds to this logic. Let us analyze it from left to right.

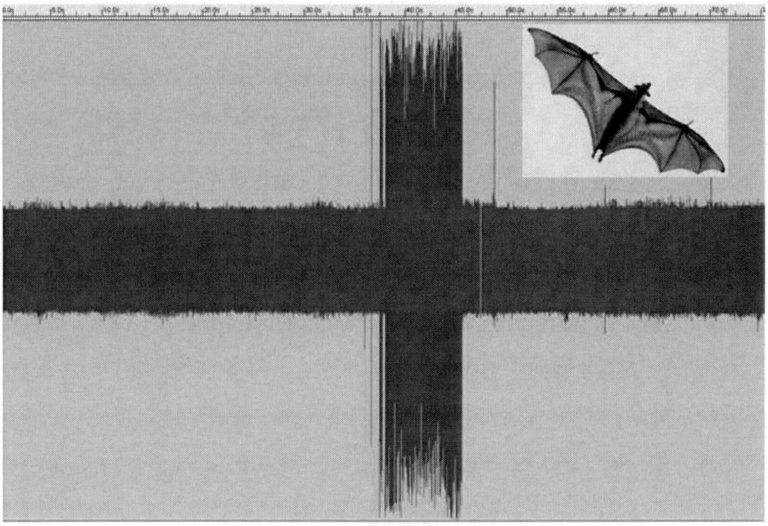

Figure 24.2.19. Fragment of the GN spectrum in Figure 24.2.18 from 0 to 80 s. We deliberately inserted a photo of a bat into the drawing to understand the analogy.

A very strong and long lasting signal (8 s) from U-2, which is directed towards the galaxy M87, is visible. The signal is to be understood as "Who are we?" It reflects the physical form of the inhabitants of Uranus. The intersection symbolizes the wings, which seems to be a little broader than the body itself. Furthermore, there are two legs at the bottom. Later UI received drawings on humanoid inhabitants from Zubow-4 (see table below). According to our previous investigations (§24.1), the inhabitants of Uranus can fly and should have "hands" and "fingers,"; they can only see well in the dark only, much like bats.

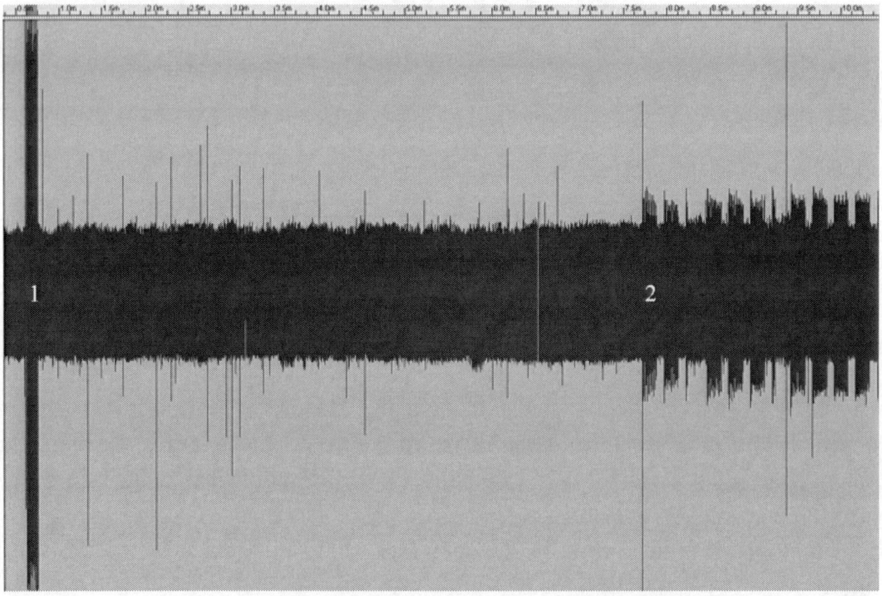

Figure 24.2.20. Fragment of the GN spectrum in Figure 24.2.18 from 0.5 to 11 minutes.

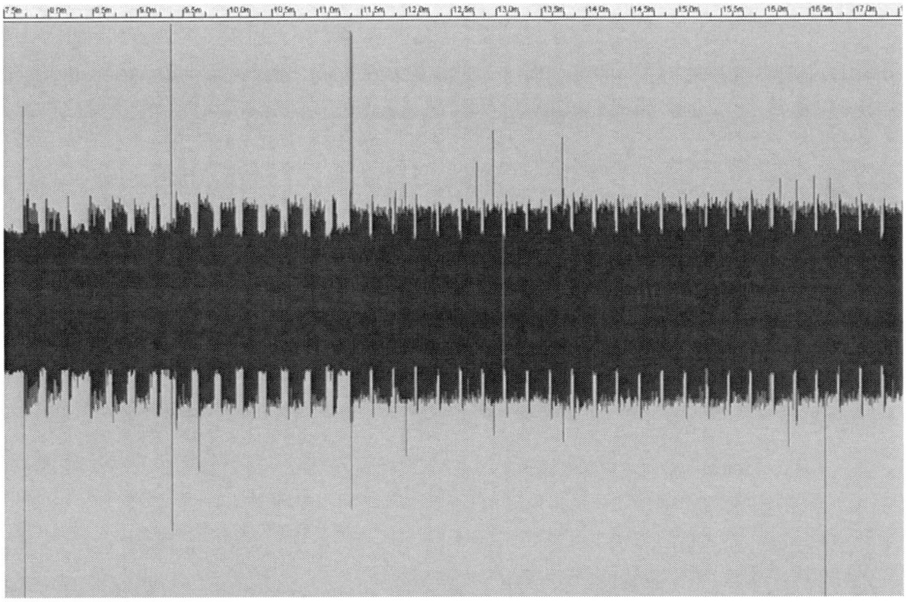

Figure 24.2.21. Fragment of the GN spectrum in Figures 24.2.18 and 24.2.20 from 7.5 and 18 minutes.

In Figure 24.2.20, there are weak information signals between the intersection (1) and the beginning of the UI information blocks (2) regarding "Where do we come from?"

We see signals from U-2, whose type has been used previously to contact us (in September 2012, §24.1). Three types of messages (I-III) are divided by very strong graser pulses.

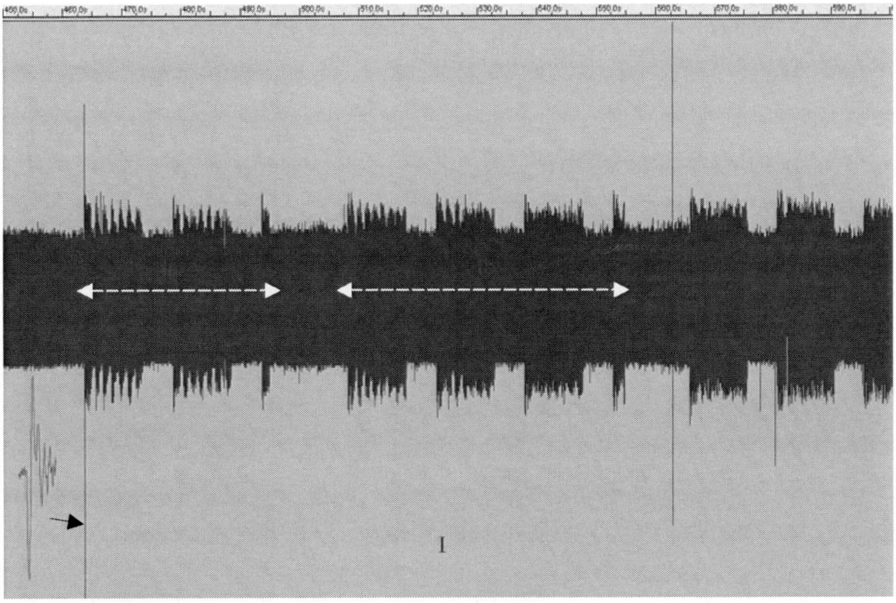

Figure 24.2.22. Expanded GN spectrum from Figure 24.2.21.

Analyzing the messages in detail, it is found that message **I** is emphasized by strong graser pulses and consists of the two blocks, **Ia** and **Ib** (Figure 24.2.22). As the expanded form of the graser pulse at the beginning of message **I** shows, it is similar to a ringing bell, which must be understood as "attention." Block **Ia** is subdivided, namely, into **A**, **B**, and **C**, where each subblock except **C** contains eight signals that are interpreted as "in our surroundings are eight planets." The message **A** is repeated by the subblock **B** (Figure 24.2.23). Block **C** is understood as the contents of this message (**A** + **B** = 2).

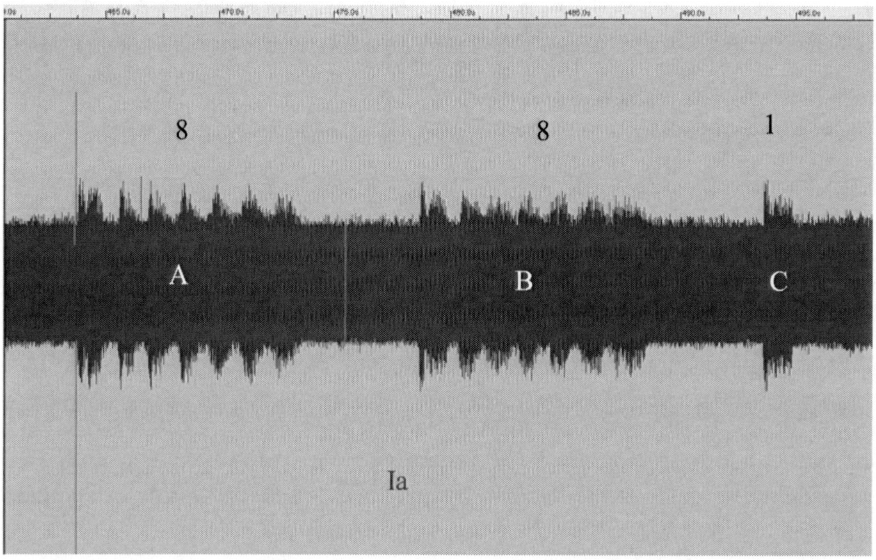

Figure 24.2.23. GN spectrum from Figure 24.2.22 and message **Ia** in expanded form.

The next message, **Ib**, is characterized by four subblocks (**D**, **E**, **F**, and **G**, Figure 24.2.24).

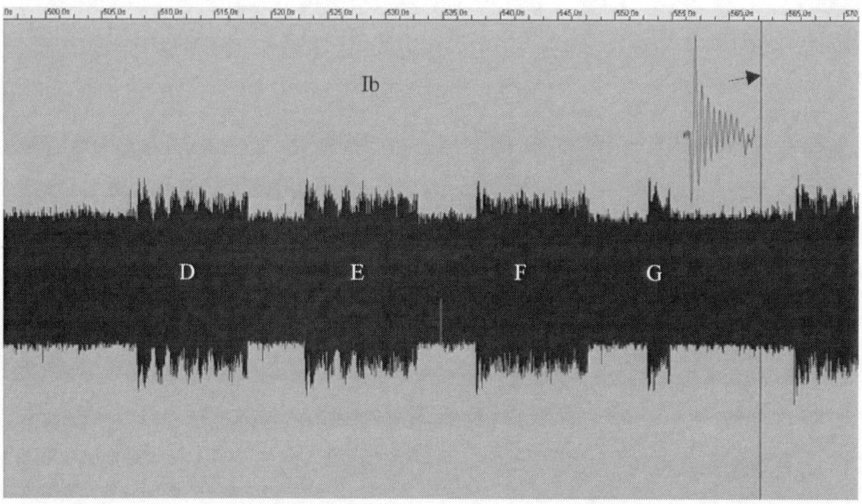

Figure 24.2.24. Fragment of the GN spectrum in Figure 24.2.22, message **Ib** in expanded form.

Subblock **D** (Figure 24.2.24) contains eight microblocks, however, some of them are represented twice or even more than twice (our sensor resolution is low, μs). **E** also consists of eight microblocks (double or more), an indication for the satellite number around each planet (8). **F** repeats **E** and **G** (triplet) is seen as the contents of the whole message. Behind this logic is the goal to transmit the information completely so that nothing gets lost. The expanded separation pulse again reminds us of a ringing bell (arrow).

Block **II** also consists of eight subblocks (Figure 24.2.25) where the division pulse is 3.2 s ahead of the first subblock; the pulse is also similar to the ringing of a bell (arrow). The number of signals in the microblocks from the first to the seventh amounts to: ~9, ~15, ~21, ~24, ~28, ~33, > 25 (due to insufficient equipment, the last two are difficult to resolve). Moreover, upon approaching the seventh, the packing density of the signals increases. Using these subblocks, U-2 is assumed to inform Zubow-3 about the planets and satellites surrounding Uranus (response on "Where we come from").

In Block **III**, a total of 35 subblocks are registered and due to technical reasons, the scanning is completed. It is possible that with the help of this block UI wish to inform the other side about all known mass concentrations in the Sun's system (planets, satellites, asteroids, etc.).

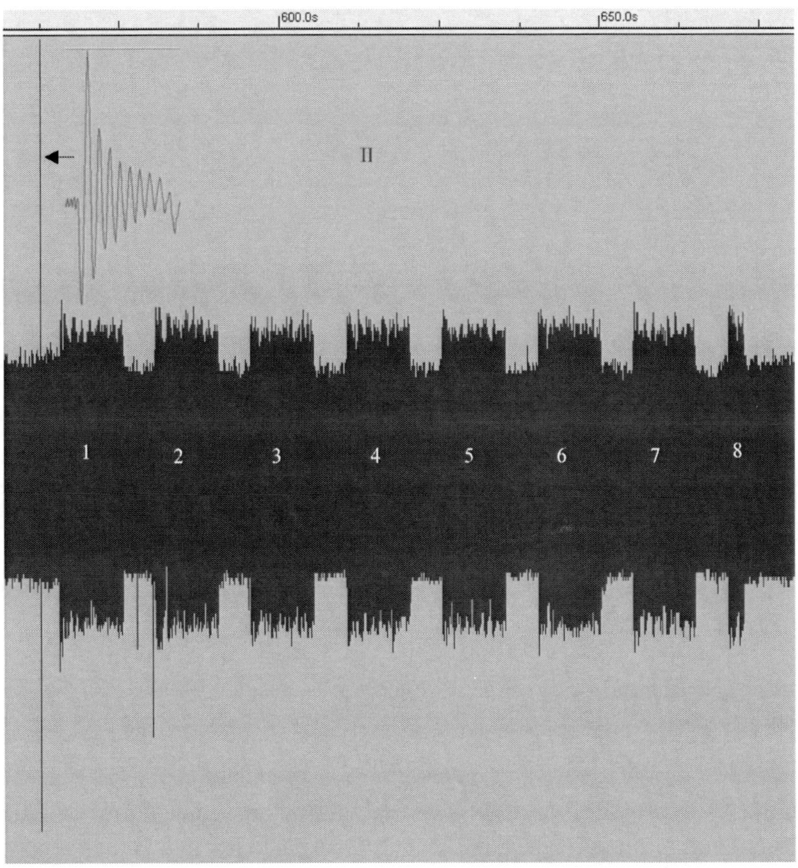

Figure 24.2.25. Fragment of the GN spectrum in Figure 24.2.21, block **II**.

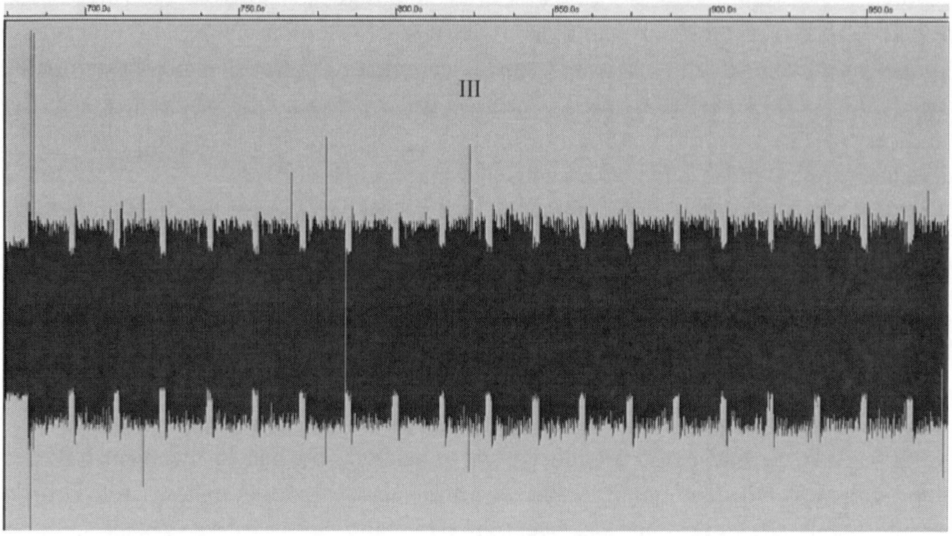

Figure 24.2.26. Expanded GN spectrum from Figure 24.2.21, block **III**.

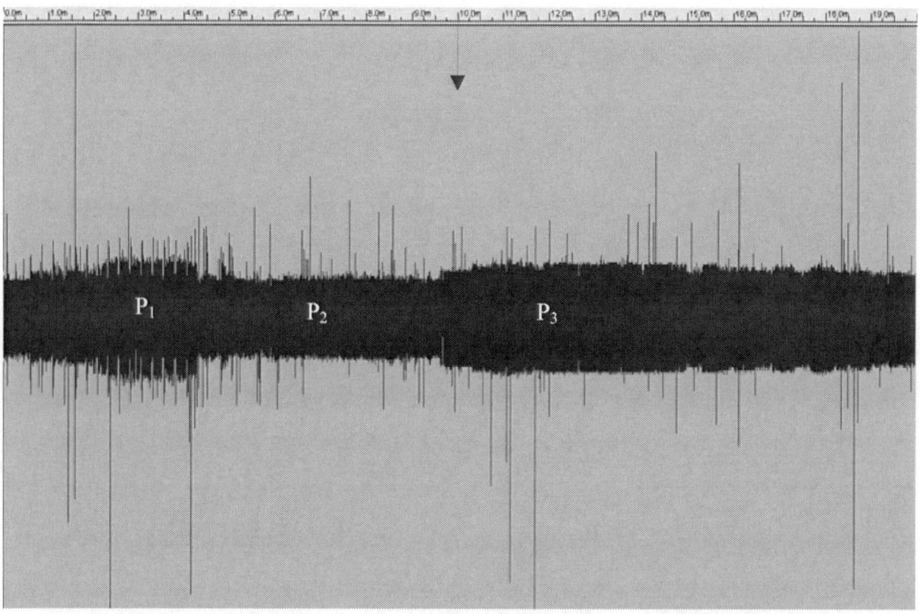

Figure 24.2.27. GN spectrum when the gravitational Uranus disk meets the SPGPR (arrow, 18:08:16, CET, altitude 28°), November 14, 2012. Agarose hydrogel. Opposition Uranus-M87 > 177.3°. Scanning from 17:58:16.

The channel of gravitational communication is concluded to be very specific and its analysis at μs level provides a discussion of the outer signal form currently without understanding of the information inside them. It is expected that ETIs use every chance of transmitting information, no matter how small; everything is important: signal and signal structure, blocks/subblocks, and their construction. The most important question in the information exchange is to understand the surroundings and principles of the living existence as well as the transmission methods of genetic information. Realizing this, U-2 generated a GN spectrum (Figure 24.2.27) giving the answers to the above questions.

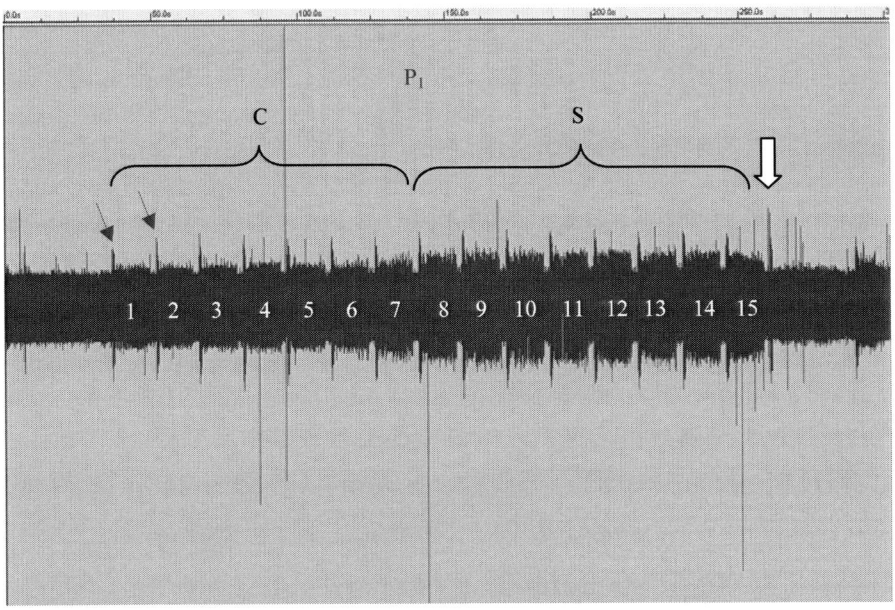

Figure 24.2.28. Expanded GN spectrum from Figure 24.2.27 between 0 and 300 s ($\mathbf{P}_1$).

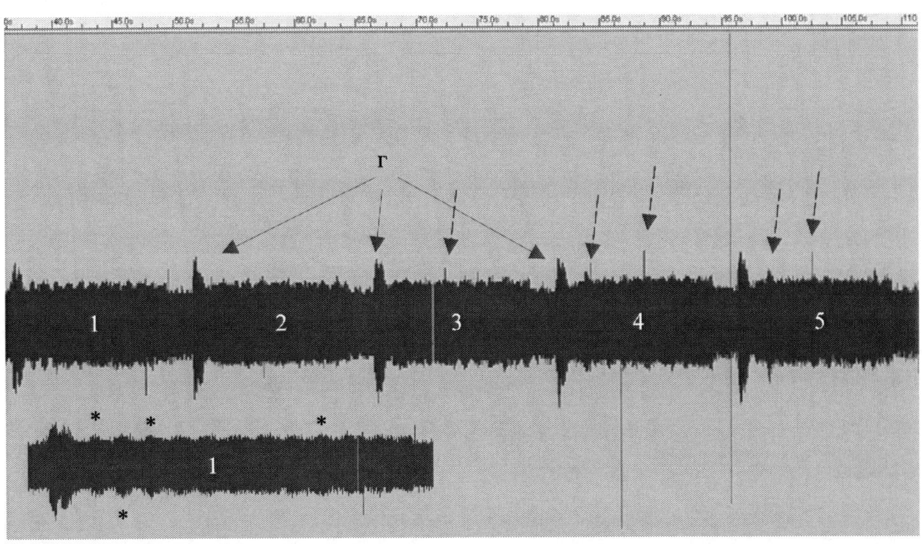

Figure 24.2.29. Fragment of GN spectrum in Figure 24.2.28 from 55 to 110 s, **r**-head belonging to the Sun. The first five information blocks are given. The first information block (1) is highlighted separately. The signals in information blocks 3-5 are marked with dotted arrows and indicate special mass concentrations.

As is shown in Figure 24.2.27, the GN spectrum is characterized by many dividing graser pulses and as typical for UI (U-2), the block starts with high-density information. The block resembles a phallus of the male sex. The form of the spectrum as a whole is intended to report a transmission method of the genetic information on Uranus. Taking into account that proteins can perform their functions in the absence of water, too [14], one can understand the formation principles of biomatrices on Uranus that they could be based on, e.g., liquid methane or methane-water azeotropes (enriched with water under high pressure and temperature). Let us consider these signals in more detail and start with $\mathbf{P}_1$.

Spectrum in Figure 24.2.28 unambiguously belongs to the inhabitants of Uranus where U-2 (§24.1) is activated. Since U-2 on Uranus; orbit is better equipped, the UI seem to be highly interested in Zubow-3. The spectrum in Figure 24.2.28 is represented by information blocks with special protocols. A total of 15 information blocks are given; the end signal is marked with a broad arrow. The 15 information blocks may correspond to the mass concentrations in our Sun's system. Due to similar levels of the signal intensities (**S**), the numbers from 8 to 15 are ascribed to planets while 1 to 7 could be large satellites (lower signal intensities, **C**).

As shown in Figure 24.2.29 the information blocks are composed of the protocol head (**r**) and the information subblocks where the signals of the head protocol are show up twice. In the first information block (1), 12 subblocks (stars) can be identified.

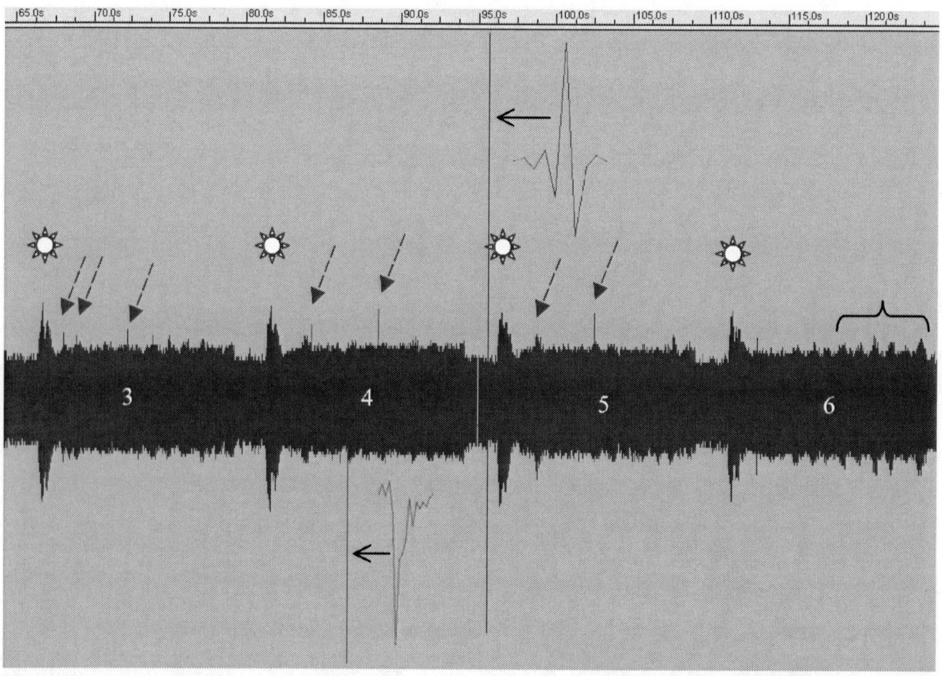

Figure 24.2.30. Fragment of the GN spectrum in Figure 24.2.28 from 65 to 125 s.

Figure 24.2.30 shows the detailed spectrum of the information blocks from 3 to 6, in which special signals (dotted arrow) are selected by the UI. Subblocks, apparently, mean mass concentrations in the Sun's system that may be of interest to Zubow-3, e.g. information on UI surroundings. We have around 11-12 large mass concentrations in our Sun's system. In information block 3, six subblocks of 12 are marked through a pulse; these could be the planets Mercury, Venus, Jupiter, Saturn, Uranus, and Neptune. The Earth is not included in this list. We are probably written off as being without perspective (§24.1). In information block 4, the subblocks with the numbers 1, 2, and 6 from the first list (Mercury, Venus, and Saturn) are emphasized while in information block 5 only the numbers 2 and 7 (Venus and Uranus) remain. Information block 6 accentuates subblocks for mass concentrations (width) that are most distant from the Sun. If the width and intensity of the signal are considered as the criterion for development then this logic is understandable only in terms of energy. UI use chemical energy from Uranus' core while the conditions are bad on Mercury (high heat and radiation gradient) but development is also possible (pulses in the information blocks 3 and 6). Development on Venus is conceivable, although the temperatures are really high (information blocks 3, 4, 5, and 6). The best conditions are assigned to mass concentrations (information block 6) that are most distant from the Sun (shown in bracket).

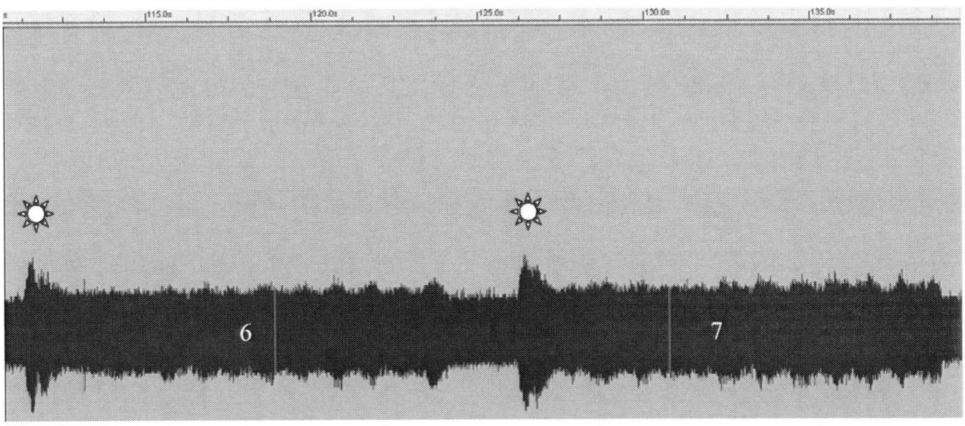

Figure 24.2.31. Resolved spectra of the information blocks 6 and 7 from Figure 24.2.28.

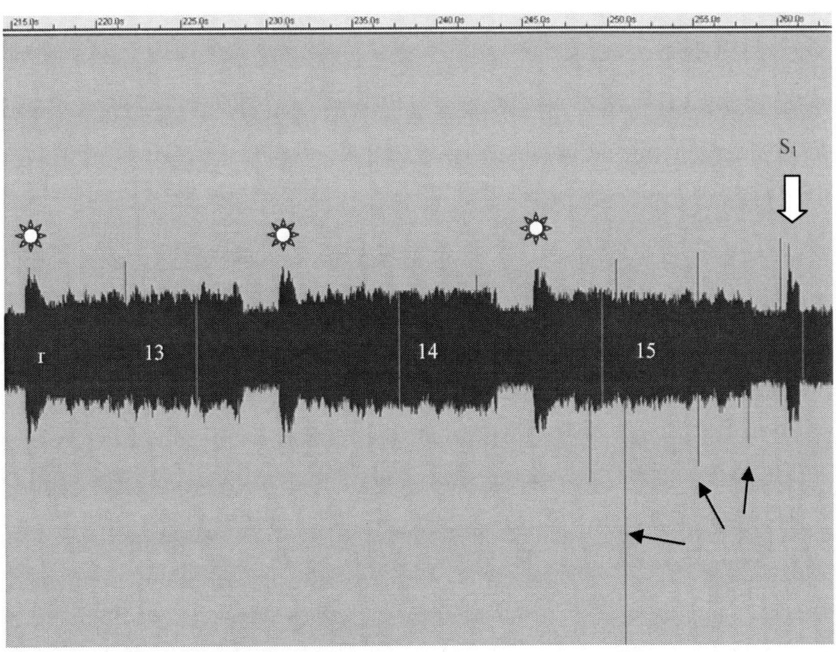

Figure 24.2.32. Fragment of the GN spectrum in Figure 24.2.27

As shown in Figure 24.2.31, upon transition from the information block 6 to 7, the signal intensities in the area of hypothetical gas giants that are far away from the Sun (the Sun is a hollow ball, double signal, see comment to Figure 24.2.47 below) become stronger. This information may reflect that U-2 understands the attractiveness of celestial bodies for their own development; it could be of interest for Zubow-3.

Additionally, the data in Figures 24.2.30, 24.2.31, and 24.2.32 indicate that UI read and write from left to right.

In information blocks 13 and 14, the signal intensities increase with distance from the protocol (**r**), while in information block 15 they decrease, followed by a strong signal that is similar to the signal protocol (star S_1, broad arrow)—the symbol of a star (Figure 24.2.32). The information block 15 is characterized by at least four dividing pulses (arrows).

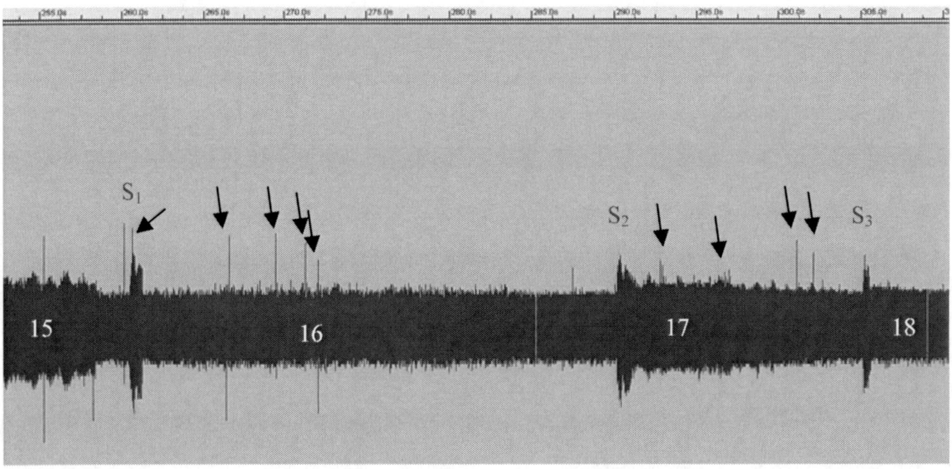

Figure 24.2.33. GN spectrum fragment from Figure 24.2.27 between 250 and 310 s.

After information block 15 follows the protocol (S_1) without any subblock signals (Figure 24.2.33). This situation can be understood as follows. The nearest star to our surroundings (S_1, α-Centauri B) has four planets (arrows); they are far away from the star itself. Very close to S_1, however, is the fifth planet. The interstellar space between S_1 and S_2 is full of dust and small asteroids. The next big star (S_2, α-Centauri A) shows mass concentrations (planets), but only a few and the most perspective ones are marked with pulses, like what was done for S_1. The weakest signal S_3 (information block 18) seems to belong to the red dwarf—Proxima Centauri; there are no planets, only a few small asteroids (http://en.wikipedia.org/wiki/Centaurus).

After 9,692 s a second scanning of the visible Uranus disk in the SPGPR is performed (Figure 24.2.34). The signals are less typical for UI, but look more like those of Zubow-3. Let us consider them in detail.

UI does not have to wait long for an answer from Zubow-3; that means the communication proceeds in real time. It confirms the theoretical and experimental views of the great Russian scientist G. L. Shipov [15] on the independence of excitation in a physical vacuum over distance. Logically, the answer of Zubow-3 should relate to its surroundings. It should be noted that Zubow-3 applies for the GN generation grasers that damp gravitational radiation (signals are directed down, Figure 24.2.34).

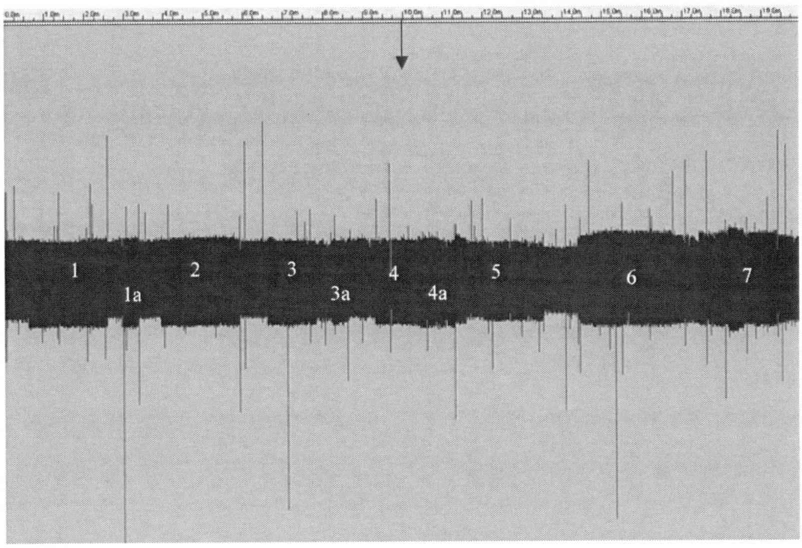

Figure 24.2.34. Zubow-3's reply on the previous message from UI (Figure 24.2.27) at the moment when the visible Uranus disk appears in the SPGPR (November 14, 2012 at 20:49:48, CET, arrow).

They are fundamentally different from the message logic of UI. Due to the very small time interval (20 minutes), the Zubow-3 surroundings in M87 cannot be fully understood. After information block 1 follows the miniblock 1a; behind information block 2 a mininblock is missing; information blocks 3 and 4 are followed by 3a and 4a, accordingly; and information block 5 stands alone, unaccompanied. Upon reading from left to right, the distance between blocks and their miniblocks becomes closer. Information blocks 6 and 7 are large compared to 1...5 and may reflect the mass of the celestial bodies surrounding Zubow-3. Then information blocks 1...5 can carry information about the planets and microblocks 1a, 3a, and 4a could be about their satellites. Concerning information blocks 6 and 7, they are assumed to be a planet set with a large number of satellites.

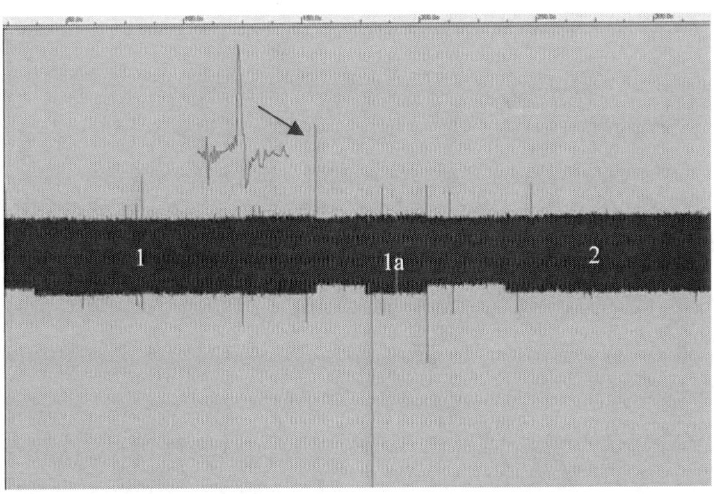

Figure 24.2.35. GN spectrum fragment from Figure 24.2.34 between 75 and 325 s

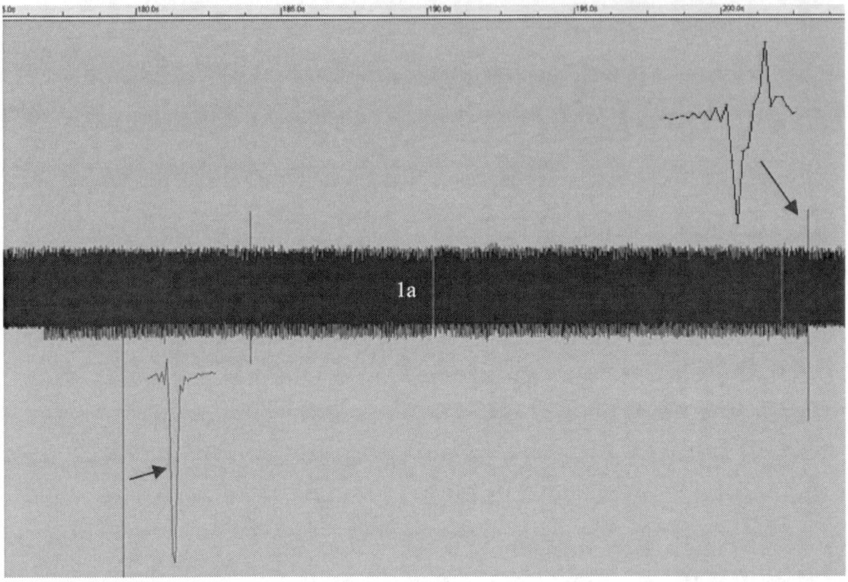

Figure 24.2.36. GN spectrum fragment from Figure 24.2.34 between 175 and 225 s, detailed spectrum of 1a. The limiting pulses are given in detail.

The spectrum in Figure 24.2.35 is typical for Zubow-3 and its signal density is higher than the frequency from U-1 and U-2. The arrow indicates the detailed signal pulse (end of information block 1) that strengthens the GN.

As is visible from Figure 24.2.36, Zubow-3 use a one-mode graser (left) damping GN as well as a two-mode variable one (right) that damps and amplifies GN.

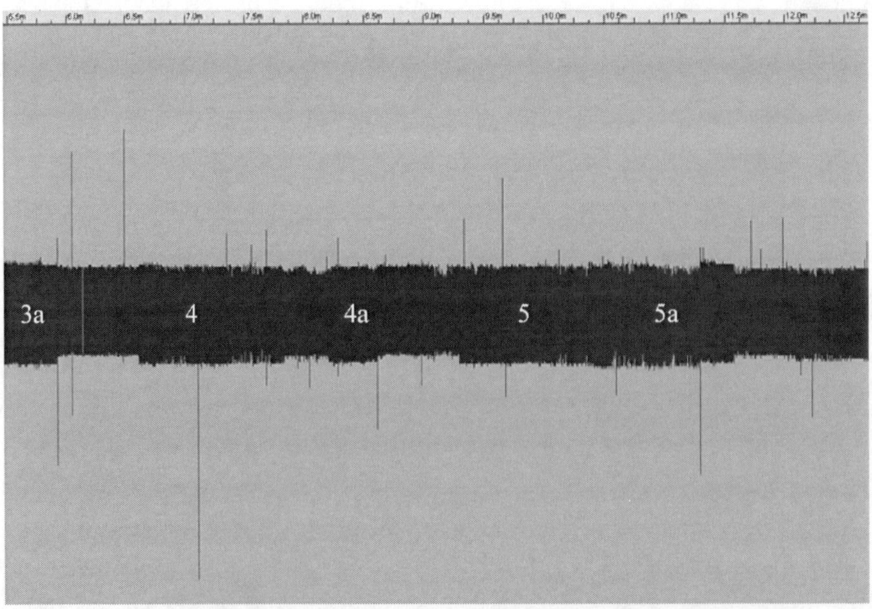

Figure 24.2.37. Fragment of the GN spectrum in Figure 24.2.34 from 5 to 13 minutes.

From Figure 24.2.37 it is evident that information block 5 has the microblock 5a, which is broad and strong, which means that the celestial body 5 holds many satellites and they are very close to this hypothetical planet.

The next GN scanning is carried out on November 20, 2012, beginning at 5:35:36 for 20 minutes (Figure 24.2.38).

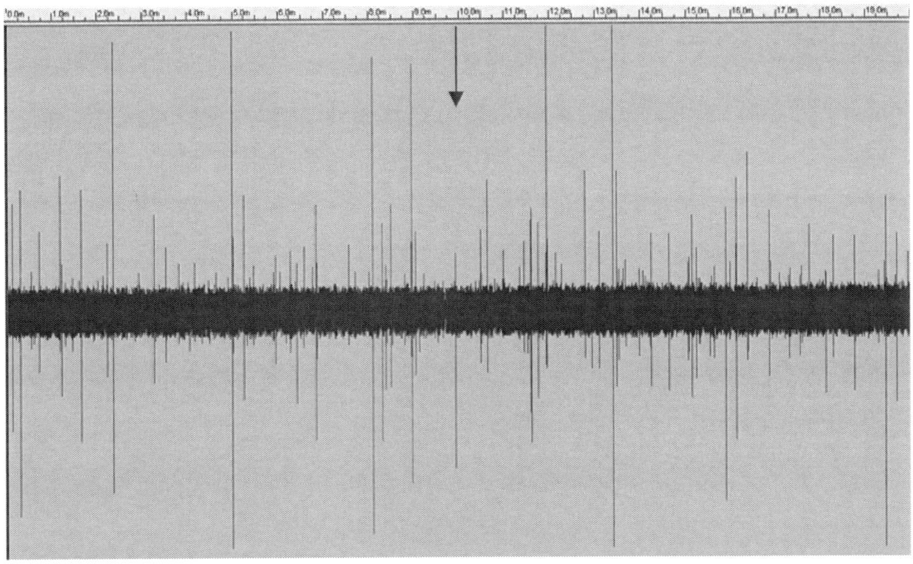

Figure 24.2.38. GN spectrum when M87 appears in the SPGPR (arrow). Opposition Uranus-M87 > 177.4°. November 20, 2012 (see http://www.zubow.de/pdf/News.pdf).

The spectrum shows a lot of dividing signals; the structure of the spectrum is new and not earlier met. Consider it in detail.

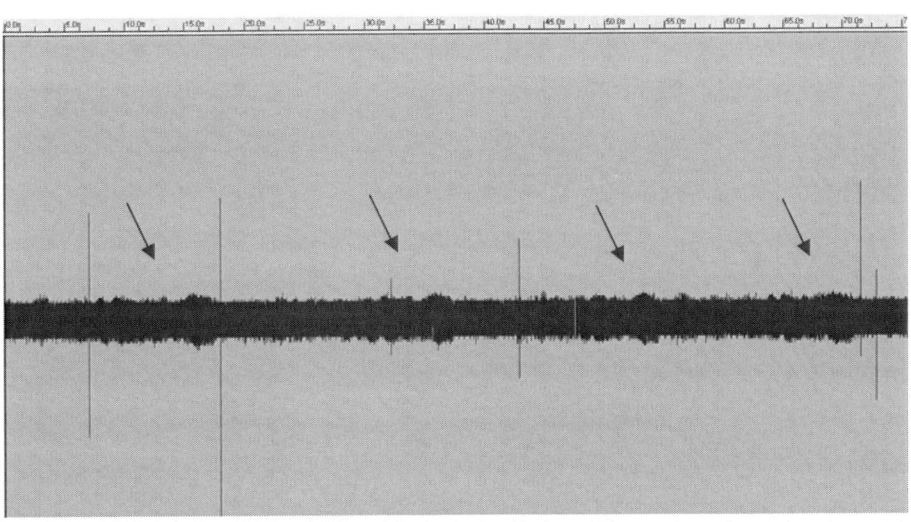

Figure 24.2.39. Fragment of the GN spectrum in Figure 24.2.38 from 0 to 75 s. Arrows denote the periodic structures of a gravitational lighthouse.

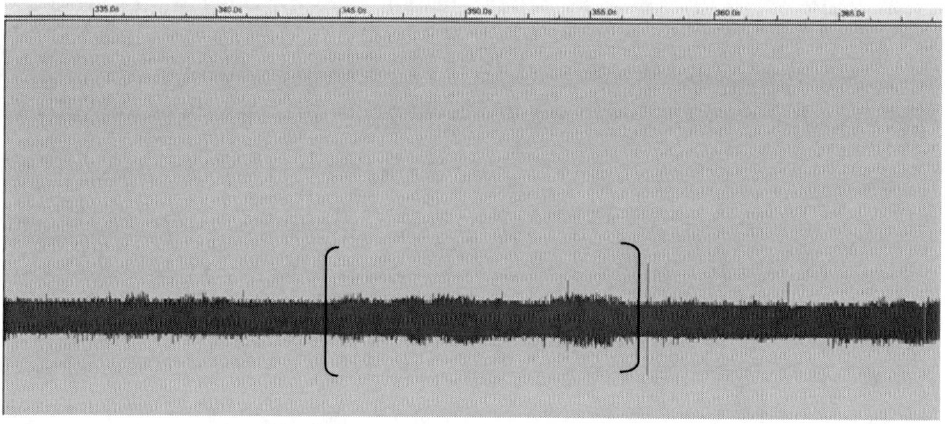

Figure 24.2.40. Fragment of the GN spectrum in Figure 24.2.38 from 330 to 370 s

The spectrum of Figure 24.2.39 resembles neither UI nor Zubow-3; it belongs to the gravitational lighthouse from a third ETI. They intercepted UI communication with Zubow-3 and called UI to contact them.

The repeating signals from the new ETI—Zubow-4 are highlighted with brackets (Figure 24.2.40). Immediately the inhabitants of Uranus answer with strong and clear signals of U-2 (Figure 24.2.41) in the direction of Zubow-4. This behavior of UI is like the targeted search for civilizations with a similar or higher level of development.

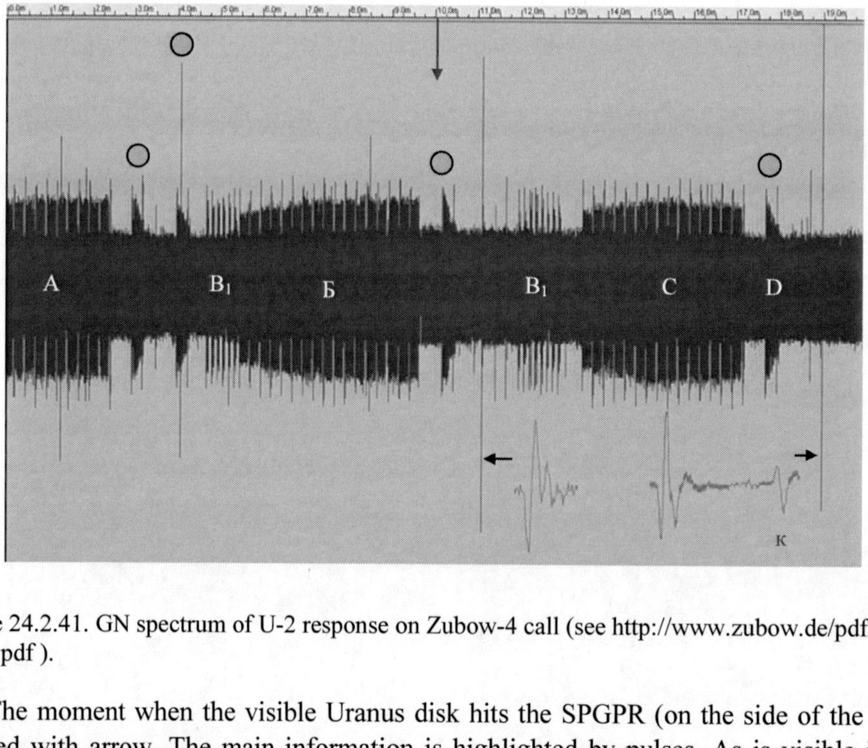

Figure 24.2.41. GN spectrum of U-2 response on Zubow-4 call (see http://www.zubow.de/pdf/News.pdf).

The moment when the visible Uranus disk hits the SPGPR (on the side of the Earth) is marked with arrow. The main information is highlighted by pulses. As is visible in Figure 24.2.41, UI switch on their best transmitter, U-2, to send the first signals to Zubow-4. Like in earlier messages, UI inform about the surroundings of Uranus and its place in the solar system. The three information blocks **A**, **B**, and **C** that are available are different!

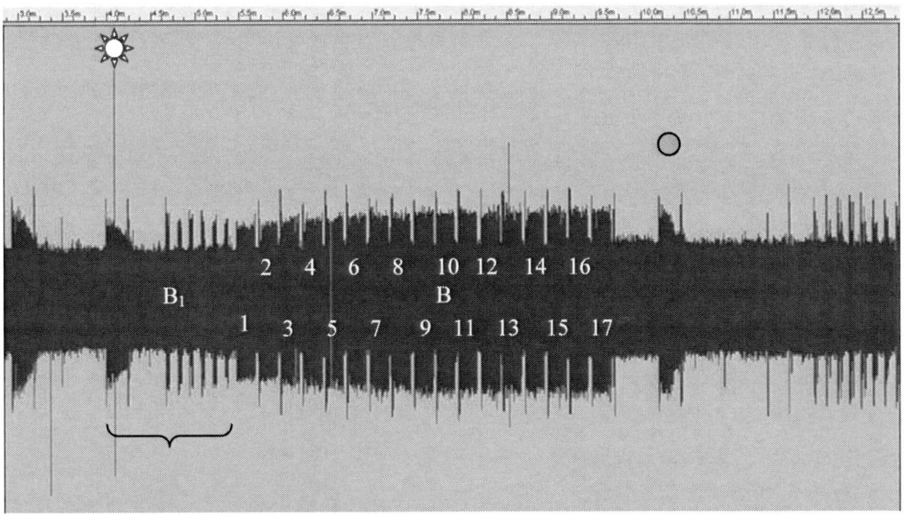

Figure 24.2.42. Spectrum of GN at the moment when M87 appears in the SPGPR (Uranus is in the SPGPR on the side of the Earth). Infoblock $\mathbf{B}_1$ and **B**, see Figure 24.02.41.

Based on early reports to U-2 the signals in block $\mathbf{B}_1$ (Figures 24.2.42 and 24.2.43) can be interpreted as the Sun and the six planets Mercury (**M**), Venus (**V**), Earth (**E**), Mars (**Ma**), Jupiter (**J**), and Saturn (**S**). The next broad signals in **B** (1…17) are ascribed to Uranus (**U**) and its main satellites.

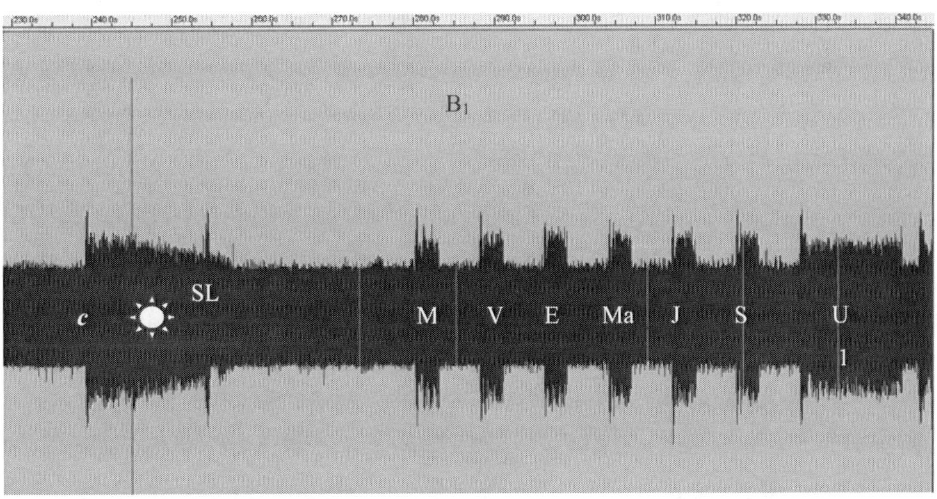

Figure 24.2.43. GN spectrum of $\mathbf{B}_1$, see Figure 24.2.42.

Here Uranus inhabitants highlight their planet by a series of signals with high intensity (**U**), and the rest of the planets do not represent any interest for the call partner. This is a classical example for a hyperbole in a dialogue.

It is interesting how the inhabitants of Uranus see the structure of the Sun. According to Figure 24.2.43, the Sun is characterized among others by an outer surface layer (**SL**) and after a radiative zone (**RZ**, Figure 24.2.44), the mass distribution monotonically decreases from the center (***C***) to the periphery.

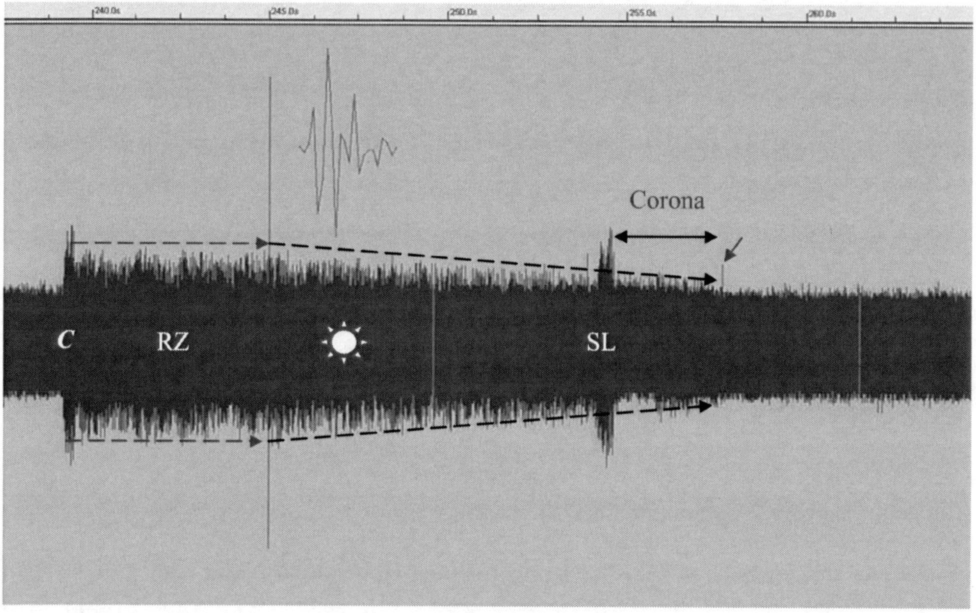

Figure 24.2.44. Structure of the Sun and mass distribution inside the star according to the UI idea: Core (***C***) – 0.325 s, **RZ** – 5.8 s, **SL** – 0.464 s, **Corona** – 3.065 s, total signal length from the nucleus (***C***) to the pulse end (slanted arrow) – 18.622 s.

The mass distribution inside the Sun, according to Figure 24.2.44, can be represented in percents: ***C*** – 1.74%, **SL** ~ 2.5%, corona – 16.4%, and **CZ** – 31.5%. The rest belongs to the photosphere with a monotonic mass decrease towards the star's periphery. Here the structure both of the Sun's center and the surface layer are interesting. Note that in modern science, the existence of **SL** is not known; unknown are the mass and length of the corona as well as the parameters of the Sun's nucleus, e.g. mass (http://en.wikipedia.org/wiki/Sun) and of the radiative transfer zone. The detailed spectrum of a dividing pulse (multi mode graser) that seems to divide the structure of the Sun into two parts is given.

Nominally, the inhabitants of Uranus emphasize 18 mass concentrations (one planet and 17 large satellites) in their surroundings. The different signal intensities are due to the numeric mass distribution in Uranus' imperium. The next $\mathbf{B}_1$ and **C** (23 blocks, Figure 24.2.41) repeat $\mathbf{B}_1$ and **B** (23 blocks) but in a different scale. UI attempted to distinguish Uranus' imperium from the remaining planets (Figure 24.2.41) to focus attention solely to it. Later (September 2013) they informed us about their habitable zone position in the planet profile (2,550 km down from the visible Uranus' surface) using a strong pulse.

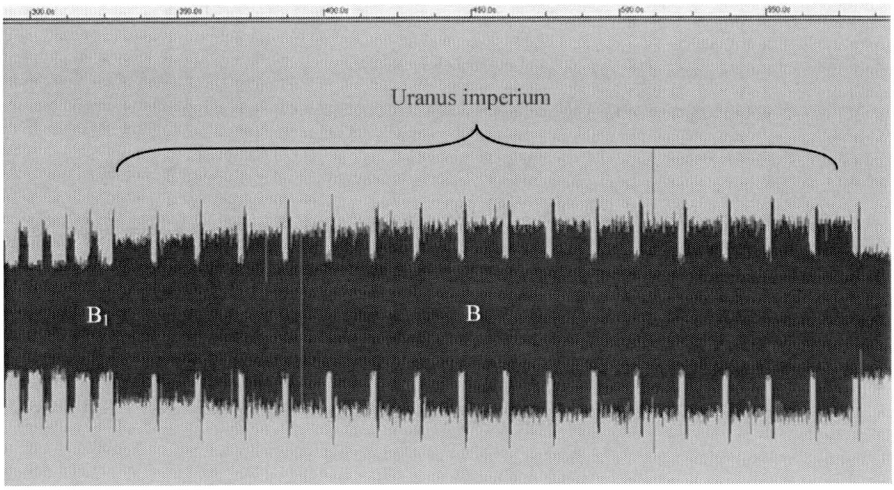

Figure 24.2.45. The imperium of the Uranus' inhabitants (17 satellites), spectrum of the block **B** in Figure 24.2.41.

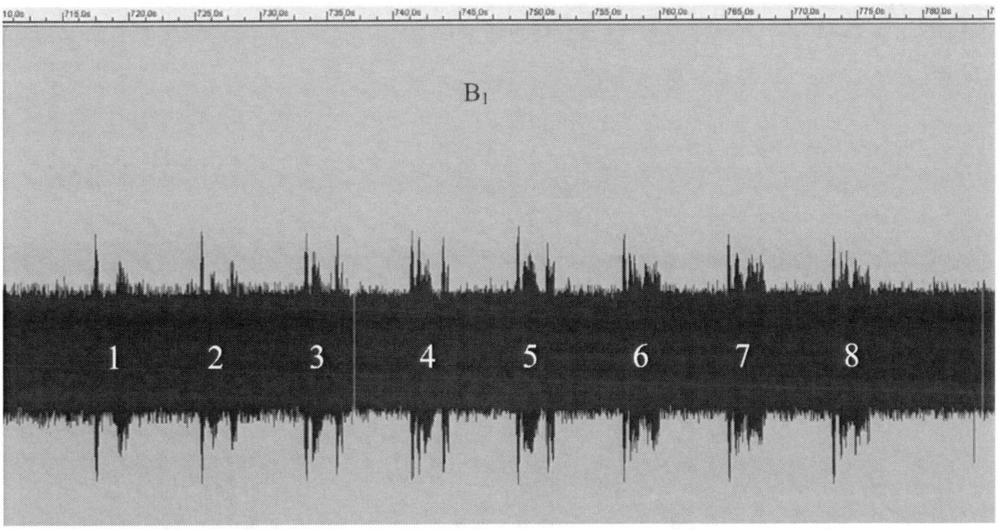

Figure 24.2.46. GN spectrum of $\mathbf{B}_1$ before block **C** from 710 to 785 s, Figure 24.2.41.

Eight equally spaced blocks (Figure 24.2.46) can be correlated with the number of the major planets and their satellites in the solar system; The structure of each block, however, is still unclear. The sequence is assumed to be started by the Earth + Moon (1), followed by Mars with two satellites (2), etc. Due to the low sensitivity of the used sensor many signals are merged. Mercury and Venus are marked with thin pulses in block $\mathbf{B}_1$ (Figures 24.2.41 and 24.2.46). They are insignificant. Furthermore, blocks 7 and 8 cannot be explained.

Perhaps blocks **C** and **D** (Figure 24.2.47) are obtained after stretching the previous blocks $\mathbf{B}_1$ and **B** to get more detailed information where **D** is the beginning of the new stretch.

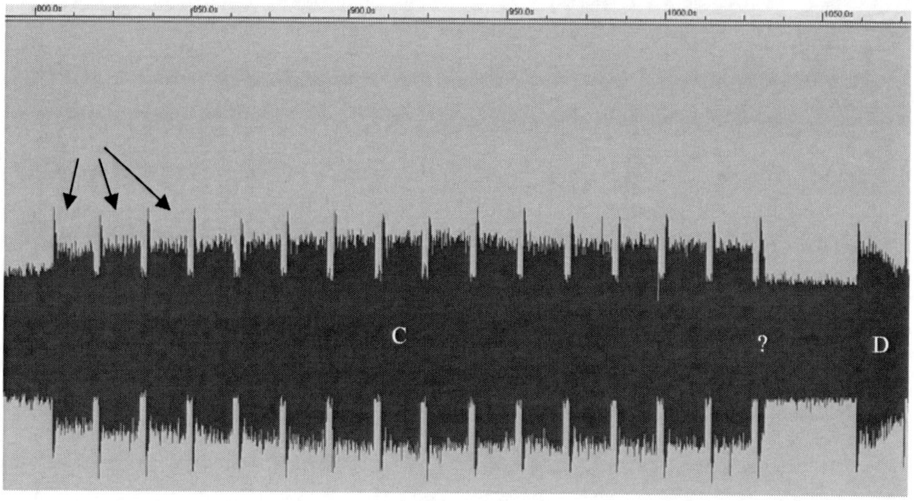

Figure 24.2.47. GN spectrum of blocks **C** and **D** (from 790 to 1,070 s, Figure 24.2.41).

An interesting message is received from another ETI, which intercepted communications between UI and Zubow-3 (details in http://www.zubow.de/pdf/News.pdf). This civilization (Zubow-4) is also located in a galaxy cluster of the Virgo constellation like M87. In Figure 24.2.48, the GN spectrum of Zubow-4 is given and in the table below the "orthography" applied by Zubow-4 is analyzed.

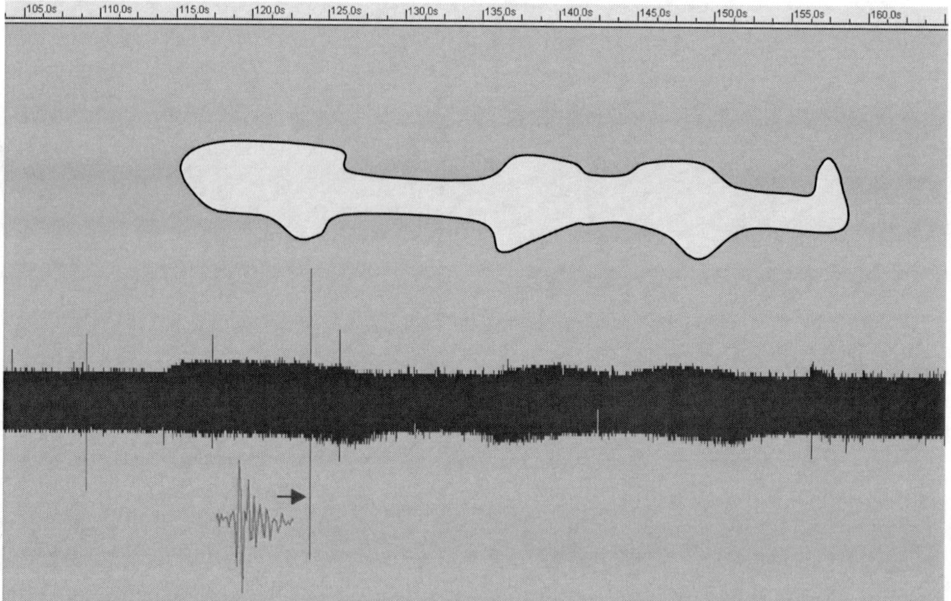

Figure 24.2.48. GN spectrum of Zubow-4 at the visible Uranus disk appearance in the SPGPR on November 19, 2012 at 20:29:45.

Figure 24.2.48 shows that the information blocks form the silhouette of a humanoid creature. A new ETI attempts again to inform UI, "Who are we?" It is difficult to transmit the form of aliens with the help of GN, nevertheless, it is possible. The detailed graser pulse (ringing of bells) at the speech organ (arrow) is an evidence for the humanoid character of Zubow-4.

Our big brother, Uranus, seems to be responsible for our safety (little brother) in the solar system. Statistically, our Earth must be destroyed every 1 million years from a cosmic boulder the size of 1 km but that has not happened for 50 million years.

Table 24.2.1. Finding the regularities in the communications of M87 and Uranus' inhabitants

Characteristic	Receiver (sample signal) Uranus inhabitants	Transmitter (sample signal). Inhabit. of Zubow-3, M87	Transmitter (sample signal). Inhabit. of Zubow-4, M87
Example	U-1* 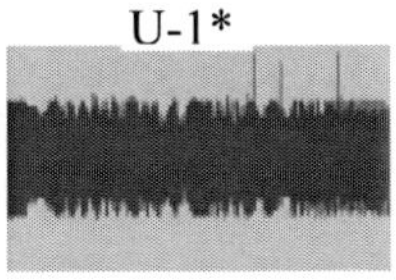	* 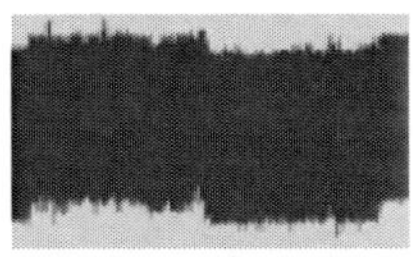	* 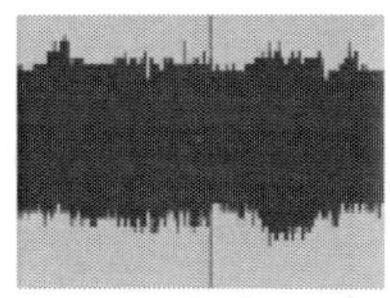
The method of signal division (example) (), (""), (!)	Figures 24.1.2; 24.1.5; 24.2.22; 24.2.24; 24.2.25 Figure 234* (brackets)	Figure173 (quotes)* 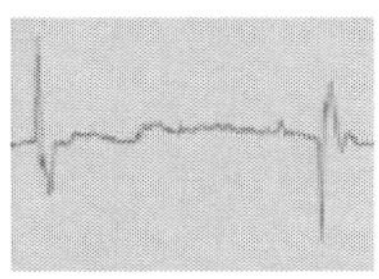	Figure264* 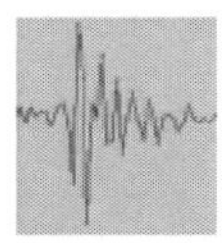
Waiting/please repeat (gravitat. Lighthouse)	Figure 25*	Figure 266*	Figures 282…284*
Figures	Figure 24.2.19 	Figure 24.2.17 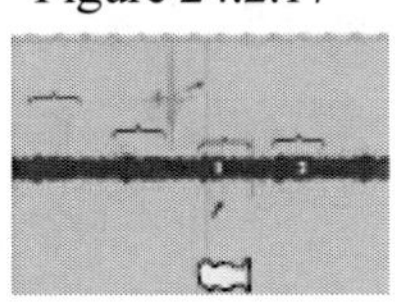	Figure 24.2.48 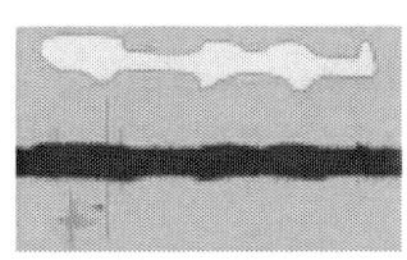
Figures	Figure 24.2.27 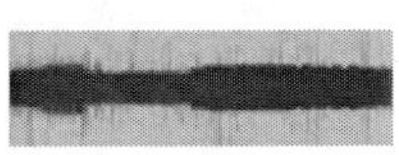		Figures 308*, 312* 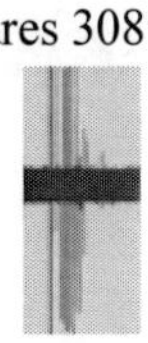

*See in http://www.zubow.de/pdf/News.pdf.

The probability that the UI visited us earlier has become greater than ever before. One thing is clear, UI are carefully monitoring our development.

Table 24.2.1 summarizes the main parts of the communication between Zubow-3 and Zubow-4 from the one side and the inhabitants of Uranus from the other side.

Conclusion

ETI widely use the technology of the generation of gravitational waves to communicate with each other (perhaps on the basis of low-energy neutrinos and axions).

A wide range of grasers are used to increase and decrease the gravitational radiation. Grasers with varying modularity are controlled by computers, which act as a typewriter (transmitter).

The main generator of high-frequency signals are different grasers working in the nano- and picosecond resolution range of signals (signal generation).

The technology of contact and the development of a relationship include progress from simple to complex matters and an understanding of the partner as well as the adaptation to its best technology.

Apparently, computers on M87 and on Uranus operate at very low temperatures, which allows for high quality signal processing.

The spelling of "text" messages includes signs used by us, such as punctuation, exclamation, accentuation marks, and so on.

Exchange of information begins with an analysis of the nature of the signals, presenting its own location within the planetary system and nearby stars, information on the physical form of a living organism, and apparently afterwards, a package of standard signal-tests to understand the level of development of the ETI and method of communication.

The inhabitants of Uranus are highly developed creatures that are capable of flying. The inhabitants of Zubow-4 are in form similar to us. There is no information on the form of life on Zubow-3. All ETIs are 200-500 years ahead in their development compared to us.

Zubow-4 is more advanced and powerful than Zubow-3. That is why Zubow-4 got the right to communicate with Uranus' inhabitants.

While highly developed civilizations in the electromagnetic spectrum are silent, they actively seek contact with civilizations of the same or even higher levels of development with the help of a gravitational communication channel.

Sponsorship

For further project development the authors are looking for sponsors. Please contact us!
viktor@zubow.de

REFERENCES TO 24

[1] Zubow K., Zubow A.V., Zubow V.A. Signals from Extraterrestrial Intelligences. Horizons in World Physics, editor, Albert Reimer. Nova Science Publishers, Inc. NY. 2013, vol. 280. https://www.novapublishers.com/catalog/product_info.php?products_id=38323

[2] Zubow K., Zubow A.V., Zubow V.A. The Phenomenon of Proton Dissolving in Vacuum and of Proton Condensation from Vacuum. Two Forms of Protons, Structure of Nuclei, Electrons and Atoms. J. of modern physics. 2010, v.1, no.1, pp.175-184.

[3] Zubow K.V., Zubow A.V., Zubow V.A. Ensemble of Clusters – New Form of Molecular Matter, Risks and Chances. Zubow Equations. In *Advances in Chemistry Research*, vol. 5, 2010, pp. 107-145. Editor J.C. Taylor. Nova Science Publishers, Inc. NY.

[4] Zubow Kr. V., Zubow A.V., Zubow V.A. Cluster structure of liquid alcohols, water and n- Hexane. Journal of Appl. Spectr., vol.72, no. 3, 2005, pp. 321-328.

[5] Zubow A., Zubow K., Zubow V.A., New Experimental Method for Determination of Distances to Celestial Bodies and their Masses, in A. Reimer (Ed.), Horizons of Physics, Nova Sci. publishers Inc., New York, 2012, vol. 279. in print. https://www.novapublishers.com/catalog/product_info.php?products_id=34674

[6] Zubow K., Zubow A.V., Zubow V.A. The Nature of Electron, Atoms and Molecules. Physical Chemistry-An Indian Journal. MS No.: Ph615210312. 3 September, 2012, pp.2-4.

[7] Zubow K., Zubow A.V., Zubow V.A. Scanning of the Sun and other celestial bodies with help of gravitation spectroscopy. Optic and Photonics Journal. 2011, № 1, pp.15-23.

[8] Bonfils X., Delfosse X., Udry S., et al. The HARPS search for southern extra-solar planets? XXXI. The M-dwarf sample. Astronomy & Astrophysics manuscript no. main c ESO 2011 November 24, 2011, arXiv: 1111.5019v2 [astro/ph.EP] 23 Nov 2011

[9] Svozil Karl. Neutrino dispersion relation changes due to radiative corrections as the origin of faster-than-the-light-in-vacuum propagation in a medium. CDMTCS preprint nr. 407/2011. arXiv:1109.5411v1 [physics.gen-ph] 25 Sep 2011.

[10] Fargion Daniele. Inconsistence of super/luminal Opera neutrino speed with SN1987A neutrinos burst and with flavour neutrino mixing. arXiv:1109.5368v1 [astro-ph. HE] 25 Sep 2011.

[11] http://news.sciencemag.org/sciencenow/2012/11/infintely-fast-light.html?ref=hp

[12] Tom Van Flandern. The Speed of Gravity What the Experiments Say. Meta Research [as published in Physics Letters A, 1998, v. 250, pp. 1-11], http://www.metaresearch.org/cosmology/speed_of_gravity.asp.

[13] Zubow K., Zubow A.V., Zubow V.A. Experimental Methods for the Determination of the Super Light Velocities of the Gravitation. Nature, Structure and Properties of Gravitation Waves. In *Horizons in World Physics*. Editor A. Reimer. NY. vol. 277. 2012. https://www.novapublishers.com/catalog/product_info.php?products_id=29310

[14] François-Xavier Gallat, Alex P. S. Brogan, Yann Fichou at al. A Polymer Surfactant Corona Dynamically Replaces Water in Solvent-Free Protein Liquids and Ensures

Macromolecular Flexibility and Activity. J. Am. Chem. Soc., 2012, vol. 134, no., 32, pp 13168–13171.

[15] G. I. Shipov, Theory of physical vacuum. Theory, Experiments, and Technologies. Moscov, Nauka, 1997. (http://publ.lib.ru/ARCHIVES/SH/SHIPOV_Gennadiy_Ivanovich/_Shipov_G.I..html).

Chapter 25

FINAL WORDS

Summarizing the work written in this monograph, some general principles of the formation of mass cluster ensembles and their properties can be derived.

Clusters are formed as a result of the energy balance between two factors: the thermal background (k*T*) and the gravitational noises. While the first is characterized to be a destructive factor, the second one is constructive. In the absence of the destructive factor the state of molecular matter can be represented by a spatial energy network in which the nodes are energy clusters of stationary gravitational waves (see Figure 3.2.1). The energy of these clusters is used by molecular matter to form organized molecular mass concentrations (1. Zubow equation, $m = 10^{11} \cdot \omega\text{-}^{2}$). A certain balance is found to exist between the energy and mass clusters. Amongst the order number of the molecular cluster (*N*) which ones are allowed to exist is determined by gravitational noises according to the 2. Zubow equation:

$$m = (250 \pm 20) \cdot N^{(2.2 \pm 0.2)}$$

Clusters exhibit properties of both waves and particles, wherein the particles are relatively stable structures and are only under the influence of weak energy fields of anthropogeneous or natural noises coming from the Earth or celestial bodies. They change their density and interaction with the surroundings. The energy capacity of a cluster ensemble serves to characterize the cluster structure of molecular matter.

It has to be understood as the internal energy of a single ensemble—as an energy unit.

Noise gravitational fields provide a new form of molecular matter shown as molecular cluster ensembles in stationary gravitational waves. The formation of stationary waves themselves (order islands) in a "sea of disorder" of white noises is known as the "Noah's Ark effect," which confuses not only scientists but also theologians. While there are many speculations about this phenomenon, there is no clear understanding of how independent random sequences can form such islands, which are not random.

Molecular clusters with a long-range order formed and enforced under the influence of white gravitational noises (disorder) are one of the most interesting directions in the natural sciences of the 21st century.

On the other hand, molecular cluster sequences reflect the heterogeneously formed structure of the physical vacuum, whose energy is beyond the real idea.

A NEW FORM OF MOLECULAR MATTER: ITS CLUSTER FORMATION AND CLUSTER ENSEMBLES

Computer structure modeling of liquids, polymers, biomatrices, gels, and the animation of defects in crystals prompted us to reconsider the principles of molecular matter formation. The use of computer programs, which were mainly based on the weak interactions of molecules and atoms, described by Lennard-Jones; potentials; densities; energies; etc., is highly limited because of the hardware, which is insufficient for large data banks. In this simulation, the molecules are discussed as systems that are isolated from external energy influences on nature, of which only vague ideas exist. This would lead to a confusing result in which infinitely many molecules stick together. It is possible to get a more or less real idea, however, if the Mie equation using empirical parameters is applied.

Now that much more information concerning small molecular clusters has been published, however, the reasons for their formation are discussed extremely gently because of our lack of ability to analyze them directly as well as because of the fact that developments in cosmology, astrophysics, and nuclear physics are stagnating. The energies combining molecules of the same type in real clusters are very small; they are in the range of energies below the thermal noise (E = kT), which is destroying organized structures. The existence of clusters is observed to cover a huge time interval. In liquids, it is in the picoseconds' range or even lower, while for polymers, the time can range between seconds and hundreds of years. These times depend on thermal energy fluctuations and can be applied for their characterization. It is assumed that clusters are formed more easily in the time of "windows" when there are no thermal energy fluctuations. To find a compromise between the destructive thermal factor kT and the process of molecules' sticking without involving a third component, however, is difficult. It is difficult due to the requirement regarding structural stability. The third force not only has "to equalize the weight categories," the destroyer and creator, but also has to constantly care on an ordered molecular structure. All these requirements are satisfied in a cluster ensemble in which each cluster is subordinated to the ensemble, while the ensemble itself acts as an energetic unit that has to determine the energy capacity of each cluster.

Through the development of GMS, we have succeeded in discovering new forms of matter. The GMS device is tested for a number of substances: liquids, proteins, synthetic polymers, starch, and processes involving these compounds, e.g. heating, dissolution, decomposition, and polymerization. Additionally, it has been analyzed how ensembles and individual clusters in these materials react on the resonance field between them and their remote analogs. The new form of molecular matter is characterized by the properties both of particles and waves. It is to be understood as a bridge between the nano- and macro-universe. This matter as the interaction product of three forces can only exist in ensembles in which the clusters are structured and harmonized by space gravitational fields. What can serve as this force, which could provide an incredibly high reserve of structured energy for the cluster formation? This structured energy (ensembles of energy clusters) is observed to be stationary gravitational waves from space. Waves are the result of the noises from baryonic and physical vacuum matter where the last one is a relatively permanent component of the white noise. Therefore, having an idea of a cluster ensemble in molecular matter, one can learn how the structure and energy of this force is built.

In water, the constant component of the white noise allows that the following cluster sequences shall be formed: $(H_2O)_{11\pm1}$, $(H_2O)_{100}$, $(H_2O)_{178}$, $(H_2O)_{280}$, $(H_2O)_{545}$, $(H_2O)_{903}$, $(H_2O)_{1,351}$, $(H_2O)_{1,601}$, $(H_2O)_{1,889}$, etc. This ensemble, however, is not stable. It is influenced by the variable component of the noise in the form of weak energy fields from: celestial bodies, shock waves, earthquakes, and human activities that results in broken clusters, for example: $(H_2O)_3$, $(H_2O)_{25}$, $(H_2O)_{45}$, $(H_2O)_{127}$, and $(H_2O)_{402}$. Similar cluster sequences are found for synthetic polymers, proteins, starch, salt solutions, acids, alkalines, ketones, alcohols, alkanes, and cellulose. Furthermore, they are observed in the water of vegetable and fruit biomatrices, heart muscle, and the sinew of ducks; they are even in the head of an earthworm.

Similar clusters and cluster ensembles of the same material are found to interact remotely, where the interaction force is directly proportional to the total cluster's mass (M) and inversely related to the distance (r) between the transmitter and receiver. For example, the remote interaction between oscillating polystyrene clusters, represented by the phenyl group dimers, is described by the following equation:

$$\Delta f_{\mathrm{d}} = 1.162 \cdot \mathrm{M}/r^2 - 0.0016$$

The interaction between remote clusters proceeds through gravitational resonance fields of similar oscillators where the gravitational field of each cluster shall be generated by proton jumps into the vacuum and out of the vacuum. The jumps of the protons are characterized by energy histograms that are individualized for each mass concentration. In these interactions, only the clusters of the same nature and substance are involved, but not those of a different nature, such as water and polystyrene. Thus, the world surrounding us is saturated by resonance fields caused by proton jumps.

This implies an understanding of the sedative effects of large water masses on the human body, harmonizing water cluster oscillation in the biomatrix and telepathic phenomena, while reflecting the same biochemical processes of thinking with the participation of oscillating water clusters and their resonance fields. Furthermore, this can be used to explain the well-known phenomenon of "bio-fields" and diffuse views in astrology, etc.

For their formation, the molecular clusters partially consume the energy that comes from the EC while upon their destruction, the earlier consumed energy will be released, leading to secondary noises. This process should be seen as an equilibrium in which the mass concentrations themselves have an influence on the energy clusters, too. One could say each substance has its individual gravitational noise and at the phase boundary of two substances, their noise fields should be overlapped. This boundary layer with differently constructed ensembles of clusters is known in technology as the "wall effect" and its origin of clusters is confirmed by us. This effect is well used, for example, to navigate the shark in the ocean depths. On the other side, the influence of molecular clusters on the energy ones could become a platform for super light communication with our surroundings.

The variable white noise component, e.g. noises from celestial bodies, Earth's rotation, movements of tectonic plates, and anthropogenic noises influence the formation of molecular clusters. Although this effect is small, biochemical processes (growth and thinking) are modified by them, resulting in changed biological reactions. Therefore, we need to look at astrology whether we like it or not.

Long-lived cluster ensembles in liquids are evidenced to be stable for a span of minutes to one year. Here, in the ensemble, there exist at least two cluster families: (i) forming a stable framework (skeletal clusters with reproducibility more than 95%) and (ii) random clusters with reproducibility worse than 20%. In the biomatrix of cucumber (95% water), the water cluster ensemble is relatively stable during the entire summer time even though in the large cluster area from time to time small changes can happen. For example, at the approach of Mars to Earth, noticeable perturbations of the cluster of 100 water molecules are recorded and when Mercury is near the Earth, there even appears a cluster consisting of 45 water molecules in a lake. Using the water cluster ensemble: $(H_2O)_{11\pm1}$, $(H_2O)_{100}$, $(H_2O)_{178}$, $(H_2O)_{280}$, $(H_2O)_{545}$, $(H_2O)_{903}$, $(H_2O)_{1,351}$, $(H_2O)_{1,601}$, and $(H_2O)_{1,889}$, a correlation connecting the frequencies of macro- and micro-universe has been observed. Parallels therefore seem to exist in the universe.

The results presented in this monograph allow us to reconsider the atomic nuclear structures and their formation mechanism as well as to discuss the stability of isotopes, atoms, and molecules by recognizing the other nature of the proton and electron with the consequence of a "fresh look" at chemical bonds.

Cluster ensembles in a consistent energy field are a new form of molecular matter. Here the ensembles interact remotely with each other and further with the energy fields of gravitational waves from celestial bodies and the physical vacuum. Hence they form spatial-temporal process events and become information carriers. The latter opens new possibilities for a new generation of super light communications, methods for the remote analysis of far galaxies, and the movement of humans in the universe.

THE INFORMATION FIELD OF A NEW MOLECULAR MATTER FORM

The natural and enforced cluster oscillations are the result of gravitational fields' excitation and the interaction of clusters with each other and with their surroundings (*f* value). The latter provides information on the energy conditions of the molecular matter on the Earth and on its involvement in various chemical, physical, and biochemical processes. This information was originally stored in the secondary gravitational waves that propagate with the speed of light (ripple). The information transfer from the secondary to the primary gravitational waves proceeds through proton jumping between a physical vacuum and the baryonic world, where in the physical vacuum a medium will be formed through which the super light waves (primary) propagate.

For the biosphere, the value *f* reflects the state of biochemical processes, growth, development, and aging while for humans and animals, it includes processes of thought, experiences, diseases, etc. The information is concentrated in clusters (domains) of super molecular structures of biomolecules. The release of this information in the surroundings is due to cluster oscillations, the effect of changed cluster forms on proton jumping in a physical vacuum, and the associated generation of weak gravitational radiation. With the help of the resonance WGR, it is able to interact with remote WGR sources and gravitational space fields. It follows that the information field generated by molecular forms of Earth's matter is in permanent contact with the gravitational fields of the solar system, our galaxy, and the physical vacuum. This field is to be understood as a "fingerprint" of Earth's molecular matter.

One could even speak of a new road map of information and a new exchange of information. The role of the road map, as the data interface of the "fingerprints" can be seen to be means that they can act as an interaction channel with the gravitational field of space. This "fingerprint" of Earth's molecular matter (energy histogram of proton jumping) transmits information about us and our biosphere into space. On the other hand, the understanding that the universe is a gigantic gravitational field consisting of many "fingerprints" from other molecular clusters leads us to the idea that this field could serve as an information accumulator and database. This information is found to be stored in the energy clusters of stationary gravitational waves of space (Figure 3.2.1). Developing the idea further, we have come to an interesting conclusion about the information flows within the archive as well as on the impact of these information flows on the initial information source, namely, on us. Erasing the information stored on such a media giant it is not possible, except in the case of a periodically collapsing universe with the subsequent explosion of a new-born universe. The universe thus may be considered a giant computer. The information field of molecular matter could be used for the development of next-generation communication channels with surroundings in real time.

THE PHENOMENON OF PROTON JUMPING IN MOLECULAR MATTER AND UNDERSTANDING THE VACUUM

In chemistry, proton jumping at significant distances is known but the explanation cannot withstand any criticism because it is difficult to understand how the proton "breaks" through a network of electron orbitals of atoms and molecules. If we extrapolate the new form of molecular matter—its cluster formation—to the level of elementary particles, then in the proton structure consisting of three quarks, there should be mass (energy) "transfusions" with the ability of them to move them into the physical vacuum (disappearance of mass clusters). According to the thermodynamic equilibrium, the mass entrance into the physical vacuum immediately requires a compensation for it in form of an equivalent mass emission in the near vacant space, which is observed as proton jumping where the emitted proton has to be understood as the twin of the vanished (dissolved) one. In this jumping model, the vacuum, as an "ambulance," takes the resource away from the thermodynamically unfavorable proton and immediately delivers it to the nearest vacant space.

THE HETEROGENEOUS STRUCTURE OF THE PHYSICAL VACUUM

The existence of molecular clusters reflects the heterogeneous structure of the physical vacuum. On the other side, the heterogeneity of the vacuum results from energy clusters of a new energy form—the energy of the gravitational field. This field arises when the proton dissolves in the physical vacuum and almost simultaneously condenses from it. With these surroundings, the gravitational energy background (noise) from similar processes is present in near and far space. The potential of the gravitational field seems to be determined by ensembles of energy clusters in the vacuum (see Figure 3.2.1) and its structure by the state of gravitational radiation flows in a given space. Our world order resembles a Russian matryoshka in which each doll (ensemble) reflects a certain energy cluster state but the ensembles themselves are considered milestones of the material world.

The role of the physical vacuum will definitely be proved by other experiments that will accelerate our understanding of the structure of the space as a saturated energy "solution" in which the baryonic matter, as the condensate of the physical vacuum matter, floats.

Cluster Ensembles in Atomic Nucleons

The energy levels (*K*) in atomic nucleons are given in §18 and they can be described by the following function:

$$N = 1.6024 \cdot K^{2.1479}$$

This equation is very similar to the 2. Zubow equation for molecular clusters. If N is proportional to the mass of these levels in the nucleons (m_N), or better yet, to the mass concentrations, then the equation and the 2. Zubow equation can be considered identical. Furthermore, it should not be discussed on the basis of levels, but rather on the basis of mass concentrations. Then the last equation would be as follows:

$$m_N = 1.6024 \cdot K^{2.1479}$$

In this monograph, the world is concluded to be a multi-level construction in which, for example, the ensembles in nucleons are also members of a giant vertical mass ensemble (planets, galaxies, etc.). With a little imagination, it should be possible to find a constant that links all the vertical levels.

Cluster Construction of the Universe

The Universe is also constructed by mass clusters: visible (galaxies) and dark matter (voids, 75%). These concentrations of masses can be seen as clusters and their formation is allowed by gravitational white noises permeating the whole Universe. The cluster formation could also be described by the Zubow equations. At the same time, the Universe has to be characterized by super cluster structures in the form of huge energy and mass concentrations (e.g. great attractor). The extrapolation of the cluster formation idea in the microcosm to that of the macrocosm harmonizes our world view and opens a new chapter in the understanding of the role of noises in the world order.

Molecules and Atoms

One of the main goals of the present monograph is the “soft” revision of the Bohr model for hydrogen. Currently one cannot conceive that an electrically charged particle moves in the field of a charged atomic nucleus without emitting some radiation. Additionally, a perpetuum mobile cannot be accepted and an eternal rotation of the electron around the nucleus must be impossible. The process of proton oscillation in a physical vacuum causes stationary shock waves, which in the 20th century were falsely termed as electron orbitals, i.e. the electron itself is not there. These stationary shock waves are observed to be quantized energy levels and to reflect the pressure of dark matter on our baryonic world. In more complex nuclei, they arise as conjugated stationary shock waves and their superpositions are those with the lowest potential energy, which will be described by chemists as the electronic orbitals *p*, sp^3, *f*, *d*, and so on. Nuclei ensembles form molecules according to the rules of the stationary shock waves' interaction (energy minimization) where the stationary shock waves themselves reflect the oscillation energy of nucleon ensembles in the nuclei. The neutrons are the only place where both dark and baryonic matter are mixed together.

A chemical bond is formed when various nuclei ensembles (chemical elements) combine to form a new ensemble (molecule) by minimizing the potential energy.

All other types of "electrons," like the electricity flow, are to be seen as a flow of shock waves with quantized energy levels.

Magnetic lines are nothing more than a superposition of stationary shock waves of atomic nuclei.

It is interesting in this context to understand the isotopes as unstable nuclei ensembles. The nuclei of stable elements are understood as harmonic oscillators but the violation of the harmony leads to nucleus decomposition and radiation.

THE GRAVITATIONAL CHANNELS OF EXTRATERRESTRIAL CIVILIZATIONS (ETI)

To contact each other, highly developed ETI use resonance GMS. Here the message shall be transferred as information about molecular clusters/domains. This information is quite sufficient to get an idea about the living conditions and living forms outside the Earth as well as on the development level of ETI.

To contact ETI and exchange information with it, the following steps have to be carried out:

1. To create grasers with the possibility to set information on the main gravitational wave between celestial bodies
2. To get a road map of the main gravitational wave in space
3. To develop knowledge in biochemistry in order to get an idea of life forms on the basis of domain structures of water and biopolymers
4. To organize a contact center with ETI that is sponsored by as many countries of the world as possible

This monograph was "The Way to ETIs."

Authors' Contact Information

Dr. Zubow Kristina
(Simferopol, Ukraine 1976)
University of Potsdam. Biochemistry
(2001, FRG)
PhD thesis, Humboldt University Berlin, 2007
Expert in biochemistry of antibodies
(http://edoc.hu-berlin.de/ dissertationen /zubow-kristina-2007-01-08/PDF/zubow.pdf)

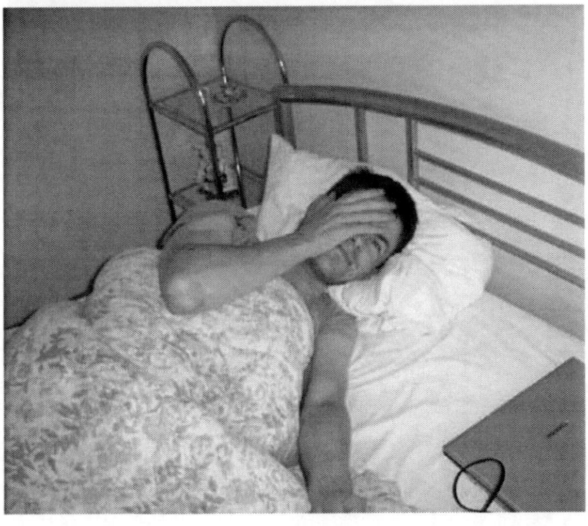

Dr. Zubow Anatolij
(Simferopol, Ukraine 1978)
Humboldt University Berlin
(2004, Berlin, FRG)
Dept. of Computer Science
PhD thesis, Humboldt University Berlin, 2008
Expert in Wireless Communication
(http://edoc.hu-berlin.de/ docviews/ abstract.php?id=29955)

Dr. Zubow Viktor Anatolievich
(Sewastopol, Russia, 1953)
Odessa Mechnikov University
Department of Chemistry
PhD thesis, Moscow Lomonosov University, 1982
Expert in Macromolecular Chemistry and Physics
(www.zubow.de)

INDEX

D

E

F

G

H

I

J

L

M

U

V

W

X

Z